WITHDRAWN
University of
Illinois Library
at Urbana-Champaign

WITHDRAWN
University of
Illinois Library
at Urbana-Champaign

CRC Handbook of Food Additives

2nd Edition

Volume II

Editor:

Thomas E. Furia

President

Intechmark Corporation

Palo Alto, California

CRC Press, Inc.

Boca Raton, Florida

Library of Congress Cataloging in Publication Data (Revised)

Main entry under title:

CRC handbook of food additives.

Includes bibliographies and index.
1. Food additives--Handbooks, manuals, etc.
I. Furia, Thomas E., ed. II. Chemical Rubber Company,
Cleveland. III. Title: Handbook of food additives.
TX553.A3C2 1972 664'.06 73-153870
ISBN 0-8493-0543-8 V.II

This book represents information obtained from authentic and highly regarded sources. Reprinted material is quoted with permission, and sources are indicated. A wide variety of references are listed. Every reasonable effort has been made to give reliable data and information, but the author and the publisher cannot assume responsibility for the validity of all materials or for the consequences of their use.

All rights reserved. This book, or any part thereof, may not be reproduced in any form without written consent from the publisher.

Direct all inquiries to CRC Press, 2000 N.W. 24th Street, Boca Raton, Florida, 33431.

© 1980 by CRC Press, Inc.

International Standad Book Number 0-8493-542-X (Volume I)
International Standard Book Number 0-8493-0543-8 (Volume II)

Library of Congress Card Number 73-153870
Printed in the United States

664
C859
1972
v. 2
cop. 2

AGRICULTURE

FOREWORD

With this volume of the CRC *Handbook of Food Additives,* a significant departure from previous handbook formats of presenting new editions has been instituted by CRC Press. In the past, new editions were published in which each chapter was updated so as to keep the reader current. Occasionally, new chapters were presented which represented essentially new topics . . . in the case of this handbook, new categories of food additives. In the current format, the notion is to publish new volumes which add, when applicable, to the body of established information to the last edition. Consequently, this publication is entitled Volume II of the 2nd Edition.

The current format is not without its fair share of problems considering the topic, Food Additives. Unlike many other disciplines where significantly chemical compositions are constantly emerging, the current regulatory climate severely restricts the approval of new additives and even more so the introduction of new categories of food additives. It is a matter of record that in the U.S., the FDA has not approved a major new food additive since the introduction of TBHQ and FD&C Red 40 nearly 10 years ago. The only area (category) where some progress in new approvals has been made is in basic flavor chemicals. The Flavor Extract Manufacturers Association (FEMA), through its own expert review committee, has been granting "FEMA GRAS" status at a fairly brisk pace to a host of new organic compounds found useful as flavor ingredients. Many are "nature identical", i.e., chemicals found in plants or as components of naturally derived flavors and flavor extractives having long histories of safe use. Others on the FEMA list are new synthetic compounds which, because of their very small (often minute) levels of use in foods, are deemed safe since they fall well below the safety factor established in animal tests. All of the new FEMA approved flavor compounds, together with a recapitulation of the previously approved ingredients, are listed in a separate volume entitled *Regulatory Status of Direct Food Additives,* and for the most part, are reviewed in the new chapter on flavor by F. Fischetti.

Official approvals notwithstanding, research goes forward and we have attempted to present an update of such happenings. For example, over the past 6 years a new concept has been introduced in the field of food additives . . . nonabsorption. The concept here is that if a functional molecule were rendered stable to processing, and after ingestion to all of the biological processes encountered while transversing the entire gastrointestinal tract, then molecular size would determine the degree of absorption. Rendering an additive both stable and sufficiently large would preclude significant absorption. As such, food additives with these chemical/physical characteristics are expected to display an enhanced safety factor. This concept has been translated via the synthesis of functional polymers into no less than three categories of potentially new food additives: (1) an oil soluble polymeric antioxidant, (2) a series of polymeric food dyes, and (3) high intensity sweeteners. The latter is as yet at an exploratory phase of development. Progress in each area is covered in the current chapters on antioxidants (Sims and Fioriti), food dyes (Noonan and Meggos), and new sweeteners (Crosby and Furia).

In addition to presenting some new concepts, progress in the more traditional categories of food additives are reviewed. The huge bibliography on saccharin and cyclamate has been updated by K. Beck. These materials are currently under regulatory review and we felt obliged to reissue the chapter and updating its contents where possible. As a separate topic, all of the activities in new sweetener research are reviewed in the chapter by Crosby and Furia.

For a number of years, enzyme technology has been looked towards with great expectations for applicability to the food-processing industry. This forecast persists and several volumes dedicated to the topic have appeared. Dr. Leland Underkofler has extracted those data of importance to the food industry and critically reviewed the details in his new chapter on enzymes. Similarly, Frank Fischetti has taken a more comprehensive approach to flavor chemistry and presents some interesting concepts to those who use flavors but are not actively engaged in creative flavor technology. Most valuable is his recapitulation of the whole new range of synthetic flavor ingredients now available for compounding into finished products.

No volume of this nature would be complete without some effort being devoted to regulatory issues. The editor considers himself fortunate to have enlisted the aid and insight of Dr. Virgil Wodicka, formerly Director of the Bureau of Foods, FDA. Readers are advised to study this chapter since it may very well set the tone for what may govern the future development of food additives.

At the onset, the editor had every intention of presenting two additional chapters of considerable importance. One would have dealt with the sort of safety tests currently deemed necessary to seek regulatory approvals for new food additives. The other chapter would have dealt with colorants of natural origins or those deemed as "nature identical". Both subjects are in a state of flux and not sufficiently finalized to be presented in a nearly complete fashion. We shall attempt to present more complete overviews on these and other topics in subsequent volumes of the CRC *Handbook of Food Additives.*

Thomas E. Furia

CONTRIBUTORS

Karl Beck
Manager, Market Research and Development, Chemical Division
Abbott Laboratories
North Chicago, Illinois

Guy A. Crosby
Director, Chemical Synthesis
Dynapol
Palo Alto, California

Joseph A. Fioriti
Senior Research Specialist
Technical Center
General Foods Corporation
White Plains, New York

Frank Fischetti, Jr.
Assistant Director, Flavor Operations
Fritzsche Dodge & Olcott, Inc.
New York, New York

Thomas E. Furia
President
Intechmark Corporation
Palo Alto, California

Harry N. Meggos
Manager, Color Service Laboratory
Warner-Jenkinson Company
St. Louis, Missouri

James E. Noonan
Senior Vice-President
Warner-Jenkinson Company
St. Louis, Missouri

Rex J. Sims
Research Scientist
Technical Center
General Foods Corporation
White Plains, New York

L. A. Underkofler
Retired
Carlsbad, New Mexico

Virgil O. Wodicka
Consulting Food Technologist
Fullerton, California

TABLE OF CONTENTS

Legal Considerations on Food Additives 1

Antioxidants as Stabilizers for Fats, Oils, and Lipid-Containing Foods 13

Enzymes 57

Nonnutritive Sweeteners: Saccharin and Cyclamate 125

New Sweeteners 187

Natural and Artificial Flavors 229

Synthetic Food Colors 339

Index 385

LEGAL CONSIDERATIONS ON FOOD ADDITIVES

V. O. Wodicka

DEFINITION

The term "food additive" is an unfortunate one. It is innately pejorative. Intrinsically, it designates something added to food. Logically, if it is something added to food, it is not itself food. If it is not food, why should it be added? The characterization of an ingredient as a food additive immediately stimulates suspicion. As a consequence, most people reserve the term for those ingredients of food that they consider unnecessary and at least possibly harmful.

In extension of this concept, many countries, including a number of the major ones, classify food ingredients as:

1. Normal ingredients
2. Processing aids
3. Food additives
4. Contaminants

Normal ingredients are often categorized as such if they can be consumed alone as food. The nature of processing aids is fairly obvious from the name. The presumption is that these aids will be totally or almost totally lost in processing, thereby not adversely affecting the properties of the food, such as color, flavor, etc., and not presenting problems of safety. Contaminants are such things as pesticide residues, substances transferred in from packaging or other food transfer surfaces, and substances produced by invading organisms, e.g., mycotoxins. (Pesticides, however, sometimes get separated out in a category of their own.) The miscellany, not covered by these categories, constitutes the food additives.

Now, consider the application of this system to flavored ice cream. The strawberries in strawberry ice cream are normal ingredients, because strawberries are eaten as food. The chocolate in chocolate ice cream is a normal ingredient, because chocolate is sometimes eaten as food. (This may be stretching the point, because chocolate is seldom consumed without admixture with sugar and usually other ingredients.) The vanilla in vanilla ice cream, however, and the oil of peppermint in peppermint ice cream are food additives, because these flavorings are not eaten alone.

Another example of the application of this system occurs with leavenings. Yeast used in making bread is a normal ingredient, because it can be and sometimes is consumed alone as food. The baking powder used in biscuits, however, is a food additive, because it would not be consumed alone. These examples could be multiplied many times over, but they will suffice to show that the categorization that appears logical in the abstract leads to some strange results in its concrete particulars. Determination of policies pertaining to the large categories becomes difficult and is obviously arbitrary.

Over 100 countries, including all the major ones, have banded together under the joint auspices of the Food and Agriculture Organization (FAO) and the World Health Organization (WHO) to establish a Codex Alimentarius Commission.[1] The major goal of the commission is to develop a set of food standards to govern international trade. The rules require, however, that any country adopting a particular standard must use it for domestic as well as international commerce. The collected volume, to be known as the *Codex Alimentarius,* has not yet been issued in its first edition, in spite of over 10 years of work, but there are over 100 standards now being circulated to member countries for adoption. When a substantial number of countries have taken a reasonable number of actions, the volume can be issued with the realization that it will require frequent updating.

Meanwhile, the work of the commission has brought together the regulatory people from most of the countries of the world and caused them to interchange views on common problems. Both formally and informally, progress is being made toward uniformity.

The definition used by the Codex Alimentarius Commission is

'Food additive' means any substance not normally consumed as a food by itself and not normally used as a typical ingredient of the food, whether or not it has nutritive value, the intentional addition of which to food for a technological (including organoleptic) purpose in the manufacture, processing, preparation, treatment, packing, packaging, transport or holding of such food results, or may be reasonably expected to result, (directly or indirectly) in it or its by-products becoming a component of or otherwise affecting the characteristics of such foods. The term does not include 'contaminants' or substances added to food for maintaining or improving nutritional qualities.[2]

This definition:

1. leaves much room for judgment regarding what is a "typical" ingredient,
2. includes only direct additives through use of the word "intentional", and
3. excludes contaminants and nutritional adjuncts, both of which can be additives under U.S. law.

As might be expected, there is also a definition for contaminants, and pesticides, also defined, constitute a separate category.

The Codex Alimentarius Commission has a Committee on Food Additives, for which the host country is the Netherlands. Its responsibility is, in part, "To endorse or establish permitted maximum levels for individual food additives and for contaminants in specific food items."[3]

The commission also has set forth a set of general principles for the use of food additives[4] which requires that they be proven safe for their intended use, that they have definitive specifications, and that they serve one or more of four defined purposes:

1. to preserve nutritional quality,
2. to provide necessary ingredients for foods for groups of consumers having special dietary needs.
3. to enhance keeping quality or improve organoleptic properties, or
4. to provide a processing aid.

Through the years, such practices as adding formaldehyde to milk as a preservative, adding borax to curing salts to preserve meat, or adding any of a variety of cheap white powders to flour to reduce the cost of a loaf of bread, led to laws in essentially every major country to control and regulate the ingredients used in preparing food for sale. Laws were prompted partly by concern for the public health, partly by aesthetic considerations, and partly by the desire to prevent economic fraud. In any specific case, it is hard to sort one from the other.

The fact that some of these now-forbidden ingredients were used has led to the feeling by most consumers and regulatory officials (who are usually trained in law rather than science) that any ingredient with which they are not familiar is probably fraudulent and used in the interest of the purveyor, not the consumer. The genesis of the term, "food additive", is understandable as is the opprobrium or suspicion it carries with it all over the world.

The terminology in the U.S. is much more complicated and confusing, in that the term, "food additive", defined legally, is only very loosely in correspondence with the popular understanding of the term. Many substances in the U.S. are legally food additives, although the general public would consider them normal ingredients; many other ingredients are not legally food additives, when most people regard them as such. This confusion in terminology will be discussed in detail later.

The fact remains that there is no better term than "food additive" to generically describe that miscellaneous collection of minor ingredients of food added to improve its functional properties in some respect, i.e., color, flavor, texture, resistance to spoilage, uniformity, nutritive value, or others. As a consequence, in spite of its innate drawbacks, the term, "food additive", will continue to be used until someone comes up with a better one.

U.S. LEGAL STATUS

Interstate commerce in the U.S. in food other than meat or poultry is controlled by the Federal Food, Drug, and Cosmetic Act, passed in 1938.[5] An amendment, added in 1958, deals with food additives. In Chapter II. Definitions, Section 201(s) says:

> The term 'food additive' means any substance the intended use of which results or may reasonably be expected to result, directly or indirectly, in its becoming a component or otherwise affecting the characteristics of any food (including any substance intended for use in producing, manufacturing, packing, processing, preparing, treating, packaging, transporting, or holding food; and including any source of radiation intended for any such use), if such substance is not generally recognized, among experts qualified by scientific training and experience to evaluate its safety, as having been adequately shown through scientific procedures (or, in the case of a substance used in food prior to January 1, 1958, through either scientific procedures or experience based on common use in food) to be safe under the conditions of its intended use; except that such term does not include —

1. A pesticide chemical in or on a raw agricultural commodity
2. A pesticide chemical to the extent that it is intended for use or is used in the production, storage, or transportation of any raw agricultural commodity
3. A color additive
4. Any substance used in accordance with a sanction or approval granted prior to the enactment of this paragraph pursuant to this Act, the Poultry Products Inspection Act (21 U.S.C. 451 and the following) or the Meat Inspection Act of March 4, 1907 (34 Stat. 1260), as amended and extended (21 U.S.C. 71 and the following)
5. A new animal drug

This is a long and complicated definition, and its meaning is by no means immediately manifest. In order to understand it, one must first strip away some of the legalisms and get down to the heart of it.

The exception from the definition of new animal drugs is necessary because Section 201(f) of the Act defines food as follows: "The term 'food' means 1. articles used for food or drink for man or other animals, 2. chewing gum, and 3. articles used for components of any such article." In other words, animal feeds are included in the classification of foods along with food for man. Since animal drugs are sometimes administered in medicated feeds, they would therefore be food additives if not explicitly excluded. Inclusion here would be redundant, because new animal drugs are regulated by other sections of the Act. Similarly, color additives are regulated by another section of the Act and do not need double coverage.

In the case of pesticides on raw agricultural products, regulation is under the Federal Insecticide, Fungicide, and Rodenticide Act, now administered by the Environmental Protection Agency, so that coverage here would cause confusion. It should be pointed out, however, that confusion is not completely avoided. If the pesticide residue on a raw agricultural commodity persists in a processed food, no further regulation is required if the level in the finished food is no higher than that in the raw agricultural commodity. In many or even most instances, however, moisture is removed from the food, thereby concentrating the residue. If the pesticide has not been removed or sufficiently reduced in some way, the residue level in the processed food will then be higher than that in the raw agricultural commodity. A regulation (21 CFR 170.19) sets forth the policy that this situation will be permitted only under a food additive regulation covering the situation.[6] These food additive regulations are issued by the Environmental Protection Agency, not by the Food and Drug Administration (FDA).

This leaves as an exception not covered elsewhere substances approved prior to the amendment of 1958. It is reasonable to set these materials aside, since the major intent of the amendment was to control new additions to the food supply and these materials had been reviewed previously for safety and appropriateness by responsible legal bodies.

This status of previous approval, commonly known as "prior sanction", presents some problems, however. First among them is that relatively few of these approvals appeared in published form. Many of them were based on laws of 1906 and 1907, long before there was a *Federal Register* or *Code of Federal Regulations;* many of them were contained in simple letters to a processor who wanted to use some substance without running afoul of the regulatory authorities. Although records of this correspondence may exist, finding them in either governmental or industrial files is almost beyond the powers of man. As a consequence, the scope of this exception is uncertain.

The second problem is that prior sanction status is accorded only to substances approved under specific laws. Substances approved by the U.S. Public Health Service for use as sanitizing agents for food-handling equipment, for water treatment, and for dairy use, are not covered by this provision of the act. Their use must rest on some other legal basis.

The third problem is that approval given in individual letters is not a matter of public knowledge. The company that has such a letter is authorized to use the substance, but no other company has any knowledge of the approval and therefore does not have the benefit of it.

Finally, the fact that these substances are specially exempted by the law makes it more than ordinarily complicated to ban them or limit their use on the basis of later safety information. Because approval is fixed by the law, it cannot be changed except by law, so it cannot be simply withdrawn by the regulatory agency. This does not bar regulatory action against the substance, but that action must be taken on the basis of other sections of the law. This makes for a difficult and cumbersome procedure.

Now, setting aside these exemptions, one can see that every substance which is or might become a component of food or which affects the properties of food is a food additive, unless that substance is generally recognized as safe for its intended use by properly qualified experts. This means that except for the few substances covered by the exemptions, the entire food supply would consist of food additives if its individual components were not "generally recognized as safe" (commonly referred to by its acronym, GRAS). It follows from the fact that the legal definition of food includes components of food (which is logical enough) and the legal definition of food additives includes all components of food (again, except for the exemptions) unless they are

GRAS. Obviously, the bulk of the food supply — the meat and potatoes, bread and milk, oats, peas, beans, and barley — is legal, because it is GRAS.

Taking a simple example to see how this definition works, suppose one wished to make mashed potatoes. One would peel some potatoes, boil them, and mash them with a bit of milk, butter, and salt. What are the legal implications of our actions? The potatoes are legal, because they are GRAS, even though they would not meet the safety criteria for food additives. The knife, which was used for peeling (thereby affecting the properties of the food), is made of stainless steel, so it did not add measurably to the content of iron, carbon, nickel, or chromium of the potatoes; however, by affecting the properties of the potatoes, its status came into question. It is GRAS. The water in which the potatoes were boiled is GRAS. The saucepan in which they were boiled was aluminum and therefore GRAS. The gas flame over which they were boiled affected their properties, but the operation is GRAS. The milk, butter, and salt are all GRAS. Now one can relax; the dish is legal.

This fact that the legal basis for consumption of most of the food supply and for conducting most kitchen operations is that all these things are GRAS is not understood by many people, including most members of Congress. There is often talk of abolishing GRAS status, which would be a difficult operation, indeed. Some substitute legal basis would have to be provided to authorize consumption of all traditional foods and use of all customary kitchen processes and utensils without a comprehensive safety review of each food, each process, and each utensil. Such a review would either be so superficial as to be farcical or so overwhelming that the paperwork alone would take long enough to cause one to starve before it was legal to eat.

GRAS STATUS

How does one know whether something is GRAS? The law does not provide a means of determining this, and it does not name a government agency to make this determination. The only way to make a definitive determination would be to poll experts qualified by scientific training and experience to evaluate safety and find the consensus. For the thousands of foods and kitchen processes and utensils used, this would be an overwhelming task, and few experts, indeed, could do more than give an offhand opinion; the sheer magnitude of the job would preclude any sort of systematic search of the scientific literature, item by item.

The government agency closest to this problem, of course, is the FDA. From 1958 until 1971, there were only two kinds of action it had taken to address this problem. One of these, an obvious one, was to write letters of opinion regarding the status of particular ingredients or practices. These letters continued all during the period indicated. This practice had the obvious disadvantage that the opinions were not a matter of general knowledge but were usually only known to the correspondent.

In the period right after the passage of the Food Additives Admendment in 1958, there were many inquiries to the FDA about minor ingredients that had then been long in use. Were they additives? This question arose particularly for ingredients that were chemically pure rather than complex mixtures of chemicals, such as foods which are the products of agriculture.

In self-defense, the FDA made up a list of substances about which they were receiving frequent questions and added other similar substances in wide use to the list. They sent the list to a large number of scientists whose judgments might qualify under the law and asked them for their opinions on the safety of each substance. Based on the replies received, the agency published the list (21 CFR 182), thereby affirming the GRAS status of these materials. Many people falsely assume that only materials on this list are officially GRAS. In fact, as indicated above, most of the food supply is GRAS, and the materials listed in this regulation are only those about which questions had arisen or might arise; the list was published as a convenience to all.

In an attempt to reduce confusion and put some order into the situation, the FDA published a regulation in 1971 (21 CFR 170.30) and revised it in 1976, construing the law and setting forth its policies in this area. The key paragraph is worth quoting verbatim:

> The food ingredients listed as GRAS in Part 182 of this chapter or affirmed as GRAS in Part 184 or §186.1 of this chapter do not include all substances that are generally recognized as safe for their intended use in food. Because of the large number of substances the intended use of which results or may reasonably be expected to result, directly or indirectly, in their becoming a component or otherwise affecting the characteristics of food, it is impracticable to list all such substances that are GRAS. A food ingredient of natural biological origin that has been widely consumed for its nutrient properties in the United States prior to January 1, 1958, without known detrimental effects, which is subject only to conventional processing as practiced prior to January 1, 1958, and for which no known safety hazard exists, will ordinarily be regarded as GRAS without specific inclusion in Part 182, Part 184, or §186.1 of this chapter.

In effect, this paragraph makes clear that there is nowhere a definitive list of substances or processes which are GRAS. Under the conditions just cited, substances and processes, used in the U.S. prior to 1958, are assumed to be GRAS unless some question arises regarding their safety. The FDA, however, reserves judgment about the status of:[7]

1. GRAS substances modified by processes developed since 1958 which may reasonably be expected to alter their composition
2. Substances, otherwise GRAS, which have had significant alteration of composition by breeding or selection after January 1, 1958, when that change may reasonably be expected to alter the nutritive value or the concentration of toxic constituents
3. Distillates, isolates, extracts, and concentrates of extracts of GRAS substances
4. Reaction products of GRAS substances
5. Substances, not of natural biological origin, including those purporting to be chemically identical to GRAS substances
6. Substances of natural biological origin consumed for other than nutrient properties.

In each of these cases, the FDA will give an opinion only after reviewing the evidence.

The critical importance of the year 1958 (the year the Food Additives Amendment was enacted) is manifest. In effect, constituents of the food supply of that year were made legal by a "grandfather" provision (GRAS status). Anything added after that date, be it ingredient or process, is subject to review. Nothing introduced after 1958 can be determined GRAS on the grounds of "experience based on common use in food"; it can only by GRAS on the basis of "scientific procedures". The interpretive regulation (21 CFR 170.30) also addresses this distinction. In paragraph (b), it says: "General recognition of safety based upon scientific procedures shall require the same quantity and quality of scientific evidence as is required to obtain approval of a food additive regulation for the ingredient. General recognition of safety through scientific procedures shall ordinarily be based upon published studies which may be corroborated by unpublished studies and other data and information." The distinction here is that a food additive petition can be supported by unpublished safety evidence, even though the FDA does make it publicly available in the Office of the Hearing Clerk.

On the other hand, a material can hardly be **generally** recognized as safe if the key safety evidence is not generally available before the question arises.

An ingredient or process newly introduced, therefore, may be either an additive or GRAS, the distinction being largely a matter of administrative convenience. Moreover, a supplier may legally introduce a new ingredient without a word to or from the FDA if he has evidence of safety in the public domain so convincing that properly qualified experts generally recognize it as safe. This situation has left many people uneasy since it leaves a large loophole in regulation of the safety of food components. It means that the FDA must monitor the food supply for new food ingredients, make judgments on their safety, and act against those for which it does not find convincing safety evidence.

The saving grace is that punitive action by the FDA would have such dire commercial consequences that few large companies would risk their public images in this way. Accordingly, suppliers find it practically, though not legally, necessary to obtain official opinions from the FDA regarding the safety of their products before they market them. The FDA, in turn, has decided that it will no longer establish policy in such matters by private correspondence but will publish its opinion in regulations. It has created a procedure (21 CFR 170.35) by which it will affirm that a substance or process is GRAS, thereby giving public notice of its opinion. It is obviously prudent and commercially valuable for an ingredient supplier to obtain such an affirmation for a new material. It protects both him and his customers.

From the foregoing discussion, it becomes apparent that what we have loosely defined as a food additive, an ingredient added in small quantities to affect the properties of the food, may be either a food additive or a GRAS substance under U.S. law. The distinction between the two is mostly administrative and psychological. Most suppliers would rather have their products GRAS, because this category includes most common natural foods. Whichever way the substance is classified, however, the FDA must be satisfied that it is safe for its intended use or it will remove the substance from the marketplace.

INDIRECT ADDITIVES

It has long been known that liquids stored in glass vessels can extract sodium and sometimes silicic acid from the glass. Liquids in "tin" cans in long storage increase in level of iron, tin, and lead. Foods in plastic packages can pick up plasticizers, antioxidants, monomers, and other constituents of the package. Foods, particularly acid foods, which are cooked in aluminum vessels show increased aluminum levels. Wines in oak barrels extract constituents of the oak. In short, foods in long contact with most surfaces pick up at least trace quantities of materials in or on the surfaces.

In many, or even most, countries, these materials extracted are defined and regulated as contaminants. In the U.S., these materials are food additives under the law. They are commonly classed as indirect additives, but they are handled in the same way as direct additives by the FDA. Just as with direct ingredients, they may be generally recognized as safe (GRAS) and thereby escape explicit regulation because that status makes them ipso facto not food additives. In general, however, regulation of these materials is more extensive and more rigorous in the U.S. than in other countries. As might be expected, packaging materials such as glass, which have been used for a long time, receive less close scrutiny than the more newly introduced materials and those materials just being proposed for introduction.

Indirect food additives — from food contact surfaces in processing or kitchen preparation, from packaging, from sanitizer residues, etc. — constitute a large and diverse family of materials that are in general quite different in chemical nature and the types of problems presented by direct additives. The governing regulations have grown in a sort of weedy fashion, and as an inevitable result, they are scheduled for review and reorganization. This large and diverse body of materials will not be discussed in this book.

FUNCTIONS OF ADDITIVES

For a long time, the U.S. Food and Drug Administration tried to control food constituents to assure safety by writing identity standards for foods with rigid control over their make-up, i.e., the so-called "recipe" standards. These standards regulated the foods, the various ingredients which could be used in them, and their levels. The *Codex Alimentarius* still takes this approach.

It soon became apparent that the American food industry was too large and creative to be controlled by the standardization process. It came out with new products faster than the FDA could write standards. The Food Additives Amendment was the reaction to this situation. It puts regulatory controls on the materials themselves rather than the foods in which they are used.

The problem with this later approach lies in the fact that any safety problems are associated with the level of intake, not just the nature of the substance. Unless the foods in which the material is used and the frequency distribution of the consumption of these foods are known to the agency, it is difficult, if not impossible, to estimate with any real confidence total intake of the ingredient in question. The FDA must estimate intake when a food additive regulation is issued; however, over a period of years, practices change, and the FDA has no systematic way of obtaining this information.

In 1970, the FDA began a survey of the safety of ingredients on the GRAS list (21 CFR 182). In order to establish intake levels, it contracted with the National Academy of Sciences to conduct a survey of the use of items on the list and a few additional GRAS items. A follow-up contract was planned for regulated additives. The first contract was also coordinated with a voluntary survey conducted by the Flavor and Extract Manufacturers Association (FEMA) on behalf of the FDA to document the use of flavoring substances, both GRAS and additive.

In conducting these surveys, it became necessary to establish classes of foods and classes of uses in order to make possible any kind of summary of the findings. Accordingly, there is now a list of functions of food additives in the larger, not the legal sense, that has had the benefit of some exposure and use and some regulatory recognition. It was published, after the usual comment period, along with a list of food classes in 21 CFR 170.3. It supplies a useful list of the major functions of additives. Stripped of definitions, the list is as follows:

1. Anticaking agents and free-flow agents
2. Antimicrobial agents
3. Antioxidants
4. Colors and coloring adjuncts
5. Curing and pickling agents
6. Dough strengtheners
7. Drying agents
8. Emulsifiers and emulsifier salts
9. Enzymes
10. Firming agents

11. Flavor enhancers
12. Flavoring agents and adjuvants
13. Flour-treating agents
14. Formulation aids
15. Fumigants
16. Humectants
17. Leavening agents
18. Lubricants and release agents
19. Nonnutritive sweeteners
20. Nutrient supplements
21. Nutritive sweeteners
22. Oxidizing and reducing agents
23. pH Control agents;
24. Processing aids
25. Propellants, aerating agents, and gases
26. Sequestrants
27. Solvents and vehicles
28. Stabilizers and thickeners
29. Surface-active agents
30. Surface-finishing agents
31. Synergists
32. Texturizers

This list is obviously much more detailed and specific than that of the *Codex Alimentarius.* It should also be obvious that many materials performing these functions are not legally food additives in the U.S., because they are GRAS.

SAFETY

Much of the public concern, world-wide, about the use of food additives relates to fears about safety and has generated some sort of regulatory structure in every major country as well as in international bodies to monitor this aspect of the field. There is a Joint Experts Committee on Food Additives, set up by the FAO and the WHO, to consider the safety of additives and set specifications and limits for them. These limits take the form of an Acceptable Daily Intake (ADI). The Codex Committee on Food Additives is required to follow the safety guidelines of the Joint Experts Committee. Its safety criteria are generally not very different from those used in the U.S., although they are not codified. In the U.S. criteria for food additives are stated in 21 CFR 170.22, **Safety factors to be considered,** and 21 CFR 170.20, **General principles for evaluating the safety of food additives.** The key sentence, which also runs through the decisions in other countries, says, "A food additive for use by man will not be granted a tolerance than will exceed 1/100th of the maximum amount demonstrated to be without harm to experimental animals." Corresponding regulations for color additives are 21 CFR 8.34, **Safety factors to be considered** and 21 CFR 8.35 **Criteria for evaluating the safety of color additives.** Provisions are essentially identical.

It is to be remembered, however, that in the U.S. these criteria apply only to substances that are legally food or color additives and by interpretive regulation to those substances that are GRAS on the basis of "scientific procedures". For those substances that are GRAS because of "experience based on common use in food", there are no rules. Decisions of safety depend on the knowledge and judgment of the "experts qualified by scientific training and experience to evaluate its safety", regardless of whether the general public would consider the ingredient to be a food additive or a normal ingredient. In most cases, the experimental evidence to which the rules would be applied does not exist, particularly for conventional foods which are often eaten as such, not just used as ingredients. The U.S. (and the rest of the world), therefore, has a double standard and realistically, will probably always have one.

One other aspect of the safety question deserves mention. In the U.S. and only in the U.S., there is a special provision of the law applicable to cancer. Introduced by Representative James Delaney of New York, it is widely spoken of as the Delaney clause. It says in part, " . . . no additive shall be deemed to be safe if it is found to induce cancer when ingested by man or animal". It is interpreted to mean that an additive is not to be permitted at any level, no matter how low, if it induces cancer at any level, no matter how high. It applies, however, only to food additives, with a corresponding clause for color additives. It is well known among scientists that many common foods contain constituents that induce cancer at high levels, but they are GRAS and therefore not subject to the exclusion.

This clause has generated much controversy because of recognition by most scientists that the continued existence of mankind is silent witness to the fact that low levels of carcinogens can be tolerated. With the tremendous progress in recent years in molecular biology, scientists are now beginning to understand some of the mechanisms responsible for this resistance to outside insult. In fact, the mechanisms must exist to protect the body from its own mistakes in cell building. The focus of the problem is that so far there has not been a generally accepted way of determining safe levels. Meanwhile, the constantly increasing capabilities of analytical chemistry to measure first milligrams, then micrograms, then nanograms, with occasional excursions into picograms and even femtograms increasingly extends the reach of this provision of the law.

The main thrust of the clause, preventing the approval of a new additive that induces cancer when eaten, is not causing the raising of objections; nobody is likely to propose such an additive. The problem that hangs like the sword of Damocles above the head of every purveyor of food is that new evidence is likely to find carcinogenesis in some long-established ingredient, thereby causing it to be blamed if it is a food or color additive or stirring great public argument if it is in one of the exempt categories, such as GRAS or prior sanction. Most major companies would eliminate the ingredient without waiting for a ban just to maintain their public images. Some disruption has already occurred along these lines, and more is virtually certain.

THE REGULATORY OUTLOOK

The crystal ball is clouded and obscures the future in regulatory trends, primarily because of uncertainties in the U.S. posture.

On the world level, there will probably be a fairly stable situation over the next few years. There are likely to be some changes in detail in the way safety evaluations are made in the Joint Experts Committee on Food Additives, but the effect on the substantive decisions is not likely to be profound. Previous decisions in this group have not been so bad that changes in techniques are likely to upset them. The primary question with reference to this group and also the Codex Committee on Food Additives has to do with the fact that the U.S. has been a strongly influential member of both groups and the position of the U.S. in food safety questions is currently unsettled. Since the U.S. is only one member, however, the work of both groups will probably carry on close to its present course, which may be characterized as conservative and sound.

If one considers Europe, even more stability can be expected. The European Economic Communities (E.E.C.) is steadily becoming more influential in regulatory matters of much common interest, such as food. Many of the same people are active here as in the Codex committees, and a similarly stable, solid activity is to be expected. This is not to say that everything will stay as it is. The safety of a number of long-used materials has been called into question, and the consequence could well be a ban or restriction on some of them. The outcome as this is written depends on the results of studies in progress, so it is essentially unpredictable. If the E.E.C. takes action, it is likely that the Codex committees and the U.S. will follow.

The situation in Japan is hard to assess from a distance. The regulatory authorities are competent and responsible but are increasingly subject to pressure from a strong consumer movement which has already forced some scientifically shaky decisions. In general, safety testing for materials to be used in Japan must be at least partly done in Japan. The trend has been toward tighter regulations than previously.

SUMMARY

Food additives may be looked upon as minor ingredients incorporated into foods to affect their properties in some desired way. Most commonly, the effects desired relate to color, flavor, texture, nutritive value, or stability to storage. There is no rigorous definition that meets all needs.

The *Codex Alimentarius,* which dominates actions in international circles, considers an additive as an ingredient "not normally consumed as a food by itself and not normally used as a typical ingredient." This obviously leaves great latitude for judgment by the committee. The Federal Food, Drug, and Cosmetic Act in the U.S. has a complex definition of food additives that comes close to any component of food introduced into U.S. commerce after 1957.

The safety of food additives is of public concern all over the world. Suspicions are aroused by the fact that these ingredients are commonly declared on label statements by their chemical names, which are unfamiliar and frightening to many people. As a consequence, most countries and the *Codex Alimentarius* make special regulatory provisions for food additives to assure safety. In general, regulations are so structured that even consumers who ingest extreme quantities will not ingest more than 1/100 of the maximum amount demonstrated to be without harm to experimental animals. (Levels are usually translated from the animals to man in terms of milligrams of additive per kilogram of body weight.)

The work of the Joint Experts Committee on Food Additives of the FAO and WHO has done much to improve communication among experts of the developed countries and consequently to cause their regulatory actions to converge. There is generally relatively good agreement among the experts, but political influences in individual countries sometimes produce differences in regulatory actions.

REFERENCES

1. Introduction, in Secretariat of the Joint FAO/WHO Food Standards Programme, *Codex Alimentarius Commission Procedural Manual,* 4th ed., Food and Agriculture Organizaton, Rome, 1975.

2. Secretariat of the Joint FAO/WHO Food Standards Programme, General principles, in *Codex Alimentarius Commission Procedural Manual,* 4th ed., Food and Agriculture Organization, Rome, 1975, 26.
3. Secretariat of the Joint FAO/WHO Food Standards Programme, Subsidiary bodies, in *Codex Alimentarius Commission Procedural Manual,* 4th ed., Food and Agriculture Organization, Rome, 1975, 74.
4. Secretariat of the Joint FAO/WHO Food Standards Programme, General principles for the use of food additives, in *Codex Alimentarius Commission Procedural Manual,* 4th ed., Food and Agriculture Organization, Rome, 1975, 71.
5. Federal Food, Drug, and Cosmetic Act as Amended, U.S. Government Printing Office, Washington, D.C., 1976.
6. Title 21, in the *Code of Federal Regulations,* Office of the Federal Register, U.S. Government Printing Office, Washington, D.C., 1976
7. Office of the Federal Register, 21 CFR 170.30(f), in the *Code of Federal Regulations,* U.S. Government Printing Office, Washington, D.C., 1979.

ANTIOXIDANTS AS STABILIZERS FOR FATS, OILS, AND LIPID-CONTAINING FOODS

R. J. Sims and J. A. Fioriti

EDITOR'S NOTE

Since the completion of the writing of this chapter in April of 1977, several important regulatory issues have emerged concerning the use of several key antioxidants employed as preservatives in food. The reader is advised to keep current with these issues since they are expected to significantly affect the use of such products in foods produced in the U.S. and other nations.

1. The U.S. Food and Drug Administration has proposed to restrict the use of butylated hydroxytoluene (BHT) as a direct and indirect additive to food pending completion of animal feeding studies necessary to resolve certain safety questions about the metabolism of BHT in man and the toxic effects seen in experimental animals *(Federal Register*, pg. 27603 to 27609; May 31, 1972). The proposal would remove BHT from the list of substances that are generally recognized as safe (GRAS). However, the proposal would allow continued use of BHT in fat-containing foods under newly established interim food additive regulations.
   On July 25, 1977, the FDA was officially petitioned in a collective action submitted by Dr. M. Jacobson, the Center for Science in the Public Interest and Nutrition Action to ban the use of BHT.
2. On June 7, 1977, the FDA announced an opportunity for public hearing on the safety of certain food ingredients to determine if they are generally recognized as safe (GRAS) or subject to a prior sanction *(Federal Register*, pg. 29105 to 29107; June 7, 1977). Among the ingredients listed for safety hearings is butylated hydroxyanisole (BHA).

T. E. F.

INTRODUCTION

Antioxidant technology is important in preserving edible fats, oils, and lipid-containing foods from development of objectionable flavors and odors, much as those resulting from oxidative rancidity and from formation of decomposition products and polymers, which may be toxic. Successful application of antioxidants requires a basic knowledge of fat and oil chemistry, the mechanism of oxidation, and the function of an antioxidant in conteracting this type of deterioration. It should be emphasized that antioxidants, to be effective, must be used with food materials of good quality. They will not protect (or mask) fat or fatty food which has already deteriorated from abusive storage or which was prepared from damaged raw materials. Best results can be expected only if the antioxidants are incorporated promptly into freshly prepared products of good quality and if these products are subsequently packaged properly and stored under correct conditions.

This chapter will include a discussion of the causes and mechanisms of oxidative deterioration in fats and fatty foods; however, more attention will be directed to a discussion of the selection and use of antioxidants as well as methods for measuring oxidative deterioration. Hopefully, this information will enable the reader to solve some of the practical problems encountered in the commercial stabilization of fats and fatty foods. References to the original literature will be included for those interested in more detailed documentation or discussions of the theoretical aspects of antioxidant usage.

COMPOSITION OF NATURAL FATS

The major components of natural, edible fats and mixed triglycerides of unsaturated and saturated long-chain fatty acids containing even numbers of carbon atoms.

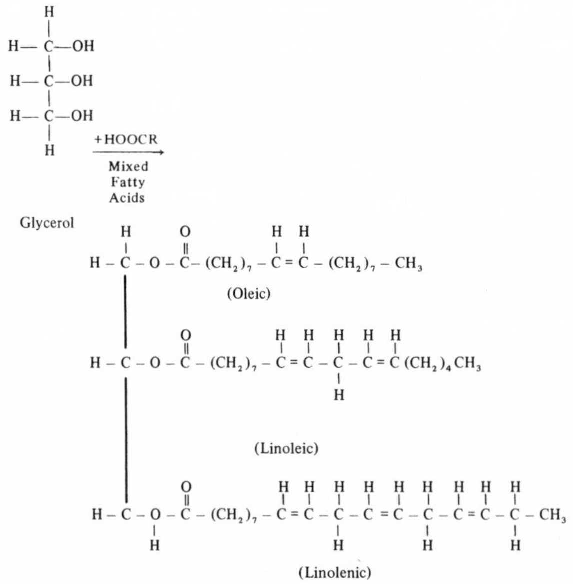

Glycerol: H–C(H)(OH)–C(H)(OH)–C(H)(OH)–H

$\xrightarrow[\text{Mixed Fatty Acids}]{+\text{HOOCR}}$

$H-\overset{H}{C}-O-\overset{O}{\overset{\|}{C}}-(CH_2)_7-\overset{H}{C}=\overset{H}{C}-(CH_2)_7-CH_3$

(Oleic)

$H-C-O-\overset{O}{\overset{\|}{C}}-(CH_2)_7-\overset{H}{C}=\overset{H}{C}-\underset{H}{\overset{H}{C}}-\overset{H}{C}=\overset{H}{C}(CH_2)_4CH_3$

(Linoleic)

$\underset{H}{H-C}-O-\overset{O}{\overset{\|}{C}}-(CH_2)_7-\overset{H}{C}=\overset{H}{C}-\underset{H}{\overset{H}{C}}-\overset{H}{C}=\overset{H}{C}-\underset{H}{\overset{H}{C}}-\overset{H}{C}=\overset{H}{C}-\underset{H}{\overset{H}{C}}-CH_3$

(Linolenic)

The diagram shows a single unsaturated triglyceride containing a glycerol moiety which is esterified with fatty acids having one, two, and three double bonds. In the natural state, virtually all the unsaturation occurs in the *cis* configuration and it is methylene interrupted. Most vegetable oils used for food, e.g., cottonseed, soybean, peanut, corn, and safflower, contain rather high levels of these unsaturated acid esters. Animal fats, e.g. lard and beef and mutton tallow, contain much higher levels of saturated fatty acid glycerides, e.g., palmitic, C_{16} and stearic, C_{18}. Since the rate of autoxidation is dependent upon the number of double bonds in the molecule, one would expect vegetable oils to be much more susceptible to deterioration than animal fats. However, vegetable oils also contain significant quantities of tocopherols which function as natural antioxidants. The alpha isomer shown here has both antioxidant and vitamin E activity.

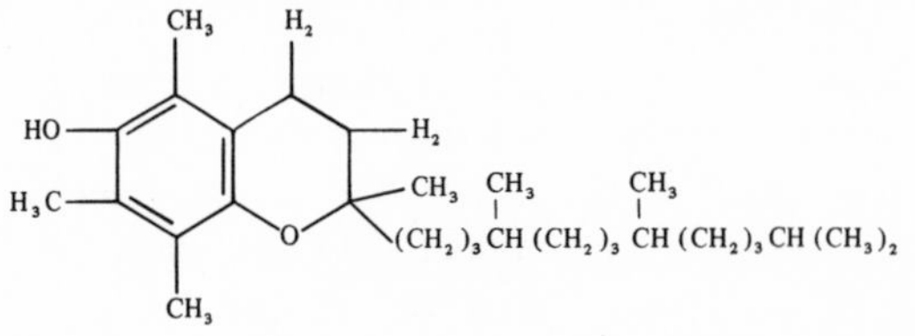

CH$_3$ H$_2$ HO H$_3$C– O –H$_2$ CH$_3$ CH$_3$

$CH_3\quad\overset{CH_3}{|}\qquad\overset{CH_3}{|}$

$(CH_2)_3CH(CH_2)_3CH(CH_2)_3CH(CH_3)_2$

CH$_3$

Since tocopherols are largely absent from animal fats, they tend to oxidize more rapidly than would be expected from their chemical compositions.

Natural fats also contain significant quantities of phospholipids. These materials consist of polyhydric alcohols, such as glycerol, which are esterified with fatty acids and also phosphoric acid. The phosphoric acid is, itself, combined with basic nitrogen compounds, such as choline and ethanolamine. These two bases are contained in the most common phospholipids, lecithin and cephalin.

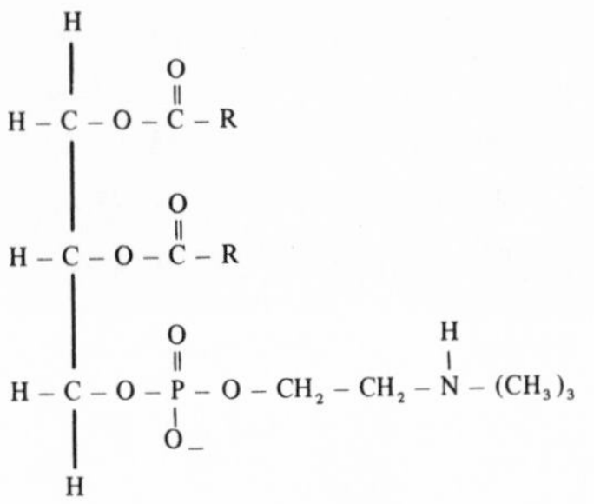

α-Lecithin

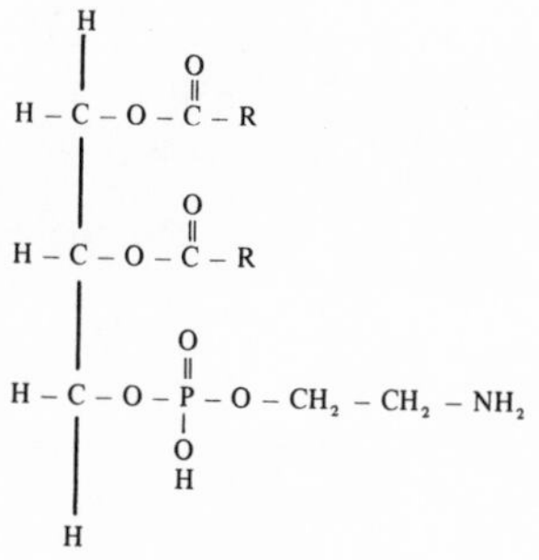

α-Cephalin

Since phospholipids also contain unsaturated fatty acid chains (shown here as R groups), they are susceptible to attack by oxygen in the same manner as the unsaturated triglycerides. Crude soybean and cottonseed oils contain 1 to 2% phosphatides, but most of these are removed during refining by a water wash. While most animal fats contain very little phospholipid, there are certain exceptions. For example, egg oil is very rich, about 20%, in phosphatides.

Mono- and diglycerides are minor components of natural fats except those which have undergone hydrolysis. Enzymatic hydrolysis of natural fats may generate substantial amounts of mono- and diglycerides as well as free fatty acids. Since monoglycerides contain the same unsaturated fatty chains as the parent triglyceride, they are also prone to oxidation. Monoglyceride concentrates are prepared synthetically for use as emulsifiers in various applications, including shortening for baking and complexing agents to modify the texture of starch.

The major pigments in edible fats are the carotenoids. β-Carotene has the structure shown below.

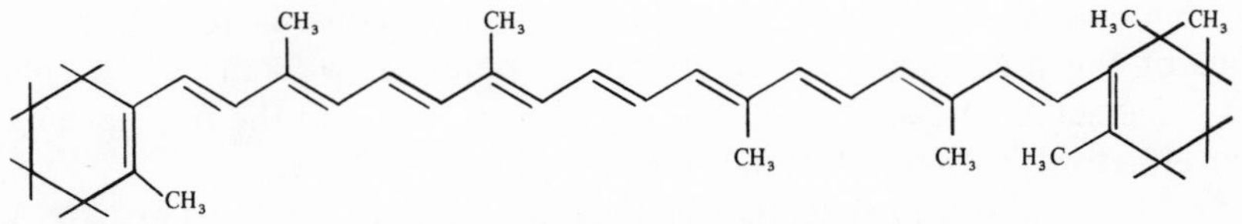

β-Carotene

Since it contains a large number of double bonds, it will oxidize quite readily. Oxidation is evident when the normal deep orange color of β-carotene bleaches to a light yellow.

Various other components occur in minor amounts in natural fats. The sterols, e.g. cholesterol and sitosterol, are not usually considered to be important factors in fat oxidative stability, since they are relatively inert. However, cholesterol, which is present in animal fats, may play a role in the origin of atherosclerosis. Squalene, a highly unsaturated hydrocarbon, is found in many natural fats. Upon oxidation, it produces sharp, rancid odors.

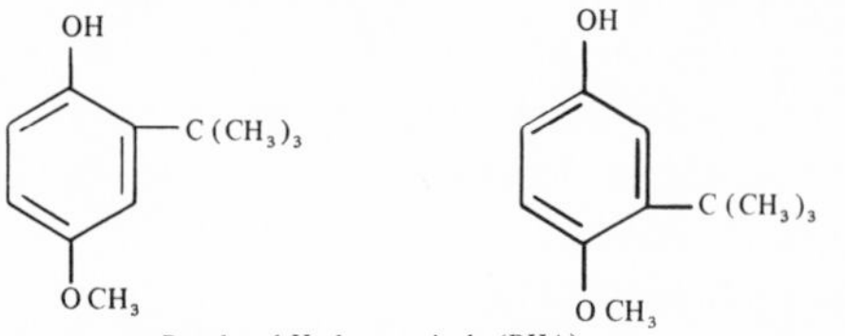

Butylated Hydroxyanisole (BHA)

3-BHA (3-*tert*-butyl-4-hydroxyanisole; 2-*tert*-butyl-4-methoxyphenol)

2-BHA (2-*tert*-butyl-4-hydroxyanisole 3-*tert*-butyl-4-methoxyphenol)

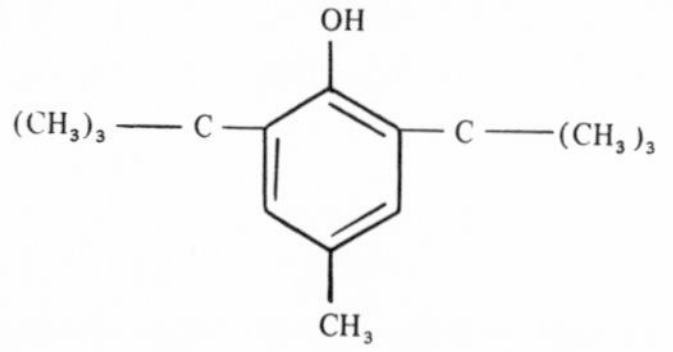

Butylated Hydroxytoluene (BHT) (2,6-di-*tert*-butyl-*o*-cresol; 4-methyl-2,6-di-*tert*-butylphenol)

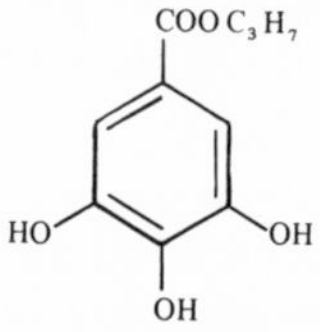

Propyl Gallate (PG) Propyl gallate (*n*-propyl-3,4,5-trihydroxybenzoate)

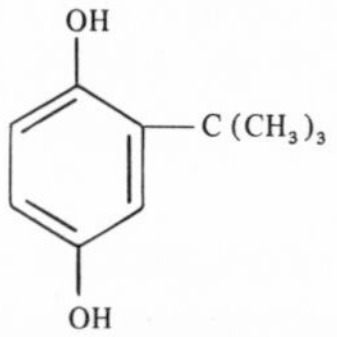

Mono-*tert*-butylhydroquinone (TBHQ)

STABILIZING FATS AND FATTY FOODS TO OXIDATION

Use of antioxidants in the U.S. for protection of foods from oxidation is limited to those additives permitted by the Food and Drug Administration. Additional requirements for a suitable additive include adequate potency at a reasonable cost and no development of objectionable color, flavor, or odor. Furthermore, the antioxidant, usually in combination with a metal scavenger, must be readily dispersible in the fatty phase. One of the problems most frequently encountered with the use of phenolic stabilizers is the achievement of uniform dispersion throughout the fat, since in order to function effectively, they must be in solution in the fatty phase.

The antioxidants most frequently used in the U.S. include: butylated hydroxyanisole (BHA), butylated hydroxytoluene (BHT), propylgallate (PG), and mono-tert-butylhydroquinone (TBHQ)

These compounds are frequently added in combination with citric acid which inactivates metals by the formation of complexes. Since PG and citric acid both have limited fat solubility, they are often incorporated by using solvent carriers, such as propylene glycol or blends of propylene glycol and glycerol monooleate (GMO). Lack of fat solubility can be particularly troublesome in the case of PG. Although it is very effective in stabilizing fats against oxidation, it will partition into any aqueous phase which may be present thereby losing some of its effectiveness. If, in addition, the system contains iron in the form of water-soluble salts, bluish black colors may develop. For this reason, the food industry has largely moved away from PG in favor of BHA, BHT, and TBHQ.

As was mentioned previously, antioxidants are effective only with fresh fats which have not been allowed to oxidize. For this reason, it is important to incorporate these additives as early as possible in the preparation process. With animal fats this should be done immediately after rendering. In fact, some authors have revealed advantages in adding the more fat-soluble antioxidants (BHA and BHT) during the rendering process itself.[1] This addition provides protection during the subsequent separation of the proteinaceous residue and clarification.

Although animal fat responds well to BHA and BHT, these phenolics are almost completely ineffective with polyunsaturated vegetable oils. As mentioned earlier, these oils, e.g., soy and cottonseed, contain significant amounts of naturally occurring tocopherols and do not benefit greatly from the addition of synthetic phenolics. However, TBHQ seems to be an exception to this rule. It outperforms PG in these highly unsaturated oils and has the added advantage of not producing off-colors in the presence of iron.

Other antioxidants, which are food approved but find little commercial use, include thiodipropionic acid and dilauryl thiodipropionate. These antioxidants are relatively ineffective and, like many sulfur-containing compounds, cause odor and flavor problems in edible fats. Tocopherols and most other available natural antioxidants are of limited benefit when used in animal fats. When they are added to vegetable oils which already contain significant amounts of natural tocopherols, little additional protection is observed.

SELECTION OF ANTIOXIDANTS FOR SPECIFIC APPLICATIONS

Processors of animal fats may add an individual phenolic antioxidant to their product. More often, they prefer combinations of two or more, usually with the addition of citric acid as a metal scavenger.* Limits on the levels permitted will be given in a later section on government regulations.

If animal fats, such as lard, are to be sold in their natural state without deodorization, it is customary to add solutions of the phenolics plus citric acid in a solvent carrier, such as propylene glycol. With agitation, BHT, which does not dissolve easily in propylene glycol, may be added directly as a concentrated solution to the heated fat (approximately 140°F). Mixing should be carried out at an elevated temperature for some time, preferably with exclusion of oxygen to assure uniform dispersion throughout the solution. Mixtures are commercially available which contain both propylene glycol and monoglycerides, e.g., GMO as carriers for occasions when it is desired to add BHA, BHT, and citric acid as a single homogenous mix. Traces of metal salts (iron and copper) act as strong pro-oxidants. Since most fat contains some of these contaminants, it is necessary to complex the metals with citric acid or some other edible metal scavenger.

* In many cases, combinations of two or more phenolic antioxidants show synergism; for example, the degree of stabilization achieved is somewhat greater than that observed with the same level of a single antioxidant.

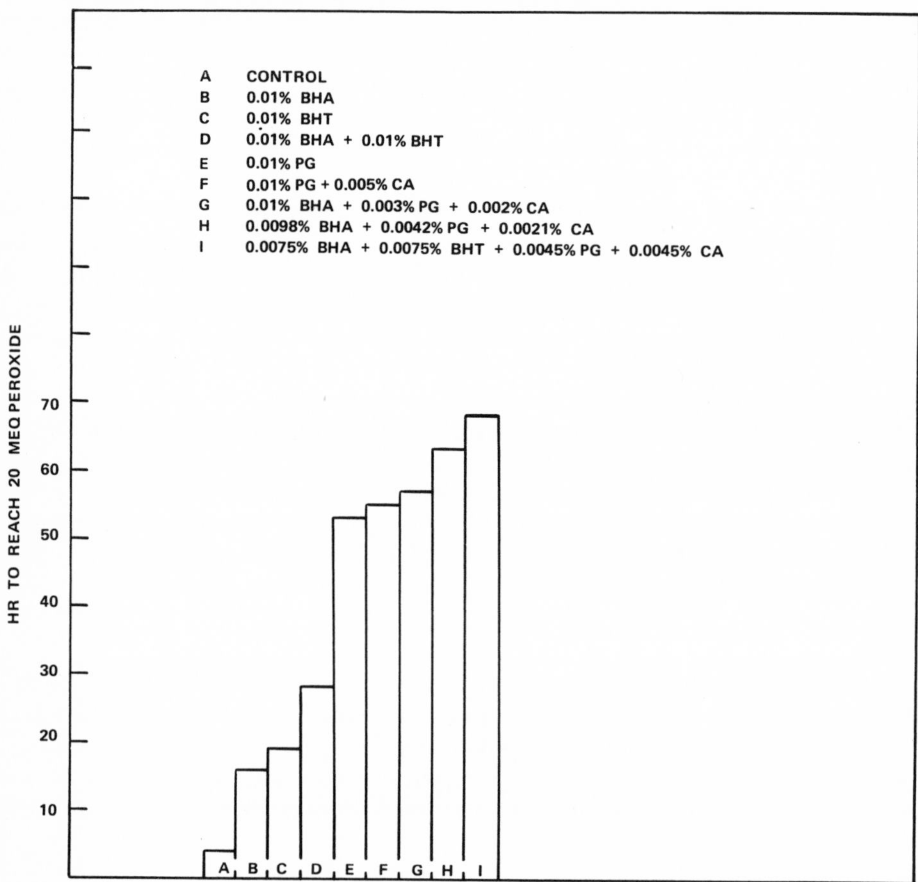

FIGURE 1. Effect of antioxidants on the active oxygen method (AOM) stability of lard. (From Stuckey, B. N., in *The Handbook of Food Additives,* 2nd ed., Furia, T. E., Ed., CRC Press, Cleveland, 1972, 205.)

Figure 1 shows the relative effects of BHA and BHT alone on the stability of lard as measured by the active oxygen method (AOM). The enhancing effect of PG over either of these is evident. From Figure 2, it is apparent that TBHQ outperforms PG in vegetable oils. Finally, the relative carry-through of these antioxidants into pastry prepared from lard is shown in Figure 3. For this particular application, BHA is still the anxioxidant of choice.

Most animal fats and all vegetable oils are steam vacuum deodorized to remove free fatty acids plus volatile flavor and odor compounds, after which as the last step in processing before the fat is formulated into the finished product, the phenolics are added. Since the phenolics are volatile under these deodorizing conditions, it is necessary to add them during the cooling cycle or after the deodorized fat has been pumped into a holding tank.

Fried foods, such as potato chips, nuts, or doughnuts, are usually cooked in vegetable oils or hydrogenated shortening. BHA gives the best carry-through in frying; for example, it protects the finished product from oxidation. Carry-through of PG, TBHQ, and BHT is not as good, since they may be steam distilled or may decompose during the frying operation. But since absorption of fat by the fried product is usually quite high, fresh fat must be added to the kettle at intervals. Consequently, fresh antioxidant is introduced to replace that lost by steam distillation.[2] Methyl polysiloxane[3]

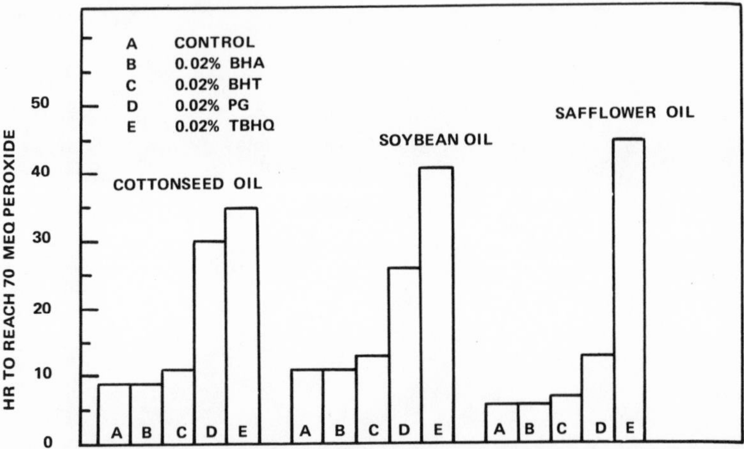

FIGURE 2. Comparison of *tert*-butylhydroquinone (TBHQ) with other antioxidants by the active oxygen method (AOM). (From Sherwin, E. R. and Thompson, J. W., *Food Technol.* (Chicago), 21(6), 106, 1967. With permission.)

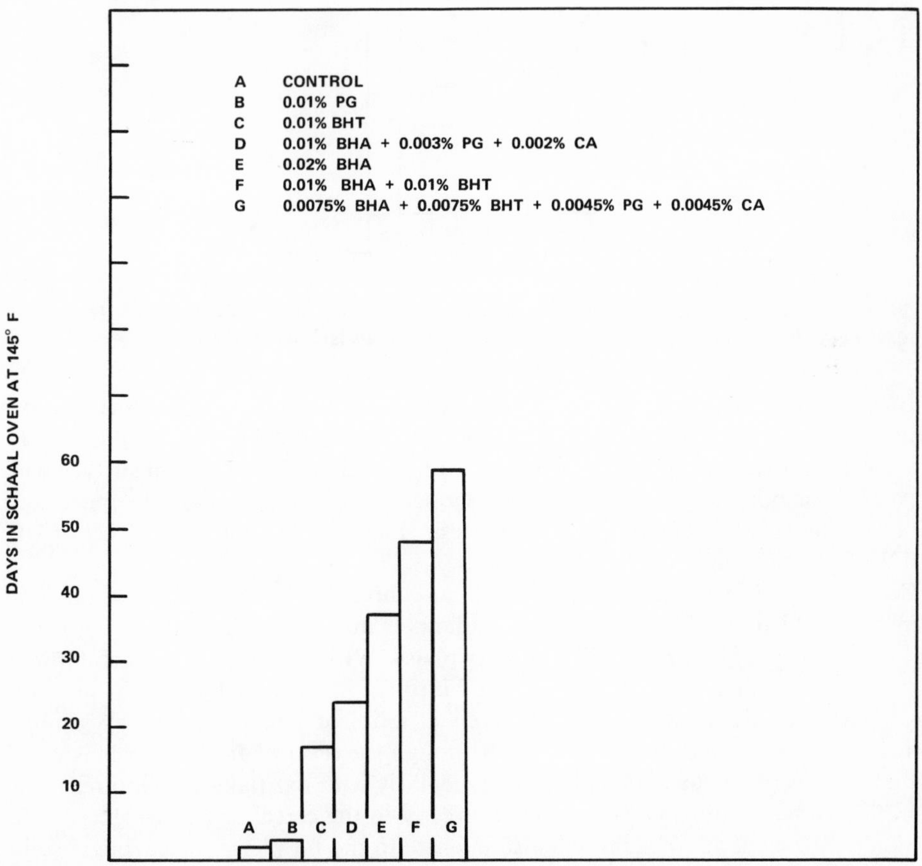

FIGURE 3. Carry-through of antioxidants in pastry prepared with treated lard. (From Stuckey, B. N., in *The Handbook of Food Additives*, 2nd ed., Furia, T. E., Ed., CRC Press, Cleveland, 1972, 205.)

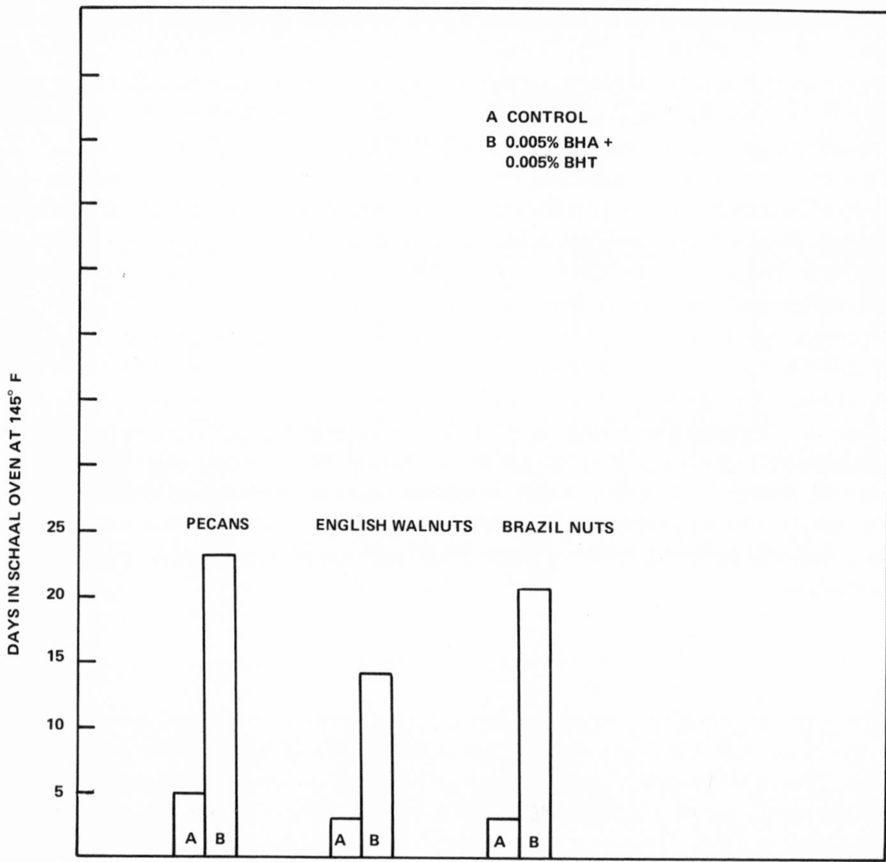

FIGURE 4. Effect of antioxidants in nuts, a high-fat food. (From Stuckey, B. N., in *The Handbook of Food Additives,* 2nd ed., Furia, T. E., Ed., CRC Press, Cleveland, 1972, 208.)

is frequently added to frying fats at levels of up to 10 ppm. Although it is not an antioxidant and has no direct effect on the stability of the finished product, it does protect the hot fat during frying by forming an insoluble film at the oil surface which prevents its exposure to oxygen. As a consequence, the processor can be more confident that the fat absorbed by the food is fresh. Figure 4 illustrates the effect of BHA and BHT on the stability of various fried nuts.

Cereals, dehydrated potatoes, and some cake mixes are examples of low fat foods. Here the problem of incorporating antioxidants is more complicated, since it is difficult to achieve proper contact with the fatty phase. With cereals, it is a general practice to incorporate a high level of BHA or BHT into the inner waxed package liner. Since these phenolics are somewhat volatile even at ambient temperature, they gradually diffuse into the product from the wax. Although the level of fat in cereals is usually quite low, it is highly unsaturated, and particularly with oat flakes, it requires protection. Sometimes, antioxidants are added directly to the cereal or potato slurry before cooking. Sufficient antioxidant will also migrate to the fat phase for adequate stabilization in this case. Spraying emulsions containing antioxidants directly on the cereal just before packaging has also been tried with some success, but uniform distribution is difficult to achieve by this technique. A recent patent to Nestle SA[4] describes the use of BHA or BHT in solution under pressure in CO_2 or dichlorodifluoromethane to achieve contact between the antioxidants and low fat foods. This method is claimed to aid in the penetration of antioxidants to the inside of the product where the fat is located.

Comminuted meats, either fresh or dried, may be stabilized by adding an antioxidant (usually BHA plus citric acid), which is dispersed on the surface of salt crystals. The antioxidant migrates to the fatty phase and gives adequate protection.

Fish products cannot be stabilized very successfully with antioxidants. The principal problem of instability is microbial rather than oxidative. Furthermore, if protection against rancidity is required, it must be recognized that fish oils are highly unsaturated making them particularly difficult to treat. Also it is not easy for antioxidants to penetrate fish skin and reach areas where fat is concentrated.

Additional edible products which often benefit from antioxidant addition include flavor oils such as orange and lemon oils and other terpene-containing flavoring materials. These products can be stabilized adequately with BHA which reduces flavor degradation. Chewing gum often becomes brittle and develops off-flavors on storage. Incorporation of phenolics into the gum base during manufacture will minimize these undesirable changes. Nuts and butter in candies may be a source of flavor problems during storage. Incorporation of antioxidants into the frying fat used to prepare the nuts and into the melted butter oil before manufacture of the candy helps to alleviate these problems.

LIPID OXIDATION MECHANISM

From what has been said earlier, it is easy to see that lipid oxidation confronts the food processor with innumerable headaches and the food technologist with a serious challenge. What is the exact nature of this deleterious process? What sets this process in motion? What keeps it going? Once we have dealt with these questions we can look at the most important question of all: what can be done about it? Because of the restricted scope of this review, treatment of this subject will be basic and limited in nature. The interested reader is referred to other papers and reviews for a fuller, more detailed picture.[5-10,20,177,178] Our primary concern here is with autoxidation, which is that process which occurs spontaneously under relatively mild conditions resulting in rancidity, reversion, and other types of off-flavors and -odors. These effects are observed almost exclusively in unsaturated, particulary polyunsaturated, fats and oils. In passing, note that under certain conditions, such as at elevated temperatures and especially in the more advanced stages of oxidation, polymerization takes place. References to this problem will be made in this review, bearing in mind, however, that for the food processor, the development of adverse organoleptic characteristics in the product is of principal concern. It should also be noted that polymerization, oxidative and thermal, is of practical interest to the protective coating and linoleum industries, but a description of these processes is beyond the scope of this review. Before getting on with the problem of autoxidation, there is another qualification: the differentiation between chemical, oxygen-produced autoxidation and enzyme-catalyzed lipid oxidation. Although quite important, particularly in the case of dairy and soybean products, this latter reaction poses a different aspect of food deterioration which will be omitted here. It is sufficient to say that enzymatic oxidation of lipids cannot be prevented by antioxidants and can only be avoided by inactivating the enzyme, usually by heat treatment.[11]

Present views on the reaction between oxygen and the double bond of a fatty acid are derived mainly from Farmer and his colleagues.[12] The major reaction is known to involve the hydroperoxidation of the methylene group adjacent (α) to the double bond. The reaction occurs by a free radical chain process and can be stated simply as follows:

Initiation	production of R• or RO_2• radicals	
Propagation	R• + $O_2 \rightarrow RO_2$•	
	RO_2• + RH → R• + RO_2H (a hydroperoxide)	
Termination	R• + R• → RR	noninitiating and nonpropagating products
	RO_2• + R• → RO_2R	
	RO_2• + RO_2• → $RO_2R + O_2$	

In the above, RH represents an olefin with H representing an α-methylenic hydrogen atom. It is an abbreviated description of an unsaturated fatty acid hydrocarbon chain. As Bateman pointed out,[13] this mechanism satisfactorily accounts for the characteristics of olefinic autoxidation. The term "auto" used in this context is indicative of a self-propagating process. The specific reaction of individual purified fatty esters has been investigated and explained in detail.[14] To date, however, there is no sure explanation of how the process of radical formation gets started. To put it another way, scientists are unsure as to the exact nature of the so-called induction period. Some credit this genesis to the presence of trace metals. Another theory postulates a special complex formation between the double bonds and atmospheric oxygen.[15] A more plausible, recent explanation lays the blame of initiation at the doorstep of singlet oxygen ($^1O_2$).[10]

Regardless of which theory is correct (indeed, they may all play a role), everyone seems to agree that the second important step involves free radical reactions as outlined above. The immediate result of these free radical reactions is the formation of hydroperoxides. As is often pointed out in the literature, pure hydroperoxides are tasteless, colorless, and unstable. It is this latter property which is responsible for the undesirable effects of autoxidation. When these hydroperoxides break down and there is evidence to show that metals, such as iron and copper accelerate the process, the result is formation of aldehydes, ketones, alcohols, and hydrocarbons which impart unpleasant flavors and odors to rancid or reverted oils.

Gas chromatography has been playing an ever increasing role in the characterization of the various off-flavor producing substances. Evans et al.[16] have determined the nature of some of these short-chain, volatile oxidation products, while Badings[17] has reported on the nature of volatile carbonyls in autoxidized linoleic acid. Other researchers have emphasized the fact that the human senses can detect very minute amounts (often lower than 1 ppm) of these identified products.[18-20]

Hoffman provides a good discussion of oxidative off-flavors produced in vegetable oils and a description of various commercial terms.[6a] Currently (1977), a special committee of the American Oil Chemists' Society is gathering an extensive list of these commercial terms and summarizing their correlations to various chemical entities. There are numerous problems connected with compiling such a list since each industry, if not each user, applies different terms to describe a similar impact. The difficulty of such an undertaking can best be illustrated by specific examples. In dairy products, for instance, Patton,[6b] although adhering to the classic hydroperoxide mechanism, outlines two distinct pathways. When water is involved, he describes the lipid oxidation flavors as "cardboardy" or "cappy"; whereas, those observed in anhydrous milk fat or dried whole milk are labeled as "oily" or "tallowy". In butter both pathways seem to proceed and the resulting flavors are described as "cardboardy," "oily tallowy," "fishy," "mushroom-like," etc. Koch[6c] has pointed out that the amount of water present can affect oxidative stability (a monomolecular layer apparently gives the best protection against oxidation) and the nature of off-flavors themselves may be altered when they interact with other food ingredients, such as carbohydrates and proteins.

It has been reported that oxidation of methyl linoleate in a model system containing protein or cellulose is accelerated in the presence of a limited amount of water. The water apparently acts as a solvent to mobilize trace metals so that they act as pro-oxidants.[21] Chelating agents, such as citric acid or ethylenediaminetetraacetic acid (EDTA), reduce oxidation rates significantly in such a system, although phenolics also have some effect. At lower moisture levels, the rate of oxidation is reduced; when a level is reached corresponding to a monomolecular layer surrounding the fat globules, maximum stability is achieved.[22] However, if moisture is removed completely from food or a model system, this protective effect is lost and rapid oxidation may result.

Selke et al. have shown that volatile decomposition products (aldehydes and methyl ketones) are generated when tristearin is heated at 192°C in air.[23] Since these carbonyl compounds are developed at levels above the taste threshold, it has been suggested that saturated fats may in part be responsible for off-flavors when shortenings are heated. However, Michalski and Hammond have demonstrated that up to 125°C, stearate does not contribute to the volatile oxidation products obtained from soybean oil.[24] ^{14}C-labeled stearic acid was inter-esterified into soybean oil, and the oil was subsequently oxidized to a high peroxide value. The radioactivity of a distillate from this oil was no higher than that of the background.

Considerable progress has been made recently in elucidating the structures of dimeric and polymeric reaction products formed during autoxidation. Artman and Smith have described the fractionation of heated cottonseed oil by methylation followed by urea treatment to separate the nonadducting fraction.[25] Branched and polymeric reaction products did not form inclusion compounds with this regeant. Subsequent distillation and gradient elution chromatography were used to obtain a cyclic dimer fraction.[25] Paulose and Chang isolated the nonvolatile decomposition products from trilinolein (185°C) as a non-urea-adducting fraction.[26] They fractionated this material further to obtain cyclic dimers, noncyclic dimers, trimers, and some material containing both dimers and trimers in the same molecule. Most of the material contained C-C linkages, but there were also some C-O-C linkages between molecules. Perkins et al. used gel permeation chromatography to separate high molecular weight components from heated fats. Dimeric fatty acids were identified by gas liquid chromatography and mass spectrometry.[27]

Fairly recently, it has been shown that photo-oxidation of unsaturated fats follows a mechanism involving singlet oxygen as the reactive intermediate.[28] The rate of this reaction is unaffected by peroxy radical scavengers, such as hindered phenols. Unlike the usual free radical chain reaction, this type of deterioration is inhibited by carotenes and is also retarded by Ni (II) chelates which quench singlet $^1O_2$. It also has been postulated that free radical intermediates are formed in the initial stages of fat oxidation and that these then stabilize the food or lipid preparation.[29]

A number of nitroxides were shown to have marked activity in unsaturated lipids. The amount of residual nitroxide can be estimated from the electron paramagnetic resonance (EPR) signal which responds to the single unpaired electron in the nitroxide free radical. For a recent review on the mechanism of free radical decomposition, the reader should consult Hiatt.[30]

Important as these recent developments may be, organoleptic deterioration is still the chief problem confronting the food processor today. Rancidity is a widely used term, covering many typical off-flavors formed by autoxidation of unsaturated fatty acids and present but detectable only at advanced stages of oxidation. Rancidity is usually distinguished from the term "flavor reversion," which is applied to the "grassy", "beany", "fishy", and "painty" flavors produced by the very early stages of oxidation of oils containing linolenic acid or even more unsaturated fatty acids, e.g., oils from soybeans, rapeseed, linseed, and marine sources. Rancidity of pork fat (lard) is rather easy to recognize. That of vegetable oils is less so since one often encounters off-odors and -flavors not usually associated with rancidity. Since the nature of the problem has been discussed at length, the role that antioxidants play in stabilizing foods against oxidation may now be examined.

MECHANISM OF ANTIOXIDANT ACTION

From what has preceded, it is obvious that one way to prevent oxidation of food lipids would be to exclude oxygen. This has already been utilized in at least one commercial product where oxidation of a reconstituted potato snack is limited by packaging under nitrogen. Since this procedure is not practical in most cases because of expense, other alternative means must be utilized. At the outset, it must be made clear that the known food antioxidants are largely ineffective in preventing flavor reversion, although in a recent publication Chang et al. claims that a natural extract from spice is effective to this end.[31] If reversion is connected with the initiation step and is due to the presence of singlet oxygen, then it is not surprising that phenolic antioxidants do not prevent it. Data from Clements et al. document this inefficacy and give a plausible explanation.[28] Singlet oxygen formation is catalyzed by light and augmented by photosensitizers, such as chlorophyll and protoporphyrin, naturally present in fats and oils. It can be inhibited only by such compounds (called quenchers) as β-carotene and nickel chelates. Recently, Cort has shown that the use of quenchers is not a satisfactory solution to the lipid oxidation problem.[32] Undoubtedly, much more work remains to be done before the food technologist can learn to cope with flavor reversion.

If antioxidants can do little to prevent reversion, they are most useful in delaying the formation of rancidity. They perform this useful function by interfering with the propagation step of the free radical reaction. Antioxidants, probably by donating a proton or an electron, act as free radical traps resulting in resonance-stabilized intermediates which increase the likelihood of chain termination. This is illustrated by Stuckey[5] as follows:

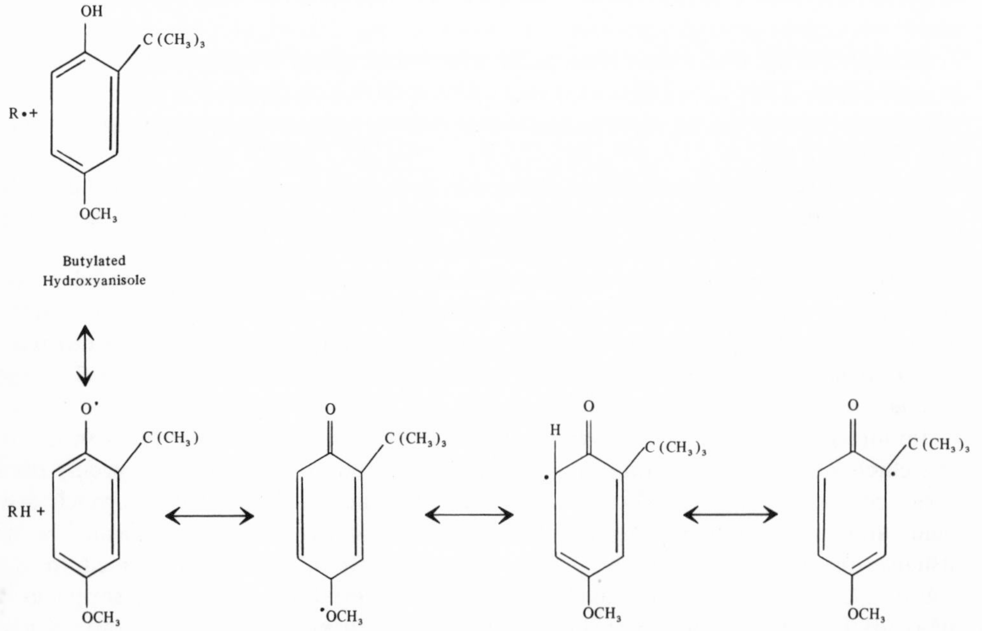

R• + (OH, $C(CH_3)_3$, OCH_3)

Butylated Hydroxyanisole

↕

RH + (O•, $C(CH_3)$, OCH_3) ⟷ (O, $C(CH_3)_3$, •OCH_3) ⟷ (O, H, $C(CH_3)_3$, OCH_3) ⟷ (O, $C(CH_3)_3$, OCH_3)

The importance of stable nitrogen-containing free radicals in antioxidant activity has recently been stressed by Lin and co-workers.[29] This stability could account for the efficacy of ethoxyquin (6-ethoxy-1,2-dihydro-2,2,4-trimethlquinoline) as a feed antioxidant.

As Stuckey points out in his review,[5] ring substitution with an alkyl group reduces the antioxidant activity of the molecule, probably because of steric hindrance. Substitution by a bulky alkyl group (tertiary butyl) as is the case with BHA above helps to stabilize the molecule against secondary reactions, renders it less volatile, more oil soluble, and, most importantly, more acceptable from the toxicological point of view. For these reasons most of the food antioxidants in use today are substituted phenols. An obvious exception to this is ascorbyl palmitate, the ester of palmitic acid with ascorbic acid (vitamin C). Whether this antioxidant also effects protection against oxidation by the radical trapping mechanism is not known. It has been shown, however, that its antioxidant activity is also due to the fact that it can react with oxygen.[32] Its activity may also be due, in part, to its known metal chelating activity. More will be said about this in a later section.

FOOD-APPROVED ANTIOXIDANTS

The food-approved antioxidants in the U.S. are relatively few in number. Only BHA, BHT, PG and TBHQ have any commercial significance. BHA is available as a mixture of two isomers, 2- and 3-tert-butyl-4-methylphenol. It was first approved by the Food and Drug Administration (FDA) in 1948. Over the years, it has been widely used both in the U.S. and in most of the industrial countries of the world. (See Table 1) Since BHA contains only one phenolic hydroxyl group, its bulky t-butyl group in the *ortho* or *para* positions makes it relatively nonpolar. The steric hindrance which this alkyl group provides may account for the poor functionality of BHA in vegetable oils, but this same structural feature probably allows it to carry through well into fats used in baked goods. By using BHA in conjunction with PG, it is possible to take advantage of the relatively high antioxidant potency of the latter while retaining the carry-through effect of the former. BHA has a strong phenolic odor, which is particularly noticeable at high temperatures during frying. Losses of BHA due to steam distillation during this process are quite substantial. As a consequence, what small protective effect BHA may have in frying oils is soon lost on repeated high temperature use. Table 2 shows the physical properties of this antioxidant and those of BHT, PG, and TBHQ.

BHT is another hindered phenol which has been used extensively in food fats. Although it has been cleared in the U.S. since 1954, in recent years questions have been raised about its safety.[51,52] Some food companies have already removed it from their products in anticipation that it may be delisted sometime in the future. BHT is weak as a stabilizer for vegetable oils. Like BHA, it was originally developed for use in lard, in which it gives excellent protection. In vegetable oils BHT can be used effectively as a component of synergistic mixtures containing BHA, PG, or TBHQ. It affords carry-through protection in baked goods much like that observed with BHA. Figure 1 illustrates the relative efficiencies of BHA, BHT, and PG in stabilizing lard to oxidative rancidity. Note that PG stands out as much more functional than either BHA or BHT. In baked goods (Figure 3), 0.02% BHA prevents rancidity for almost 40 days in the Schaal oven test, whereas PG shows no significant effect over the control.

During the early 1940s, propyl gallate and several other alkyl gallates were allowed for use in foods in a number of countries. Since gallic acid contains three phenolic hydroxyl groups, it is a rather potent antioxidant; however, the propyl ester has a significant degree of water solubility (Table 2). As a consequence, it tends to partition

Table 1
PRIMARY ANTIOXIDANTS AND SYNERGISTS FOR VEGETABLE OILS IN VARIOUS COUNTRIES

Countries	Primary antioxidants												Synergists									
	Tocopherols	Gum guaiac	Propyl gallate	Butyl gallate	Octyl gallate	Dodecyl gallate	NDGA (nordihydro-guaiaretic acid)	BHA (butylated hydroxyanisole)	BHT (butylated hydroxytoluene)	THBP (2,4,5-trihydroxy-butyrophenone)	4-Hydroxymethyl-2,6-di-*tert* butylphenol	TBHQ (*tert*-butylhydroquinone)	Citric acid	Isopropyl citrate	Phosphoric acid	Thiodipropionic acid	Didodecyl or dilauryl thiodipropionate	Dioctadecyl thiodipropionate	Ascorbic acid	Ascorbyl palmitate	Tartaric acid	Lecithin
Australia			X		X	X		X														
Austria	X	X	X	X	X	X	X	X	X													
Belgium	X		X		X	X		X	X											X		
Brazil			X		X	X		X	X			X										
Canada	X	X	X					X	X				X	X					X	X	X	X
Ceylon			X		X	X		X	X													
Czechoslova-kia			X																			
Denmark	X	X	X		X	X	X	X	X				X						X	X	X	X
Finland	X		X		X	X	X	X	X				X						X			
France			X		X	X		X	X										X			
Britain			X		X	X		X	X													
Greece			X		X	X																
Haiti			X				X	X	X													
Hong Kong			X		X	X		X	X													
India	X	X	X		X	X	X	X	X				X						X		X	X
Italy	X		X		X	X		X												X		
Jamaica			X		X	X		X	X			X										
Japan			X					X	X													
Korea			X					X	X													
Malaysia			X		X	X		X	X													
Mexico	X	X	X				X	X					X			X						

Morocco			X	X	X			X													
Netherlands	X		X	X	X		X	X				X						X	X		X
New Zealand			X	X	X		X	X													
Nicaragua	X	X	X			X	X					X		X	X						X
Norway	X		X	X	X	X	X	X										X			
Pakistan	X	X	X	X	X	X	X								X	X					
Peru			X	X	X		X	X			X										
Poland	X																	X	X		
Republic of South Africa	X	X	X	X	X		X					X						X		X	X
Rumania						X															
Spain			X	X	X		X	X													
Sweden	X	X	X	X	X		X	X													X
Switzerland	X		X	X	X		X					X						X			X
Turkey	X		X	X	X		X	X													
Taiwan			X	X	X		X	X			X										
U.S.	X	X	X				X	X	X	X	X	X	X	X	X	X	X	X	X	X	X
U.S.S.R.					X		X	X													
Wales			X	X	X		X	X													
Yugoslavia			X	X	X	X	X					X						X		X	

From Sherwin, E. R., *J. Am. Oil Chem. Soc.*, 53, 430, 1976. With permission.

Table 2
PHYSICAL PROPERTIES OF COMMONLY USED ANTIOXIDANTS

BHA
Butylated Hydroxyanisole
(a mixture of 3-*tert*-butyl-4-hydroxyanisole and 2-*tert*-butyl-4-hydroxyanisole)

PHYSICAL PROPERTIES

Structural formula (mixture of 2 and 3 isomers)

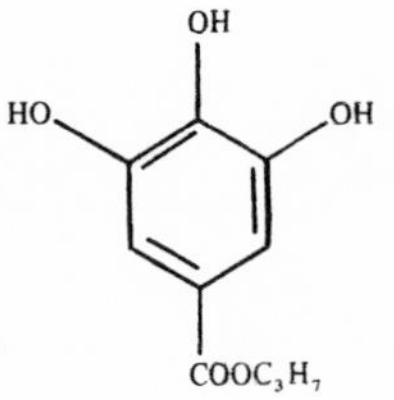

Molecular weight	180
Physical appearance	White, waxy tablets
Boiling range, 733 mm (°C)	264—270
Melting range (°C)	48—55
Odor	Slight

SOLUBILITY IN COMMON SOLVENTS

Solvent	Temperature (°C)	Solubility, approx. (%)
Water	0—50	Insoluble
Glycerol	25	1
Ethyl alcohol	25	50 +
Diisobutyl adipate	25	60
Propylene glycol	25	70
Glyceryl monooleate	25	50
Coconut oil	25	40
Cottonseed oil	25	30
Corn oil	25	40
Peanut oil	25	40
Soya oil	25	50
Lard	50	50
Yellow grease	50	50
Paraffin	60	60 +
Mineral oil	25	5

PHYSICAL PROPERTIES

Structural formula

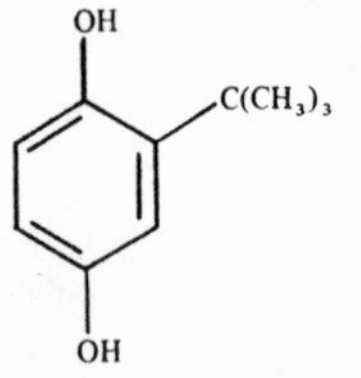

Molecular weight	212
Physical appearance	White, crystalline powder
Boiling nt (°C)	Decomposes above 148
Melting range (°C)	146—148
Odor	Very slight

SOLUBILITY IN COMMON SOLVENTS

Solvent	Temperature (°C)	Solubility approx. (%)
Water	25	<1
Glycerol	25	25
Ethyl alcohol	25	60 +
Diisobutyl adipate	25	15
Propylene glycol	25	55
Glyceryl monooleate	25	5
Cottonseed oil	25	1
Corn oil	25	Insoluble
Peanut oil	25	<1
Soya oil	85	2
Lard	50	1
Yellow grease	50	1
Mineral oil	25	<1

BHT
Butylated Hydroxytoluene
(2,6-di *tert*-butyl-*o*-cresol)

PHYSICAL PROPERTIES

Structural formula

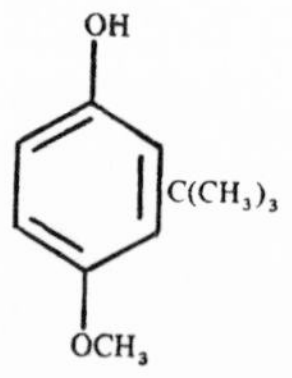

Molecular weight	220
Physical appearance	White, granular crystal
Specific gravity, 20/4°C	1.048
Boiling point, 760 mm (°C)	265
Melting point (°C)	69.7

Table 2 (continued)
PHYSICAL PROPERTIES OF COMMONLY USED ANTIOXIDANTS

PHYSICAL PROPERTIES

Vapor pressure @ 100°C (mm)	2
Vapor pressure @ 195°C (mm)	253
Odor	Very slight

SOLUBILITY IN COMMON SOLVENTS

Solvent	Temperature (°C)	Solubility, approx. (%)
Water	0—60	Insoluble
Glycerol	25	Insoluble
Ethyl alcohol	25	25
Diisobutyl adipate	25	40
Propylene glycol	0—25	Insoluble
Glyceryl monooleate	25	15
Coconut oil	25	30
Cottonseed oil	25	30
Corn oil	25	30
Olive oil	25	25
Peanut oil	25	30
Soya oil	25	30
Lard	50	40
Yellow grease	50	40
Paraffin	60	60 +
Mineral oil	25	5

TBHQ
Mono-*tert*-Butylhydroquinone

PHYSICAL PROPERTIES

Structural formula

OH
$(CH_3)_3C$ $C(CH_3)_3$
CH_3

Molecular weight	166.22
Physical appearance	White to light tan crystals
Melting range (°C)	126.5—128.5
Odor	Very slight

SOLUBILITY IN COMMON SOLVENTS

Solvent	Temperature (°C)	Solubility, approx. (%)
Water	25	<1
Water	95	5
Ethanol	25	60
Ethyl acetate	25	60
Propylene glycol	25	30
Glyceryl monooleate	25	10
Cottonseed oil	25	10
Corn oil	25	10
Safflower oil	25	5
Soybean oil	25	10
Lard	50	5

Data from Tenox® Food Grade Antioxidants, Publication ZG-109C, Eastman Chemical Products, Inc., Kingsport, Tenn., 1976, 4. With permission.

into any water phase which may be present and to complex with iron salts to give bluish black colors. Also, its relatively poor resistance to heat degradation makes it ineffective for baking and frying applications. The higher alkyl gallates (butyl and octyl) are more fat soluble but have not been approved for use in this country.

In 1963, 2,4,5-trihydroxybutyrophenone (THBP) was cleared for use in foods.

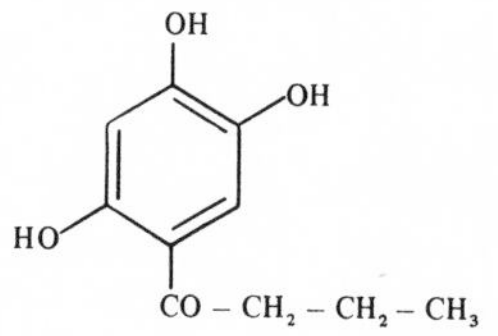

OH
OH
HO
$CO - CH_2 - CH_2 - CH_3$

2,4,5-Trihydroxybutyrophenone (THBP)

It is effective and apparently safe for use based on adequate toxicity testing, but it suffers from the same deficiencies as PG, i.e., lack of carry-through into baked and fried foods and discoloration in the presence of iron salts. It has achieved no significant commercial usage.

Another phenolic antioxidant, 4-hydroxymethyl-2,6-di-tert-butylphenol has been available since 1967. The objective in changing the BHT structure to include a hydroxymethyl

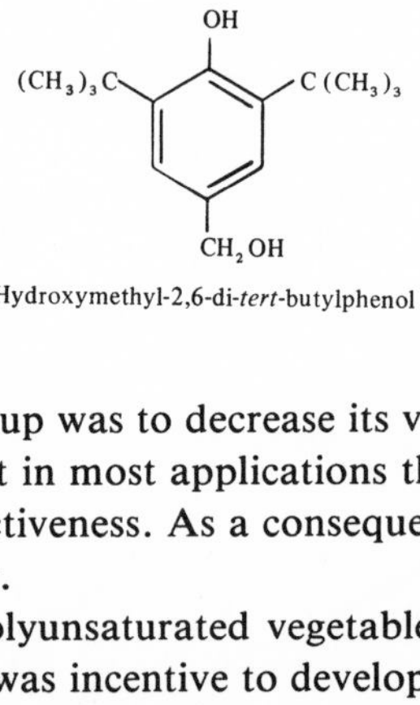

OH
$(CH_3)_3C$ $C(CH_3)_3$
CH_2OH

4-Hydroxymethyl-2,6-di-*tert*-butylphenol

rather than a simple methyl group was to decrease its volatility and help in the stabilization of oils during frying, but in most applications this small structural change had little effect on antioxidant effectiveness. As a consequence, this material has not been used commercially to any extent.

Since the consumption of polyunsaturated vegetable oils has been on the increase during the past 15 years, there was incentive to develop new, more functional antioxidants for this application. TBHQ has been identified as an answer to this need. It was cleared for food use in the U.S. in 1972. Figure 2 illustrates its effectiveness in prolonging the stability of vegetable oils as measured by the AOM test. It shows significantly better protection than PG, especially with soybean and safflower oils, which are highly unsaturated and particularly difficult to stabilize.

TBHQ was shown (Figure 5) to function more effectively than other food-grade antioxidants in preventing the development of peroxides in crude oil during storage.[34,35] It was particularly good with the highly unsaturated oils, such as safflower and sunflower. This property is important to a processor in making sure that his oil can be delivered to the food manufacturer in a fresh condition. Refined palm oil and hydrogenated cottonseed oil were treated with BHA, BHT, or TBHQ; only TBHQ gave a significant reduction in rate of hydroperoxide formation during bulk storage.[36]

The recent literature contains many references to the advantages of TBHQ. It has been found to retard the development of rancid flavors and odors in many difficult-to-stabilize unsaturated fats. TBHQ shows no discoloration in the presence of iron

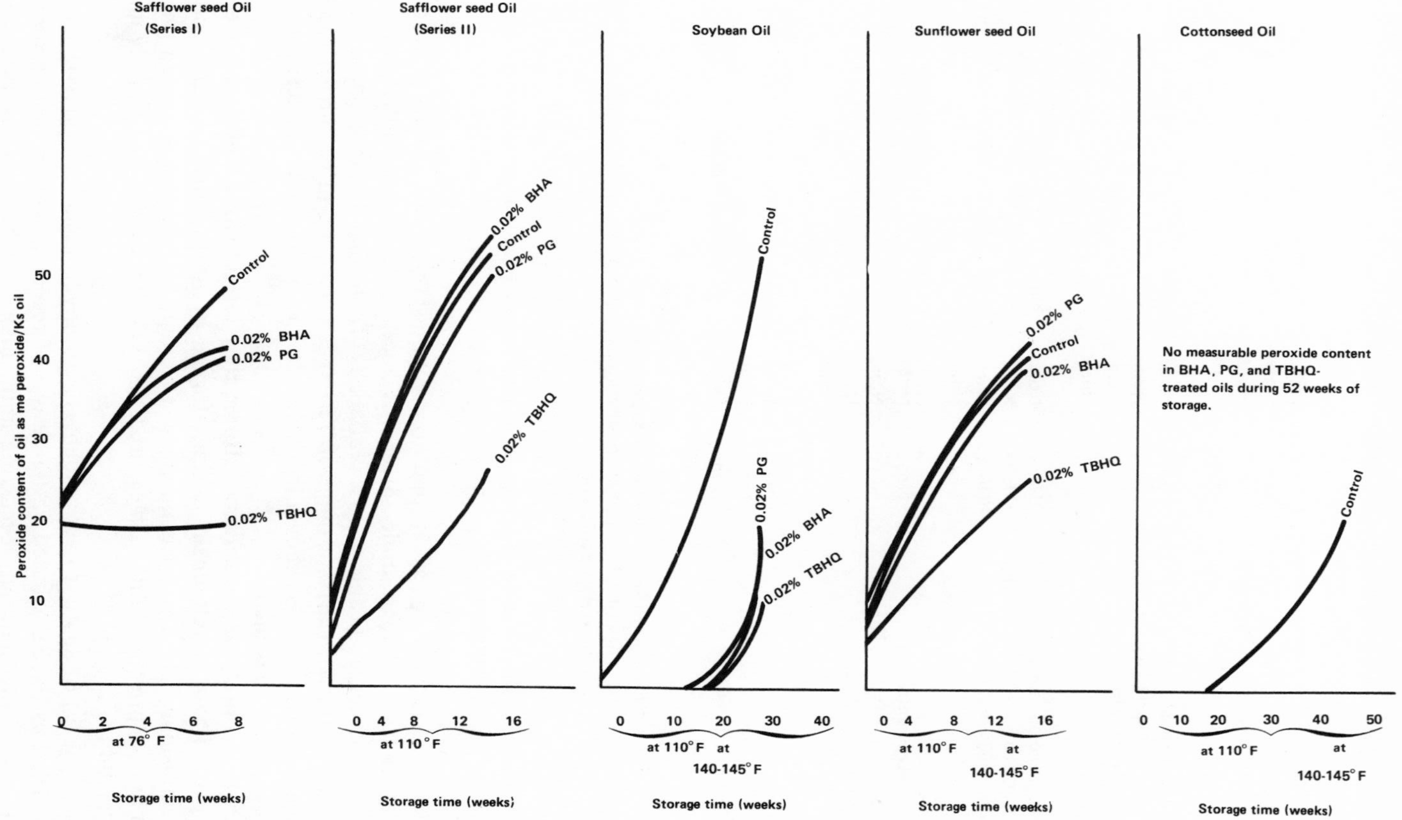

FIGURE 5. Peroxide formation in crude vegetable oils during storage. (From Hartman, K. T. and Rose, L. C., *J. Am. Oil Chem. Soc.*, 47, 19, 1970. With permission.)

Table 3
COMPARISON OF IRON DISCOLORATION TENDENCIES OF TBHQ vs. PG BY THE IRON STRIP AND SCRAMBLED EGG TESTS[a]

Type of fat or oil and treatment (wt. %)	Iron strip test — color development in 14 days in the: Oil phase	Water phase	Scrambled egg test — discoloration of eggs prepared with lard or oil
Lard, untreated	None	None	None
Lard + 0.02% TBHQ	None	None	None
Lard + 0.02% PG	None	Dark blue	Severe discoloration[b]
Cottonseed oil, untreated	None	None	None
Cottonseed oil + 0.02% TBHQ	None	None	None
Cottonseed oil + 0.02% PG	None	Dark blue	Severe discoloration[b]
Soybean oil, untreated	None	None	—[c]
Soybean oil + 0.02% TBHQ	None	None	—[c]
Soybean oil + 0.02% PG	None	Dark blue	—[c]
Safflower oil, untreated	None	None	None
Safflower oil + 0.02% TBHQ	None	None	None
Safflower oil + 0.02% PG	None	Dark blue	Severe discoloration[b]

[a] TBHQ = *tert*-butylhydroquinone and PG = propyl gallate.
[b] Difficult color to describe, but normal bright yellow of the eggs was completely masked by a grayish green discoloration.
[c] Dash indicates test not conducted.

Reprinted from Sherwin, E. R. and Thompson, J. W., *Food Technol.* (Chicago), 21, 912, 1967. With permission.

(Table 3); whereas, PG under the same conditions gave a severely discolored egg product. Solubility in fats and food-grade solvents is good. Studies show TBHQ to be effective in breakfast cereals (Table 4), in essential oils (Table 5) and in food packaging materials (Tables 6 and 7). Carry-through of TBHQ is very good in deep-fried food products, such as potato chips as shown in Table 8.[33] The safety of this material as a food additive is well documented.[37]

In the authors' laboratory, it was demonstrated that TBHQ has a significant effect in preventing off-flavor development during storage of unsaturated vegetable oils alone or in emulsions.[38] This effect is in contrast to many other food approved phenolics which may improve stability as measured by AOM but do not significantly improve flavor stability.[39]

Missing from the list of approved antioxidants is nordihydroguiaretic acid (NDGA), which was delisted by the FDA based on the results of recent Canadian feeding studies.[40] The history of this naturally occurring antioxidant can be found in a review article by Oliveto,[41] who states that most of the previous studies on NDGA were carried out on the natural material which may have varied considerably in composition. Now

Table 4
COMPARISON OF TBHQ WITH BHA AND BHT IN BREAKFAST CEREALS[a]

Cereal treatment (wt %)	Storage life (days) at 145°F until development of rancid odor in:		
	Cornflakes	Wheat flakes	Oat cereal
Control (no antioxidant added)	12	10	10
0.001 TBHQ	19	18	13
0.005 TBHQ	56	54	38
0.001 BHA	18	16	17
0.005 BHA	36	24	34
0.001 BHT	24	12	12
0.005 BHT	65	20	43

[a] TBHQ = *tert*-butylhydroquinone, BHA = butylated hydroxyanisole, and BHT = butylated hydroxytoluene.

Reprinted from Sherwin, E. R. and Thompson, J. W., *Food Technol.* (Chicago), 21, 912, 1967. With permission.

Table 5
EFFECTIVENESS OF TBHQ IN CITRUS AND ESSENTIAL OILS[a]

Antioxidant treatment of oil (wt %)	Oven storage life, (days to develop rancid odor) — Lemon		Orange		Peppermint	
	85°F (29.4°C)	100°F (37.8°C)	85°F (29.4°C)	100°F (37.8°C)	85°F (29.4°C)	100°F (37.8°C)
Untreated (control)	59	60	40	39	94	20
TBHQ						
0.005	126	127	82	73	122	25
0.010	145[b]	125	75	103	124	133
0.020	—[c]	—[c]	91	65	107	72
BHA						
0.005	120	87	118	93	68	42
0.010	133	96	217	103	42	—[c]
0.020	—[c]	—[c]	366	142	64	33
PG						
0.005	72	116	104	93	138	39
0.010	64	118	102	103	86	129
0.020	—[c]	—[c]	95	105	121	125

[a] TBHQ = *tert*-butylhydroquinone, BHA = butylated hydroxyanisole, and PG = propyl gallate.
[b] Samples not rancid at end of test period.
[c] Dash indicates that tests were not conducted.

From Tenox® TBHQ Antioxidant for Oil Fats and Fat-containing Foods, Publication ZG-201D Eastman Chemical Products, Inc., Kingsport, Tenn., 1976, 8. With permission.

Table 6
EFFECTIVENESS OF TBHQ AS A TREATMENT FOR WAXED GLASSINE LINERS (CEREAL PACKAGING)[a]

Antioxidant treatment of wax (wt %)	Concentration of antioxidant on waxed glassine (wt %)	Oven storage life (days to develop rancid odor)					
		Corn cereal		Oat cereal		Wheat cereal	
		85°F (29.4°C)	145°F (62.8°C)	85°F (29.4°C)	145°F (62.8°C)	85°F (29.4°C)	145°F (62.8°C)
Untreated (control)	0	280	30	68	5	125	6
TBHQ							
0.2	0.038	365[b]	79	365[b]	18	282	13
0.3	0.056	365[b]	101	365[b]	38	271	20
0.5	0.094	365[b]	175	365[b]	96	365[b]	38
BHA							
0.2	0.038	365[b]	55	365[b]	30	229	16
0.3	0.056	365[b]	95	365[b]	35	338	17
0.5	0.094	365[b]	150	365[b]	46	365[b]	13

[a] TBHQ = *tert*-butylhydroquinone and BHA = butylated hydroxyanisole.
[b] Samples not rancid at end of test period.

From Tenox® TBHQ Antioxidant for Oil Fats and Fat-containing Foods, Publication ZG-201D, Eastman Chemical Products, Inc., Kingsport, Tenn., 1976, 8. With permission.

Table 7
EFFECTIVENESS OF TBHQ AS A TREATMENT OF GLASSINE (LARD PACKAGING)[a]

Antioxidant treatment of glassine (wt %)	Oven storage life of lard, days to develop rancid odor at 145°F (62.8°C)
Untreated (control)	20
TBHQ	
0.2	57
0.3	106
0.5	120
BHA	
0.2	91
0.3	109
0.5	128

[a] TBHQ = *tert*-butylhydroquinone and BHA = butylated hydroxyanisole.

From Tenox® TBHQ Antioxidant for Oil Fats and Fat-containing Foods, Publication ZG-201D, Eastman Chemical Products, Inc., Kingsport, Tenn., 1976, 8. With permission.

that a completely synthetic, high-purity product is available for the first time, it may be appropriate to reevaluate this potent antioxidant.

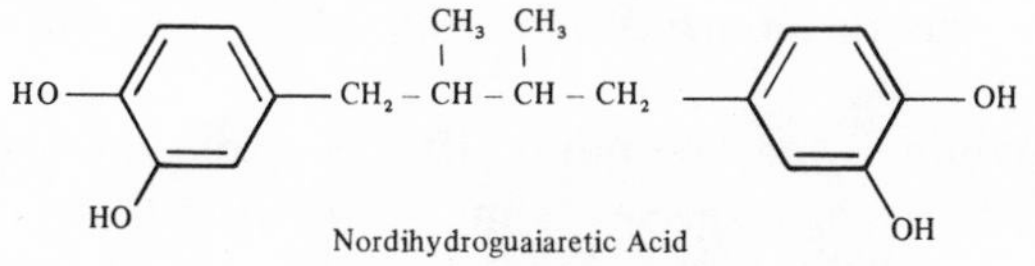

HO, HO, CH_3 CH_3, $CH_2 - CH - CH - CH_2$, OH, OH

Nordihydroguaiaretic Acid

NDGA [β, γ-dimethyl-α, δ-bis (3, 4-dihydroxphenyl) butane; 4, 4-(2, 3-dimethyltetrarnethylene) dipyrocatechol]

It should be noted, however, that NDGA resembles PG in that it forms a colored complex with iron salts and does not carry through into baked or fried food products.

As mentioned previously, antioxidants function well only with fresh, well-refined fats of good quality. This point is demonstrated by a recent publication in which free fatty acids were deliberately added to olive oil.[42] The acids, added at levels as low as 0.5%, reduced the protective effect of BHA, NDGA, and ascorbyl palmitate. Their catalytic action appears to be independent of unsaturation and is, largely, a function of the carboxyl group. It was also shown that phenolic antioxidants are more effective in retarding hydroperoxide development when oils are stored in the dark than when they are irradiated with direct sunlight.[43] In the section on Mechanism of Lipid Oxidation, it is noted that irradiation involves singlet oxygen ($^1O_2$) and that a phenolic would not be expected to function well in this situation.

In view of recent concerns about the toxicity of many food additives, e.g., the delisting of NDGA, a number of workers have returned to the natural vegetable oil tocopherols. In cottonseed oil, the best level of total tocopherols for maximum stability was 40 to 70 mg%.[44] Above this range, oxidation was actually enhanced. Other workers report that with sunflower oil, mixed tocopherols showed increased antioxidant activity at levels of up to 800 mg%.[45] In frying experiments, it was found that tocopherols provided satisfactory protection to the oil only when the system was protected from air by a metal float during periods when batches of food were not being prepared.[46] Otherwise, there was excessive thermal oxidation with a rapid decrease in tocopherol content. In the U.S., most producers of frying fats include 1 to 2 ppm of methyl polysiloxane. This additive forms a monolayer at the oil-air interface and effectively excludes oxygen.[47]

The antioxidant effect of various commercial tocopherol concentrates (including α, β, γ, and δ was assessed in olive oil.[48] Although there was some increase in AOM stability, no improvement in flavor stability was observed. With rapeseed oil, the best oxidative stability was obtained with 10 to 60 mg% mixed tocopherols (α, γ, and δ).[49] Losses of tocopherol from sunflower seed oil during refining were reported by Morrison.[50] Snacks, fried in oil which retained the most tocopherol during refining, also had the best shelf life. Cort has reviewed the antioxidant activity of various tocopherols and has discussed the mechanism of their action.[32] A more recent review gives a comprehensive account of the tocopherol content of foods.[53]

It is considered beyond the scope of this chapter to critically evaluate the toxicological evidence for safety of food antioxidants. But perhaps a general description of the tests required by the FDA might give the reader a better understanding of the complexity of this field. For more details, the review by Johnson is recommended.[185] Safety of antioxidants may be determined by acute toxicity tests to determine the LD_{50}, i.e., dose lethal to 50% of the experimental animals. Longer term feeding experiments may be run using lower dose levels with subsequent histopathological examination. Also, the metabolic pathway of the additive may be studied, and the biochemical effects of these metabolites determined.

To be considered safe, a food additive must have an LD_{50} of not less than 1 g/kg of animal body weight. In addition, it must not have any significant effect on growth when fed to experimental animals for up to 2 years at a level 100 times that proposed for human consumption. Animals recommended for this testing are the rat, rabbit, and dog.

Since enzyme pathways in animals may be quite different from those in humans, the metabolism of an additive in a single species cannot be used to predict the results in man. Furthermore, results may differ from one test animal to another. As a consequence, interpretation of animal feeding experiments is often difficult and leads to controversy. It seems obvious that the most reliable results will come from human feeding tests.

ANTIOXIDANTS FROM NATURAL SOURCES

With the recent delisting of NDGA and the questions which have been raised about the toxicity of BHT,[51,52] there has been renewed interest in the development of natural, edible replacements for the synthetic phenolics. There are many reports which show that amino acids and proteins have some antioxidant activity. Since these materials are quite polar, they have limited solubility in fats and consequently show weak activity. Recently, ethyl, lauryl, and benzyl esters of several amino acids were prepared and found to have enhanced activity, probably due to their improved solubility.[54] Petroleum ether extracts of textured vegetable protein were also shown to have good antioxidant activity for linoleate.[55] A recent patent has claimed that the stabilizing effect of cystine on fats is greatly enhanced if this amino acid is heated briefly in the fat at 140°C or higher.[56] Sims and Fioriti have shown that a similar effect obtained with methionine is the result of thermal decomposition to yield methional (β-methylmercaptopropionaldehyde).[57]

The efficacy of methional as a polyunsaturated oil antioxidant is illustrated in Figure

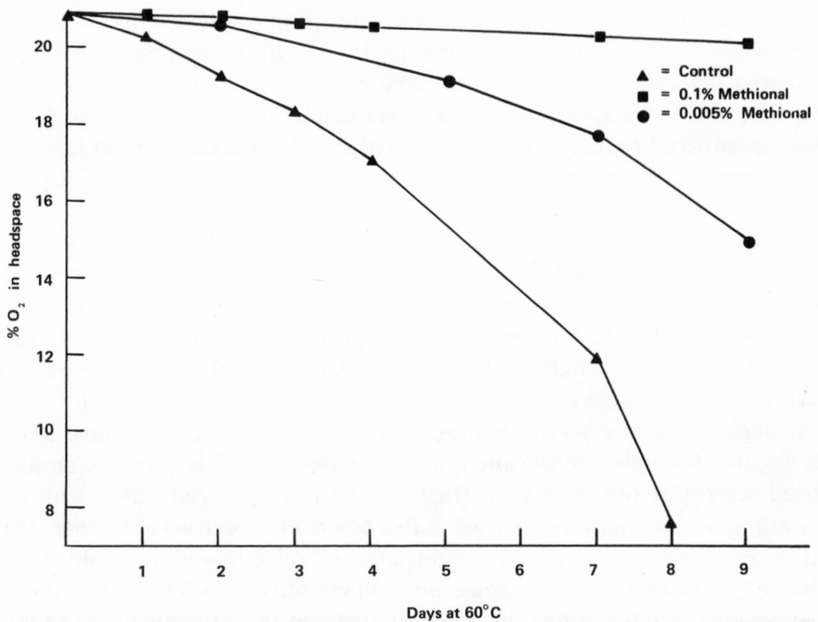

FIGURE 6. Stabilization of safflower oil with methional. (From Sims, R. J. and Fioriti, J. A., *J. Am. Oil Chem. Soc.*, 54, 4, 1977. With permission.)

6. A more detailed study of commercial methional revealed the presence of a dimer and a trimer, which are shown below along with the monomer.

$CH_3-S-CH_2-CH_2CHO$ Monomer

```
CH3 - S - CH2 - C - CHO          Dimer
                ||
                CH
                |
                CH2
                |
                CH2
                |
                S - CH3
```

```
                          O
                        /   \
CH3S - CH2 - CH2 - CH         CH - CH2 - CH2 - SCH3     Trimer
                   |           |
                   O           O
                     \       /
                        CH
                        |
                        CH2
                        |
                        CH2
                        |
                        SCH3
```

Testing showed that all three species have similar antioxidant activity. It is believed that methional and its polymers function as antioxidants by conversion of the sulfide linkage to sulfoxide. Methional has a GRAS status and as such can be used in various food preparations right now; unfortunately it has such a strong odor and flavor, even at low concentrations, that its use in food products may be rather limited.

There has also been considerable activity in the area of browning reaction products (Maillard) formed by heating simple sugars with amino acids. It was shown that the most effective antioxidants are formed in the early reaction stages and do not involve typical brown-colored pigments.[58] All combinations of amino acids and sugars gave almost the same antioxidant activity.[59] Browning reaction products from reducing sugars and amino acids were found to act synergistically with tocopherol in margarine.[60] They compared favorably with BHA and BHT when used at equal weights in linoleic acid[61] and were more effective than the various tocopherol isomers.[62] Mastich, a resin from *Pistacia lentiscus L.*, contains terpenolic acids which act as antioxidants similar to BHA. Egyptian farmers have used mastich for centuries as a butter oil preservative.[63] It is claimed that oils, such as soybean, cottonseed, corn, safflower, and lard, can be protected from autoxidation by adding Tempeh oil. Tempeh is obtained by fermentation of cooked, dehulled soybeans with *Rhizopus oligosporus*; the oil is extracted from the dried Tempeh with hexane and alcohol (2/1, v/v).[64] This protective effect seems overstated since peroxide values are poor indicators of oxidative stability, especially in the presence of high levels of free fatty acids, such as are present in Tempeh extracts.

Sugar alcohols, such as glycerol, xylitol, sorbitol, and maltitol, inhibited oxidation of lard in biscuits, but only xylitol functioned in linoleic acid.[65] A nontoxic antioxidant of improved activity is obtained by extraction of sunflower seed husks with boiling water, filtration, and evaporation to a gel.[66] Petroleum ether extracts of white cabbage, black radish, parsnips, celery, chicory, kohlrabi, and olive leaves were added to pork fat. The olive leaf extract showed strong antioxidant activity at 90°C while the others all showed positive results.[67] Phenolic compounds from the aqueous phase of crushed olive pulp were obtained by extraction of the NaCl-saturated solution with acetone-ethyl acetate. In olive oil, they had activity similar to the synthetic phenolics.[68]

Several articles describe the value of pigments. Yellow, brown, and green pigments isolated from teas were used in margarine, vegetable oil, and bakery products at 0.1 to 0.2% basis fat. They functioned as well as BHT at the 0.02% level.[69] The color of the red pepper, *Capsicum annum*, is well preserved in the dried state. The active components of this vegetable were separated and identified as α-tocopherol and capsaicins.[70] Extracts of *Aspergillus niger* were fractionated by gel permeation chromatography to give brown and yellow pigments. The yellow pigment was found to have a naphthopyrone structure. Both of these materials had strong antioxidant activity in lard.[71] Extracts of dried cranberry leaves containing a tannincatechin complex were found to be effective in preventing oxidation of butter, milk, margarine, and hydrogenated fats. The basis of this effectiveness was improvement in flavor stability and a reduction in the accumulation of hydroperoxide and the secondary products of oxidation as measured by benzidine and thiobarbituric acid color reactions.[72] Bishov and Henick claim that autolyzed yeast protein enhances the antioxidant properties of BHA, BHT, and α-tocopherol.[73]

One of the recently reported natural products, Gemini® base (Fritzsche®-Dodge Olcott, Inc.), is a green-colored, spicy-flavored powder which, although sold as a flavoring agent for sausages and other cured meats, has been reported to have considerable antioxidant activity. At 0.03 to 0.08% concentrations, this spice extractive has been found to give acceptable flavor and shelf life to nitrate- and nitrite-free frankfurters.[76] It is GRAS and has been approved for use by the USDA Consumer Marketing Service. It is currently used in the preparation of various meat products. Because of its mint-like flavor, its use would be restricted to spicy foods.

In 1973, the Campbell Soup Co. obtained a patent (U.S. 3,732,111) on the manufacture of a spice antioxidant principle, which is also intended for use in the manufacture of processed meat products. In this case the ground spice, e.g., sage, is extracted with a heated animal or vegetable oil. The oil-soluble extract containing volatile and nonvolatile components is separated from the spice solids. The extract is then heated to 175°C under vacuum conditions and simultaneously sparged with steam to obtain a deodorized oil extract containing an oil-soluble, nonvolatile spice antioxidant princi-

Table 8
COMPARISON OF TBHQ WITH BHA, BHT, AND PG IN VEGETABLE OILS BY AOM AND OVEN STORAGE TESTS[a]

Oil treatment (wt % conc)	AOM stability (hr to reach 70 meq peroxide content)			Storage life at 115°F of potato chips prepared with oils (days)		
	Cottonseed	Soybean	Safflower	Cottonseed	Soybean	Safflower
Control (no antioxidant)	9	11	6	11	8	4
0.010 TBHQ	24	29	29	17	—[b]	—[b]
0.020 TBHQ	34	41	40	23	25	23+
0.030 TBHQ	42	53	77	33	—[b]	—[b]
0.010 BHA	9	12	8	9	—[b]	—[b]
0.020 BHA	9	10	6	10	9	7
0.030 BHA	9	10	7	7	—[b]	—[b]
0.010 BHT	10	12	10	12	—[b]	—[b]
0.020 BHT	11	13	7	7	10	9
0.030 BHT	13	15	8	15	—[b]	—[b]
0.010 PG	19	21	13	12	—[b]	—[b]
0.020 PG	30	26	10	14	20	7
0.030 PG	37	31	12	17	—[b]	—[b]

[a] TBHQ = *tert*-butylhydroquinone, BHA = butylated hydroxyanisole, BHT = butylated hydroxytoluene, PG = propyl gallate, and AOM = active oxygen method.
[b] Dash indicates test not conducted.

Reprinted from Sherwin, E. R. and Thompson, J. W., *Food Technol.* (Chicago), 21, 914, 1967. With permission.

ple. The oil obtained is reported as bland and having little or no aroma. The patent shows that at 2.5% level is increases the AOM stability of pork fat by a factor greater than 50. It is also reported to impart good flavor stability to pork sausage patties. The authors are not aware of a commercial application of this product at this time.

In a related development, Chang and co-workers described a spice isolate from rosemary and sage with strong antioxidant properties.[31] As shown in U.S. 3,950,266, the approach is somewhat similar to that described above. Chang and his colleagues have investigated the field more thoroughly, however, stopping just short of identification of the active compound(s). Various solvents, including hexane, benzene, ethyl ether, chloroform, dioxane, and methanol, were investigated as spice extractants. Because of yield, volatility, and toxicity considerations, ethyl ether was used in most experiments. As shown in Table 9 this crude extract is effective in inhibiting peroxide formation in fats and oils. Although active, the extract has a greenish color and a strong mint flavor. The color may be removed by bleaching with activated charcoal, and the odor and flavor by dispersion in a commestible oil and vacuum steam-distillation or molecular (thin-film) distillation. Alternatively, it may be purified by silicic acid column chromatography. Active fractions are reported to be white, odorless, and tasteless. At the 0.02% level, they show good carry-through stability in potato chips fried in sunflower oil. They also delay the development of reversion flavor in soybean oil, a feat, as mentioned earlier, beyond the capacity of antioxidants on the market today. Because this is rather unique, the reader is invited to look at the data taken from the patent and presented in Table 10. In this case also, perhaps the most extensively studied area of natural antioxidants, there is not, as yet, any commercial utilization.

Table 9
ANTIOXIDANT EFFECT OF ROSEMARY EXTRACT IN CHICKEN FAT, SUNFLOWER OIL, AND CORN OIL[31]

Sample	Additive	Peroxide value meq/kg after days at 60°C 0	1	3	6	8
Chicken fat	None	0.3	6.4	25.2	—	—
	0.02% Rosemary extract	0.3	5.1	9.8	—	—
Sunflower oil	None	0.4	—	5.4	32.3	63.3
	0.02% Rosemary extract	0.4	—	3.2	15.8	19.7
	0.02% P-A prod.	0.4	—	3.5	32.4	42.9
Corn oil	None	0.9	—	3.2	9.8	22.1
	0.02% Rosemary extract	0.9	—	2.8	4.3	9.6
	0.02% P-A prod.	0.9	—	3.1	6.2	15.2

Table 10
IMPROVEMENT OF FLAVOR STABILITY OF SOYBEAN OIL WITH THE USE OF ROSEMARY ANTIOXIDANT[31]

	Flavor stability after days at room temp.							
	0		14		21		35	
Sample	Flavor score[a]	Peroxide value	Flavor score	Peroxide value	Flavor score	Peroxide value	Flavor score	Peroxide value
Soybean oil	8.5	0.3	3.5	1.7	3.0	—	3.0	5.1
Soybean oil + 0.02% rosemary antioxidant	8.5	0.3	5.5	0.6	5.6	—	5.5	1.9

[a] Hedonic scale 0 to 10; the higher the score, the better the flavor.

NEW ANTIOXIDANTS

A number of new synthetic chemicals have been tested for antioxidant activity. None of these is, as yet, food approved. An ester of glycerol with gallic acid, 1-galloyl glycerol, was found to have much greater activity than PG.[74] The 5-hydroxybenzimidazoles were found to possess strong antioxidant activity.[75] Phenothiazine derivatives were active antioxidants for methyl linoleate.[77] Triphenylphosphine was also found to stabilize linoleate.[78] Ikeda and Fukugumi have found that ion-exchange resins and ethylenimine polymers are antioxidants for methyl linoleate.[79] In this heterogeneous system, the mechanism involves donation of a hydrogen atom to the peroxy radicals.

Certain vegetable oil unsaponifiables protect safflower oil from oxidative polymerization when it is heated at frying temperature.[80] The fraction responsible is apparently sterol in nature. Although the common sterols are inactive, those having a 4-α-methyl group function well. Those sterols from *Vernonia anthelmintica* oil which contain no 4-α-methyl group are also active. The isofucosterol side chain may be the structural feature required. Also, several anionic surfactants were found to function much like these sterols in preventing oxidative polymerization at higher temperature.[81] The most effective additive was the sodium salt of phosphated mono- and diglycerides. It is suggested that these surfactants behave in a manner similar to methyl polysiloxane, as oxygen barriers at the oil-air interface.

The preparation and testing of 6-hydroxy-2,5,7,8-tetramethyl-chroman-2-carboxylic

Table 11
COMPARATIVE ANTIOXIDANT ACTIVITY[a]

Antioxidants	Oxygen analyzer, RT (% NDGA)[d]	Thin layer (days at 45°C)[b,c] Soybean oil	Chicken fat	AOM (hr at 98°C)[c] soybean oil
Trolox	100	17—36	35 +	27
BHT	85	9	15	6.5
BHA	140	8	20	5
PG	85	15	—	22
AP	58	11	10	11.5
TBHQ	85	20—37	34	32
NDGA	100	17	25	22
α-Tocopherol	7	6	13	5
γ-Tocopherol	9	7	29	5
None	0	5	5	5

[a] RT = room temperature, NDGA = nordihydroguaiaretic acid, BHT = butylated hydroxytoluene, BHA = butylated hydroxyanisole, PG = propyl gallate, AP = ascorbyl palmitate, and TBHQ = *tert*-butylhydroquinone.

[b] To reach peroxide values of 70 meq/kg in vegetable oil and 20 meq/kg in animal fat.

[c] Thin layer and active oxygen method (AOM) antioxidants at 0.02%.

[d] Oxygen analyzer, 100 μg antioxidant per test.

From Cort, W. M., Scott, J. W., Araujo, M., Mergens, W. J., Cannalonga, M. A., Osodca, M., Hardley, H., Parrish, D. R., and Pool, W. R., *J. Am. Oil Chem. Soc.*, 52, 174, 1975. With permission.

acid (called Trolox C by Hoffman-LaRoche,® Inc.) represents one of the more serious, recent attempts to introduce a new, synthetic food antioxidant into the market place.[82,83] As shown below, Trolox C resembles α-tocopherol structurally; its 2-position has a carboxyl group replacing the hydrocarbon chain.

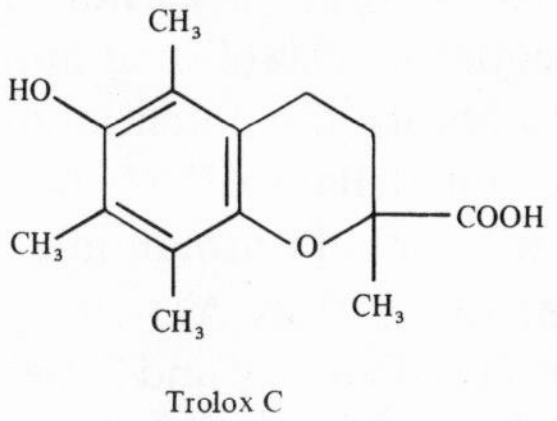

Trolox C

This gives it limited (0.05%) solubility in water, while that in common vegetable oils is about 0.2%. As shown in Table 11, its activity compares favorably to the antioxidants in use today. In vegetable oils and animal fats, it has two to four times the antioxidant activity of BHA and BHT and is more active than PG, NDGA, ascorbyl palmitate, and the tocopherols. In most cases, its activity is close to that of TBHQ, except in those tests where higher temperatures are required, e.g., the AOM test. The decomposition rate of Trolox C is significant above 100°C making it of doubtful use in frying operations. This high temperature loss can be mitigated by addition of ascorbyl palmitate with which it acts synergistically. Its LD_{50} and other preliminary toxicity tests show it to cause no physiological problems; however, Trolox C has not been released for human consumption and its price (considerably higher than that of antioxidants now on the market) may well keep it on the shelf permanently. Cort has also reported on ascorbyl palmitate (below) as a highly effective antioxidant which has been cleared for food use with no restirctions on the level employed.

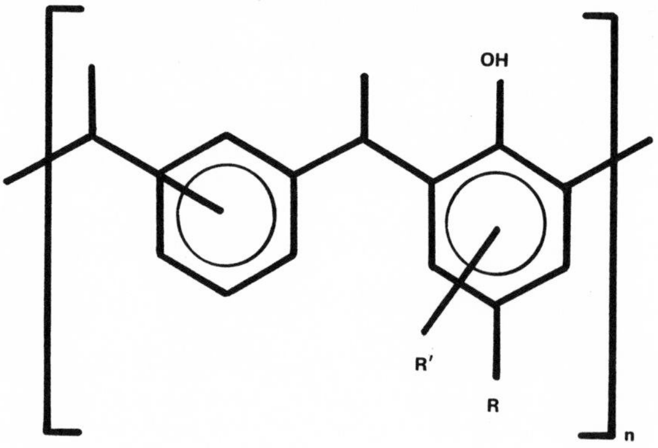

FIGURE 7. The general structural formula of Poly AO-79®. R = −OH, −OMe, or alkyl group and R′ = H or alkyl group. (From Furia, T. E. and Bellanca, N., *J. Am. Oil Chem. Soc.*, 54, 239, 1977. With permission.)

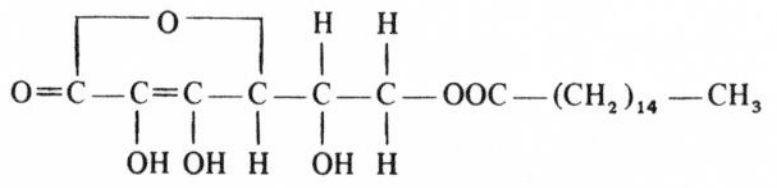

```
  ┌──O──┐   H   H
  |     |   |   |
O=C—C=C—C—C—C—OOC—(CH2)14—CH3
    |  |  |   |   |
   OH OH  H  OH   H
```

Ascorbyl Palmitate

It has been shown that in soybean oil and in potato chips at the 0.01% level, it is more effective than 0.02% TBHQ.[32]

The last development in antioxidants to be described in this section is also the latest and appears to be the most promising. It does not involve entirely new antioxidant structures but a new way of rendering established antioxidants nonabsorbable to the gastrointestinal tract in order to reduce the potential toxic risk to man over a life span of use. As described by Furia and Bellanca,[84,84a] the functional compounds in these novel compositions are covalently bound onto an inert, polymeric matrix (backbone). The general structure of one of these, Poly AO-79®, is shown in Figure 7. It is prepared by polycondensation of divinylbenzene and a blend of various sterically hindered phenols and hydroquinone. Although the data indicate that the product has good antioxidant activity in stabilizing vegetable oils, including essential oils, e.g., lemon essence oil (Figure 8), its main virtue lies in high temperature applications, such as frying, where it is not depolymerized at temperatures well above 300°C. Also, its carry-through into foods from frying oils is nearly quantitative. Results of comparative studies using ^{14}C-radiolabeled antioxidants (Table 12) illustrate this remarkable property. Additional radiolabeled studies indicate Poly AO-79® is virtually unabsorbed and is without toxic affect when fed to animals at very high levels. Further applications and more extensive toxicological studies now underway should bring this new development into commercial use in the near future.

SYNERGISM AND CHELATION

Tartaric acid and substituted methyl tartaric acids were shown to deactivate added iron salts (as Fe^{++} and Fe^{+++}) in oil-in-water emulsions containing soybean oil.[85] Rates of oxidation were measured using a polarographic oxygen analyzer. Unsubstituted tar-

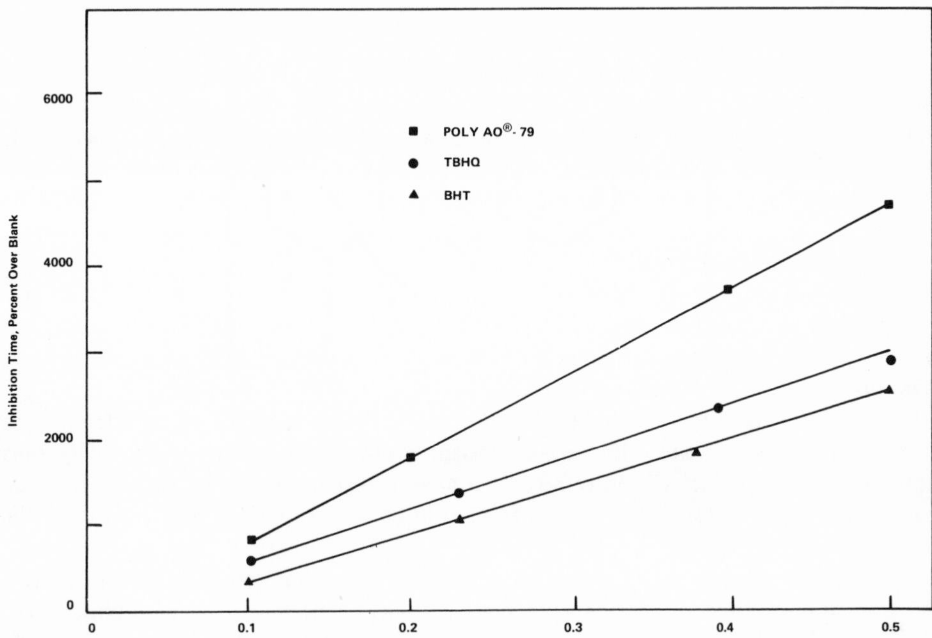

FIGURE 8. Comparative activity of butylated hydroxytoluene (BHT), *tert*-butylhydroquinone (TBHQ), and Poly AO-79® in lemon essence oil. Oxygen bomb test at 100°C and PSIG O_2. (From Furia, T. E. and Bellanca, N., *J. Am. Oil Chem. Soc.*, 54, 239, 1977. With permission.)

Table 12
CARRY-THROUGH OF ^{14}C-RADIOLABELED ANTIOXIDANTS ONTO POTATOES FROM FRYING OIL AT 190°C (375°F)[a]

	Percent carry-through of antioxidant[b]		
No. of fryings	^{14}C-BHA	^{14}C-TBHQ	^{14}C-Poly AO-79®
1	79—84	80—85	96—99
2	83—87	78—81	96—98
3	87—90	83—85	95—96
4	83—85	80—82	95—97
12	—	80—83	—
20	78—82	—	96—98

[a] BHA = butylated hydroxyanisole and TBHQ = *tert*-butylhydroquinone.

[b] Range of 3 samples per frying.

From Furia, T. E. and Bellanca, N., *J. Am. Oil Chem. Soc.*, 54, 239, 1977. With permission.

taric acid was more effective than 2-methyl tartaric acid, which was, in turn, more effective than 2,3-dimethyl-tartaric acid.

Synergism between citric acid and BHT[86] was reported in soybean oil when stored in the dark or irradiated in direct sunlight. Under both conditions, $CuSO_4$ promoted increases in peroxide values. In this case, it is unclear whether citric acid functions as a true synergist or a metal deactivator alone. Similarly, citric acid is described as having a strong synergistic effect with red pepper.[87] Again, this may be principally or entirely metal deactivation.

Linow and Mieth have reported the synergistic action of phosphates with tocopherol using methyl linoleate as substrate.[88] Crude phosphatides were shown to act in part as metal deactivators, but chromatographically pure phosphatides, such as phosphatidyl choline and phosphatidyl ethanolamine, appear to show true synergism with tocopherol and have no metal-capturing properties. Melanoidin, the browning reaction pigment formed when glycine is heated with glucose, was shown to function principally as a strong metal deactivator.[89]

Synergism is also claimed for mixtures of ascorbic acid with 6-hydroxy-2,5,7,8-tetramethyl-chroman-2-carboxylic acid.[90] This combination is shown to be quite functional in fats and oils and also in various food products, such as margarine, butter, cheese, salad dressings, potato chips, and sausages.

Lunde et al. has shown that metal ions (Ni^{+2} and Fe^{+3}) can be transferred from water to an oil phase when the oils contain phospholipids.[91] This may explain why these materials have sometimes been reported to have pro-oxidant and, at other times, antioxidant properties. Unless they are strong enough to complex all the metal ions, they may appear to catalyze oxidation. Since Fe^{+2} and Cu^{+2} below 5 ppm were shown to increase the oxidation rate of meat lipids,[92] it is evident that complete metal deactivation is most important. Chelating agents, such as citric acid and ethylenediaminetetraacetic acid (EDTA), were effective in retarding the oxidation rate of peanut butter.[93] Best results were observed when they were added as a concentrate in water.

Several papers have appeared recently describing new and more sensitive methods for measuring the levels of trace metals in oils.[94] In assessing the synergistic effect of antioxidant combinations, it is important to be aware of the complicating factor of metal contamination.

ANALYTICAL METHODS

A literature survey on the detection and determination of food antioxidants has been published recently.[95] What follows is a synthesis of this report with some of the most relevant features of the previous article in this series[5] and some new developments in this field. From these and other publications on this subject, it immediately becomes clear that when discussing the analysis of antioxidants in foods there are two recognizable and distinct steps: (1) separation and (2) quantitative measurement. Separation includes extraction, distillation, and to a lesser extent, chromatography. Quantitative measurements involve spectrophotometric, chromatographic, and lately, polarographic techniques. Each of these topics will be discussed below.

Separation Techniques

By far, extraction is the most widely applied separation technique. In her review, Endean[95] further subdivides the procedure according to its application to fats, high-fat foods, and low-fat foods. PG can be extracted from a fat solution in petroleum ether, hexane, or heptane by aqueous extraction[96] or with aqueous ammonium acetate.[97] The most useful solvents for extracting most food antioxidants are acetonitrile and alcohol. Acetonitrile was used by Aczel et al. to extract BHT directly from lard.[98] A freezing step following extraction removes the concomitant fat by precipitation. More commonly, the fat is dissolved in petroleum ether or hexane, and the antioxidant, extracted with acetonitrile or hexane-saturated acetonitrile.[99-103] This type of extraction is subject to high blank values and gives poor recovery with BHT. For this reason, many workers prefer to use aqueous ethanol or methanol,[104-106] which avoids the extraction of fat thus minimizing the interference problems in the succeeding analytical step. Its disadvantage is that, when detecting low levels of antioxidant, loss of BHA and BHT by volatilization may be encountered. According to Alessandro et al.[107] this loss can be minimized by using dry methanol as the extractant. This has been confirmed by Endean and Bielby[108] who found that all of the common food antioxidants, except ascorbyl palmitate, can be extracted with dry methanol at 40°C. Ascorbyl palmitate requires acetonitrile for extraction. Antioxidants can be recovered from both extracts in good yields by vacuum and low-temperature evaporation. Phipps found that the best way to quantitatively recover BHA and BHT from an oil is to dissolve it in heptane and extract it with four aliquots of dimethyl sulfoxide (DMSO). The combined DMSO extracts are mixed with water and 2*M* NaCl, and the antioxidants are then back-extracted with petroleum ether.[109]

Extraction is also used quite frequently to separate antioxidants from high-fat foods. In this case, petroleum ether soxhlet extraction of the food material is usually applied after the product is dried and ground with sand. Losses by volatilization are avoided by drying with anhydrous sodium sulfate. Even with the best of care, some workers have found that extractions are incomplete either because of the presence of phospholipids[110] or a high sugar and protein content.[111] Groebel and Wessels have extracted fat from soups, sauces, sausages, nougat, marzipan, and other similar material in the cold with peroxide-free ether.[112] Extraction is preceded by a step in which the food is ground with sodium sulfate. Perhaps the most useful method for high-fat foods is that of Endean and Bielby[113] which utilizes chloroform/methanol (1/1, v/v) as the extracting medium. Antioxidant evaporation losses are minimized by drying the extract over sodium sulfate prior to solvent evaporation.

Most of the procedures mentioned above can be utilized with varying degrees of success to obtain antioxidants from low-fat foods. The aforementioned low-temperature ether extraction of Groebel and Wessells[112] seems applicable to low-fat foods, specifically cereals and potato products. The Canadian Food and Drug Laboratories recommends the use of benzene for extracting BHT from polished rice.[114] Takahashi[115] makes use of carbon disulfide to obtain BHA and BHT in cereals, while Daniel and Cage prefer diethyl ether.[116] Good recovery from both cereals and potato granules is reported. Needless to say, applications of extraction to low-fat foods are numerous, even if most of them are not completely satisfactory. The most widely used method is that developed by Anglin et al.,[117] which was modified by Filipic and Ogg[118] and Sloman et al.[119] In this method, BHA and BHT are steam distilled from a food and water mixture containing added salts (usually magnesium oxide) to improve antioxidant recovery. The method seems to work well for less than 10 ppm BHA and BHT in dried potatoes and similar levels for desserts and beverage mixes[119] but poorly for breakfast cereals.[120] Even with its limitation, distillation still stands as a primary means of obtaining the more volatile antioxidants from foods.

Chromatographic methods, usually column, are commonly employed to separate antioxidants from food samples. Occasionally, they are used alone; more often, they are ancillary to extraction and distillation, an additional step in the purification of the sample for further analysis. Silicic acid and alumina columns have been used to separate various phenolic antioxidants from food products. Caldwell et al. describe this application by illustrating the separation of BHT from various cereal products and paperboard.[121] Florisil containing 4% water, for example, was found most useful in isolating BHT from fat.[113] Alumina,[122] acid alumina,[103] cellulose,[98] silica gel,[123] and silver nitrate containing silica gel[124] have been used at one time or another to separate antioxidants from fats. A precolumn of siliconized glass wool can be used to aid in the assay of BHA and BHT by gas chromatography (GC). A more typical application of column chromatography is a cleanup step for antioxidant extracts and distillates; thus, a deactivated acidic alumina column has been used to purify the acetonitrile-extracted antioxidants from meat.[126] Acid alumina, Brockman activity grade I, has been used to clean up methanol extracts of BHA and BHT from fats and oils.[127]

Other authors have shown that polyamide powders are also useful in cleaning up antioxidant-containing extracts.[99,102,128] As mentioned previously, these purification procedures are an essential intermediate step to quantitative determination, which will be discussed next.

Quantitative Measurements

Most of the quantitative methods for antioxidant detection fall broadly into two categories: spectrophotometry and chromatography. Polarography, the latest entry in this field, will be briefly mentioned at the end of this section. Historically, spectrophotometry was first and, in all likelihood, is still the most widely used method for quantitative determination of antioxidants.

Spectrophotometry

Most of the spectrophotometric methods make use of the fact that except for ascorbyl palmitate, all food antioxidants are phenolic in nature. This characteristic means that they exhibit absorption in the ultraviolet and infrared regions of the spectrum.[153] Because of the minute amounts of antioxidants involved and the lack of specificity due to the presence of interfering substances, very few methods have been able to capitalize on this fact. One is described by Kaszyka[129] in which infrared spectrophotometry is used to detect propyl and octyl gallates in lard. In a more typical spectrophotometric determination, the antioxidant is reacted with a phenol reagent, which results in a color absorbing in the visible region. A good example is the widely used Emmerie-Engels method which Mahon and Chapman adapted for the determination of antioxidants in fats.[130] In this instance, the ferric chloride reagent is reduced to the ferrous state by the antioxidant with the concomitant formation of a reddish purple ferrous-bipyridyl complex which is measured spectrophotometrically. These same authors used Gibbs reagent (2,6-dichloroquinonechloroimide) in developing a colorimetric test, specific for BHA.[131] Similar tests, specific for propyl gallate[117] and BHT,[132] have also been developed. Today, most spectrophotometric tests are being replaced by more specific and sensitive chromatographic methods.

Chromatography

Chromatography in all its forms, paper, thin-layer, gas, and liquid, has been applied to the antioxidant determination in foods. In the past, many applications of paper chromatography (both normal and reverse phase) have been cited in the literature.[95] Although an occasional reference to this technique, such as the one by Dumitrescu et al. on the detection of dodecyl gallate in foods,[133] is still encountered, it has been, for the most part, replaced by the more rapid thin-layer chromatography (TLC). TLC has been used by Sahasrabudhe to separate BHA, BHT, and propyl gallate from fats.[134] Here, quantitation was done by elution and spectrophotometry; whereas, Kimura and Terada[135] used densitometry to detect BHA on thin-layer plates. With the most common food antioxidants, BHA, BHT, propyl gallate, and ascorbyl palmitate, the best separations are obtained by using activated silica gel layers,[113] particularly those incorporating a chelating agent, such as citric acid.[136] Various review articles, detailing solvent systems and visualizing reagents and separation characteristics of most of the food antioxidants, have been published[95,100,137,138] and indicate continued, extensive interest in this subject.

From the early work on breakfast cereal by Buttery, and Stuckey on potato granules[139] to the latest antioxidant determination in vegetable oils by Karleskind and Valmalle,[143] one of the most favored methods for quantitizing the presence of antioxidants in foods has been gas-liquid chromatography (GLC). GLC's precision, rapidity, and sensitivity still make it the method of choice for many antioxidant analyses today, particularly BHA and BHT. Hence, Anderson and Nelson[140] have used it to determine BHA and BHT in cereals; Choy, Quattrone, and Alicino,[141] to detect the presence of these antioxidants in vitamin formulations; Singh and Lapointe, in meat products;[126] Stuckey and Osborne in rice and potato flakes;[110] Groebel and Wessels in soups, sauces, and sausages;[112] and Schwecke and Nelson in breakfast cereals, potato granules, and packing materials,[142] Numerous applications of GLC to the detection of antioxidants in fats and oils have been reported.[127,143-150] Among the most successful methods in this area and one of the widest in scope is the work of Stoddard[151] who describes the assay of BHA, BHT, PG, TBHQ, and NDGA. After the hexane solution is partitioned by shaking against 80% aqueous ethanol and acetonitrile, the antioxidant-rich phase is cleaned up on a florisil column, derivatized with a silylating reagent, and then injected into a gas chromatograph equipped with a flame ionization detector. The glass column, 6' × ¼", is packed with 3% JXR silicone on Gas-Chrom® Q and is programmed from 105 to 250°C at the rate of 7.5°C/min. Good separation and recovery of 76 to 100% is indicated. With several new applications being reported every year, it is clear that GLC will continue to play a leading role in the quantitative analysis of antioxidants.

Column chromatography, on the other hand, has played a relatively minor role in the determination of antioxidants in foods. Sporadic applications have been recorded like the one by Pokorny et al. who used gel permeation chromatography to determine antioxidants in edible fats and oils.[152] The advent of high-performance liquid chromatography (HPLC) is rapidly changing this situation, thus, Endean in her review[95] gives five references for HPLC in antioxidant analysis, with more successful applications being reported in increasing numbers. As previously noted and as reported by Seher,[153] all food antioxidants (except ascorbyl palmitate) are phenolic in nature and exhibit ultraviolet absorption in the range of 240 to 300 nm. These facts favor detection by HPLC which offers the added advantage of being nondestructive and readily amenable to preparative requirements. With new substrates and more sensitive detectors becoming available every year, this field is ripe for additional development.

Of the various miscellaneous methods reported for antioxidant analysis, the most promising seems to be polarographic. It has been known for a long time that phenolic antioxidants can be determined by anodic oxidation at a rotating graphite electrode.[154] McBride and Evans have found that linear sweep voltammetry can be used to estimate added antioxidants in tocopherol-containing vegetable oils.[155] More recently, Waltking et al.[156] have repeated and expanded the method, demonstrating its sensitivity and accuracy. Both reports note the interfernce of δ-tocopherol in BHT determination, and the technique is still far from being a routine laboratory method.

MEASUREMENT OF ANTIOXIDANT EFFECTIVENESS AND OXIDATIVE STABILITY

Techniques for measuring oxidative stability of a fat or fatty food generally include storage tests run under either normal or accelerated conditions. Since the exact point at which a product becomes rancid is often difficult to determine and depends upon subjective judgment, a number of objective tests have been developed to determine this end point. Many of these methods are only partially satisfactory since the manufacturer is concerned primarily with the projected storage life of his product, before it develops an odor or flavor which is objectionable to the consumer. This point is usually reached long before the product might be described as typically rancid. Consequently, some of the tests most frequently used are aimed at measuring the induction period, e.g., the time required for the product to begin oxidizing rapidly, which coincides fairly closely with the beginning of off-flavor development.

AOM

Figure 1 shows data on the number of hours required for lard to reach a hydroperoxide content of 20 meq/kg during aeration at 210°F. It illustrates that the induction period of lard, i.e., that portion of the curve which is nearly horizontal, can be extended dramatically by the addition of an antioxidant. Although vegetable oils have a somewhat longer induction period than lard, they do not respond well to the use of an antioxidant because of the naturally occurring tocopherols in vegetable oils which are absent in lard.[157] Almost maximum stabilization has been obtained from these natural antioxidants in the vegetable oils, and further antioxidant addition has little effect.

The AOM test is the most frequently used method in industry to estimate the rate of hydroperoxide development in fats.[158] Dry, filtered air is bubbled at a constant rate through the liquified fat, held at 210°F. Periodically, samples are removed and titrated for peroxide content iodometrically. The American Oil Chemists' Society method specifies 100 meq/kg as the end point of this determination. However, it is known that animal fat becomes rancid under these test conditions at much lower peroxide values (20 to 60 meq/kg); whereas, most vegetable oils and shortenings do not become noticeably rancid until values between 70 and 150 are reached. As a consequence, in some laboratories it is customary to observe the time for the effluent air to develop a rancid odor and to use this subjective evaluation as the end point.

Since hydroperoxides are quite unstable at temperatures above those used in the AOM test, it is necessary to follow directions closely when carrying out this method. Good correlation with actual shelf life at ambient storage temperatures are usually not obtained, particularly with vegetable oils which have been hydrogenated and contain added antioxidants. Strong off-flavor development often occurs as the result of hydroperoxide decomposition and before high levels of peroxides accumulate. The flavor responses given by tasters are "painty, cardboard-like, fishy, melony, buttery, metallic," rather than typically "rancid". Thus, the OAM method can be used to compare the relative resistance of various fat samples to oxidation, but it is not necessarily a good indicator of their flavor stabilities. Furthermore, its use is restricted to fats and oils since it cannot be applied to food systems.

2-Thiobarbituric Acid Test (TBA)

A colorimetric method, based on the reaction of TBA with the decomposition products of fat hydroperoxides to form a red complex, has been used by many research workers.[159] It is known that test conditions have an effect on the results obtained.[160] Consequently, it is necessary to be cautious in considering the significance of the measurements and comparing them with organoleptic evaluation or findings by other objective methods. It has been shown that the pigment obtained absorbs at 532 nm and involves the reaction of a three-carbon compound (malonaldehyde) with the TBA reagent. Marcuse and Johansson have suggested that absorption at 450 nm be used in assessing lipid oxidation,[161] but this latter wavelength may also reflect the degree of browning in a food system.[159] In any case, application of the TBA test to measure aldehydes during various stages of sample incubation has proven quite useful in monitoring lipid oxidation with a variety of food products.

Free Fatty Acids

The level of free fatty acids has been used for many years as a measure of oil quality and is determined by titrating a weighed sample in neutralized alcohol to the phenolphthalein end point with a sodium hydroxide solution. Since both fatty acids generated by hydrolysis and those formed by thermal oxidation are titrated collectively, this test does not accurately measure oxidative rancidity.

Shelf Storage Test

The most reliable method to determine stability of a food product is to evaluate it organoleptically at intervals during actual shelf storage. This is a slow process often requiring many months to complete. The usual procedure for following oxidative breakdown in shelf storage tests is by organoleptic panels,[162] which vary from one or two trained people to large numbers of panelists. Statistical methods are often used to analyze the data. Although the statistical approach gives better reliability, it is not always possible to use such a large panel. Adequate results can often be obtained with three to four trained people. In some cases, this data may be supplemented with peroxide value determinations. With animal fats, this technique has proven quite valuable since rancidity development is rapid and positive. With food products, it is necessary to separate the fat using a solvent, such as hexane or a mixture of chloroform with methanol in those cases where the fat is bound to protein. As a consequence, the procedure may be quite time consuming.

Oven Storage Test

The oven storage test is merely a technique for reducing the storage time required to make an evaluation. Samples are held at several temperatures above ambient temperature and evaluated periodically by an organoleptic panel or various chemical tests for rancidity. The most commonly accepted method is the Schaal oven test.[163] In this method, samples are stored in loosely capped jars at 140 to 145°F and evaluated daily until they become rancid. This technique has been used widely for evaluating shortenings in cookies and crackers and for evaluating potato chips and other fried foods.

Oxygen Bomb Test

The oxygen bomb test is based on the fact that there is measurable absorption of oxygen during the oxidative deterioration of fats. The material is tested in a stainless steel bomb, which is connected to a pressure recorder. To accelerate the reaction, the bomb is charged with oxygen under pressure and is heated in a bath. When oxidation occurs, the pressure in the bomb is reduced through absorption and measured on the chart of the pressure recorder. Stuckey et al. have shown this technique to be quite accurate for various fats and fatty foods.[164] With this technique,[165] Wintermantel et al. have shown the effectiveness of antioxidants in various cereal products. The oxygen bomb test has three advantages: reproducibility, no requirement for skilled technicians, and performance with a minimum of operator time.[166] Petroleum bombs, now being used, are not too satisfactory, since considerable difficulty is found in maintaining a leak-free system. One disadvantage of the bomb method is the high cost of the equipment. Consequently, only a limited number of samples can be run simultaneously. Also, the reaction must be accelerated by relatively high temperatures, thus giving results which cannot be expected to correlate well with shelf storage organoleptic data.

Recent Objective Methods — Correlation with Organoleptic Evaluation

Fioriti et al. have subjected samples of various fats and spray-dried emulsions to accelerated storage conditions, accelerated by higher temperatures, 37.8 and 60°C. Flavor panel results on these fats and on the reconstituted wet emulsions were compared with data obtained by several objective test methods.[167] Good to excellent correlations were observed with oxygen absorption measurements following the method of Bishov and Henick,[168] which is a rapid GLC headspace technique (approximately 2 min per sample). It can be used to evaluate a large number of samples on a routine, daily basis. Since the storage temperature (50 to 60°C) is considerably lower than that customarily used with the oxygen bomb test, the time required for a complete evaluation is longer; however, the data should be inherently more meaningful since these conditions more closely approximate actual shelf storage.

Good to excellent correlations were also found between organoleptic data and the peroxide and carbonyl values, the pentane levels, and the carbon dioxide in the headspace. Total carbonyl values were obtained by steam distillation of the reconstituted, spray-dried emulsions, which was followed by a reaction of the distilled carbonyl compounds with 2,4-dinitrophenylhydrazine. This process produces red-colored phenylhydrazones, which can be determined colorimetrically. This method for measurement of total carbonyls was found to be superior to the TBA test, which measures malonaldehyde levels. Decomposition of hydroperoxides also generates both pentane and carbon dioxide. Pentane values were obtained by GLC using the method of Evans et al.[169] Carbon dioxide levels were determined on headspace gas by GLC.[167]

Anisidine has also been used in reactions with the carbonyls generated by hydroperoxide decomposition.[170] The optical density of the colored reaction product, when applied to soybean oil autoxidation, gave good correlation with flavor scores. Because of the carcinogenic nature of benzidine, anisidine has replaced benzidine in most analyses today.

Gas-liquid chromatography has recently been used to measure the level of total and individual flavor-related volatile compounds in vegetable oils and shortenings.[171] Flavor scores of aged samples were shown to correlate highly with the level of volatiles obtained; thus, the flavor characteristics of fats and oils can be obtained rapidly and efficiently by use of this instrumental technique.

Recently, the rate of disappearance of dissolved oxygen in oil-in-water emulsions has been measured using an oxygen electrode (polarograph). Results were shown to correlate with the oxidative stability of oil as measured by the AOM test.[172] Also, during the early stages of oxidation, there was good correlation with the determined peroxide values of the oil.[173] This is quite an effective technique, particularly when a pro-oxidant, such as hemoglobin, is added for rapid screening of antioxidant effectiveness.[174] In milk that is rich in linoleic acid and obtained from cows fed on a formaldehyde-treated casein and safflower supplement,[175] disappearance of oxygen coincided with the development of oxidized flavors.

GOVERNMENT REGULATIONS

Use of antioxidants in edible fats and oils and food products is regulated b the FDA. Meat and poultry products are covered separately by the Meat Inspection Bureau of the Department of Agriculture. Generally, food-approved primary antioxidants and synergists are permitted up to 0.02%, based on the level of fat in a food. In meat and poultry products, the level of each individual antioxidant is restricted to 0.01%, but two or more may be added if the total does not exceed 0.02%.

Table 1 lists the primary antioxidants and synergists which are permitted in the U.S. and various industrial countries of the world.[176] There are no limits on the levels of tocopherols permitted in the U.S. The level of gum guaiac is restricted to 0.1% in edible fats and oils, but to our knowledge, there is no longer any significant commercial use of this material. The U.S. alone permits the usage of 2,4,5-trihydroxybutyro-

Table 13
ADDITION LIMITS OF ANTIOXIDANT(S) TO VARIOUS FOODS IN PPM BASED ON TOTAL WEIGHT OF FOOD[a]

Food	BHA	BHT	PG	Total permissible[b]
Beverages and desserts prepared from dry-mixes	2	—	—	2
Cereals, dry breakfast	50	50		50
Chewing gum base	1000	1000	1000	1000
Dry mixes for beverages and desserts	90	—	—	90
Emulsion stabilizers for shortenings	200	200	—	200
Fruit, dry glaceed	32	—	—	32
Meats, dried	100	100	100	100
Potato flakes	50	50	—	50
Potato granules	10	10	—	10
Potato shreds, dehydrated	50	50		50
Rice, enriched	—	33	—	33
Sausage, dry	30	30	30	60
Sausage, pork, fresh	100	100	100	200[c]
Sweet potato flakes	50	50	—	50
Yeast, active dry	1000	—	—	1000

[a] BHA = butylated hydroxyanisole, BHT = butylated hydroxytoluene, and PG = propyl gallate.
[b] Combination of lawful antioxidant.
[c] Based on fat content of sausage.

From Stuckey, B. N., in *The Handbook of Food Additives,* 2nd ed., Furia, T. E., Ed., CRC Press, Cleveland, 1972, 215.

phenone, 4-hydroxymethyl-2,6-di-*tert*-butylphenol, and dioctadecyl thiodipropionate, but none are used commercially to any significant extent.

Table 13 lists the levels of addition of antioxidants permitted in various food products.[5] Details as to the levels of TBHQ permitted in food products are not available at this time. Most likely these levels would be comparable to those of the other antioxidants listed. It is suggested, however, that the manufacturer (Eastman®) be consulted with reference to specific applications for this relatively new additive.

Foreign regulations, as shown in Table 1, vary considerably. Some countries, such as the Federal Republic of Germany, permit no antioxidants at all. Since technology is so advanced in the U.S., the number of additives has proliferated. Currently, all additives on the GRAS list are being reviewed; consequently, the situation as it exists at the moment may be quite different from that in the near future. In addition, a joint conference of representatives from the Food and Agricultural Organization of the United Nations (FAO) and the World Health Organization (WHO) has resulted in the formation of an expert committee on food additives and the Codex Alimentarius Commission whose task is to reconcile differences in the food laws of the various countries. Ultimately, these bodies may achieve simplification and unification of regulations, which would stimulate world trade in fats, oils, and fatty foods.

PRESENT NEEDS AND FUTURE RESEARCH

Having discussed problems caused by oxidation and the role that antioxidants play in reducing these problems, it seems fitting to conclude this review by looking at present needs and observing some new developments looming on the horizon, which might fulfill some of these needs. In regard to needs, the shopping list would still include those items mentioned in recent reviews and publications on this subject.[5,20,177-179] Order or priority might vary according to what one personally believed to be the most pressing (or attractive?) problem. Lindsay and colleagues,[180] for instance, have chosen to investigate various methods of introducing antioxidants into food. Their studies appear to be both interesting and useful since not much is known about the actual distribution of antioxidants in foods. Equally interesting is the approach of Wiegeland and colleagues[181] who limit off-flavor production by enzymatic action.

Most agree that consumer activity has lent new impetus to the search for inexpensive, effective, and safe natural antioxidants. Bearing safety and novel approaches in mind, one cannot but applaud the efforts of those involved in the development of nonabsorbable, polymeric antioxidants.[84] Along with a better delivery system and new antioxidants, better objective methods for detecting oxidation, since it affects the organoleptic properties of foods, are required. Both chemiluminescence[182] and hexanal determination[183] promise real progress in this area. A recent symposium[184] on objective methods for food evaluation summarizes some of the successes and needs in this very complex field. New approaches will, no doubt, make our efforts both more rewarding and interesting.

REFERENCES

1. **Sims, R. J. and Hilfman, L.,** *J. Am. Oil Chem. Soc.,* 33, 381, 1956.

1a. **Dugan, L. R., Jr., Postby, L. M., and Wilder, O. H. M.,** *J. Am. Oil Chem. Soc.,* 31, 46, 1954.

2. **Sims, R. J. and Stahl, H. D.,** *Baker's Dig.,* 44, 50 1970.

3. **Martin, J. B.,** U.S. Patent 2,634,213, 1953.

4. **Nestle SA, Swiss Patent 1,446,140,** 1976.

5. **Stuckey, B. N.,** Antioxidants as food stabilizers, in *The Handbook of Food Additives,* 2nd ed., Furia, T. E., Ed., CRC Press, Cleveland, 1972, 185.

6. **Schultz, H. W., Day, E. A., and Sinnhuber, R. O., Eds.,** *Symposium on Foods: Lipids and Their Oxidation,* AVI Publishing, Westport, Conn., 1962.

6a. **Hoffman, G., in** *Symposium on Foods: Lipids and Their Oxidation,* Schultz, H. W., Day, E. A., and Sinnhuber, R. O., Eds., AVI Publishing, Westport, Conn., 19)2, chap. 12.

6b. **Patton, S.,**in *Symposium on Foods: Lipids and Their Oxidation,* Schultz, H. W., Day, E. A., and Sinnhuber, R. O., Eds., AVI Publishing, Westport, Conn., 1962, chap. 10.

6c. **Koch, R. B.,** in *Symposium on Foods: Lipids and Their Oxidation,* Schultz, H. W., Day, E. A., and Sinnhuber, R. O., Eds., AVI Publishing, Westport, Conn., 1962, Chap. 13.

7. **Lundberg, W. O.,** *Autoxidation and Antioxidants,* Vol. 1, Interscience, New York, 1961.

8. **Shelton, J. R.,** Off. Dig., *(J. Paint Technol.,* formerly), p.1, June 1962.

9. **Scott, G.,** *Atmospheric Oxidation and Antioxidants,* American Elsevier, New York, 1966, 115.

10. **Rawls, H. P. and Van Senten, P. J.,** *J. Am. Oil Chem. Soc.,* 47, 121, 1970.

11. **Johnson, C. J.,** A critical review of the safety of phenolic antioxidants, *Crit. Rev. Food Technol.,* 2, 267, 1971.

12. **Farmer, E. H., Bloomfield, G. F., Sundraligan, A. and Sutten, D. A.,** *Trans. Faraday Soc.,* 38, 348, 1942.

13. **Bateman, L.,** *Q. Rev.,* 8, 147, 1954.

14. **Gunstone, F. D.,** *An Introduction to the Chemistry and Biochemistry of Fatty Acids and Their Glycerides,* 2nd ed., Chaucer Press, Suffolk, England, 1967, 105.

15. **Campbell, I. M., Caton, R. B., and Crozier, D. N.,** *Lipids,* 9(11), 916, 1974.

16. **Evans, C. D., Frankel, E. N., Cooney, P. M., and Moser, H. A.,** *J. Am. Oil Chem. Soc.,* 37, 452, 1960.

17. **Badings, H. T.,** *Ned. Melk Zuiveltidscher.,* 24, 147, 1970.

18. **Grosch, W. Z.,** *Lebensm. Unters. Forsch.,* 157, 70,1975.

19. **Hoffman, G. and Meijboom, P. W.,** *J. Am. Chem. Soc.,* 46, 620, 1969.

20. **Labuza, T. P.,** Kinetics of Lipid Oxidation in Foods, in *Crit. Rev. Food Technol.,* 2, 355, 1971
21. **Lubuza, T. P., Heidelbaugh, N. D., Silver, M., and Karel, M.,** *J. Am. Oil Chem. Soc.,* 48, 86, 1971.
22. **Salwin, H. and Slawson, V.,** *Food Technol.* (Chicago), 13, 715, 1959.
23. **Selke, E., Rohwedder, K., and Dutton, H. J.,** *J. Am. Oil Chem. Soc.,* 52, 232, 1975.
24. **Michalski, S. T. and Hammond, E. G.,** *J. Am. Oil Chem. Soc.,* 48, 92, 1971.
25. **Artman, N. R. and Smith, D. E.,** *J. Am. Oil Chem. Soc.,* 49, 318, 1972.
26. **Paulose, M. M. and Chang, S. S.,** *J. Am. Oil Chem. Soc.,* 50, 147, 1973.
27. **Perkins, E. G., Taubold, R., and Hseih, A.,** *J. Am. Oil Chem. Soc.,* 50, 223, 459, 1973.
28. **Clements, A. H., Van Den Engh, R. H., Frost, D. J., Hoogenhout, K., and Nooi, J. R.,** *J. Am. Oil Chem. Soc.,* 50, 325, 1973.

28a. **Carlson, D. J., Suprunchuk, T., and Miles, D. M.,** *J. Am. Oil Chem. Soc.,* 53, 656, 1976.

29. **Lin, J. S., Smith, V., and Olcott, H. S.,** *Agric. Food Chem.,* 22, 682, 1974.
30. **Hiatt, R. R.,** *Crit. Rev. Food Sci. Nutr.,* 7, 1, 1975.
31. **Chang, S. S., Ostrich-Matijasevic, B., Huang, C. L., and Hsieh, A. L.,** U.S. Patent 3,950,266, 1976.
32. **Cort, W. M.,** *J. Am. Chem. Soc.,* 51, 321, 1974.
33. **Anon.,** *Food Process.,* 34(5), 35, 1973.
34. **Luckadoo, B. M. and Sherwin, E. R.,** *J. Am. Oil Chem. Soc.,* 49, 95, 1972.
35. **Chahine, M. H. and McNeill, R. F.,** *J. Am. Oil Chem. Soc.,* 51, 37, 1974.
36. **Fritsche, C. W., Weiss, V. E., and Anderson, R. H.,** *J. Am. Oil Chem. Soc.,* 52, 517, 1975.

36a. Tenox® Food Grade Antioxidants, Publication ZG-109C, Eastman Chemical Products, Inc., Kingsport, Tenn., 1976, 4.

36b. **Sherwin, E. R. and Thompson, J. W.,** *Food Technol.* (Chicago), 21, 912, 1967.

36c. Tenox® TBHQ Antioxidant for Oil Fats and Fat-containing Foods, Publication 2G-201D, Eastman Chemical Products, Inc., Kingsport, Tenn., 1976, 8.

37. **Astill, B. D. et al.,** *J. Am. Oil Chem. Soc.,* 52, 53, 1975.
38. **Sims, R. J.,** unpublished results, March 1977.
39. **Slowikowska, J. Jakubowski, A., Pilat, K., and Rudzka, Z.,** *Zesz. Probl. Postepoa Nauk Roln.,* 136, 235, 1973.
40. **Goodman, T. et al.,** *Lab Invest.,* 23, 93, 1970.

40a. **Grice, H. C. et al.,** *Food Cosmet. Toxicol.,* 6, 155, 1968.

41. **Oliveto, E. P.,** *Chem. Ind.* (London), 17, 677, 1972.
42. **Catalano, M. and DeFelice, M.,** *Riv. Ital. Sostanze Grasse,* 47(10), 484, 1972.
43. **Yoon, S. U. and Kim, D. H.,** *Korean J. Food Sci. Technol.,* 5(1), 42, 1973.

43a. **Han, D. B. et al.,** *Korean J. Food Sci. Technol.,* 5(2), 129, 1973.

44. **Khafizov, R. H. H.,** *Izv. Vyssh. Uchebn. Zaved. Pishch. Tekhnol.,* 4, 37, 1975.
45. **Peredi, J.,** *Olaj Szappan Kozmet.,* 22(4), 110, 1975.
46. **Yuki, E.,** *Yakagaku,* 20(8), 488. 1973.
47. **Freeman, I. P., Padley, P. B., and Sheppard, W. L.,** *J. Am. Oil Chem. Soc.,* 50, 101, 1973.
48. **Gracian, J. and Arevalo, G.,** *Grasas Aceites,* 22(3), 177, 1972.
49. **Pilat, K.,** *Pr. Inst. Lab. Badaw. Przem. Spozyw.,* 22(4), 565, 1973.
50. **Morrison, W. H.,** *J. Am. Oil Chem. Soc.,* 52, 522, 1975.
51. **Capps, N. K.,** *Food Cosmet. Toxicol.,* 17, 367, 1974.
52. **Cumming, R. B. and Walton, M. F.,** *Food Cosmet. Toxicol.,* 11, 547, 1973.
53. **Bauernfeind, J. C.,** *Crit. Rev. Food Sci. Nutr.,* 8(4), 337, 1977.
54. **Yuki, E., Ishikawa, Y., and Yoshiva, T.,** *Yakagaku,* 23(11), 714, 1974.
55. **Phillip, F.,** *Diss. Abstr. Int. B,* 35(11), 5475, 1975.
56. **Kawasaki-Shi, H. E.,** U.S. Patent 3,585,223, 1971.
57. **Sims, R. J. and Fioriti, J. A.,** U.S. Patent 3,957,837, 1976.

57a. **Sims, R. J. and Fioriti, J.,** *J. Am. Oil Chem. Soc.,* 54, 4, 1977.

58. **Hwang, C. I. and Kim, D. H.,** *Korean J. Food Sci. Technol.,* 5(2), 84, 1973.
59. **Fujimaki, M., Morita, M., and Kato, H.,** Conf. Proc. Int. Atomic Energy Agency, University of Tokyo, 1974, 95.
60. **Yamaguchi, N. and Fujimaki, M.,** *J. Food Sci. Technol.,* 21(6), 280, 1974.
61. **Yamaguchi, N. and Fujimaki, M.,** *J. Food Sci. Technol.,* 21(1), 6, 1974.
62. **Yamaguchi, N. and Fujimaki, M.,** *J. Food Sci. Technol.,* 21(1), 13, 1974.
63. **Abdel-Rahman, A. H. Y. and Youssef, S. A. M.,** *J. Am. Oil Chem. Soc.,* 52, 423, 1975.
64. **Gyorgy, P., Murata, K., and Sugimoto, Y.,** *J. Am. Oil Chem. Soc.,* 51, 377, 1974.
65. **Yamaguchi, N., Oshima, K., and Murase, M.,** *J. Food Sci. and Technol.,* 21(4), 131, 1974.
66. **Kashevatskaya, L. A., Alekseeva, Y. K., Matygina, L. M., Minasyan, N. M., Proskvrina, V. L., and Tsygankova, G. E.,** U.S.S.R. Patent 502,012, 1976.
67. **Mihelic, F.,** *Hrana Ishrana,* 16(314), 143, 1975.

68. **Cantarelli, C. and Montedoro, G.,** *Food Sci. Technol. Abstr.,* 11, J1685, 1975.
69. **Soboleva, M. I.,** *Khlebopek. Konditer. Prom.,* 9, 24, 1974.
70. **Fujimoto, K., Seki, K., and Kaneda, T.,** *J. Food Sci. Technol.,* 21(2), 86, 1974.
71. **Zaiko, L. L. and Smith, J. L.,** *J. Sci. Food Agric.,* 26(9), 1357, 1975.
72. **Soboleva, M. I. and Sirokhman, I. V.,** *Izv. Vyssh. Ucheb. Zaved. Pishch. Tekhnol.,* 1, 44, 1975.
73. **Bishov, S. J. and Henick, A. S.,** U.S. Patent 3,852,502, 1974.
74. **Takasago, M., Horikawa, K., and Masuyama, S.,** *Yakagaku,* 25(1), 16, 1976.
75. **Cole, E. R., Crank, G., and Salam-Sheckh, A.,** *J. Agric. Food Chem.,* 22(5), 918, 1974.
76. **Eiserle, R. J.,** *Food Prod. Dev.,* p. 70, October 1971.
76a. **MacNeil, J. H. and Mast, M. G.,** *Food Prod. Dev.,* March 1973, 36.
77. **Fukuzumi, K., Ikeda, N., and Egawa, M.,** *J. Am. Oil Chem. Soc.,* 53, 623, 1976.
78. **Morita, M., Mukunoki, M., Okubo, F., and Tadokoro, S.,** *J. Am. Oil Chem. Soc.,* 53, 489, 1976.
79. **Ikeda, N. and Fukugumi, K.,** *J. Am. Oil Chem. Soc.,* 53, 618, 1976.
80. **Sims, R. J., Fioriti, J. A., and Kanuk, M. J.,** *J. Am. Oil Chem. Soc.,* 49, 298, 1972.
81. **Sims, R. J., Fioriti, J. A., and Kanuk, M. J.,** *Lipids,* 8, 337, 1973.
82. **Scott, J. W., Cort, W. M., Harley, H., Parrish, D. R., and Saucy, G.,** *J. Am. Oil Chem. Soc.,* 51, 200, 1974.
83. **Cort, W. M., Scott, J. W., Araujo, M., Mergens, W. J., Cannalonga, M. A., Osodca, M., Hardley, H., Parrish, D. R., and Pool, W. R.,** *J. Am. Oil Chem. Soc.,* 52, 174, 1975.
84. **Furia, T. E. and Bellanca, N.,** *J. Am. Oil Chem. Soc.,* 53, 132, 1976.
84a. **Furia, T. E. and Bellanca, N.,** *J. Am. Oil Chem. Soc.,* 54, 239, 1977.
85. **Horikawa, K. and Masuyama, S.,** *Yakagaku,* 21(3), 155, 1972.
86. **Lee, H. B.,** *Korean J. Nutri.,* 8(2), 95, 1975.
87. **Yang, K. S., Yu, J. H., Hawang, J. I. and Yang, R.,** *Korean J. Food Sci. Technol.,* 6(4), 193, 1974.
88. **Linow, F. and Mieth, G.** *Nahrung,* 20(1), 19, 1976.
89. **Gomyo, T. and Horikoshi, M.,** *Agric. Biol. Chem.,* 40(1), 33, 1976.
90. **Cort, W.,** U.S. Patent 3,903,317, 1975.
91. **Lunde, G., Landmark, L. H., and Gether, J.,** *J. Am. Oil Chem. Soc.,* 53, 207, 1976.
92. **Ke, P. J. and Ackman, R. G.,** *J. Am. Oil Chem. Soc.,* 53, 636, 1976.
93. **St. Angelo, A. J. and Ory, R. L.,** *J. Am. Oil Chem. Soc.,* 52, 38, 1975.
94. **Black, L. T.,** *J. Am. Oil Chem. Soc.,* 52, 88, 1975.
94a. **Olejko, J. T.,** *J. Am. Oil Chem. Soc.,* 53, 480, 1976.
94b. **Farhan, F. M. and Pazandeh, H.,** *J. Am. Oil Chem. Soc.,* 53, 211, 1976.
95. **Endean, M. E.,** The Detection and Determination of Food Antioxidants — A Literature Survey, Scientific and Technical Survey No. 91, British Food Manufacturing Industries Research Association, Leatherhead, Surrey, Great Britain, 1976, 57 pp.
96. **Vos, H. J., Wessels, H., and Six, C.H.T.,** *Analyst* (London), 82, 362, 1957.
97. **Mahon, J. H. and Chapman, R. A.,** *Anal. Chem.,* 23(8), 1116, 1951.
98. **Aczel, A., Selmeci, G., Noske, O., and Marik, M.,** *Husipar,* 21(5), 222, 1972; *Food Sci. Technol. Abstr.,* 5, 3N143, 1975.
99. **Lehman, G. and Moran, M.,** *Z. Lebensm. Unters. Forsch.,* 145(6), 344, 1971.
100. **Van Dessel, L. and Clement, J.,** *Z. Lebensm. Unters. Forsch.,* 139(3), 146, 1969.
101. **Schwien, W. G. and Conroy, H. W.,** *J. Assoc. Off. Agric. Chem.,* 48, 489, 1965.
102. **Takeshita, R., Sakagomi, Y. and Itoh, N.,** *J. Hyg. Chem.,* 15(2), 77, 1969; *C.A.,* 71, 79803m, 1969.
103. **Johnson, D. P.,** *J. Assoc. Off. Anal. Chem.,* 50(6), 1298, 1967.
104. **Amato, F.,** *Ind. Aliment.* (Pinerolo, Italy), 7(12), 81, 1968; *C.A.,* 71, 48361M, 1969.
105. Analysis of Phenolic Antioxidants, Eastman® Chemical Products, Inc., Kingsport, Tenn., Technical Data Report G101, 1957.
106. International Association of Broth and Soup Industry, Analytical Methods for the Food Industry, Association of Soup Manufacturers, Berne CH 5/1, 1973.
107. **Alessandro, A., Mazza, P., and Donati, C. G.,** *C.A.,* 75, 34043C, 1971.
108. **Endean, M. E.,** The Quantitative Extraction of BHT from Fat and Its Colorimetric Determination, Technical Circular No. 601, British Food Manufacturing Industries Research Association, Leatherhead, Surrey, Great Britain, 1975.
109. **Phipps, A. M.,** *J. Am. Oil Chem. Soc.,* 50, 21, 1973.
110. **Stuckey, B. N. and Osborne, C. E.,** *J. Am. Oil Chem. Soc.,* 42, 228, 1965.
111. **Pearson, D.,** *The Analysis of Food,* J. & A. Churchill, London, 1970, chap. 1.
112. **Groebel, W. and Wessels, A.,** *Dtsch. Lebensm. Rundsch.,* 62(12), 453, 1973.
113. **Endean, M. E. and Bielby, C. R.,** Detection of Antioxidants by TLC, Technical Circular No. 587, British Food Manufacturing Industries Research Association, Leatherhead, Surrey, Great Birtain, 1975.

114. Qualitative Test for BHT on Polished Rice, Method No. FA-52, Canadian Food and Drug Labs, Ottawa, Canada, 1965.
115. **Takahashi, D. M.,** *J. Assoc. Off. Anal. Chem.*, 53(1), 39, 1970.
116. **Daniel, J. W. and Cage, J. C.,** *Food Cosmet. Toxicol.*, 3, 405, 1965.
117. **Anglin, C., Mahon, J. H., and Chapman, R. A.,** *J. Agric. Food Chem.*, 4, 1018, 1956.
118. **Filipic, V. J. and Ogg, C. L.,** *J. Assoc. Off. Agric. Chem.*, 43, 795, 1960.
119. **Sloman, K., Romagnoli, R.K., and Cavagnol, J. C.,** *J. Assoc. Off. Agric. Chem.*, 45, 76, 1962.
120. **Takahashi, D. M.,** *J. Assoc. Off. Agric. Chem.*, 48, 694, 1965.
121. **Caldwell, E. F., Nehring, E. W., Postweiler, J. E., Smith, G. M., and Wilbur, C. T.,** *Food Technol.* (Chicago), 18(3), 125, 1964.
122. **Wolff, J. P.,** *Rev. Fr. Corps Gras*, 5, 630, 1958.
123. **Ratto, G.,** *C. Ab.*, 69, 11613C, 1968.
124. **Vigneron, P. Y. and Spicht, P.,** *Rev. Fr. Corps Gras*, 17(5), 295, 1970; *C.A.*, 73, 75740g, 1970.
125. **Hartman, K. T. and Rose, L. C.,** *J. Am. Oil Chem. Soc.*, 47(1), 7, 1970.
126. **Singh, J. and LaPointe, M. R.,** *J. Assoc. Off. Anal. Chem.*, 57(4), 804, 1974.
127. **Schwein, W. G., Miller, B. J., and Conroy, H. W.,** *J. Assoc. Off. Anal. Chem.*, 49, 809, 1966.
128. **Copius-Pereboom, J.,** *Nature (London)*, 204, 748, 1964.
129. **Kaszyka, A.,** *Rocz. Panstw. Zakl. Hig.* 27, 171, 1976; *Food Sci. Technol. Abstr.*, 10, 51692, 1976.
130. **Mahon, J. H. and Chapman, R. A.,** *Anal. Chem.*, 23(8), 1116, 1951.
131. **Mahon, J. H. and Chapman, R. A.,** *Anal. Chem.*, 23(8), 1120, 1951.
131a. **Mahon, J. H. and Chapman, R. A.,** *Anal. Chem.*, 24, 534, 1952.
132. **Szalkowski, C. R. and Garber, J. B.,** *J. Agric. Food Chem.*, 10, 490, 1962.
133. **Dumitrescu, H., Barduta, Z., and Dumitrescu, D.,** *Igiena*, 24(2), 111, 1975; *Food Sci. Technol. Abstr.*, 10, 466, 1976.
134. **Sahasrabudhe, R. M.,** *J. Assoc. Off. Agric. Chem.*, 47, 888, 1964.
135. **Kimura, S. and Terada, S. J.,** *Food Hyg. Soc. Jpn.*, 14(1), 94, 1973; *Food Sci. Technol. Abstr.* 6, 7T403, 1974.
136. **Scheidt, S. A. and Conroy, W. H.,** *J. Assoc. Off. Anal. Chem.*, 49(4), 807, 1966.
137. **Van der Heide, R. F.,** *J. Chromatogr.*, 24(1), 239, 1966.
138. **Copius-Pereboom, J.,** in *Thin-Layer Chromatography, A Laboratory Handbook*, 2nd ed., Stahl, E., Ed., George Allen and Unwin, London, 1969, 630.
139. **Buttery, R. G. and Stuckey, B. N.,** *J. Agric. Food Chem.*, 9(4), 283, 1961.
140. **Anderson, R. H. and Nelson, J. P.,** *Food Technol.* (Chicago), 17(7), 95, 1903.
141. **Choy, T. K., Quattrone, J. J., and Alicino, N. J.,** *J. Chromatogr.*, 12, 171, 1963.
142. **Schwecke, W. M. and Nelson, J. H.,** *J. Agric. Food Chem.*, 12(1), 86, 1964.
143. **Karleskind, A. and Valmalle, G.,** *Rev. Fr. Corps Gras*, 23, 431, 1976.
144. **Hartman, K. T. and Rose, L. C.,** *J. Am. Oil Chem. Soc.*, 47, 7, 1970.
145. **Cucks, L. V. and van Rede, C.,** *Laboratory Handbook for Oil and Fat Analysis*, Academic Press, London, 1966, 170
146. **Melchert, H. U.,** *Chem. Mikrobiol. Technol. Lebensm., M.*, 2(3), 94, 1973; *Food Sci. Technol. Abstr.*, 5, 12N627, 1973.
147. **Miethke, H.,** *Dtsch. Lebensm. Rundsch.*, 57, 170, 1966.
148. **Wachs, W. and Gerhardt, K.,** *Dtsch. Lebensm. Rundsch.*, 62(7), 202, 1966.
148a. **Wachs, W. and Gerhardt, K.,** *Dtsch. Lebensm. Rundsch.*, 66(2), 37, 1970.
149. **McCauley, D. F., Fazio, T., Howard, T. W., DiCurio, F. M., and Ives, J.,** *J. Assoc. Off. Anal. Chem.*, 50(2), 243, 1967.
150. **Wolfe, J. and Audiau, F.,** *Bull. Soc. Chim. Fr.*, 10, 2662, 1964; *Anal. Abstr.*, 13, 979, 1966.
151. **Stoddard E. E.,** *J. Assoc. Off. Anal. Chem.*, 55(5), 1081, 1972.
152. **Pokorny, S., Coupek, J., and Pokorny, J.,** *J. Chromatgr.*, 71, 576, 1972.
153. **Seher, A.,** *Fette Seifen Anstrichm.*, 60, 1144, 1958.
154. **Barendrecht, E.,** *Anal. Chim. Acta*, 24, 498, 1961.
155. **McBride, H. D. and Evans, D. H.,** *Anal. Chem.*, 45(3), 446, 1973.
156. **Waltking, A. E., Kiernon, M., and Bleffert, G. W.,** *J. Assoc. Off. Agric. Chem.*, 1977, in press.
157. **List, C. R., Evans, C. D., and Moser, H. A.,** *J. Am. Oil Chem. Soc.*, 49, 287, 1972.
158. Free Fatty Acids, American Oil Chemists' Society, Official and Tentative Methods, Ca5a-40, American Oil Chemists' Society, Champaign, Ill., 1972.
158a. Peroxide Value, American Oil Chemists' Society, Official and Tentative Methods, Cd8-53, American Oil Chemists' Society, Champaign, Ill., 1960.
158b. Fat Stability — Active Oxygen Method, American Oil Chemists' Society, Official and Tentative Methods, Cd12-57, American Oil Chemists' Society, Champaign, Ill., 1972.
159. **Tarladgis, B. G., Pearson, A. M., and Dugan, L. R.,** *J. Sci. Food Agric.*, 15, 602, 1964.
160. **Patton, S.,** *J. Am. Oil Chem. Soc.*, 51, 114, 1974.

161. **Marcuse, R. and Johansson, L.,** *J. Am. Oil Chem. Soc.,* 50, 387, 1970.
162. **Moser, H. A., Dutton, H. J., Evans, C. D., and Cowan, J. C.,** *Food Technol.,* (Chicago), 4, 105, 1950.
163. **Joyner, N. T. and McIntyre,. J. E.,** *Oil Soap* (Chicago), 15, 184, 1938.
164. **Stuckey, B. N., Sherwin, E. R. and Hannah, F. D., Jr.,** *J. Am. Oil Chem. Soc.,* 35, 581, 1958.
165. **Wintermantel, J. F., New, D. J., and Ramstad, P. E.,** *Cereal Sci. Today,* 6, 186, 1961.
166. **Blakenship, B. R., Holaday, C. E., Barnes, P. C., Jr., and Pearson, J. L.,** *J. Am. Oil Chem. Soc.,* 50, 377, 1973.
167. **Fioriti, J. A., Kanuk, M. J. and Sims, R. J.,** *J. Am. Oil Chem. Soc.,* 51, 219, 1974.
167a. **Fioriti, J. A., Stahl, H. D., Cseri, J. and Sims, R. J.,** *J. Am. Oil Chem. Soc.,* 52, 395, 1975.
168. **Bishov, S. J. and Henick, A. S.,** *J. Am. Oil Chem. Soc.,* 43, 477, 1966.
169. **Evans, C. D., List, G. R., Hoffman, R. L., and Moser, H. A.,** *J. Am. Oil Chem. Soc.,* 46, 501, 1969.
170. **List, G. R., Evans, C. D., Kwolek, W. F., Warner, K., Boundy B. K., and Cowan, J. C.,** *J. Am. Oil Chem. Soc.,* 51, 17, 1974.
171. **Jarvi, P. K., Lee, G. D., Erickson, D. R., and Butkus, E. A.,** *J. Am. Oil Chem. Soc.,* 48, 121, 1971.
171a. **Dupuy, H. P., Fore, S. P., and Goldblatt, L. A.,** *J. Am. Oil Chem. Soc.,* 50, 340, 1973.
171b. **Dupuy, H. P., Rayner, E. T., and Wadsworth, J. I.,** *J. Am. Oil Chem. Soc.,* 53, 628, 1976.
172. **Berner, D. L., Conte, J. A., and Jacobson, G. A.,** *J. Am. Oil Chem. Soc.,* 51, 292, 1974.
173. **Ozawa, T., Nakamura, Y. and Hiraga, K.,** *J. Food Hyg. Sci. Jpn.,* 13, 205, 1972.
174. **Cort, W. M.,** *Food Technol.* (Chicago), 28(10), 60, 1974.
175. **Sindhu, G. S., Brown, M. A., and Johnson, A. R.,** *J. Dairy Res.,* 42(1), 185, 1975.
176. **Sherwin, E. R.,** *J. Am. Oil Chem. Soc.,* 53, 430, 1976.
177. **Kochar, S. P. and Meara, M. L.,** A Survey of the Literature on Oxidative Reactions in Edible Oils as It Applies to the Problem of "Off-flavors" in Foodstuffs, Scientific and Technical Survey No. 87, British Food Manufacturing Industries Research Association, Leatherhead, Surrey, Great Britain, 1975, 22 pp.
178. **Kochar, S. P., Meara, M. L., and Gilburt, D. J.,** Oxidative Reactions in Edible Oils, Research Report 252, British Food Manufacturing Industries Research Association, Leatherhead, Surrey, Great Britain, 1976, 84 pp.
179. **Berger, K. G.,** The use of antioxidants in foods, part II, *Chem. Ind.* (London), March 1, 1977.
180. **Lindsay, R. C., and Lund, D. B., and Branen, A. L.,** Investigation of Methods for Introducing Antioxidants in Foods, Paper 183, in 36th Annu. Meet. Inst. Food Technol., Anaheim, Cal., June 1976.
181. **Wiegeland, E., Niemann, W., Kohler, M., and Gernhardt, C.,** German Patent 2,306,824, 1973.
182. **Mendenhall, G. D. and Nathan, R. A.,** *J. Am. Oil Chem. Soc.,* 53(7), 456A, 1976.
183. **Fritsch, C. W. and Gale, J. A.,** *J. Am. Oil Chem. Soc.,* 54, 225, 1977.
184. *Proc. Symp. Objective Methods for Food Evaluation,* Newton, Mass., November 1974, National Academy of Science, Washington, D.C., 1976.
185. **Johnson, F. C.,** *Crit. Rev. Food Technol.,* 2(3), 267, 1971.

ENZYMES

L. A. Underkofler

INTRODUCTION

Enzyme technology has been called a solution in search of problems. It has yet to reach the scope of its potential, but even so, in recent years several enzymes have achieved importance in new commercial processes. The major purposes of a revision of this chapter are inclusion of these new enzyme processes and a summarization of improvements made in older ones. Of particular importance are a new tabulation of enzyme products marketed in the United States for food applications and the general section on immobilized enzymes. Specific potential or actual applications of such fixed enzymes are also considered along with the uses of individual enzymes. Other new material also includes some discussion of debranching enzymes, a new use of α-galactosidase (melibiase) in the refining process for beet sugar, the potential use of cellulase in the profitable utilization of wastes, the new multimillion dollar high-fructose syrup industry employing glucose isomerase, and the development of successful microbial rennin products.

Enzymes are important factors in food technology because of the roles they play in the composition, processing and spoilage of foods. Enzymes occur naturally in many food raw materials and can affect the processing of foods in many ways. Sometimes the presence of natural enzymes is advantageous; for example amylases in sweet potatoes assist in curing to give desirable texture and flavor. In other cases, natural enzymes may produce undesirable reactions, such as rancidity produced by lipases, or browning reactions due to polyphenol oxidases. Sometimes, tests for natural enzymes in foods are used as an index of whether heat treatment has been sufficient, such as detection of phosphatase in milk or cheese as evidence of inadequate pasteurization or of catalase or peroxidase in vegetable products as evidence of inadequacy of blanching. Recognition of the function of enzymes and their usefulness in bringing about desirable changes has led to their large-scale use as modifiers of food ingredients. Such commercial enzyme applications have grown from a relatively insignificant role to one of the most important aspects of food processing during the past quarter of a century. Several books have been published recently relating to enzymes in food technology.[190, 191, 281, 283, 286]

The practical application of enzymes to accomplish certain reactions has been conducted for centuries. The fermentation of foods — beer, wine, bread, and cheese — is older than recorded history. All fermentations are conversions produced enzymatically through the metabolism of living organisms; hence, these food fermentations are examples of enzymatic modifications. Where a number of enzymes, constituting an enzyme system, are necessary to produce desired changes, it is advantageous to employ intact cells in fermentation processes. Where a single enzyme or a system of only two or three enzymes is involved in a desired reaction, isolated enzyme products are preferable to intact cells. A detailed treatment of fermentations is beyond the scope of this chapter.

Examples of ancient uses of enzymatic action other than fermentations are malt in brewing, stomach mucosa in milk clotting for cheese making, papaya juice for meat tenderization, and dung for leather bating. Of course, the ancient peoples using these natural products did not know there were such entities as enzymes; however, crude enzyme preparations extracted from animal tissues, such as pancreas and stomach mucosa, or from plant tissues, such as malt and papaya fruit, became articles of commerce.

The usefulness of crude enzyme preparations was established, and when the biocatalytic enzymes responsible for their action became recognized, a search began for better, less expensive, and more readily available sources of such enzymes. Advances in enzymology and microbiology have resulted in more progress in enzyme production and application during the past 75 years than in the preceeding 5000 years. Early pioneers who laid the foundations for the production and use of industrial enzymes were Takamine,[236] Rohm,[197] and Wallerstein.[271] The development of methods for large-scale production of enzymes, along with knowledge for controlling enzymatic processes and applications, has resulted in a sizeable number of industrial enzyme products. These have been listed by deBecze by trade name, giving for each the producer, enzymes present, primary action on substrate, uses, and recommended pH and temperature ranges.[58, 59]

Today, commercial enzyme preparations obtained from plant, animal, and microbial sources are widely used by industry. Malt amylase and the proteases, papain, ficin, and bromelain, from tropical plants are the best known enzymes from plant sources. Proteases, amylases, and lipases from the pancreas, pepsin and rennet from stomach mucosa, and catalase from liver are useful commercial enzymes from animal tissues. Certain microorganisms have been found to produce abundant amounts of useful enzymes and have become a major source for commercial enzymes.

ENZYME THEORY

Before considering the specific applications for enzymes in present day food industries, it is necessary to consider briefly enzyme theory — what enzymes are, how they work, and factors affecting their action — in order to understand and control the practical applications of enzymes. For an exhaustive treatment of enzyme theory, a modern textbook, such as that of Dixon and Webb[64] or Whitaker,[281] should be consulted.

Chemical Nature of Enzymes

Enzymes belong to the broad class of substances which the chemist calls catalysts. A catalyst is a substance which influences the velocity of a reaction without being used up in the reaction. Catalysts take part in reactions but reappear in their original form, describing a cycle. Theoretically, a catalyst can convert an unlimited amount of reacting substance. Enzymes are very special kinds of catalyst with very distinctive properties.

The most commonly accepted definition is that an enzyme is a soluble, colloidal, organic catalyst produced by a living cell; even more simply, an enzyme is a biocatalyst produced by a living cell. Enzymes can be produced only by living cells to accomplish specific metabolic needs. But fortunately, enzymes can be separated readily from the cells which produce them and perform their catalytic activities entirely apart from the cells; hence, they are available for useful applications. Since all enzymes are either simple or conjugated proteins, another definition quite widely used is that an enzyme is a protein with catalytic properties due to its power of specific activation.

Enzymes, like all catalysts, affect the rates of chemical reactions, but not the extent of the chemical change concerned. Enzymes accelerate reactions which are, in themselves, thermodynamically possible; that is reactions attended by losses of free energy. However, in order for the reaction to occur a certain amount of resistance must be overcome; that is the molecules must be activated by supplying a certain amount of activation energy. One can describe enzyme action by stating that the enzyme lowers the amount of activation energy required by the reaction; hence, enzymes are able to bring about, under mild conditions near room temperature, reactions which without enzymes would require drastic conditions of high temperature or other high-energy conditions.

All of the highly purified enzymes which have been isolated have been found to be proteins. In some enzymes, nonprotein prosthetic groups are also present, such as a specific metal like zinc or calcium, an organic heme, a flavin, or other group.

Although the beginnings of enzymology can be traced back to the early nineteenth century, the development of enzymology as a science has come mainly during the last 70 years. Enzyme activities were fairly well understood by 1920, and about this time serious attempts at purification of enzymes began. The first enzyme to be prepared in crystalline form was urease by Sumner in 1926. Even 25 years ago, there were very few purified enzymes; whereas, now the number of pure and crystalline enzymes is over 400, more than 1500 have been somewhat purified, and over 2000 have been studied to some extent. Just how many enzymes may exist is unknown, but there are probably 10,000 or more.

Enzymes differ from other catalysts in several respects. These differences are, of course, due to the protein nature of the enzymes. Being proteins, enzymes are denatured and inactivated when subjected to unphysiological conditions, such as heat or strong chemicals. Hence, one of the distinctive properties of enzymes, in contrast to ordinary chemical catalysts, is their thermal lability and sensitivity to acids and bases. But the difference between enzymes and other catalysts is most clearly displayed in one respect: enzymes generally have extremely specific actions. Whereas acids, that is hydrogen or hydronium ions, may catalyze hydrolysis of many kinds of substances, such as esters, acetals, glycosides (sugars), or peptides (proteins), separate and different enzymes are necessary for the hydrolysis of each of these and even of specific members of each class. Thus, esterases and lipases can split only esters or fats without hydrolyzing glycosides or peptides. Carbohydrases act specifically on glycosides and cannot attack ester or peptide bonds. Individual carbohydrases are necessary for their hydrolysis. The enzyme lactase, which hydrolyzes lactose, has no effect whatever on sucrose or maltose and so on. In enzymology two types of enzyme specificity are recognized, substrate specificity and reaction specificity. Examples of substrate specificity are the hydrolysis of lactose by lactase and the oxidation of glucose by glucose oxidase. Examples of reaction specificity are the actions of proteinases, which are capable of splitting particular peptide bonds of proteins. These peptide bonds are amide linkages between carboxyl and amino groups of constituent amino acids. Usually, specific proteinases have narrow reaction specificity as to the exact peptide bonds they can split. For example, the action of pepsin is restricted largely to hydrolysis of peptide linkages to which the carboxyl group has been furnished by phenylalanine or tyrosine and has little or no effect on peptide bonds formed by other amino acids.

A good scientific definition is that an enzyme is a protein with catalytic properties due to its power of specific activation. Active research is beginning to give information regarding the exact composition of enzymes. They are proteins with molecular weights of individual enzymes ranging from about 13,000 to over a million. The amino acid compositions of a few enzymes have been determined and the amino acid sequence in some; but there is nothing in the amino acid analysis or sequence which differentiates enzymes from other proteins. Since enzyme proteins possess specific catalytic functions, special structures must be responsible. All proteins contain reactive groups, such as free amino, carboxyl, hydroxyl, sulfhydryl, and imidazole groups, and frequently nonprotein prosthetic groups. However, the mere presence of reactive groups, or combining sites, in protein molecules does not insure enzyme activity. Enzymes must be so arranged that the reactive groups of the substrate may fit them and be held for further action. The long peptide chains of native protein molecules are known to be folded and arranged in exact positions, so in an enzyme the combining sites are suitably arranged to make up active centers necessary for substrate binding and catalytic activity.

Naming and Classification of Enzymes

Enzymes bring about changes in specific compounds. The compound upon which an enzyme acts is known as its substrate. Up until recently, there have been no completely systematic methods for naming and classifying enzymes. Names of early enzymes were completely unsystematic — diastase, emulsin, pepsin, ptyalin, trypsin, and catalase. Later, it became customary, where possible, to name an enzyme after the substrate upon which it acts, with the ending "-ase"; thus, peptidase, esterase, urease, lactase, amylase. In other cases, enzymes were named from the reactions they catalyzed, such as dehydrases, dehydrogenses, transferases, phosphorylases.

In 1961, modified slightly in 1964 and again in 1972, the Commission on Enzymes and the Commission on Biochemical Nomenclature of the International Union of Biochemistry and the International Union of Pure and Applied Chemistry published an entirely systematic method for classifying and naming enzymes based on substrate and type of reaction.[11] The enzymes are divided into six main classes, oxidoreductases, transferases, hydrolases, lyases, isomerases and ligases (synthetases). Each class is further divided into a number of subclasses and sub-subclasses, according to the nature of the chemical reaction catalyzed, and is coded on a four-number system intimately connected with this system of classification. Oxidoreductases catalyze oxidations or reductions. Transferases catalyze the shift of a chemical group from one donor substrate to another acceptor substrate. Hydrolases catalyze hydrolytic splitting of substrates. Lyases remove groups or add groups to their substrates (not by hydrolysis). Isomerases catalyze intramolecular rearrangements. Ligases (synthetases) catalyze the joining together of two substrate molecules. Both systematic and trivial names are recommended for the enzymes. The systematic name is formed in accordance with definite rules and will identify the enzyme and indicate its action as precisely as possible. In general the systematic name consists of two parts; the first part names the substrate, and the second, ending in "-ase," indicates the nature of the process. The systematic rules are quite extensive and they are difficult to apply if the substrate composition or the enzyme reaction are not fully understood. The trivial name is sufficiently short for general use, and in the majority of cases is the name already commonly employed. To illustrate, the familiar β-amylase (trivial name) has the systematic name: 1,4-α-D-glucan maltohydrolase. The name indicates that the substrate is a glucan (starch), a glucose polymer in which the glucose molecules are joined by α-1,4 linkages, and the reaction is a hydrolytic splitting off of maltose.

How Enzymes Act

Enzyme-substrate Combination

It has been shown that an enzymatic reaction proceeds in two stages. The enzyme (E) unites with substrate (S) to form a labile intermediate complex:

$$E + S \rightleftharpoons ES$$

The result is that the substrate molecules become more chemically reactive by some intramolecular changes, i.e., the substrate is activated by the enzyme. The unstable intermediate complex reacts with a reactant (R) and breaks down with formation of the end products (P) of the reaction and regeneration of the enzyme:

$$ES + R \rightleftharpoons E + P$$

The idea that enzymatic action is due to the formation of an intermediate complex between enzyme and substrate was expressed by Michaelis and Menton[155] and demonstrated by Chance.[39] Recently, an enzyme-substrate complex was isolated in crystalline form.[291] The mass action equilibrium constant for the formation and dissociation of enzyme and substrate, as given below has come to be known as the Michaelis constant.

$$\frac{(E - ES) \times (S - ES)}{(ES)} = K_m$$

Without going into details of how K_m may be determined, it will suffice here to point out that it is a fundamental constant in enzyme work since its value reflects the affinity of an enzyme for its substrate or substrates. The lower the value of K_m, the higher is the affinity of the enzyme for its substrate.

Enzymes are effective in extremely small amounts because the enzyme-substrate complex is very reactive and is present for a very short time. With enzymes, a useful term to describe the amount of substrate converted in unit time by a given quantity of enzyme is the molecular activity or turnover number, which represents the number of moles of substrate converted per mole of enzyme per minute. The molecular activities of different enzymes vary widely (100 to 5,000,000).

Factors Affecting Enzyme Action

There are numerous factors which affect enzyme activity, and these must be taken into account in the use of the enzyme. Among the most important are 1. concentration of enzyme, 2. concentration of substrate, 3. time, 4. temperature, 5. pH, and 6. presence or absence of activators or inhibitors.

For most enzymatic reactions, the rate of the reaction is directly proportional to the concentration of enzyme, at least during the early stages of the reaction. With very low substrate concentration enzymatic reaction velocity is proportional to substrate concentration. In many practical enzymatic applications, where the substrate is present in considerable excess during the early stages, the reaction follows zero-order kinetics, and the amount of product formed is proportional to time:

$$dP/dT = k_0$$

For such reactions the amount of end product is doubled, if reaction time is doubled.

However, during the course of an enzymatic reaction, there is a continuing decrease in substrate concentration which results in slowing down of the reaction with time as shown in Figure 1. Most enzymatic reactions follow the kinetics of a first-order reaction:

$$dP/dT = k_1 \times (S - P)$$

where k_1 is the first-order reaction constant and (S − P) is the concentration of substrate remaining at any given time. The rate of the reaction is directly proportional to the remaining substrate concentration. Equal fractions of the remaining substrate are transformed in equal time intervals. For example, if 50% of the substrate is converted

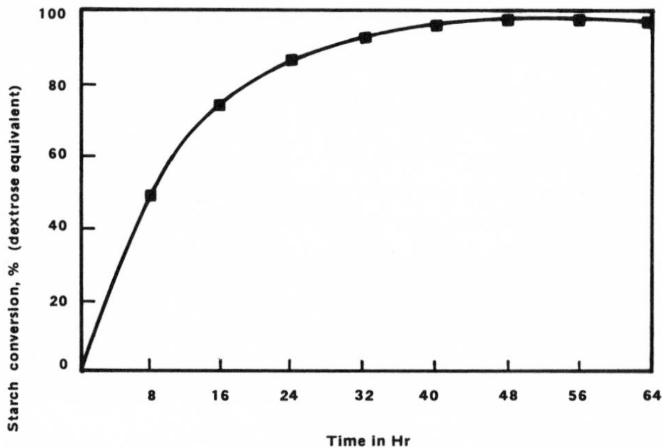

FIGURE 1. Enzymatic hydrolysis of starch to dextrose by glucoamylase. Initial starch, 30%, glucoamylase, 80 units/1b starch, 60°C, pH 4.0.

in 30 min, an additional 25% of the original substrate (50% of the remining substrate) will be converted in the next 30 min, and so on.

Time is a very important factor in practical enzymatic applications. As mentioned above, initial reaction velocities are proportional to enzyme concentration. However, as a reaction proceeds, the rate diminishes as shown in Figure 1. The decreased velocity may be due to many reasons. Most important usually are exhaustion of substrate and inhibition of a reaction by its end products. First-order kinetics postulates the slowing up of a reaction as available substrate diminishes in amount. For practical uses, because of this inherent behavior of enzymes, sufficient time must be allowed for enzyme reactions to approach completion. They may involve reaction periods of several hours or even days, for example, 48 to 96 hr in the industrial enzymatic process for dextrose production from starch.

Heat may affect enzymes in two ways. One effect is inactivation since high temperatures cause denaturation of the enzyme protein resulting in a loss of catalytic properties. Actual temperatures at which heat inactivation is substantial vary a great deal depending upon the particular enzyme. For many enzymes useful in food processing, inactivation becomes appreciable and rapid at temperatures above about 50°C. Some enzymes are much more resistant to heat; for example, glucoamylase is effectively employed at 60°C, and bacterial amylases, at 80°C or even higher.

The second effect of temperature on enzymatic reactions is on their rate. Like most chemical reactions, enzyme-catalyzed changes are increased in rate by raising the temperature. A rough rule for chemical reactions, including those catalyzed by enzymes, is that every 10°C increase in temperature approximately doubles the rate; that is, the temperature quotient, Q_{10}, is about 2. Measurements of individual enzyme reactions have given Q_{10} values in the range of 1.2 to 4.

As temperature is increased in an enzyme reaction, however, thermal inactivation of the enzyme may take place so rapidly as to more than offset the increased rate of reaction due to higher temperatures. The so-called optimum temperature is that point of maximum activity above which the rate of reaction decreases because of thermal inactivation. Optimum temperature values must be interpreted with caution, since such factors as substrate concentration and particularly time of an enzymatic reaction have considerable effect on optimum temperature. This is apparent from the temperature-activity curves of Figure 2 which show apparent temperature optima differing by 10°C

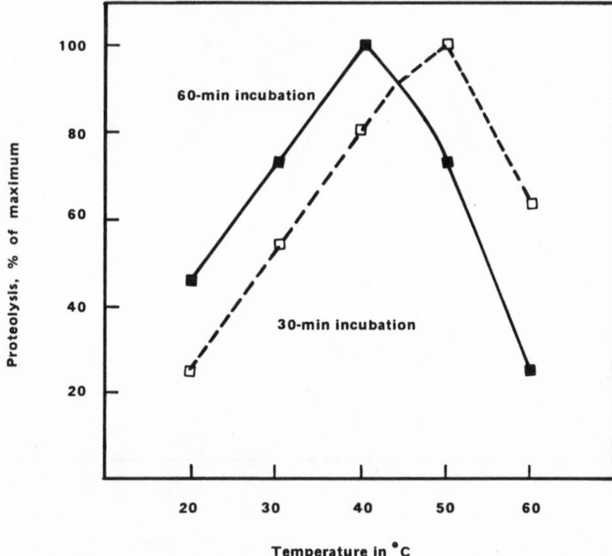

FIGURE 2. Temperature-activity curves for bacterial proteinase. Gelatin, pH 7.0. (From Underkofler, L. A., in *Production and Application of Enzyme Prepartions in Food Manufacture, Soc. Chem. Ind.* (London), Monograph Macmillan, New York, 11, 1961, 48.

depending upon whether 30- or 60-min incubations were used. This difference is, of course, due to greater inactivation at higher temperatures with longer incubation.

The pH of the system also has a profound effect on enzyme activity. Each enzyme in the presence of its substrate has a characteristic pH at which its activity is highest, known as the optimum pH. For some enzymes, the optima are quite sharp; for others, there are rather broad optimum pH ranges. Typical pH-activity curves are shown in Figure 3. When pH is changed, activity decreases rapidly on both sides of the optimum range until the enzyme is completely inactive. Change of pH toward the optimum will reactivate the enzyme, but change of pH above or below the levels of temporary inactivation may gradually denature the enzyme to permanent inactivity. The pH optimum values for various enzymes vary widely. For example, the optima for the two proteolytic enzymes, pepsin and trypsin, are about 2.0 and 8.0, respectively. There is also a pH range for best enzyme stability, which is not necessarily the same as for optimum activity.

Some enzymes need cofactors or activators for maximum effectiveness. Frequently metal ions, such as those of calcium, magnesium, or manganese, are necessary enzyme activators, either for maximum activity or stability or both.

Enzyme inhibition is a very important area of enzymology. Enzymes are inhibited by a number of conditions, such as lack of moisture, since all enzyme reactions take place in aqueous systems. However, the term "enzyme inhibitor" usually refers to a substance or chemical which causes inhibition of an enzyme reaction. There are reversible and irreversible, specific and nonspecific inhibitors.

Since enzymes are proteins, groups, such as free carboxyl, amino, and sulfhydryl, will be common to many enzymes, and blocking of these groups will, of course, give rise to a rather nonspecific inhibition. If enzyme activity is restored by removal of the inhibitor, as by dialysis, the inhibition is reversible. Irreversible inhibition is produced by some poisons which cause irreversible denaturation or destruction of the enzyme.

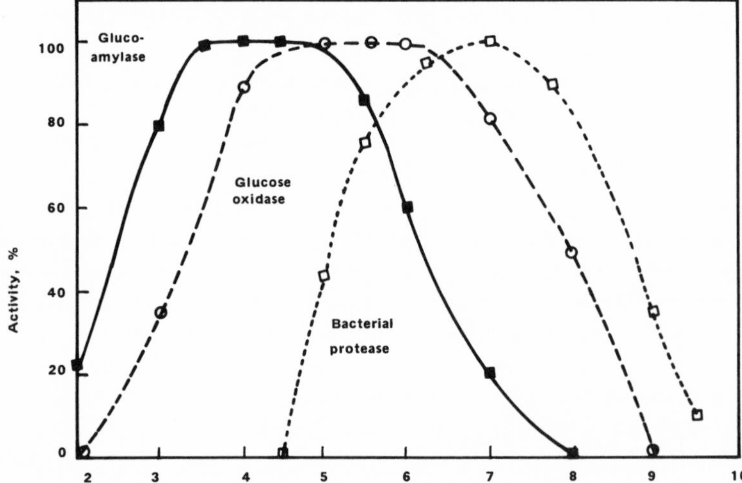

FIGURE 3. pH-activity curves for three commercial enzymes.

Specific inhibitors are those substances which block groups conferring specificity on an enzyme; that is to say, which react with combining sites of the active centers. A study of these reversible and specific inhibitions has shown that two types may be distinguished, competitive and noncompetitive.

Competitive inhibition is shown by substances with structural similarity to the normal enzyme substrate, including frequently products of the enzyme action. The enzyme is capable of combining with such substances but cannot activate them, and inhibition of the enzyme reaction results from hindered access of the normal substrate to the active centers.

In noncompetitive inhibition, there is no competition between substrate and inhibitor. Instead, the inhibitor combines with groupings not essential for the formation of the enzyme-substrate complex but necessary for substrate activation. There are naturally occurring inhibitors of this kind, such as trypsin inhibitors found in serum, eggs, and soybeans. Many poisons involve inhibition of a single enzyme involved in a main metabolic chain, thus rendering the whole chain inoperative and having a profound and even fatal effect upon the organism. The toxicity of cyanide is due to its inhibition of cytochrome oxidase. Recognition of poisons as enzyme inhibitors is of fundamental importance to pharmacologists and toxicologists and has practical applications. An increasing use of enzyme inhibitor compounds as insecticides is now the basis of a large and expanding industry. The "nerve gases" of military importance are essentially specific enzyme inhibitors.

Methods of Enzyme Assay

An important problem of the enzymologist is to quantitatively measure the activity or concentration of an enzyme.[26,59,258] This measurement cannot be done by the ordinary means of analytical chemistry, because the only means of detecting an active enzyme is by what it does to its specific substrate. Hence, the only way of measuring the activity or quantity of an enzyme is by determining how fast it changes its substrate under controlled conditions. For example, one can assay the potency of an amylase by adding a solution containing a known weight of the enzyme preparation to a starch solution and allowing it to react for a given time. Careful control of temperature and pH during the digestion period is necessary, since these factors so markedly affect the action of the enzyme. Likewise, the time chosen must be within the range of linear proportionality of reaction rate with enzyme concentration. The reaction is then stopped by adding a suitable reagent, and the amount of starch hydrolyzed is determined. In the case of the amylase reaction, the change in viscosity of the starch paste may be measured, the amount of starch not changed may be determined (starch-iodine color reaction), or the amount of sugar produced from the starch may be determined. In other words, enzymes are assayed by determining the amount of substrate changed under standardized conditions of the procedure, by the change in physical nature of the substrate, or by the amount of a reaction product produced. There are numerous ways of expressing the results of enzyme assays, a common method being in terms of arbitrary units, which are defined for the specific enzyme and substrate used. For amylase, for example, enzyme potency might be expressed in saccharification units, a unit being defined as the amount of enzyme which produces 1 g of maltose when it acts on a 2% starch solution for 30 min at 30°C. The Commission on Enzymes of the International Union of Biochemistry originally recommended that for all enzymes, uniform units be employed and defined as: One unit of any enzyme is the amount which will catalyze the transformation of 1 μmol of substrate per min under specified conditions of temperature, pH, concentration, etc. More recently,[11] the Commission on Biochemical Nomenclature recommended that the unit in which enzymatic activity is expressed is that amount of activity that converts 1 mol of substrate per sec. The new unit is named the katal (symbol *kat*). Enzymatic activities expressed in the former enzyme units (U) may be interconverted to *katals* by the relationship:

$$1 \text{ kat} = 1 \text{ mol/sec} = 60 \text{ mol/min} = 60 \times 10^6 \ \mu\text{mol/min} = 6 \times 10^7 \text{ U}$$

MANAGEMENT OF NATURAL ENZYMES AND ENZYMATIC ACTION

Food materials are of natural plant or animal origin. Since enzymes are biocatalysts produced by living cells, enzymes of many different kinds are present in most fresh food materials. These natural enzymes may contribute to the desirable characteristics of food or may be an important factor in the deterioration or spoilage of food. An important aspect of food enzymology is therefore management of natural enzymes and enzymatic action.

In a general way spoilage of food from enzyme action may be as varied as that caused by microorganisms; in fact, all microbial food spoilage is caused by enzymes contained in or elaborated by these organisms. However, microbial food spoilage is a distinct problem in itself, beyond the scope of this chapter.

Many food products contain diastatic enzymes which, when conditions are right, can change starches to sugars. Thus, if stored at too high temperature, potatoes become sweet and soft. Most living cells contain autolytic enzymes that can act on proteins. When animal or plant tissues are stored, these enzymes can bring about extensive changes, which in some cases may be desirable and in other cases considered as spoilage.

Studies on frozen and dehydrated foods have made it apparent that naturally occurring enzymes can spoil foods. In spite of freezing or removal of moisture, products may undergo marked color or flavor changes. A variety of enzymes may be involved in these undesirable changes. Some act on fats, some on proteins, others are pectolytic, and some are phenolases causing enzymatic browning reactions. These undesirable enzymatic changes can often be avoided by blanching, which consists of preheating the product to the point of enzyme inactivation.

The use of heat in pasteurization or sterilization not only destroys microorganisms but also is a convenient way of inactivating enzymes. Enzyme inactivation by heat often takes more time than the attainment of complete sterility from microorganisms. In some cases, however, such as in milk, a test for the absence of phosphatase enzyme is a good indication of adequate pasteurization since the enzyme is completely inactivated at a temperature high enough to destroy undesirable organisms including resistant pathogens, such as *Mycobacterium tuberculosis.*

Situations are known in food technology when, although they have been inactivated, certain enzymes may be regenerated and their activity restored after some time. Such enzyme regeneration has been observed with the phosphatase of milk, pectolytic enzymes in citrus juices, and especially peroxidases in fruits and vegetables. In general, the shorter the time of heat treatment, the greater are the chances for enzyme regeneration. It is believed that this regeneration of enzymes is due to only partial denaturation of the enzyme proteins, which have been somewhat unfolded or uncoiled by the heat treatment but find themselves after a time partially or completely recombined by the hydrogen or sulfhydryl bonds and their activity thus restored.

Dairy Products

All dairy products are, of course, based upon milk. Milk itself is one of our most important foods; moreover, butter, cheese, and ice cream are derived directly from milk.

Milk contains a large number of natural enzymes. Among these are α-amylase, lipase, peroxidase, phosphatase, *p*-diamine oxidase, and xanthine oxidase.[219, 290] There seems no conceivable purpose for the presence of some of the enzymes (α-amylase for example) as far as the digestibility of the milk by the suckling animal is concerned. Some of the enzymes, lipase and phosphatase, are destroyed by pasteurization; others are not even completely destroyed by sterilizing at 115°C for 15 min. Peroxidase and xanthine oxidase are inactivated by this sterilization, but 5% of the original α-amylase and 12% of the *p*-diamine oxidase remain after sterilization.

The presence of phosphatase in milk is used in quality control to determine whether the milk has been adequately pasteurized. In making the test, a portion of the milk is added to a substrate containing phenyl phosphoric compound and incubated for a short time. An indicator is then added. If the phosphatase has not been inactivated, it causes liberation of phenol which forms a colored dye with the indicator.[115] The test is sensitive enough to indicate the addition of 0.1% raw milk to pasteurized milk, a drop of 2°C in the pasteurizing temperature, or a slight shortage in the holding period.

The natural enzymes in milk seem to have little importance in the quality of dairy products. Milk lipases may lead to undesirable rancidity, if freshly drawn milk is cooled too rapidly, if raw milk is homogenized and agitated, or if foaming or great temperature fluctuations occur. Since lipase is destroyed by pasteurizing, important lipolytic actions occurring in cheesemaking come from enzymes elaborated by starter cultures or by added pregastric lipases, which will be discussed later.

Meat

In the American way of life, meat is a most desired food. The flesh of many types of animals serves as excellent protein food sources. Wild game as well as seafood are favored items of our diet. However, meats from cattle, hogs, sheep, and poultry have the most importance from the standpoint of quantities consumed. Since flesh from which meat is derived is composed of animal cells, a large number of enzymes, particularly of metabolic types, are present. Some of these enzymes have considerable importance.

A period of time usually elapses between the dispatching of an animal and the consumption of a part of the carcass as food. Normally, much of the length of this period is determined by the necessities of commercial practice in making meat available to the consumer. In addition, however, there may be deliberate periods of aging, which have as their purpose the improvement of quality particularly as to tenderness and flavor. Such aging which allows for favorable enzymatic action is most commonly employed only for beef in the fashionable quest for the most tender steaks.

Lean meat is essentially voluntary muscle tissue. Muscles are made of a mass of muscle fibers bound together by connective tissue composed of collagen and elastin. The more connective tissue present, the tougher the meat. After an animal carcass is dressed, it is chilled. There is solidification of the fat and, simultaneously, a development of rigor mortis. In rigor mortis, muscles harden and shorten and make the meat quite tough. This phenomenon is accompanied by the formation of lactic acid and other acids from tissue glycogen through the action of the glycolytic enzyme system. After reaching a maximum in about 24 hr, there is a slow diminution of rigor, and the muscles again become soft and flaccid. This softening occurs mainly inside the muscle fibers and is catalyzed by native proteolytic enzymes present in the meat and known as cathepsins. The cathepsins are present in much greater amount in glandular organs, such as spleen, kidney, and liver, but have been demonstrated and their action studied in muscle tissues also. The rate of tendering is a function of temperature. Usually, meat is held just above the freezing point in order to discourage microbial growth and spoilage for from 1 to 4 weeks. For choice cuts of beef, longer aging periods are used which results in greater disintegration of protein, including connective tissue. Beef aged to this degree has been tendered. Such aging is expensive not only because of the expenses involved in holding the meat for such extra long periods under refrigeration but also because of losses due to shrinkage, loss of bloom, and microbial growth. Bloom describes the appearance of the surface and is related to the condition of the pigments, connective tissue, and fat. Surface microbial growth occurs necessitating trimming. It is believed by some that growth of *Thamnidium* mold on the surface of the meat contributes materially to a better flavor in properly age-tendered steaks.

The autolytic processes of normal aging can be greatly speeded up if temperatures higher than normal refrigeration at 2 to 4°C are used. In the Tenderay® process, meat is hung at 15°C for 3 days in rooms flooded with ultraviolet light to minimize microbial growth. A comparable increase in tenderness can also be obtained by aging for 24 hr at 40°C in which case antibiotics or irradiation must also be employed to protect the meat from spoilage.[285]

Fish and seafood products, of course, also contain many enzymes. After death, the glycogen in fish muscle is converted into lactic acid, and rigor mortis sets in. Autolytic enzyme action may occur in fish and is undesirable, but most fish spoilage is due to microbial action. Fish may be protected from spoilage by freezing, canning, smoking, salting, or drying. Quick freezing with a protective ice glaze and cold storage have made quality seafood products available throughout the country and throughout the year, even though there are alterations during storage, particularly due to oxidation, which affect color and flavor. Enzymatic or microbial factors are of little importance.

Pork and poultry meat products undergo the same enzymatic changes; however, these meats are sufficiently tender after the long, high-temperature cooking times employed, so that like fish, tenderness is usually no problem. Microbial spoilage of poultry is now almost always controlled by quick freezing and low-temperature storage.

Cereal Products

Important agricultural crops in which seeds are utilized for foods belong to the grass or legume families. The members of the grass family which are grown for their edible seeds are called cereals or cereal grains, the most important of which are wheat, corn (maize), rice, rye, barley, oats, and sorghum. Cereal products constitute a major part of the food consumed in the world.

Unlike meat, just discussed, which is dead tissue, the cereal grains, as harvested and stored until processed, are alive and continue to undergo enzymatic metabolism, such as respiration although, of course, at slower rates. Hence, cereal grains contain a host of enzymes necessary for living organisms, such as the metabolic enzymes necessary for respiration, glycolysis, protein metabolism, etc., as well as the hydrolytic enzymes, of which the amylases, proteases and lipases are of greatest interest in cereal technology.

Ungerminated cereals have considerable β-amylase activity, but relatively little α-amylase. The proteolytic enzymes which have been extensively studied, are of the plant enzyme or papain type as indicated by the inhibitory action of oxidizing agents and the powerful activating effect of sulfhydryl compouonds. At the beginning of germination, marked cytological changes occur in the epithelial cells of the scutellum. The scutellar epithelium seems to synthesize enzymes, such as cytase, amylases, and proteinases.

Most cereal products, such as prepared breakfast foods, are heat processed, and the enzyme content of the cereal ingredients are of little importance. A very important exception is in flour milling. The enzyme content of flour for bread baking is of great importance and must be properly controlled. The important enzymes are β-amylase, α-amylase, and proteases. β-Amylase is present in unmilled wheat and, hence, present in excess in flour. This excess does no harm, since the degree of saccharification it can bring about is also dependent upon the presence of α-amylase. The amount of α-amylase is controlled by adding malted wheat or malted barley flour to the regular flour to increase its diastatic power to a certain level and is measured by gas production tests or by so-called maltose number, the number of milligrams of maltose produced in 1 hr at 30°C by 10 g of flour in suspension.

Control of α-amylase is even more important in the baking of bread made largely from rye flour. Where there is insufficient α-amylase, the bread is characterized by a dry, brittle crumb, and the crust becomes cracked and torn upon cooling; whereas, an excess results in bread with a wet, soggy crumb which frequently falls away from the crust leaving large hollow spaces in the bread. The effect of excess enzyme is due to the thermostability of cereal α-amylase so that considerable dextrin formation may take place during baking before heat inactivation of the enzyme takes place. As baking proceeds the starch swells and undergoes partial gelatinization with a consequent increase in its hydration capacity; whereas, coagulation of proteins results in a decrease in their water-holding capacity. Control of the quantity of α-amylase in rye flours is, therefore, very important in Europe and is generally measured on the Brabender® amylograph, which provides a continuous automatic record of the changes in viscosity of a flour-water suspension as temperature is increased at a uniform rate. Variations in amylase activity are considerably greater in northern Europe and Scandanavia, where there is considerable sprout damage as a result of wet harvest weather.

Proteases in flour directly influence the quality or elasticity of gluten proteins in dough. If too much proteolytic action is present, too much breakdown of gluten will occur and a "sticky" dough will result. If too little protease activity occurs, gluten proteins are tough and nonelastic and the baker calls the dough "bucky." When the miller adds malted flour to increase diastatic activity, he also introduces proteases. It is, therefore, difficult to maintain a protease quantity which is just in balance to give elasticity and still not be sticky. The amount of protease action is controlled indirectly by means of oxidation; that is, the addition of oxidizing agents. Wheat proteases, being of the papain type, are sensitive to oxidation; however, its protease activity can be controlled so that normal elastic doughs can be obtained.

Vegetables and Fruits

The foods known as vegetables are derived from different parts of plants. Some are botanically bulbs; others are fruits, roots, shoots, tubers, leaves, stems, and flowers. Fresh vegetables and fruits are living and respiring tissues and have many different enzymes present. Because of this active enzyme content, special precautions are necessary in harvesting, handling, and storing vegetables to be marketed in the fresh condition or in processing for marketing in frozen, dried, or canned conditions.

Fresh fruit and vegetables continue to respire after harvest. The respiratory enzymes continue their actions at sometimes greater and sometimes lesser rates than when attached to the plant before harvest. In general, storage of such fruits and vegetables must be under conditions which minimize respiration and other enzymatic changes. Proper control of temperature, humidity, ventilation, etc., is of paramount importance for successful storage of such commodities as potatoes and apples, for example. Conditions chosen must permit slow respiration; otherwise, the tissues die, and spoilage becomes very rapid.

Besides respiratory enzymes, most fruits and vegetables contain a great many other enzymes, which commonly include carbohydrases (amylases, invertase), proteases, lipases, pectinases, lipoxidases, tyrosinase, peroxidases, catalase, polyphenol oxidases, chlorophyllase, and ascorbic acid oxidase, all of which may be responsible for deterioration in the products. In addition, there are a host of other enzymes which may be of little or no significance. The carbohydrases may cause an undesirable breakdown of starch or sucrose. The proteases can produce autolytic action on proteins. The lipases and lipoxidases can cause undesirable rancidity or flavor changes in the lipid components. Pectinases can cause undesirable softening or viscosity changes in fruit products. Tyrosinase, peroxidases, polyphenol oxidases, and chlorophyllase can be responsible for undesirable color changes. Ascorbic acid oxidase may destroy valuable vitamin C.

Food technologists have long been aware of the structural considerations in the preservation of fruits and vegetables. When cells remain intact, enzymes and substrates are often mechanically separated from each other. When the structure is damaged, as in bruising of fruits or upon freezing, various enzymatic reactions proceed at an accelerated pace.

Fresh fruits and vegetables play important roles in diet. Much attention has been given to the proper time for harvest and proper storage conditions to ensure as long a storage life as possible. Many items are best harvested while still immature or green. Storage conditions are chosen to permit artificial ripening, which may involve not only the many complex physiological respiratory and metabolic enzymatic reactions but also the disappearance of green chlorophyll by the action of chlorophyllase to allow the ripe color of the fruit or vegetable to show.

Because of the perishable nature of fresh fruits and vegetables even under the best devised systems of storage, other methods are commonly used for preserving these food products, including canning, dehydration, and freezing. These preservation methods prevent spoilage by quite different means. Canning involves cooking and sterilization which effectively prevent further enzymatic or microbial changes. The texture and flavor of the cooked products are, of course, usually quite different from those of the fresh products but still quite desirable in most cases. Dehydration reduces moisture content to that point where microbial spoilage cannot occur. Sometimes, if permitted to remain, native enzymes can continue their action in dehydrated food products producing undesirable changes, and certain enzymes are particularly bad if present while the foods are being dried. Frozen fruits and vegetables frequently retain much of the flavor of their fresh counterparts and are protected from spoilage by low storage temperatures. Enzyme action can be very important, if allowed to be present during the freezing operation as well as after freezing during storage. Action of enzymes at freezer temperatures will be slow but will proceed, causing flavor and aroma changes.

In all of these preservation methods, blanching has become the established method for destroying native enzyme systems, which might adversely affect the quality of the product, in canning during the relatively slow rise to cooking temperature, in drying during the extended period of reducing moisture content, and in freezing during the freezing and storage periods. Blanching is carried out in various ways, by heating in boiling water or with steam. The length of time of heating and the temperature required depend upon the subsequent process — canning, dehydration, or freezing, the medium used, temperature of the medium, temperature of the product, rate of circulation of the medium, and the size of the vegetable or pieces of the vegetable. Shelled peas, being small, need only about 60 sec in boiling water or live steam; whereas, ears of sweet corn require 8 to 10 min in the same medium before freezing.

In general, processes employed in freezing may be summarized as follows. Vegetables or fruit of proper variety and maturity are selected and harvested. The freshly harvested product is carefully washed, cleaned, inspected, and prepared as for cooking. It is then blanched for a period long enough to inactivate all catalase content and immediately cooled to 60°F or lower in cold water. The product is given a final inspection, then packed in moisture- and vaporproof cartons, and quick-frozen. After freezing, packages are placed in corrugated fiberboard, shipping containers and stored at 0°F or lower.

For canning fruits and vegetables, preparation is similar to that for freezing through the blanching step. The hot product is then immediately sealed into cans and processed under steam pressure.

For dehydration, the fruit or vegetable must be harvested at the ideal stage for eating and then handled and processed as carefully and rapidly as possible. The freshly harvested product is thoroughly washed, cleaned, and inspected and may be peeled. Fruits to be dried whole are dipped in weak alkaline solutions. Fruits and many vegetables are sulfured, fruits usually by placing them in compartments containing sulfur dioxie from burning sulfur and vegetables usually by immersion in sulfite solutions or by spraying. Cabbage and potatoes are routinely sulfited prior to dehydration; other vegetables may also be so treated after blanching. Most vegetables and cut fruits are blanched, the adequacy of which is judged by measurements for inactivation of catalase or peroxidase. They are then sulfured and dried in counterflow dehydrators. Whole fruit is frequently sun-dried.

It may be noted that adequacy of blanching has usually been determined by ascertaining the absence of peroxidase or catalase, since these enzymes are universally present in fruits and vegetables. In many vegetables, peroxidase has been found to undergo self-reactivation. It may be shown to be absent after a heating cycle but 24 hr later will have in part or wholly recovered its activity. Testing for presence of catalase is easily done with hydrogen peroxide. In the case of vegetables blanched sufficiently to inactivate the catalase, the frozen pack can be expected to keep reasonably well, but a more severe treatment than that necessary to inactivate catalase is required to insure the best storage properties.

An interesting example of controlling natural enzyme content is in the preparation of tomato juice. Desirable juice of high viscosity with little tendency for suspended solids to separate is dependent upon the pectin content of the tomatoes. If cold pressed, the juice has low viscosity and suspended solids separate because of the action of the natural pectin methylesterase. Hence, the hot break method is employed, in which the fruit is heated to a temperature of about 165°F. This heat inactivates the pectinase enzyme, and the juice is then pressed.

A great deal of attention has been given to controlling enzymatic browning in cut fruits and vegetables.[268] Darkening of plant tissues, when exposed to air, is due to the oxidation of *o*-dihydroxyphenol derivatives, such as catechol, protocatechuic acid, caffeic acid, and hydroxygallic acid, which are abundant in nature. This oxidation occurs through the action of enzymes originally called oxygenases. The preferred name now is polyphenol oxidase, although the terms phenolase and polyphenolase also are found in use. This entire class of enzymes has copper as its prosthetic group and comprises a number of different enzymes distinguished by their specific substrates, such as tyrosinase, catecholase, laccase, etc.

The fruits, apples, peaches, apricots, bananas, cherries, grapes, pears, strawberries, and figs, and vegetables, potatoes and red beets, contain polyphenol oxidases and substrate compounds which cause enzymatic browning when the cut tissues are exposed to air. For the browning to occur, three components must be brought together, enzyme, substrate, and oxygen. If any of the three is missing or prevented from reacting by some means, oxidation and browning will not take place. Natural control exists in many fruits and vegetables, as in cantaloupe and tomato which have neither enzyme nor substrate, in the unique Sunbeam peach which has the enzyme but no substrate, and in most of the berries and citrus fruits which do not have the enzyme.

Polyphenol oxidases are quit sensitive to heat, and the blanching process is a means of controlling enzymatic browning. Besides heat, these enzymes can be inactivated by heavy metals, halogens, sulfites, and cyanides. The heavy metals and cyanides cannot be used in foods because of their toxicity, but halogen salts, ascorbic acid, and sulfites are very frequently used in the food industry. Sulfuring with sulfur dioxide or sulfites is the most common method employed with most fruits and many vegetables before dehydration to inhibit the browning reaction. Sodium chloride brines are used for a similar purpose for dipping sliced potatoes before dehydrating. The influence of pH in reducing the rate of browning is widely applied in the food industry. Apricots or peaches, after lye peeling, are immersed immediately into citric acid solution to reduce their pH far below the sharp optimum for polyphenol oxidase activity. A solution of 0.5% citric acid and 0.03% ascorbic acid may be used to prevent browning in cut fruits intended for freezing. In this case, the ascorbic acid acts as an antioxidant.

Not all cases of enzyme-catalyzed browning are undesirable. In the manufacture of tea, coffee, and cocoa, these reactions are essential to the manufacturing processes. Also, enzymatic browning is responsible, at least in part, for the characteristic colors of certain dried fruits, such as prunes, dates, and raisins.[268]

PRODUCTION OF INDUSTRIAL ENZYMES

Industrial enzymes are produced from animal tissues, plant tissues, and microorganisms. For example, commercial animal enzymes include pancreatin, trypsin, chymotrypsin, and lipase obtained from pancreas, pepsin and rennet from stomach mucosa, and catalase from liver. Commercial plant enzyme products include ficin from fig latex, bromelain from pineapple, papain from papaya, and malt amylase from barley malt. A wide range of enzymes of all classes is obtained from a variety of species and strains of molds, yeasts, and bacteria. Microbial production processes afford a degree of control of efficacy, quality, and quantity, which is difficult to achieve with plant or animal sources. For this reason, together with the greater number of enzymes readily available from microorganisms, microbial enzymes are assuming increasingly predominant roles as industrial enzyme products.

Enzyme preparations are obtained from animal and plant sources by collecting or pressing out juices or extracting tissues with water. Microbial enzyme preparations are produced by cultivating selected organisms. Species and strains of special molds, bacteria, and yeasts are selected, developed, and cultivated to produce maximum yields of the desired enzymes. Research is going on in three areas to improve microbial enzyme production: environmental manipulation, mutation, and genetic engineering. Publications on production methods are available.[25, 256, 257]

Crude enzyme solutions obtained from animal or plant tissues or from microbial fermentations are clarified by filtration or centrifugation. Frequently, these solutions are concentrated by vacuum evaporation at relatively low temperatures. To obtain solid products, salting-out or spray-drying procedures may be used, but most commonly, precipitation by acetone or aliphatic alcohols is employed. Precipitated enzyme concentrates are recovered by filtration or centrifugation and dried in atmospheric or vacuum driers. The resulting liquid and solid enzyme concentrates are the basic materials for formulating commercial enzyme products. They represent concentrated, but not highly purified, products. For many commercial applications further purification is not necessary. For some uses, it may be undesirable to have contaminating enzymes or other substances present, and procedures, such as dialysis and chromatographic adsorption methods, are employed. Frequently, enzymes used for specific analytical purposes, for some pharmaceutical uses, and for research purposes must be highly purified. High-purity enzymes are almost invariably expensive and are used only where high cost is not significant relative to the need they satisfy.

Liquid or solid enzyme concentrates are assayed for potency and may be sold as produced by the manufacturer on the basis of their potencies. More commonly, they are diluted to standard activities. Liquid products may require the addition of stabilizers, such as benzoate, glycerol, propylene glycol, sorbitol, or sodium chloride, to prevent microbial growth or loss of enzyme activity during storage. Solid products are adjusted to standard potencies by addition of such diluents as starch, sucrose, lactose, flour, salts, and gelatin. Frequently, buffers and other salts are also used in the formulation of either liquid or solid enzyme products to ensure favorable pH conditions, enzyme activity, and stability. Frequently, manufacturers supply the same basic enzyme standardized to different potency levels or with different diluents depending upon its intended use.

It has been reported that the total enzyme market in the U.S. in 1971 was $36 million, in 1975, $51 million; for 1980, $67 million[226] has been estimated. These figures do not include the very considerable quantities of enzymes produced for their own use by major manufacturers of glucose, fructose, and other products. Probably the value of the world-wide use of enzymes is about three to four times that of the U.S.

A listing of the commercial enzyme products marketed in the U.S. for applications in food industries is presented in Table 1. For each enzyme product, the brand name, enzymes present, source, and company are given. Not included in this list are the specialized and high-purity enzymes and the companies marketing them. They are available at a relatively high cost for analytical, medical, and research purposes.

ENZYME APPLICATIONS

Advantages in Using Enzymes

Enzymes have several distinct advantages for use in industrial processes:

Table 1
MAJOR FOOD ENZYMES MARKETED IN THE UNITED STATES

Brand name	Enzymes	Source	Company
Bud® Chips	α-amylase Protease	*Aspergillus oryzae*	Anheuser-Busch®[a]
Diastatic supplement	Amylases	Barley malt	Anheuser-Busch
Irgazym® BA-20	α-Amylase	Bacterial	Ciba®-Geigy®[b]
Irgazym BF-20	α-Amylase β-Glucanase Protease	Bacterial	Ciba-Geigy
Irgazym BS-10	Glucoamylase	Fungal	Ciba-Geigy
Irgazym BS-15	Glucoamylase	Fungal	Ciba-Geigy
Irgazymn BP-5	Protease	Fungal	Ciba-Geigy
Irgazym BP-3	Protease	Bacterial	Ciba-Geigy
Irgazym CP	Protease	Plant	Ciba-Geigy
Irgazym M-10	Pectinase	*Aspergillus* sp.	Ciba-Geigy
Irgazyme® 100	Pectinase	Fungal	Ciba-Geigy
Ultrazym® 100	Pectinase	Fungal	Ciba-Geigy
American® rennet	Rennin	Calf stomachs	Dairyland[c]
Emporase®	Rennin	*Mucor pusillus*	Dairyland
Quikset®	Rennin, pepsin	Bovine and porcine stomachs	Dairyland[c]
Regulase®	Rennin, pepsin	*M. pusillus* Porcine stomachs	Dairyland
Italase®	Lipase	Calf oral glandular tissue	Dairyland
Capalase®	Lipase	Kid or lamb oral glandular tissue	Dairyland
Liberase®	Lipase	Animal glandular tissue	Dairyland
Dawe's catalase	Catalase	*A. niger*	Dawe's®[d]
Dawe's glucose oxidase	Glucose oxidase Catalase	*A. niger*	Dawe's
Bromelain	Protease	Pineapple	E.M.®[e]
Diastase	Amylases		E.M.
Invertase	Invertase	Yeast	E.M.
Papain	Protease	Papaya	E.M.
Pepsin	Pepsin	Porcine stomachs	E.M.
Asperzyme®	α-Amylase	*A. oryzae*	Enzyme Development[f]
Enzopharm®	Amylase, protease	*A. oryzae*	Enzyme Development
Adjuzyme®	Glucoamylase	*A. niger*	Enzyme Development
Pulluzyme®	Pullulanase	*Klebsiella aerogenes*	Enzyme Development
Enzobake®	Protease	*A. oryzae*	Enzyme Development
Liquipanol®	Protease	Papaya	Enzyme Development
Panol®	Protease	Papaya	Enzyme Development
Cookerzyme®	Amylase	*Bacillus subtilis*	Enzyme Development
Fungal alpha-amylase	α-Amylase	*A. oryzae*	Fermco[g]
Fungal amylase VAC	α-Amylase	*A. oryzae*	Fermco
Bacterial amylase F	α-Amylase	*B. subtilis*	Fermco

Table 1 (continued)
MAJOR FOOD ENZYMES MARKETED IN THE UNITED STATES

Brand name	Enzymes	Source	Company
Bacterial amylase F	α-Amylase	*B. subtilis*	Fermco
Fermvertase®	Invertase	*Saccharomyces* yeast	Fermco
Beta-glucanase	Carbohydrases	*A. niger*	Fermco
Pentosanase	Carbohydrases	*A. niger*	Fermco
Extractase®	Pectinase	*A. niger*	Fermco
Fermcozyme®	Glucose oxidase Catalase	*A. niger*	Fermco
Ovazyme®	Glucose oxidase Catalase	*A. niger*	Fermco
Fermcolase®	Catalase	*A. niger*	Fermco
Fermlipase PL	Lipase	Pancreas	Fermco
Fungal lipase	Lipase	Fungal	Fermco
Bactamyl®	Amylase	*B. subtilis*	GB® Fermentation[h]
Maxamyl®	α-Amylase	*B. subtilis*	GB Fermentation
Maxilact®	Lactase	*Saccharomyces* yeast	GB Fermentation
Maxinvert®	Invertase	*Saccharomyces* yeast	GB Fermentation
Maxazyme NP	Protease	*B. subtilis*	GB Fermentation
Maxazyme P	Protease	*B. subtilis*	GB Fermentation
Lipase	Lipase		GB Fermentation
Pectinase	Pectinase		GB Fermentation
Microbial rennet	Rennin		GB Fermentation
Glucanase GV	β-Glucanases	Fungal	Grindsted[i]
Pektolase L-60	Pectinase	Fungal	Grindsted
Rennet	Rennin	Calf stomachs	Chr. Hansen's[j]
Microbial rennet	Rennin	Fungal	Chr. Hansen's
Lipase	Lipase	Glandular tissue	Chr. Hansen's
Catalase	Catalase	Liver	Chr. Hansen's
Pepsin	Pepsin	Porcine stomachs	Chr. Hansen's
Fungal alpha-amylase	α-Amylase	*Aspergillus* sp.	Henley[k]
Catalase	Catalase	*Aspergillus* sp.	Henley
Cellulase	Carbohydrases	Basidiomycetes	Henley
Glucox®	Glucose oxidase	*Aspergillus* sp.	Henley
Lipase	Lipase	*Rhizopus arrhizus*	Henley
Panzyme®	Pectinase	*Aspergillus* sp.	Henley
Fungal protease	Protease	*Aspergillus* Papaya	Henley
Bromelain	Protease	Pineapple	Henley
Bromelain	Protease	Pineapple	Dr. Madis[l]
Cellulase	Cellulase	Fungal	Dr. Madis
Ficin	Protease	Fig latex	Dr. Madis
Papain	Protease	Papaya	Dr. Madis
Meer papain	Protease	Papaya	Meer[m]
Clarase®	α-Amylase (protease)	*A. oryzae*	Miles®[n]
HT-Amylase®	α-Amylase	*B. subtilis*	Miles
Takamine® fungal amylase	α-Amylase	*A. oryzae*	Miles
Tenase®	α-Amylase	*B. subtilis*	Miles
Dextrinase® A	Amylases	Fungal	Miles
Takamyl®	α-Amylase	*A. oryzae*	Miles

Table 1 (continued)
MAJOR FOOD ENZYMES MARKETED IN THE UNITED STATES

Brand name	Enzymes	Source	Company
Milezyme® 8X	Alkaline protease α-Amylase	*B. subtilis*	Miles
Kinase ® K	α-Amylase	*A. oryzae*	Miles
Kinase M	α-Amylase	*B. subtilis*	Miles
Brewnzyme®	α-Amylase	β-Glucanase	Protease *B. subtilis* Miles
Hemicellulase CE-100®	Carbohydrases	*A. niger*	Miles
Takamine® pancreatin	Amylase lipase Protease	Porcine pancreas	Miles
Diazyme®	Glucoamylase	*A. niger*	Miles
Miles fungal lactase	Lactase	*A. niger*	Miles
Takamine cellulase	β-Glucosidases	*A. niger*	Miles
Spark-L®	Pectinase	*A. niger*	Miles
HT-Proteolytic®	Protease	*B. subtilis*	Miles
Milezyme® AFP	Acid fungal protease	Fungal	Miles
Tendrin®	Proteases	*A. oryzae* bromelain	Miles
Takamine fungal protease	Protease	*A. oryzae*	Miles
Takamine bromelain	Protease	Pineapple stem	Miles
Takamine papain	Protease	Papaya	Miles
Chilco®	Protease	Papaya	Miles
DeeO®	Glucose oxidase Catalase	*A. niger*	Miles
Takamine catalase L	Catalase	Bovine liver	Miles
Takamine lipase powders	Esterase	Pregastric animal tissue	Miles
Takamine pancreatic lipase	Lipase	Porcine pancreas	Miles
Marschall rennet	Rennin	Calf stomachs	Miles
Marzyme®	Rennin	*M. miehei*	Miles
Marla-Set®	Pepsin	Porcine stomachs	Miles
Chymo-Set®	Rennin pepsin	Calf stomachs Porcine stomachs	Miles
Marsin®	Rennin pepsin	*Mucor miehei* porcine stomachs	Miles
Bacterial amylase Novo®	α-Amylase	. **subtilis**	Novo°
BAN 120L	α-Amylase	*B. subtilis*	Novo
BAN 360S	α-Amylase	*B. subtilis*	Novo
Termamyl®	α-Amylase	*B. licheniformis*	Novo
Fungamyl®	α-Amylase	*A. oryzae*	Novo
Novozyme C	α-Amylase	*B. licheniformis*	Novo
Amyloglucosidase Novo 150 AMG 150L	Glucoamylase	*A. niger*	Novo

Table 1 (continued)
MAJOR FOOD ENZYMES MARKETED IN THE UNITED STATES

Brand name	Enzymes	Source	Company
Neutrase®	Protease	*B. subtilis*	Novo
Cereflo® 200L	β-Glucanase	*B. subtilis*	Novo
Pectinex®	Pectinase	*A. niger*	Novo
Sweetzyme®	Glucose isomerase	*B. coagulans*	Novo
Rennilase®	Rennin	*M. miehei*	Novo
Paniplus MLO	Protease (amylase)	*A. oryzae*	Paniplus[p]
Paniplus amylase tablets	Amylase (protease)	*A. oryzae*	Paniplus
Papain	Protease	Papaya	S. B. Penick[q]
Mycozyme®	α-Amylase	*A. oryzae*	Pfizer®[r]
Cerevase®	Protease	Papaya	Pfizer
Mashase®	Protease	Plant	Pfizer
Pfizer rennet	Rennin	Calf stomachs	Pfizer
Metroclot®	Pepsin	Porcine stomachs	Pfizer
Econozyme®	Rennin pepsin	Calf stomachs Porcine stomachs	Pfizer
Morcurd®	Rennin	*M. miehei*	Pfizer
Hei-Pep®	Rennin pepsin	*M. miehei* Porcine stomachs	Pfizer
Sure-Curd®	Rennin	*Endothia parasitica*	Pfizer
Rhozyme® H-39	α-Amylase	*B. subtilis*	Rohm and Haas®[s]
Rhozyme	α-Amylase	*A. oryzae*	Rohm and Haas
Rhozyme HP	Carbohydrases	Fungal	Rohm and Haas
Pectinol®	Pectinase	*A. niger*	Rohm and Haas
Rhozyme 41	Protease	*A. oryzae*	Rohm and Haas
Rhozyme P-11	Protease	*A. flavusoryzae* group	Rohm and Haas
Rhozyme P-53	Protease	*B. subtilis*	Rohm and Haas
Rhozyme 54	Protease	*B. subtilis A. flavusoryzae* group	Rohm and Haas
Lipase	Lipase	Glandular tissue	Scientific Protein[t]
Pepsin	Pepsin	Porcine stomachs	Scientific Protein
Sucrovert®	Invertase	*S. cerevisiae*	SuCrest[u]
Pancreatin NF	Amylase Lipase Protease	Porcine pancreas	Viobin®[v]
Pancrelipase NF	Lipase Protease Amylase	Porcine pancreas	Viobin
Dex-lo®	α-Amylase	*B. subtilis*	Wallerstein[hhw]
Enzyme W	α-Amylase	*B. subtilis*	Wallerstein
Enzyme WC-8	α-Amylase	*B. subtilis*	Wallerstein
Fresh-N®	α-Amylase	*B. subtilis*	Wallerstein
Hazyme®	α-Amylase	*A. niger*	Wallerstein
Enzyme 4511-3	Amylase Protease	*B. subtilis*	Wallerstein
Mylase®	α-Amylase Glucoamylase Carbohydrases	*A. oryzae*	Wallerstein
Fermex®	α-Amylase (protease)	*A. oryzae*	Wallerstein

Table 1 (continued)
MAJOR FOOD ENZYMES MARKETED IN THE UNITED STATES

Brand name	Enzymes	Source	Company
Mycolase®	α-Amylase	*A. oryzae*	Wallerstein
Malt amylase PF	α-Amylase β-Amylase	Barley malt	Wallerstein
Lactase LP	Lactase	*A. niger*	Wallerstein
Convertit®	Invertase	*S. cerevisiae*	Wallerstein
Cellzyme®	Cellulase	*A. niger*	Wallerstein
Klerzyme®	Pectinase	*A. niger*	Wallerstein
Prolase® EB-21	Protease	*B. subtilis*	Wallerstein
Prolase RH	Protease	*B. subtilis*	Wallerstein
Wallerstein papain	Protease	Papaya	Wallerstein
Prolase® 300	Protease	Papaya	Wallerstein
Enzyme 201	Proteases	*B. subtilis*	Wallerstein
Fromase®	Rennin	*M. miehei*	Wallerstein
Esterase lipase	Esterase	*M. miehei*	Wallerstein

[a] Anheuser-Busch, Inc., Industrial Products Division, Bechtold Station, P.O. Box 1810, St. Louis, Mo. 63118.

[b] Ciba-Geigy Corporation, P.O. Box 11422, Greensboro, N.C. 27409 (Ciba-Geigy Ltd., Basel, Switzerland).

[c] Dairyland Food Laboratories, Inc., 620 Progress Ave., Waukesha, Wis. 53186.

[d] Dawe's Laboratories, Inc., 450 State Street, Chicago Heights, Ill. 60411.

[e] E. M. Laboratories, Inc., 500 Executive Blvd., Elmsford, N.Y. 10523. (E. Merck-Darmstadt).

[f] Enzyme Development Corporation, 2 Penn Plaza, New York, N.Y. 10001.

[g] Fermco Biochemics, Inc., 2638 Delta Lane, Elk Grove Village, Ill. 60007.

[h] GB Fermentation Industries, 1 North Broadway, Des Plaines, Ill. 60016 (As of June 1977, Gist-Brocades, N.V., Delft, Holland, purchased the enzyme manufacturing facilities and enzyme products, except for the brewing enzymes, papain and invertase, from Wallerstein Company, and established their U.S. division as GB Fermentation Industries.)

[i] Grindsted Products, Inc., 2701 Rockcreek Parkway, North Kansas City, Mo. 64116.

[j] Chr. Hansen's Laboratory, Inc., 9015 West Maple Street, Milwaukee, Wis. 53214.

[k] Henley and Company, Inc., 750 Third Avenue, New York, N.Y. 10017 (distributors for enzymes produced by C. H. Boehringer Sohn, Ingelheim, West Germany, and John and E. Sturge Company, Birmingham, England, a wholly-owned Boehringer company).

[l] Dr. Madis Laboratories, Inc., 375 Huyler Street, South Hackensack, N.J. 07606.

[m] Meer Corporation, 9500 Railroad Avenue, North Bergen, N.J. 07047.

[n] Miles Laboratories, Inc., Marschall Division, 1127 Myrtle Street, Elkhart, Ind. 46514.

[o] Novo Laboratories Incorporated, 59 Danbury Road, Wilton, Conn. 06897 (Novo Industri A/B, DK-288 Bagsvaerd, Denmark).

[p] Paniplus Company, 3406 East 17th Street, Kansas City, Mo. 64127.

[q] S. B. Penick and Company, 1050 Wall Street West, Lyndhurst, N.J. 07071.

[r] Pfizer, Inc., 4215 North Port Washington Avenue, Milwaukee, Wis. 53212.

[s] Rohm and Haas Company, Independence Mall West, Philadelphia, Pa. 19105.

[t] Scientific Protein Laboratory Division, Oscar Mayer® Company, P. O. Box 1409, Madison, Wis. 53701.

[u] SuCrest Corporation, Specialty Products Division, 120 Wall Street, New York, N.Y. 10005.

[v] Viobin Corporation, Monticello, Ill. 61856.

1. They are of natural origin and nontoxic.
2. They have great specificity of action; hence, they can bring about reactions not otherwise easily carried out, especially without unwanted side reactions.
3. They work best under mild conditions of moderate temperature and near neutral pH, thus not requiring drastic conditions of high temperature, high pressure, high acidity, and the like, which necessitate special expensive equipment, and may cause undesirable side reactions.
4. They act rapidly at relatively low concentrations, and the rate of reaction can be readily controlled by adjusting temperature, pH, and amount of enzyme employed.
5. They are easily inactivated when a reaction has gone as far as is desired.

Broadly, enzyme technology deals with the production, isolation, purification, and use of enzymes for a variety of industrial, medical, and analytical[83] purposes. Soluble enzymes can be used once and discarded, or they can be reused by fixing them in some fashion by immobilization.

Immobilized Enzymes

Recent developments in the production and use of immobilized enzyme systems, that are insoluble in the reaction medium, further advance the inherent advantages of enzyme catalysis. The literature on immobilized enzymes has become so extensive that references are given here for only five reviews from which many individual references may be obtained.[66, 81, 222, 275, 296]

In spite of the inherent advantages of the use of enzyme catalysts, a number of potentially useful industrial applications have not developed because: 1. the cost of enzyme isolation and purification is high, making the process uneconomical; 2. the isolated enzymes are too unstable when removed from the living cell; and 3. it is difficult and costly to recover or to remove soluble enzymes from the reaction mixtures after completion of the enzymatic conversions. Enzymes of enhanced stability immobilized on water-insoluble carriers, which permit easy recovery and repeated reuse, should significantly reduce or eliminate these problems. Also, requisite enzymes may not be on the Generally Recognized as Safe (GRAS) list of the U.S. Food and Drug Administration, and the cost of obtaining food additive clearance is prohibitive. Immobilization of these non-GRAS enzymes could permit application of them, otherwise unavailable, to the food processing industry.

Immobilized enzymes achieve their effect very efficiently, even though the enzyme is not truly in solution. The problem of separating the enzyme from the substrate at the end of the reaction is minimal since the solid phase and attached enzyme are readily removed from the reaction mixture by simple filtration. The enzyme is then available for reuse in subsequent reactions.

Often, it is possible to use the immobilized enzyme in a continuous reactor system. A solution of the material to be subjected to the enzymatic action may be passed continuously through a bed of the enzyme, and the desired reaction takes place as the solution flows through. The degree of reaction can be controlled by the rate at which the solution passes through the bed of fixed enzyme. A high concentration of enzyme on the solid material may reduce the residence time necessary to attain the desired degree of reaction in the solution, resulting in high rates of production in relatively small reactors. Exposure of the substrate to the conditions of temperature and pH required for the reaction may thus be of low duration, reducing undesirable side reactions. A high concentration of immobilized enzyme also permits the carrying out of continuous reactions at relatively low temperatues with reasonably sized reactors.

These factors result in a greater useful life for the enzyme, which in turn, decreases the overall cost of enzyme per unit weight of product made.

Four principal methods have been used for preparing water-insoluble immobilized enzyme preparations.

1. Adsorption on inert carriers or synthetic ion exchange resins — This has inherent limitations because the adsorbed enzyme is weakly bound and is rather easily lost during use; however, it has been successfully applied, for example, in resolving DL-amino acid mixtures by acylase.
2. Entrapping enzyme in gel lattices, the pores of which are too small to allow the enzyme to diffuse but large enough to allow the passage of substrate and product. This method has been used mainly for analytical purposes and in the form of microencapsulated products might have medical significance where attention must be paid to immunological responses to enzyme foreign protein or where the enzyme is very fragile and unstable.
3. Covalent binding to a wettable, but water-insoluble carrier via functional groups, nonessential for biological activity.
4. Covalent cross-linking of the enzyme protein by an appropriate bifunctional reagent or by copolymerization — The covalent linking techniques have been most studied and seem to have great potential; however, enzymes immobilized on ion exchange materials have had greatest industrial applications so far.

The inert carrier materials used in immobilizing enzymes include glass beads, diazotized cellulose particles, polyaminostyrene beads, polyacrylamide, DEAE-cellulose, DEAE-Sephadex®, and other synthetic polymers or copolymers. Various physical properties, such as mechanical stability, hydrophilic or hydrophobic nature, swelling characteristics, as well as electric charge, are considered in the selection of suitable carriers for immobilization of specific enzymes for particular applications.

Goldstein[81] and Weetall[274, 275] have briefly considered some of the important factors relating to the properties of immobilized enzymes, including stability, kinetic behavior, and its physical state, such as beads, membranes, sheets, etc. As mentioned above, the insolubilized enzymes often have advantages over soluble ones: 1. They may be chemically and thermally more stable; 2. they are easier to recover and reuse; and 3. they provide a choice of supporting materials that can be tailored to enhance the specific reaction and reactor design. However, there are factors which adversely affect the use of immobilized enzymes: 1. Loss of enzyme from the support; 2. lower reaction rate; and 3. blocking of enzyme sites and support structure by chemical reaction, air, or contaminating solid materials.

Already considerable use has been made in the laboratory of immobilized enzymes. Examples of their use are in the study of protein structure by obtaining well-defined fragments of complex proteins, such as immunoglobulins; in the isolation of enzyme inhibitors, such as trypsin inhibitors from crude animal organ extracts; in continuous separation of synthetic DL-amino acid mixtures by columns of immobilized aminoacylase; and in preparations of polynucleotides by columns of immobilized polynucleotide phosphorylase and immobilized enzyme columns for analytical applications. Among the most promising areas for applications of immobilized enzymes are analysis of specific enzyme substrates, e.g., glucose and urea, and analytical and process instrumentation.[36, 276]

The first commercial use of an immobilized enzyme system was in Japan employing aminoacylase immobilized by ionic binding to a support of DEAE-Sephadex for resolution of DL-amino acid mixtures in continuous, automatically controlled preparations of the amino acids L-methionine, L-phenylalanine, L-tryptophan and L-valine.[42] Production costs by the immobilized technique are only 60% of those for the conventional batch process using soluble enzyme.[226]

The newest multimillion dollar industrial enzyme application, isomerization of glucose to fructose, employs immobilized glucose isomerase. Most other large-scale, industrial processes employing enzymes continue to use soluble enzymes, but it seems probable that in due course immobilized enzymes may also take their place in some of these industries. Weetall[277] indicated that the major reason so little progress has been made in commercialization of immobilized enzyme processes is cost. Developing and scaling up an immobilized enzyme process can cost as much as half a million dollars. It is difficult and time consuming and requires specially skilled personnel. Weetall[277] lists numerous design parameters and system properties important for any scale-up. Among these are the type of enzyme used; carrier cost, size, and composition; reactor type and dimension; enzyme immobilization techniques and time; and operating conditions, such as temperature, pressure, and pH.

MAJOR CLASSES OF INDUSTRIAL ENZYMES AND THEIR APPLICATIONS

Carbohydrases

The carbohydrases are enzymes which hydrolyze polysaccharides or oligosaccharides. Among the carbohydrases are found the most investigated as well as the most widely used enzymes both for laboratory and industrial applications.

Starch-splitting Enzymes

Of all the commercial enzymes, amylases, the enzymes which act on starch, have the most numerous applications. Various amylases from plant, animal, fungal, and bacterial sources have been in use for many years. There are three types of amylases which hydrolyze starch in different manners:

$$\text{Starch} \xrightarrow{\alpha\text{-amylase}} \text{Dextrins + Maltose}$$

$$\text{Starch} \xrightarrow{\beta\text{-amylase}} \text{Maltose + Dextrins}$$

$$\text{Starch} \xrightarrow{\text{glucoamylase}} \text{Glucose}$$

The terms "liquefying" and "saccharifying" amylases are general terms denoting the two principal types of enzyme action. An α-amylase (1,4-α-D-glucan glucanohydrolase, EC 3.2.1.1) hydrolyzes α-(1→4) linkages in large starch molecules in random manner, thereby liquefying starch rapidly but also producing extensive saccharification on prolonged action. α-Amylases from different sources — animals, higher plants, fungi, and bacteria — vary widely in saccharifying ability, in thermal stability, and in their extent of hydrolysis. Since α-amylases cannot hydrolyze α-(1→6) branching linkages in starch, the ultimate products from high saccharifying α-amylase are maltose, small quantities of other malto-oligosaccharides, a small amount of the trisaccharide, panose, which contains the original α-(1→6) linkages of the branched starch fraction, and a little glucose. Considerable amounts of low molecular weight dextrins are generally formed by low saccharifying or dextrinizing types of α-amylases.

β-Amylase (1,4-α-D-glucan maltohydrolase, EC 3.2.1.2) is produced by higher plants, particularly cereals and sweet potatoes. It is a saccharifying enzyme, producing maltose as its only sugar by splitting maltose units progressively from the nonreducing ends of starch chains. When acting on the branched amylopectin fraction of starch, the action of β-amylase ceases when it reaches an α-(1→6) linkage, leaving so-called "beta limit dextrins."

Glucoamylase (1,4-α-D-glucan glucohydrolase, EC 3.2.1.3), frequently also called amyloglucosidase, seems to be formed mainly by fungi. It is a saccharifying enzyme producing only glucose by progressive hydrolysis of glucose units from the nonreducing ends of starch chains. Studies by Pazur and co-workers with highly purified glucoamylase have shown that it acts preferentially on longer chains and that it also hydrolyzes α-(1→6) and α-(1→3) linkages, although more slowly than the α-(1→4) linkages.[174, 175, 176] Hence, this enzyme is capable of converting starch completely to glucose.

Some important commercial uses for amylases are indicated in Table 2. Amylases have a variety of uses in food processing and also other applications as shown. For example, the first industrial manufacture of the fungal enzyme, Taka-Diastase®, was for a pharmaceutical digestive aid, which continues to be its major application. A recent publication has reviewed the use of enzymes in starch processing.[22]

An extremely important use for amylases is in the production of sweet syrups. The hydrolysis of starch to sugars by acids and enzymes has been known and practiced since the early 19th century. The first plant for production of syrups by acid conversion was built in France in 1814. When starch slurries are acidified with hydrochloric acid to about pH 1.8 and heated under pressure, random hydrolysis occurs yielding glucose and glucose oligosaccharides of various degrees of polymerization. For any given extent of acid hydrolysis of starch, as measured by the copper-reducing value and expressed as dextrose equivalent (DE), there is only one composition that will occur.

For example, regular acid conversion syrup of 42 DE has a composition of about 22% glucose, 20% maltose, 20% tri- and tetrasaccharides, and 38% dextrins.[136] Furthermore, acid catalysis at the necessary high temperatures produces recombination or reversion products and decomposition products. Acid conversion syrups, above about 50 DE, are found to have an objectionable, bitter taste, amber color, and a tendency to crystallize. Since enzyme hydrolysis is characterized by specificity, it was found that use of commercially available fungal amylases on syrups, acid-hydrolyzed to within a range of 40 to 50 DE, made possible 62- to 65-DE syrups of superior flavor and sweetness. Such high-conversion acid-enzyme syrups have been marketed since the early 1940s.[55] A typical high-conversion syrup of this kind has the composition: 63 DE, 38% glucose, 34% maltose, 16% tri- and tetrasaccharides, and 12% dextrins.[136]

Within recent years, manufacturers have learned to take advantage of different commercially available enzyme preparations containing various proportions and combinations of α-amylase, β-amylase, and glucoamylase either simultaneously or successively with either acid-liquefied or bacterial amylase-liquefied starch to produce syrups of widely differing compositions, from very low DE-high dextrin syrups to very high DE-high sugar syrups, to meet the demands of different food industries.[260] For example, processes have come into commercial use for producing syrups of about 70 DE and of over 80% yeast fermentability with a glucose content below about 44%, the limit for noncrystallizing high-solids syrup. One method employs acid hydrolysis to about 20 DE, converting with malt to about 52 DE and completing the conversion with fungal amylase.[74] Another method uses new fungal enzyme systems which will produce, starting with 15- to 20-DE, acid-thinned starch, syrups of 67 to 69 DE, of 80 to 82% fermentability and containing approximately equal amounts (about 40%) of glucose and maltose.[260]

So-called high-maltose syrups, low in glucose, are also produced commercially for the hard candy industry. By use of malt extract or commercial, concentrated malt

Table 2
SOME COMMERCIAL USES OF AMYLASES

	Bacterial	Fungal	Plant
Syrup manufacture	x	x	x
Dextrose manufacture	x	x	
Baking	x	x	x
Saccharification of fermentation mashes			
Distillery	x	x	x
Brewery	x	x	x
Food dextrin and sugar products	x	x	x
Dry breakfast foods	x	x	x
Chocolate and licorice syrups	x	x	
Starch removal from fruit extracts and juices and from pectin		x	
Scrap candy recovery	x	x	
Starch modification in vegetables	x	x	
Textile desizing	x		
Starch coatings for paper and fabrics	x		
Cold water-dispersible laundry starch	x		
Wallpaper removal	x		
Pharmaceutical digestive aid		x	

enzyme preparations, either low DE, acid- or enzyme-thinned starches, are converted to syrups of about 40 to 42 DE without materially increasing the level of glucose in the final syrup. For example, incubation of 20 DE, acid-thinned starch with 0.009% of a 1500° Lintner malt enzyme preparation for 48 hr at pH 5.0 and 55°C gives a DE of about 41 and a carbohydrate composition of about 6% glucose, 47% maltose, 9% triose, and 38% higher saccharides.[260] A commercial fungal amylase designed for high-maltose syrup production also is in use. It may be satisfactorily employed with either the enzyme-liquefied or acid-liquefied starch substrate, the former being preferred because it results in lower glucose and higher maltose contents. With 15 to 20 DE, acid-liquefied corn syrup containing 40% solids at pH 5.3, incubation for 24 to 28 hr at 54°C using 0.02% of the enzyme gives a DE of about 43 and a composition of 7.5% glucose and 47% maltose. Starting with a 30% starch slurry liquefied with 0.1% bacterial amylase at 88°C in a jet cooker, incubation with 0.06% of the fungal enzyme at pH 5.3 for 40 hr at 53°C gives a syrup of about 45 DE with 3% glucose and 58% maltose.

The most recent extensive application of amylolytic enzymes is the use of fungal glucoamylase for the production of crystalline dextrose from starch.[61, 88, 233, 260] Over a billion pounds of dextrose are produced annually in the U.S. All major dextrose manufacturers world-wide, now employ glucoamylase instead of the classical acid conversion process. It has permitted more than a doubling of the concentration of starch in the conversion slurry, has simplified handling, including evaporation, decolorization, and crystallization, and has very materially increased yields of recovered dextrose.

Production of dextrose from starch by conversion with glucoamylase was proposed before 1950. However, commercialization was not possible until some 10 years later when the glucoamylase became available at economical cost and essentially free from transglucosylases because of the discovery of very high yielding strains of fungi which produced minimal amounts of transglucosylases.[15, 33] The presence of transglucosylases is undesirable because of their production of glucose polymers having α-(1→6) linkages, which reduce dextrose yields and interfere with dextrose crystallization. Several methods have also been patented for the removal of transglucosylases from glucoamylase preparations.[52, 99-101, 114, 127]

Commercial operations employ starch concentrations of 30 to 40%. Glucoamylase gives essentially quantitative conversion to dextrose at low starch concentrations with progressively less complete conversion as starch concentrations are increased.[260] This is due to "back polymerization" catalyzed by the glucoamylase, forming reversion sugars, mainly isomaltose. This polymerizing reaction restricts the maximum concentration of starch which can be efficiently converted. Considering all factors, such as dextrose yield and plant throughput, about 30% starch is usually considered the most practical level.

Prior to enzyme conversion, the starch must be gelatinized to be susceptible to enzyme attack and liquefied in order to be handled. Thinning may be accomplished either by heating with dilute acid or by use of thermostable bacterial amylase. Following thinning, the solution is then adjusted to the optimum of about pH 4.0 for the glucoamylase which is then added, and saccharification, allowed to proceed to completion in tanks maintained at about 60°C. Conversion times-employed vary between 48 and 96 hr depending upon the enzyme level used. Usually, the enzyme concentration employed requires about 72 hr for completion of the conversion. After the conversion, the material is filtered, carbon- and ion exchanges-treated, evaporated, and the dextrose crystallized.

Experience has shown that by the acid-enzyme process, a maximum DE of about 95 and about 92% dextrose based on total solids can be obtained, and the practical yields are about 100 lb anhydrous dextrose per 100 lb of starch. The double enzyme process gives about 97 DE and about 95% dextrose, with practical yields of 105 lb of dextrose per 100 lb of starch. A practical disadvantage which has hindered general adoption of the double enzyme process is the slow filtration rates for the converted starch solutions, which is due to undigested residual solids. One method which obviates this difficulty is to employ a steam jet heater for the bacterial liquefaction step, with a prior addition of optimum amounts of calcium and sodium salts for maximum thermostability of the enzyme.[260] The very rapid heating of the starch slurry in the jet heater causes almost instantaneous gelatinization of the starch and prevents retrogradation to material not susceptible to glucoamylase action. The amount of residual insolubles is greatly decreased, and filtration rates are markedly improved by this method.

Although the traditional liquefying amylases derived from *Bacillus subtilis* are quite thermostable and have been successful, modern technology, involving high-temperature, continuous jet cookers, has led to a search for enzymes which perform well at higher temperatures. A liquefying α-amylase derived from *Bacillus licheniformis* is said to be useful in starch slurries up to 115°C.[22, 215, 226]

A patented process permits the use of crude starch sources, such as corn flour, for dextrose production instead of the much more costly separated starch usually employed in the industry.[32] With crude starch sources, enzyme thinning must be used since thinning with acids would produce soluble, contaminating, nitrogenous and other compounds which would make purification and crystallization of the dextrose liquors difficult or impossible. In this process, the starch in the substrate is gelatinized and thinned with bacterial amylase using a steam jet heater and then converted with glucoamylase, which must be essentially free from protease, lipase, and transglucosylase. Following conversion, the insoluble solids are separated by filtration and marketed after drying as a premium quality, high-protein, gluten livestock feed. The solution is purified by carbon and ion exchange treatment; the dextrose can then be recovered in pure form by evaporation and crystallization. One plant operates this process on a very large scale for the production of dextrose in solution which is then used for producing citric acid by fermentation.

The use of glucoamylase immobilized on various matrices, such as DEAE-cellulose, synthetic resins, and glass, has been described.[172, 227, 275, 278] The immobilized glucoamylase may be used in a continuous flow reactor. Under the conditions advocated, the enzyme loses activity very slowly, and continuous saccharification offers the potential advantages of a continuous process with efficient use of enzyme. However, much remains to be learned about the preparation and use of immobilized glucoamylase. Under practical operating conditions, it has been difficult to obtain the same degree of conversion of preliquefied starch to dextrose with the immobilized enzyme as can be obtained with soluble glucoamylase under comparable conversion conditions. The interplay of hydrolytic and transferase reactions in the immobilized system where concentration of the enzyme is extremely high is probably of significance and is not directly related to experience with batch operation with soluble enzymes, which are generally at much lower concentrations in reaction zones. Further research on the fundamental aspects of the mechanisms of reactions catalyzed by immobilized enzymes may be expected to generate the new concepts necessary to provide the basic structure for more advanced immobilized enzyme technologies in the manufacture of crystalline dextrose.

Pitcher and Weetall[181] have estimated the cost of saccharification by immobilized glucoamylase on the basis of present knowledge. They concluded that in its present state this system appears to be of interest mainly for new facilities rather than as a replacement for existing ones.

The "debranching" enzymes, pullulanase (pullulan 6-glucano-hydrolase, EC 3.2.1.41) and isoamylase (glycogen 6-glucanohydrolase, EC 3.2.1.68), have been receiving considerable attention both from theoretical and practical standpoints.[5,129] These enzymes catalyze the hydrolysis of α-(1→6) glucosidic bonds of unmodified glycogen, amylopectin, and their partial degradation products. The main difference between the microbial pullulanases and isoamylases is the inability of the isoamylases to degrade the linear polysaccharide, pullulan. Isoamylases have a strict requirement for the α-(1→6) linkage to constitute a true branch point rather than a linkage in a linear chain. This makes isoamylase a valuable tool in studies on the structure of carbohydrates.

Known microbial sources for pullulanase are *Aerobacter aerogenes* (*Klebsiella aerogenes*) and *Streptococcus mitis*, and for isoamylase, *Pseudomonas* and *Cytophaga* species. Pullulanase cannot significantly act upon native glycogen but can hydrolyze partially degraded glycogen. Simultaneous action of β-amylase and pullulanase on amylopectin gives nearly quantitative yields of maltose. Isoamylase is the only known type of enzyme which will totally debranch glycogen.[90,295] However, it is unable, as opposed to pullulanase, to remove the 2- and 3-glucose units of the side chains of beta- and alpha-limit dextrins of oligosaccharides.

Probably, the most important industrial use of debranching enzymes will be the production of maltose or high-maltose syrups. The production of maltose and high-maltose syrups can be accomplished by various techniques, which include simultaneous reaction of starch with pullulanase and β-amylase or fungal α-amylase[94] and production of amylose using isoamylase followed by β-amylase treatment. The production of a linear polysaccharide, high-amylose product may also offer advantages in both food and industrial processes, such as in making edible films.

The unique properties of high-maltose syrups and amylose offer advantages in both food and industrial applications. Hence, debranching enzymes should play an important role in the future of starch technology.[5]

Another major food industry using amylases is the baking industry.[6,54,111,254] Two types of amylases are recognized as important in the baking industry, α- and β-amylase. Flour, milled from sound wheat, contains a relatively high content of β-amylase and very low level of α-amylase.[125] During malting of wheat or barley, the β-amylase increases only a little, but the α-amylase increases several thousandfold. Microbial amylases are mainly α-amylases. Differences in the thermostability of cereal, fungal, and bacterial α-amylases are very important considerations in determining their usefulness in baking processes. The principal function of supplementation with either malt or fungal amylase is to increase the α-amylase content of the flour.

Amylase supplementation affects fermentation and bread quality. Panary fermentation by yeast requires sugar both for growth and formation of carbon dioxide. The amount of sugar normally present in flour is quite small. Hence, the production of bread depends upon added sugar in the dough and upon maltose formation by amylases present in the starch of flour. The enzymes act during fermentation only on the so-called damaged starch granules which constitute a rather small and variable percentage of the total starch content of flours.

During fermentation, the β-amylase, which is able to act only up to the points of branching, acts upon dextrins to produce maltose. When flour is supplemented with α-amylase during fermentation, this enzyme causes dextrinization of the damaged starch, producing more chain ends for β-amylase action. The combined action of excess β-amylase in flour and the α-amylase contributed by enzyme supplementation results in rapid and complete saccharification, the major part of the conversion being achieved during early stages of fermentation. The importance of this increased sugar as a yeast substrate for increased gas production depends upon the level of sugar added to the dough. Where added sugar is limiting, the maltose from starch hydrolysis is particularly important during the proofing period and the first few minutes in the oven when the supply of other sugars may be insufficient for adequate gas production.

Addition of α-amylase not only increases the rate of fermentation but produces enough sugar to increase the sugar content in baked bread. This is important because of its effect on the flavor of bread, crust color, and toasting characteristics of bread slices.

Another role of amylase during fermentation is its effect on dough consistency. Damaged starch granules have high water-absorbing and- holding capacity; when this hydrated starch is broken down by amylase, the water released causes softening or a decrease in the consistency of the dough.[112]

The formation of fermentable sugars by the amylases is not their only function in bread baking. α-Amylase has a considerable effect on the viscosity of doughs, the formation of dextrins, the grain and texture of bread, and the compressibility of bread crumb. Although the action of amylases during fermentation is limited by the content of damaged starch, as the temperature of the dough increases during the baking period above about 60°C, the starch is gelatinized and becomes susceptible to enzyme action. For a few minutes in the oven, until the heat inactivates the amylase, dextrinization and saccharification of the starch can be rapid. It may result in increased gas production, improved crust color, improved moisture and keeping quality of the crumb due to dextrins, and additional sugar which contributes to the flavor and caramelized sugar in the crust. The extent of amylase action and starch breakdown depends primarily upon the thermostability of the α-amylase used.[6,157] The most notable and important differences between α-amylases from different sources are their thermostabilities, as shown in Figure 4. Fungal amylase is quite labile, being destroyed rapidly at temperatures above 60°C. Temperatures of 70 to 75°C are required for rapid inactivation of malt α-amylase. Bacterial amylase is the most stable of all and shows little inactivation

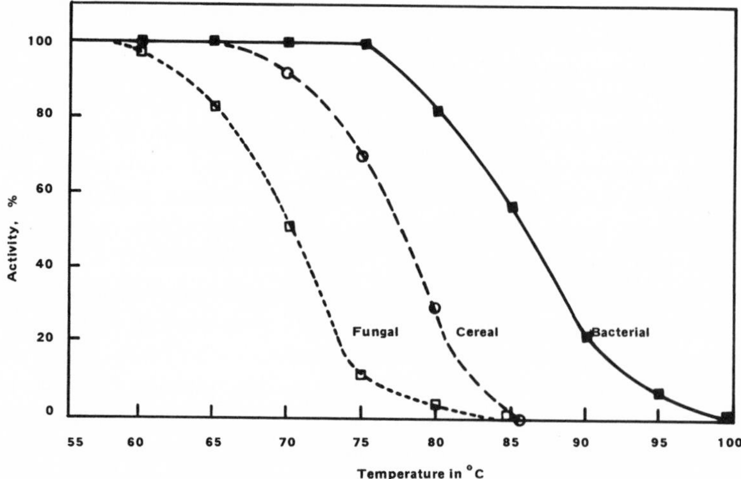

FIGURE 4. Thermal stability of α-amylases.

at temperatures up to 85 and 90°C in the presence of starch. Dextrins resulting from the action of cereal α-amylase in the oven have a very important effect on the texture of bread crumb. Small amounts of the dextrins are desirable from the standpoint of bread quality and retardation of staling, but larger amounts cause a gummy, sticky crumb and loaf fragility.[24, 54] Since the α-amylase of wheat and barley malt is quite stable over the temperature range for starch gelatinization, it is important that any excess of cereal α-amylase not be present to cause excess dextrinization. On the other hand, baking temperatures inactivate fungal amylase before any appreciable quantities of starch are gelatinized. Supplementation with fungal amylase, therefore, offers a considerable margin of safety against overdextrinization and stickiness in the crumb. More than twenty times the level of fungal amylase necessary for a desirable baking response does not adversely affect baking results in any way.

Early work led to the conclusion that bacterial amylase had no place in bread making because of its high thermostability, which led to gummy crumb. However, quite recent work has shown that very low levels of bacterial amylase retard staling without harmful effects on the quality of the bread.[202, 223] Lack of tolerance to a slight overdosage of bacterial amylase does, however, represent a hazard to its practical use in the bread baking industry.

The effect of bacterial amylase in dextrinizing flour starch can be used to advantage when a moist, soft, and somewhat tacky crumb is desirable, as in fruit cakes.[231] There is also a place for bacterial amylase in the production of specialty products having soft consistencies like pie fillings.

Enzyme supplementation to flour is practiced both at the mill and by the baker. Most countries now permit the addition of both malt flour and fungal amylase to flour at the mill. The α-amylase content of almost all commercial flour in the U.S. is increased by supplementation at the mill with wheat or barley malt flours. The baker also usually adds more fungal enzyme depending upon the formula, equipment, and process used.

Another use for amylases, probably the oldest of all enzyme usages, practiced since antiquity, is the use of malt for alcoholic fermentation of grains. There are numerous excellent books which should be consulted for the details of malting and the brewing and distilling industries. In brief,[47,180,186,187,261] malt is produced by controlled sprouting of grains, usually barley, followed by drying. During malting, high levels of α-amylase are formed, and some increase in β-amylase occurs. The principal difference in the three general classes of barley malts is in the α-amylase content. Brewers' malt assays at 25 to 30 units, distillers' malt at 50 to 60 units, and gibberellin malt at 80 to 100 units. About 100 million bushels per year of malt are produced in the U.S., of which about 6% is used by the distilling industry, 3% by the food industry, and 91% by the brewing industry. The malt is usually ground and used in this form for its enzyme content, but it also provides carbohydrates, proteins, and flavoring constituents which are useful as yeast nutrients or as constituents of a product, such as beer. The active enzymes are water soluble and can be extracted by water into solutions of about 25% malt. By evaporating such extracts, concentrates can be prepared. These are marketed as high-diastatic malt extracts with considerable maltose content for many purposes, such as brewery and bakery adjuncts.

In brewing, malt alone, or usually along with starchy adjuncts from rice or corn, is mixed with water and heated to boiling to effect gelatinization and liquefaction of the starch. It is then cooled somewhat, and additional malt is added to effect saccharification. After a suitable holding period, hops are added and the mash is boiled for a period sufficient to inactivate the enzymes, extract the hops, and sterilize the mash. The hot mash is then filtered to remove solids, yeast is added to the wort, and fermentation allowed to go to completion. Finishing operations include final filtration, addition of chillproofing protease preparations, carbonation, and filling into containers. Bottled and canned beer are pasteurized; draft beer is not and subject to microbial spoilage. A recent innovation is millipore filtration instead of pasteurization.

In distillery practice starchy raw materals are employed: potatoes, malt, maize, and rye in Europe; potatoes, rye, and wheat in Russia; corn, milo, malt, and rye in the U.S., rice and sweet potatoes in the Orient; and cassava in tropical countries. The ground material is mixed with water, and the mixture, heated under pressure to effect gelatinization of the starch and sterilization of the mash. The hot mash is cooled to about 60°C, ground malt equal to approximately 10% of total grain weight, added, and the mash, cooled and pumped into fermentors where it is inoculated with pure yeast cultures. A relatively quick, partial conversion of starch to sugars occurs during the brief period at 60°C after malt addition. Sufficient maltose is produced to promote development of the yeast and strong fermentation. As fermentation proceeds, maltose is used up by the yeast and dextrins are continually converted to sugars by the active amylase present until conversion and fermentation continues to completion. At the end of fermentation, the mash is distilled to produce industrial alcohol or beverage products, such as whiskey or grain neutral spirits. The amount of industrial alcohol produced by alcoholic fermentation of grains is now quite insignificant in the U.S., the grain alcohol distilleries in this country exist almost solely for the production of beverage alcohol products; however, in numerous other parts of the world, production of industrial alcohol by fermentation continues to be important.

Because of their high thermostability, bacterial amylases have found some use in the liquefaction of brewery and distillery mashes. Fungal amylase preparations, containing α-amylase and glucoamylase, are now employed commercially in the U.S. instead of malt for saccharification of fermentation mashes for the production of grain neutral spirits and light whiskey.[89] In most countries which produce industrial alcohol by fermentation, fungal amylase preparations are employed extensively. The use of fungal enzymes has contributed to increased yields of alcohol per bushel of grain, to reduced operating time, and to overall reduced costs.[180]

Considerable investigation has been made concerning the use of microbial enzymes in partial or full replacement of malt enzymes in brewery mashing.[140] The use of microbial enzymes has been investigated along two lines: 1. the preparation and use of a "barley syrup" to replace wort in the brewhouse,[62, 284] and 2. substitution of microbial enzymes for all, or part, of the malt in a "conventional" brew house operation. Evaluation of "barley beers" varies considerably. Most investigators agree that barley beers are different from malt beers but are, in many instances, at least equally acceptable. However, there is no consensus of opinion regarding the technical feasibility, economic advantage, or other criteria for judging the essential, commercial utilization of malt replacement enzymes in brewing.

Other uses of amylases in food industries are quite minor with respect to amounts of enzymes involved but are quite important and frequently almost indispensible to the manufacturers they serve.[257, 259] Amylases are utilized in processing cereal products for food dextrin and sugar mixtures and in processing baby foods and some breakfast foods. They are indispensible for the preparation of free-flowing chocolate and licorice syrups since they liquefy the starch present and so keep the syrups from congealing. They are used for recovering sugars for reuse from scrap candy of high-starch content. Fungal amylases are also used for starch removal from fruit juices and extracts, from flavoring extracts, and in preparing clear, starch-free pectin. For the latter use, of course, special amylase preparations completely free from pectinase are requisite. Microbial amylases also have limited use in modifying starch in vegetable purees and in treating vegetables for canning.

In addition to their important applications in food processing, large amounts of bacterial amylases have other important industrial uses, among which are textile desizing, preparing modified starch coatings for paper and fabrics, and making cold water-dispersible laundry starches. Detailed discussion of these applications is beyond the scope of this chapter.

Disaccharide Splitting Enzymes

Several carbohydrases, invertase (sucrase) or β-fructofuranosidase (β-D-fructofuranoside fructohydrolase, EC 3.2.1.26), lactase or β-galactosidase (β-D-galactoside galactohydrolase, EC 3.2.1.23), maltase or α-glucosidase (α-D-glucoside glucohydrolase, EC 3.2.1.20), and melibiase or α-galactosidase (α-D-galactoside galactohydrolase, EC 3.2.1.22) hydrolyze disaccharides and have considerable importance. These enzymes hydrolyze their corresponding disaccharides with the formation of two molecules of monosaccharide as indicated.

$$\text{Sucrose} \xrightarrow{\text{invertase}} \text{Glucose} + \text{Fructose}$$

$$\text{Lactose} \xrightarrow{\text{lactase}} \text{Glucose} + \text{Galactose}$$

$$\text{Maltose} \xrightarrow{\text{maltase}} \text{Glucose} + \text{Glucose}$$

$$\text{Melibiase} \xrightarrow{\text{melibiase}} \text{Glucose} + \text{Galactose}$$

All these enzymes may be obtained from various microorganisms, with selected strains of yeast most commonly used for commercial invertase and lactase. Maltase is not marketed as such. Melibiase, a galactosidase, also hydrolyzes the trisaccharide, raffinose, into sucrose and galactose.

Invertase has several important applications (Table 3).[43,102,162] It can be employed in manufacturing artificial honey and particularly for invert sugar which is much more soluble than sucrose. Hence, a large use of crude invertase is to prevent crystallization in high-test molasses. The high solubility of invert sugar is important in the manufacture of confectioneries, liqueurs, and frozen desserts where high sucrose concentrations

Table 3
SOME USES FOR INVERTASE

Artificial honey
Invert sugar (noncrystallizing jams, etc.)
Soft cream candy centers
Prevention of crystallization in high-test molasses
Removal of sucrose from foods

Table 4
SOME USES FOR LACTASE

Upgrading of milk products
- Sweeter milk
- Stabilization frozen milk concentrates
- Prevention of crystallization in ice cream
- Prevention of lactose intolerance in man and poultry

Upgrading whey
- Hydrolysis for animal feed
- Growing of single cell protein

Bread making

would lead to crystallization. Invertase is also used in the preparation of chocolate-coated, liquid-center candies. Molding and coating are carried out, while the centers are firm, after which the invertase that has been added to the cast centers acts to yield a smooth, stable cream.[102,106-109]

The solubility of lactose is low, compared to other sugars, and it has little sweetness. Hydrolysis of lactose by lactase, a β-D-galactosidase, forming glucose and galactose, increases sweetness, solubility, and the osmotic pressure of products containing lactose. Lactose is in abundant supply, particularly in by-product cheese whey. Some 1.5 billion pounds of lactose are available from the 30 billion pounds of liquid whey produced annually by U.S. cheese industry. Some whey is used for the production of crystalline lactose and the heat-coagulable lactalbumin. Much whey is concentrated or spray-dried for use in animal feeds and human food products, such as baked goods, candies, prepared foods, and infant formulas. However, there is still a very large amount of cheese whey which is discarded and presents a serious disposal problem to the cheese industry.

In spite of the apparent potential for lactase in upgrading whey and various dairy products, its actual commercial utilization has been quite limited. A recent review should be consulted by those interested in lactase potentials.[183]

One use for lactase is in preventing lactose crystallization in ice cream, such crystallization causing "grainy" or "sandy" ice cream. Small-scale commercial use of lactase has been in hydrolysis of lactose in whey concentrate for animal feeds. The proteins in whey have high biological value, but the presence of lactose limits the amount of whey which can be used in feeds. Hydrolysis of lactose improves digestibility and palatability and also prevents crystallization and setting-up in liquid whey concentrates.

Frozen milk concentrates have been produced on a limited scale for reconstitution as a fluid milk beverage, mostly for use on ocean-going vessels and by the armed services. Crystallization of the lactose in frozen concentrates is a problem from the standpoint of difficulty in reconstitution but more importantly because of its effect on the stability of the frozen concentrate. The presence of crystals causes flocculation and coagulation of casein, so that after the frozen concentrate had been stored for several weeks, the reconstituted milk is not acceptable. Lactose hydrolysis prevents formation of lactose crystals and, hence, results in protein stability.[230,251] For the domestic market, lactase seems to have real potential in the development of frozen whole milk concentrates of acceptable flavor when reconstituted. Such a product would have many convenience factors for the distributor, retail market, and the housewife from the standpoints of keeping qualities, storage space necessary, and frequency of purchase.[183]

It has been suggested that applications involving lactase activity may be aided by the use of immobilized enzyme technology. Pilot plant runs with lactose solutions and whey have been successful, and the concept is being commercially applied.[169, 289] The use of immobilized lactase would seem to have great large-scale industrial potential — for the conversion of the lactose in milk to reduce human rejection of milk due to lactose intolerance and for the conversion of lactose in whey to increase utilization of this nutritious by-product. Only about one third of the 30 billion pounds of fluid whey produced annually is used in food or animal feeds. Lactose hydrolysis would permit use of higher amounts of whey with its protein of high biological value in foods and feeds; the hydrolyzed lactose, itself, as a cheap by-product could find even greater use in foods as a sweetener.

Another potential use for lactase is in bread making.[184,185] Skim milk solids are used in bread, contributing about 1.5% of lactose, based on weight of flour used. This lactose is not yeast fermentable and contributes hardly any sweetness to the bread. Hydrolysis by lactase to glucose and galactose would make glucose available for fermentation, and galactose would still contribute to the formation of crust color.[185] It might be more practical and more economical to use milk products in which the lactose was hydrolyzed prior to its incorporation in the dough than attempt to hydrolyze lactose during the dough fermentation.

One of the potentially most important, recent enzymatic processes to come out of Japan is the use of α-galactosidase (melibiase, raffinase) in beet sugar-refining, to hydrolyze raffinose to galactose and sucrose. The process is commercial[213, 234] in Japan and the U.S., and perhaps in other countries.

This process is of particular importance since raffinose builds up in concentration as the mother liquor from crystallization of beet sugar is recycled. When the concentration reaches about 8%, it has been necessary, in the past, to discard the molasses since the raffinose interferes with crystallization of the sucrose and results in too fine needlelike crystals of different appearance.

The enzyme is produced by growing the organism *Mortierella vinacea* var. *raffinose-utilizer*, ATCC 20034. The separated mycelium is used as a crude enzyme; the enzyme mycelium "pellets" are suspended in a horizontal vat divided into several chambers, each agitated and separated by a screen that prevents passage of the mycelium from one chamber to the other. The molasses, rich in raffinose, is fed into one end of the trough and allowed to move through the screen from chamber to chamber, while the mycelial pellets containing the enzyme are maintained in suspension within each chamber. On hydrolysis, approximately 65% of the raffinose is transformed into sucrose and is, therefore, available for crystallization as well. More importantly, crystallization of the sucrose is not interfered with, and large suitable crystals in proper yield are thus obtained. In operation, the juice travels by gravity through the agitated chambers at a temperature of 45 to 50°C at 30° Brix and at pH about 5.2. Loss of enzyme in the chambers due to attrition necessitates the addition of more enzyme to the first chamber periodically. Retention time is only about 2 hr in the troughlike device. In the Steffen process subsequently employed, the sucrose (and raffinose, if present) precipitates as calcium saccharate; whereas, the galactose does not and is thus eliminated. Experience in Japan indicated that plant throughput increased by at least 5% and recovery of sucrose increased, too.

Table 5
NOMENCLATURE OF PECTIC ENZYMES

Acting on pectin	
Pectin methylesterases (PME)	
Polymethylgalacturonases (PMG)	Pectin lyases (PL)
Endo-PMG	Endo-PL
Exo-PMG	Exo-PL
Acting on pectic acid	
Polygalacturonases (PG)	Pectic acid lyases (PAL)
Endo-PG	Endo-PAL
Exo-PG	Exo-PAL

Pectic Enzymes

Pectic enzymes are those which take part in the degradation of pectic substances. These pectic substances are found widely in plant tissues, particularly in fruits, and include a multitude of compounds of complex and variable nature.[17, 65, 119] The pectins are polymers made up of chains of galacturonic acid units joined by α-(1→4) glycosidic linkages. In native pectins, approximately two thirds, more or less depending upon the source, of the carboxylic acid groups are esterified with methanol. These highly esterified pectins yield semi-solid gels with sugar and acid of the kind familiar in jellies and jams. Partial hydrolysis of the methyl esters give low methoxyl pectins which form gels with small amounts of calcium ions. Complete hydrolysis of the methyl esters gives pectic acids. Since crude natural pectins upon complete hydrolysis give small amounts of several different sugars and acetic acid along with galacturonic acid, it was once thought that other carbohydrates, such as arabans and galactans, were present as impurities. It now seems more probable that rhamnose, arabinose, galactose, and traces of other sugars are integral parts of the pectin molecules, and some of the free hydroxyl groups of the galacturonic acid are acetylated.[149]

The pectic enzymes may be classified into pectin methylesterases (PME, pectinesterase) which hydrolyze methyl ester linkages, polygalacturonases which split glycosidic bonds between galacturonic acid molecules, and pectin transeliminases or lyases. The latter enzymes[3] bring about nonhydrolytic cleavage of α-(1→4) linkages forming unsaturated derivatives of galacturonic acid. Some of the enzymes previously considered to be polygalacturonases may turn out to be lyases, since both show the same effect of reducing the viscosity of pectin and producing one molecule of reducing sugar for each α-(1→4) linkage broken. Further divisions exist with the pectic enzymes depending upon whether they act on the methylated or free polygalacturonic acid substrates and whether they attack internal linkages. A scheme which shows the nomenclature of the pectic enzymes and their common abbreviations is given in Table 5.[163]

Those enzymes of Table 5 accepted by the Enzyme Commission[11] are: Pectin methylesterase (PME) (pectinesterase), [pectin pectylhydrolase, EC 3.1.1.11]; endopolygalacturonase (Endo-PG), [poly(1,4-α-D-galacturonide) glycanohydrolase, EC 3.2.1.15]; exopolygalacturonase (Exo-PG), [poly(1,4-α-D-galacturonide) galacturonohydrolase, EC 3.2.1.67]; pectin lyase (PL), [poly(methoxygalacturonide) lyase, EC 4.2.2.10]; endopolygalacturonate lyase (Endo-PAL) (pectate lyase), [poly(1,4-α-D-galacturonide) lyse, EC 4.2.2.2]; and exopolygalacturonate lyase (Exo-PAL), [poly(1,4-α-D-galacturonide) exo-lyase, EC 4.2.2.9].

Pectic enzymes occur rather widely in many plants, including fruits, and in many microorganisms. Commercial pectinases are derived from fungi, most commonly strains of the *Aspergillus niger* group. These commercial pectinases, widely employed in processing fruit products, are mixtures of several of the pectic enzymes, particularly pectin methylesterases, polygalacturonases, and pectin lyases.[60,146,198] Differences in

Table 6
SOME USES FOR PECTIC ENZYMES

Increase in fruit juice yields
Fruit juice clarification
Reduction in viscosity of purees and concentrates
Jelly making
Low-sugar jellies
Better extraction of juice and color for wines
Peel digestion
Citrus oil recovery

the effectiveness of different commercial preparations with different fruits and under different conditions are undoubtedly due to variations in the kinds and amounts of particular pectic enzymes present. Many attempts have been made to correlate the effectiveness of different enzymes in clarifying fruit juices with the contents of the various pectic enzyme fraction but with little success.[14,113,160] In a brilliant work by Endo in Japan with a commercial pectic enzyme derived from *Coniothyrium diplodiella*, three endopolygalacturonases, one exopolygalacturonase and two pectin methylesterases were isolated in highly purified form.[69-72] None of the individual fractions clarified apple juice satisfactorily. Only combinations of at least one of the endo-PG fractions with at least one of the PME fractions effectively clarified apple juice. When the isolated fractions were all mixed in proportions in which they were present in the crude mixture, the apple juice — clarifying activity was identical with that of the crude preparation itself.[73] This work has demonstrated that other enzymes, besides polygalacturonases and pectin methylesterase, are not required for clarifying apple juice. More recently, it has been reported[103,104] that highly purified pectin lyase can also clarify apple juice and juice from some varieties of grapes. A mixture of pure pectin lyase and endopolygalacturonase was required for clarification of juice from some other grape varieties.[105]

Freshly pressed fruit juices contain pectin, the hydrophilic colloidal nature of which makes them viscous and hold dispersed solids in suspension. This effect is highly desirable in some cases, as in tomato, orange, and apricot juices; for such juices, the natural pectic enzymes must be destroyed by heat in order to maintain the desired stable cloud and viscosity. In other fruit juices, such as apple and grape juices, a clear product is usually desired. To obtain such clear juices, a commercial pectinase is used. The major application for pectic enzymes is in the production of brilliantly clear apple juice, grape juice, other fruit juices, and wines.

The appropriate commercial pectinase added to crushed apples results in higher yields of the juice. The enzyme added to the juice reduces viscosity and degrades the soluble pectin-protective colloid, permitting finely divided insoluble particles to flocculate. Considerable variation exists in the practical use of pectic enzymes in clarifying apple juice depending upon the season, the varieties of apples, and the factors of time, temperature and enzyme concentration employed. The floc formed may be allowed to settle or may be removed by centrifugation or filtration to yield crystal clear, brilliant juice. When this juice is pasteurized, it is free from the boiled taste, characteristic of untreated pasteurized juices. The pasteurized juice can be marketed the year around.[2,58] Vinegar and jelly produced from depectinized juice are superior in brilliance, color, and aroma.

For Concord grapes, pectic enzymes are indispensible in the hot press process, now common in the industry, for increasing the yield of grape juice and improving color extraction. Destemmed and crushed grapes are mixed with the pectinase preparation and heated to 60 to 65°C. This process permits most of the juice to be taken off as free-run grape juice with the remaining juice recovered by a continuous screw press.[37] The enzyme action is sufficient to permit high juice yields but not extensive enough to clarify the juice completely or to lower viscosity unduly.[178]

In making wines, pectic enzymes present a number of processing advantages at different stages of the operation. The addition of pectinases to the crushed grapes will increase the yield of free-run juice, reduce pressing time, and increase total juice yield. Addition of pectic enzyme increases the extraction of color when grapes are extracted hot or fermented on the skins. When pectic enzymes are added before or during fermentation, the yeast sediment will be more compact at first racking, giving a better yield and clearer wine. Enzyme addition to the fermented wine facilitates filtration and reduces the requirement of bentonite, giving superior clarity and color.[31, 53] Pectic enzymes are also widely used in making wines from berries, apples, pears, peaches, and other fruits.

Prune juice is commonly produced by hot extraction of dried prunes, and pectic enzymes are not usually employed. For production of juice from fresh Italian prunes, however, addition of pectic enzymes to the pulp doubles the yield of juice.[270]

In the U.S., citrus, apple, grape, and prune juices make up the major part of juice production. However, juices from cranberries, blueberries, strawberries, black currants, cherries, plums, and many other fruits are prepared for beverage use or jelly manufacture, and pectic enzymes are almost universally employed.[78, 265] Fruit juices for jelly manufacture are commonly depectinized completely in order to obtain clear juices and make uniform jelly possible by adding back a standard amount of apple or citrus pectin when the jelly is made. The variable quality and quantity of the natural pectins in the original juices do not then interfere.[58]

Pectic enzymes are indispensible for making high-density fruit juice concentrates or purees.[28, 269] Enzyme depectinization is required for apple juice concentrate of about 72° Brix, which is made on a large scale, since, otherwise, a gel would result rather than the desired liquid concentrate. Pectic enzymes are useful in treating orange, grape, and prune pulps prior to vacuum puff drying.[166]

A few miscellaneous uses for pectinases exist. One is production of D-galacturonic acid, di- and trigalacturonic acids, and higher polymers from pectin.[16] Employing pectin methylesterase alone allows production of low methoxyl pectins. Fruit juices may be depectinized with PME and used for making low-sugar jellies for diabetics. Sorbitol, as a preservative and nonnutritive sweetener, may be added, then calcium ions which with low-methoxyl pectin causes the formation of a stable gel.[58] Pectic enzymes are also used for the recovery and stabilization of citrus oils from lemon and orange peels.[182] Proposals have been made for using immobilized pectinases for clarification of apple and other fruit juices. These enzymes would have apparent advantages of economy (reuse of the enzyme) and purity (no additives); however, no commercialization has yet occurred.

Naringinase[161]

Numerous flavonoids are found in citrus fruits, the highest amounts of which are found in the bitter naringin and tasteless hesperidin. Naringin, the 7-(2-rhamnosido-β-glucoside) of 4′5,7-trihydroxyflavanone (naringenin), predominates in grapefruit and bitter oranges. Bitter naringin can be rendered tasteless by naringinase preparations, which contain two glycosidases, one a rhamnosidase which hydrolyzes naringin to rhamnose plus prunin and a second β-glucosidase which hydrolyzes prunin to naringenin and glucose. Commercial naringinase preparations free of pectinase activity to prevent alteration of the citrus cloud are available. They are of particular use in Japan to treat the bitter orange, (*Citrus aurantium,* natsudaidai).

Table 7
SOME USES FOR CELLULASES

Production of glucose from cellulosic wastes for
  Dextrose
  Single cell protein
  Alcohol or other fermentation products
Agar-agar recovery from seaweeds
Animal feeds
Digestive aid
Enzymatic drain cleaner ingredient
Paper making
Dehydrated vegetables and fruits
Tenderization of fibrous vegetables
Removal of soybean seed coat

Cellulases and Hemicellulases

The most abundant carbohydrate in nature is cellulose, the principal structural material of plants. Each year plants produce 100 billion tons of cellulose, about 150 lb/day for each of the 3.9 billion people on the earth. Cellulose is a water-insoluble β-1,4-glucan, a glucose polymer having β-(1→4) linkages. Cotton is almost pure cellulose; whereas, structural parts of plants — wood, straw, stalks, etc. — contain about 50% cellulose. In these plant structural materials, cellulose is bound together by lignin; there are also present varying amounts of other carbohydrates, hexosans, and pentosans, besides cellulose. These other carbohydrate polymers associated with cellulose in nature are of ill-defined composition and may be classified as hemicelluloses, which occur in almost all plant tissues — wood, straw, stalks, hulls, seeds, etc. — from relatively low to relatively high concentrations. Some, the gums, are water soluble or dispersible; others are insoluble. Mannose, galactose, xylose, and arabinose are the most common sugars found in the various hemicelluloses, but other sugars also often occur. Frequently, more than one kind of sugar is present in common hemicelluloses. For example, the principal component of guar gum is a complex galactomannan.

Cellulosic materials are degraded in nature by certain microorganisms. These microorganisms obviously produce enzymes which actively degrade cellulose, lignin, and hemicelluloses. Active cellulase enzymes have long been sought because of the availability of huge amounts of cellulosic plant wastes as potential sources of glucose for the chemical and food industries, because of the interest by food technologists in upgrading fibrous food products, and because of their potential as a digestive aid for man and animals and in waste disposal systems. Although commercial cellulase products are on the market, even the best of these are seriously lacking in high activity against native cellulose. A recent qualitative comparison of the efficacy of attack by several fungi on cellulose fibers with the efficacy of cell-free cultures of the same fungi showed almost no correlation between the two activities.[171] One of the most puzzling problems in the field of applied enzymology is why cell-free cultures or isolated cellulases from microorganisms, the living cultures of which rapidly digest cellulose, attack cellulose so slowly and so poorly.

The solution to this problem is most important because of the very wide use to which highly active cellulases could be put in the food industry and the potential production of glucose from cellulosic waste materials. Once obtained, the glucose could be used as either animal or human food, converted to fructose, used as a growth medium for single-cell protein, or fermented to chemical fuels and solvents, like ethanol and acetone. Numerous laboratories are investigating the potential of cellulase enzymes. Of particular importance is the work at the U.S. Army Natick Development Center in

Massachusetts.[226] Active cellulolytic enzyme systems are being produced from *Trichoderma viride* which will totally hydrolyze milled newspaper in 48 hr, in a prepilot plant facility handling 1000 lb of cellulose per month. They believe large-scale enzymatic conversion of cellulose to glucose is technically feasible and practically achievable by 1980 in a plant which could handle 500 tons of trash per day.[226]

A large amount of work has been conducted during the past few years with β-glucanases or cellulases derived from several sources, including *Aspergillus niger, Trichoderma viride, T. koningii, Myrothecium verrucaria,* various wood rot fungi, and various strains of rumen bacteria. Those especially interested will need to consult the voluminous original literature.[84,86,87,120,121,138,139,165,173,193,244,247,287]

In brief it seems evident from the extensive work on cellulase fractionation and characterization that the cellulase complex contains the following components, listed in the order in which their action occurs:[122]

1. C_1 enzymes (cellobiohydrolases) required for the hydrolysis of insoluble, highly oriented solid cellulose by β-1,4-glucanases — This enzyme probably functions by an endwise mechanism removing cellobiose units.[68,173,287]
2. β-1,4-glucanases (C_x enzymes) which are hydrolytic [cellulase = endo-1,4-β-glucanase = 1,4-(1,3;1,4)-β-D-glucan 4-glucanohydrolase, EC 3.2.1.4] — These include various fractions differing significantly in pH optima and temperature stability; some are endoenzymes and act in random manner, but there are also exoenzymes.
3. β-glucosidases or cellobiases [cellobiase = β-glucosidase = β-D-glucoside glucohydrolase, EC 3.2.1.21]

The available commercial cellulases have generally been derived from *Aspergillus niger* and split cellulose to cellulodextrins and glucose. They have rapid action on soluble cellulose derivatives but act much more slowly on insoluble native cellulose. Cellulolytic enzyme preparations from other microorganisms, particularly, *Trichoderma viride* and *T. koningii,* have been developed with better activity against native cellulose.[86,120,139,165,173,244,247,287]

Commercial applications for cellulases have been slow in developing, largely because of the poor activity of the available products on the insoluble substrate. There is some use in enzymatic digestive aids. Cellulase has shown some promise in removing vegetable fibers from wool and in enzymatic drain cleaner formulations. Although there has been some success in increasing the strength of paper by cellulase treatment to dissolve more easily digested constituents in wood pulp, the cost of the enzyme and length of treatment have prevented commercialization. Increase of fibrillation in papermaking usually achieved by extensive beating can be obtained by use of cellulase.[34]

In the food industry, cellulase treatment has been tried with citrus products prior to concentration and in the extraction of flavoring materials. The cellulolytic tenderizing of fibrous vegetables and treatment of otherwise indigestible plant materials for foods or feeds may assume increasing importance. For example, fiber digestion may be accomplished by cellulase treatment of coconut cake or in common garden vegetables.[153,189,245,247]

Dehydrated vegetable materials which have been pretreated with cellulase enzyme can be rehydrated more readily in water with a return to their original shape.[244,247] It has also been suggested that cellulolytic enzymes be used for predigestion of animal feeds and for the isolation of starch from potatoes.[245,247]

Several poorly characterized hemicellulases have been tested for various food applications; moreover, several hemicellulase and "gumase" products are commercially available. Various individual hemicellulases hydrolyze specific types of hexosans and pentosans, including more or less complex mannans, galactans, xylans, arabans, etc. One type of hemicellulase has been found useful commercially in processing products containing galactomannans, such as guar and locust bean gums. It has also been found effective in improving the digestibility of guar gum meal for animal feeding.[266] A hemicellulase has been used in preventing gelation in coffee concentrates.[194] Increased yields of high-quality starch from wheat are obtained by use of a pentosanase.[224]

The use of hemicellulases in baking is another area of application which appears to have a good chance of practical development. Wheat endosperm contains about 2.4% hemicellulose, mainly pentosans containing xylose and arabinose. The insoluble pentosan fraction in wheat flour has a harmful effect on bread, reducing loaf volume and giving coarser texture.[130, 131] A suitable pentosanase should be useful in baking technology.[128]

Dextranase

This enzyme, (1,6-α-D-glucan 6-glucanohydrolase, EC 3.2.1.11) produced by *Penicillium funiculosum*,[38] received interest during World War II in the processing of dextran solutions for possible use as a blood plasma extender. More recently, it has been proposed as a possible means of reducing dental caries by removal of the dextran plaque produced on teeth by *Streptococcus mutans*.[288] Formation of dextrans by the action of *Leuconostoc mesenteroides* in "stale" or "sour" sugar cane causes difficulty in refining because of the increased viscosity of the juice.[241] Degradation of dextran by dextranase reduces this viscosity and improves the processing rate.[216]

Lysozyme

This enzyme (mucopeptide *N*-acetylmuramoylhydrolase, EC 3.2.1.17) hydrolyzes mucopolysaccharides in bacterial cell walls. It is naturally present in numerous animal secretions, such as tears and milk. Its best practical source is egg white. It is available commercially in crystalline form or in the form of egg white. It has been employed industrially for releasing catalase from cells of *Micrococcus lysodeikticus* and is widely used in Europe for the "humanization" of cow's milk.[214] Human milk has a higher content of lysozyme, and addition of lysozyme to cow's milk is claimed to make it more suitable for infant feeding.

Glucose Isomerase

A very recent commercial development has been the application of glucose isomerase (D-glucose ketol-isomerase, EC 5.3.1.18) for converting glucose to fructose for large-scale production of high-fructose sweet syrups. This has become a multimillion dollar industry since its commercialization in the early 1970s. Approximately 1.1 billion pounds of high-fructose syrup were produced in the U.S. in 1975, and it captured 6% of the nutritive sweetener market.[12, 226] It has been estimated that production by at least eight companies in the U.S. will soon reach an annual production of 4.4 billion pounds of high-fructose syrups,[12,22,226] requiring about $6 million worth of glucose isomerase. The potential market in the U.S. for high-fructose syrup is in the range of 6 to 7 billion pounds.[22] In the U.S., some 24 billion pounds of sucrose are consumed per year, about half coming from imported raw sugar. With the advent of the process for conversion of fructose from glucose by glucose isomerase, for the first time, a product which is made from corn starch can compete economically with sucrose on a sweetness per unit weight basis. Thus, tremendous amounts of starch, efficiently obtained from our major farm crop (maize), can be converted to products that will supply an important part of our food needs, satisfaction of which to a great extent, depends upon foreign sources.

In 1957, Marshall and Kooi[148] reported that *Pseudomonas hydrophila* cultivated in a culture medium containing xylose produced an enzyme that isomerized glucose to fructose, as well as xylose to the corresponding ketose xylulose. Marshall[147] patented the use of enzymes showing xylose isomerase activity for conversion of glucose to fructose.

Sweet, "invert sugar" syrups containing fructose are readily and cheaply obtained by acid or enzymatic hydrolysis of sucrose. Even though glucose is cheaper than sucrose, the high cost of the enzyme has prevented the process of converting to fructose by glucose isomerase from being competitive. Although other source organisms giving higher isomerase yields were discovered,[249,250,293] all of these also required expensive xylose in the growth medium as an inducer of the glucose (xylose) isomerase. Apparently, the process became practical only after the discovery by Japanese workers of a strain of *Streptomyces* sp. in which the isomerase could be induced by crude xylan sources, such as wheat bran or corn cobs, as well as by separated xylan and by xylose.[237, 238] Recently, it was found that certain strains of *Arthrobacter* can form glucose isomerase in good yield in the absence of xylose and with glucose as the principal carbohydrate in the fermentation media.[137] *Bacillus coagulans* is also a source for commercial immobilized glucose isomerase. No details of the exact procedures employed industrially have been published, although immobilized glucose isomerase preparations are universally employed in industrial processes.[225,226,242]

Publications by Takasaki, Kosugi, and Kanabayashi[237] and by MacAllister, Wardrip, and Schnyder[143] give considerable general information regarding the production and utilization of the isomerase as well as the isolation, crystallization, and properties of the enzyme. The discussion and flow sheet[237] indicate that the medium contained 3% wheat bran, 2% corn steep liquor, and 0.024% $CoCl_2 \cdot 6H_2O$. After incubation at 30°C with aeration for maximum enzyme production in 25 to 30 hr, the wheat bran residue is removed by screening and the *Streptomyces* cells are recovered by filtration and washed. By heat treating the cells filtered from the primary fermentation broth at about 65°C, the intracellular enzyme is rather firmly fixed in the cells. These enzyme-fixed cells can then be used as immobilized enzyme preparations for glucose isomerization, either batchwise being filtered off for reuse or by a continuous column operation. In the batch system, the loss of perhaps 10% of the enzyme activity in each cycle can be made up by the addition of a requisite amount of fresh enzyme-fixed cells.

Other preferred methods for preparation of fixed enzymes include adsorption on DEAE-cellulose[225, 242] or covalent bonding.[232] In preparing such immobilized enzyme preparations, the enzyme must be in solution before the fixation operation. Glucose isomerase, which occurs intracellularly, can, in some cases, be readily solubilized by sonication, by the action of lysozyme, or by treatment with a surfactant, such as cetylpyridinium chloride.[237] Under well-controlled conditions, a high proportion of the enzyme present in the primary fermentor broth can be recovered readily in the immobilized form. In the properly fixed form, the isomerase is quite stable and can be stored, shipped, and transferred to isomerization reactors with relative ease and little loss of activity. The fixed glucose isomerase enzyme systems have many advantages with respect to enzyme use efficiency, ease of handling, and adaptability to continuous reactor operation.

For glucose isomerization,[237] glucose isomerase is employed with 50% glucose (starch enzyme hydrolyzate) solution containing 0.005M magnesium sulfate and 0.001M cobalt chloride with pH maintained at about 7 and temperature between 65 to 70°C. After isomerization is complete, cells are filtered off; the solution is decolorized with carbon, subjected to ion exchange purification, and evaporated to about 75% solids content. The sugar composition of the finished syrup is 45 to 50% fructose and 55 to 50% glucose.

One industrially used, patented process[242] employs glucose isomerase fixed on DEAE-cellulose is shallow beds, 1 to 5 in. deep, placed in several reactors arranged in series. Each small reactor may contain several hundred pounds of immobilized enzyme at a different stage of enzyme lifetime. During the semicontinuous process carried out from about 20 to 80°C over a pH range of about 6 to 9, a 93% glucose solution is isomerized to about 42% fructose.[242]

One of the advantages of the fixed isomerase system is its very high actvity per unit weight.[143] A large amount of enzyme can be packed into a relatively small reactor making it possible to process large quantities of substrate through a small reactor in a short time. The very high effective concentration of enzyme per unit volume of substrate in the reaction zone results in high reaction rates and such short residence times that undesirable color and off-flavor developments, which would, otherwise, occur at the reaction temperatures and pHs employed, are avoided.[143]

Proteases

Next to the carbohydrases, proteases have the greatest variety of applications. Because of the large number of different amino acids present, the resulting complex nature and high molecular weights of proteins, and the variety of proteases of differing specificity, the action of proteases from a practical standpoint is much more complicated than that of the carbohydrases. A greatly oversimplified picture of proteolytic action may be illustrated as follows:

Endopeptidase

HO—H

NH_2CHCO ⁞ NH CHCO ···· NH CHCO ⁞ NH CHCO ·· NH CHCO ⁞ NHCHCOOH

R_1 R_2 R_3 R_4 R_5 R_6

HO — H HO — H

Exopeptidase (Aminopeptidase) Exopeptidase (Carboxypeptidase)

The proteases break down proteins and their degradation products, polypeptides and peptides, by hydrolyzing the —CO—NH—peptide linkages. Individual proteolytic enzymes are of two types. Endopeptidases hydrolyze internal peptide linkages along the protein chain and do not usually attack terminal units; they produce peptides. Exopeptidases hydrolyze the peptide linkages which join terminal amino acid residues to the main chain, thus producing amino acids. Aminopeptidases act upon the ends of protein chains having terminal amino groups and carboxypeptidases on the ends having terminal carboxyl groups.

A considerable number of commerical proteases of plant, animal, and microbial origin are marketed. These commercial preparations usually contain several individual proteolytic enzymes. The individual proteases vary widely in their specific activities, their pH optima and ranges, and heat sensitivity. All highly purified proteases which have been studied show specificity for certain peptide bonds and have little or no action on other peptide bonds. For example, trypsin will rapidly hydrolyze only those bonds linking the carboxyl group of the two basic amino acids, lysine and arginine, to the amino group of any other amino acid. Depending upon the individual enzymes present in commercial proteases, they may range from quite specific to fairly general in their

Table 8
USES OF PROTEASES IN FOOD INDUSTRIES

Use	Enzymes employed
Oriental condiments	Fungal
Dairy — cheese	Animal, bacterial, fungal
Baking — bread, crackers	Fungal, bacterial
Brewing — chillproofing	Plant, fungal, bacterial
Meat — tenderizing	Plant, fungal
Fish — solubles	Bacterial, plant

actions on proteinaceous substrates. Hence, in practical applications of proteases in food processing, it is necessary to select the appropriate protease complex or combination of enzymes. Usually, this selection can be determined only by empirical methods. With proper selection of enzymes and appropriate conditions of time, temperature and pH, either limited proteolysis or nearly complete hydrolysis of most proteins to amino acids can be brought about. For some uses, proteases of different origin are directly competitive; in other cases, the special properties of one may give it an advantage over the others.

The important uses for proteases in the food industries are shown in Table 8.

Fungal protease has been used for centuries in the Orient for production of soy sauce, tamari sauce, and miso, a breakfast food. The application is a rather complex art, although simple in operation. Proteolytic strains of *Aspergillus oryzae* are grown on rice or soybeans to produce a heavily sporulated koji which is added to vats containing roasted soybeans, cereal grains, and salt brine. Over a period of many months, proteolysis, solubilization, and extraction occurs. The soy sauce is then recovered by straining and pressing out the liquid with hydraulic presses.

Rennin

One of the oldest uses of an enzyme is that of rennet for coagulating milk in cheesemaking. Most proteolytic enzymes will clot milk but upon further contact will dissolve the curd by hydrolysis of the insoluble proteins to peptides and amino acids. Rennin (chymosin, EC 3.4.23.4) is a unique enzyme which after clotting milk has no further, rapid or extensive action on the curd. It seems quite clear, however, that rennin acts as a protease, splitting peptide bonds in one particular casein fraction. The mechanism of curd formation has been postulated to involve the limited proteolysis of κ-casein by rennin forming soluble glycopeptide and para-κ-casein. Paracasein precipitates and leads to precipitation of other casein fractions which have been exposed to calcium ions released as a result of the κ-casein hydrolysis. This theory is probably an oversimplification of the process, and alternate ones have been proposed.[75, 145]

The production of cheese is, still, very much an art. Procedures vary depending upon the type of cheese and somewhat from plant to plant for the same type of cheese. In general, milk is heat treated or pasteurized, and then a lactic starter culture is added. After a period of about an hour to allow the culture to develop, rennet is added and the curd, allowed to form. The curd is then cut into small pieces, and the mass "cooked" at a temperature a little over 100°F. After the whey is drained off, the slabs of curd are piled and turned and finally salted and pressed. The cheese is then aged or ripened by keeping it under controlled, moderate temperatures. During the ripening processes, enzymatic reactions occur producing desirable and characteristic flavors in the cheese. Proteolytic activity during the ripening of cheese is extensive. This proteolysis is in part due to enzymes produced by the lactic starter cultures and in part by the continuing, slow proteolytic action of the rennet-coagulating enzyme system.[292] The proteolytic breakdown in cheese results in production of peptides and free amino acids.[152]

A search for substitutes for rennet for milk coagulation in cheesemaking has been underway for many years. In certain countries, such as India, ritual requirements make a "vegetable" rennet desirable. Also, because the source of commercial rennet is the stomach mucosa of young calves, rennet has become increasingly expensive and in short supply. While there is a continuing increase in demand for cheese and hence rennet, there is at the same time a decrease in the number of young calves being slaughtered and thus a shortage of rennet. Sardinas[204] recently reviewed new sources for rennet.

Milk can be clotted by most of the known proteases, but with the exception of rennin and pepsin, proteolysis proceeds beyond the slight rise in nonprotein nitrogen (NPN) observed with rennet. In an investigation with eight different proteases, Tsugo and Yamauchi[248] reported that rennin and pepsin (EC 3.4.23.1) produced the largest amount of curd after 15 min and no loss of yield after 60 min in contact with the whey. All the other proteases produced less curd, and continuing proteolysis led to further loss on holding in the whey.[248] Continued proteolysis also had a profound effect on the strength of the curd. Rennin gave the firmest curd, while pepsin and protease from *Pseudomonas myxogenes* gave firmer curds than the other enzymes. Pepsin has, in fact, come into commercial use in combination with rennet for cheesemaking. There are differences in the behavior of the two enzymes, but pepsin, when properly used, has proved to be a satisfactory extender for rennet. The effect of rennet extract and a commercial pepsin preparation has been followed in great detail throughout the entire production period of cheddar cheeses by Melachouris and Tuckey.[152] The amount of NPN formed during the ripening of cheeses was slightly greater with rennet.

Veringa reported earlier work in India where enzymes from *Ficus carica* (ficin) and *Streblus asper*, a wild shrub, yielded curds only slightly less in weight than rennet curd. The curd made with ficin was slightly softer than the rennet curd, and that with the *S. asper* enzyme was much softer.[264] Of all plant enzymes, ficin has been used more often than any of the other proteases for cheesemaking.

Recently, research efforts to obtain rennet substitutes have shifted to microbial sources. Milk-clotting enzymes for cheesemaking (fungal acid protease, EC 3.4.23.10) from the fungi *Mucor pusillus, M. miehei,* and *Endothia parasitica* have been patented[13,40,203] and are currently marketed. Preparations from all three organisms are now used in many countries for making a variety of cheeses. Proteolysis by the currently used fungal proteases and calf rennet has been compared.[262] These new microbial enzyme preparations, particularly those from *M. miehei*[188, 229] compare favorably with rennet in cheesemaking with respect to curd yield and firmness. There have been reports, however, that some of them lead to more bitterness in young cheese. The bitterness may remain, increase, or decrease, during prolonged aging of the cheese, depending upon the enzyme preparation and conditions of use and aging. Nevertheless, suitable "microbial rennets" are now widely accepted in the cheese industry, and it is estimated that 50% of the rennet market has been taken over by microbial products during the past 4 years.[226]

Immobilized coagulant enzymes, chemically bonded to resins or glass beads, have been suggested to help alleviate the rennet shortage. Treatment of milk through a column of fixed enzymes at low temperatures would allow their repeated use and permit different enzymes to be employed for clotting and for curing. No commercial applications of this idea have been developed, however.

One of the greatest uses of microbial proteases is in baking bread and crackers. The unique and variable properties of bakery doughs are due largely to the insoluble or gluten proteins of flour. When wheat flour is mixed with water, a complex dough system is formed. Because of the hydrated gluten, the dough has extremely viscous flow properties and is therefore said to be extensible. However, it is at the same time elastic and resists deformation. With protease enzyme supplementation, gluten hydrolysis can be controlled to achieve better properties in the dough.

Cereals normally contain a very low level of protease. Protease content does somewhat increase when wheat or barley is malted. However, malts are still poor sources of protease, and cereal proteases provided by malt supplementation of flour have relatively little effect on dough properties.

The special viscoelastic properties of doughs are provided by the wheat gluten. The so-called strength of flours is determined by the quantity and quality of the gluten. Strong flours result in doughs which have good tolerance to extensive mixing. The use of oxidizing agents — bromates, iodates, and peroxides — as dough improvers has a strengthening effect, while reducing agents have a weakening effect.[179] The effect of proteolytic enzymes resembles in some respects that of reducing agents. Mixing time and viscosity of doughs are reduced, and doughs are more pliable and extensible. In practice, combined use of oxidizing agents and proteolytic enzymes has shown considerable benefits. In present-day baking, flours and doughs which have not been treated with oxidizing dough improvers are very rare. Probably, fungal protease derived from *Aspergillus oryzae* (*Aspergillus* protease, EC 3.4.21.15) is added by the baker for more than two thirds of the bread baked in the U.S. Recently, the use of bromelain and papain has been permitted in bread baking, but data on their effectiveness are not available.

A major benefit of adding fungal protease to sponge doughs is the marked reduction in mixing time.[45,192] Proper use of fungal protease reduces mixing time by about one third and improves the handling properties of doughs. Protease increases the extensibility of doughs and is, thus, valuable in controlling the pliability of doughs, eliminating buckiness, and ensuring proper machinability. Improved loaf characteristics, including better volume, greater symmetry, and improved grain, texture, and compressibility are also observed. These qualities probably come as secondary benefits resulting from the greater extensibility and better machinability; the primary benefits are, unquestionably, reduction in mixing time and better handling properties of the dough.[192]

The action of protease seems to be limited to the mixing and fermentation periods, since very low inactivation temperatures (50 to 60°C) preclude any appreciable action during baking.[156] The initial process in dough formation must be hydration of the flour particles to bring out the cohesive properties of gluten. The amount of mixing required depends upon the gluten quality, more mixing development being required with strong flours than weak ones. Bucky doughs from very strong flours may require excessively long mixing times for proper development. By judicious use of fungal protease supplementation, the progressive softening induced by the enzyme can effectively substitute for a large part of the mixing development normally required.[45] The baker can, thus, regulate mixing time and adapt flours of differing qualities to his standard bakeshop practices. The optimum amount of protease required is quite critical. It depends not only upon flour quality but also on bakeshop conditions, particularly mixing and fermentation time and temperature. Excess protease activity, however, is very undesirable since it results in doughs becoming progressively more sticky and slack with poor rather than better loaf characteristics.[112]

Fungal protease preparations from different strains of *A. oryzae* or from the same strain cultivated in different manners vary considerably in their effect on the baking performance of flours.[156,192] The modified Ayre-Anderson hemoglobin method is the generally accepted procedure for standardizing the level of fungal protease for breadmaking.[8] It has been shown, however, that different fungal preparations of equal hemoglobin assay values may vary widely in their effect on doughs.[192] The only satisfactory way for evaluating different protease preparations is by dough handling and baking test methods. The strain of the organism and the conditions of manufacture must be rigidly controlled, as they are, of course, by enzyme manufacturers to insure consistent baking performance. Under these conditions, the hemoglobin assay procedure is a useful and reliable quality control method for standardizing an individual product.

The largest application of fungal protease supplementation has been in conventional breadmaking. It is also extensively used in the brew-conventional process. In this process, the preferment or brew contains all of the water and water-soluble ingredients; that is all the ingredients except flour and shortening. After a short 2- to 4-hour fermentation time, the flour and shortening are mixed with the brew, and the doughs, processed through conventional makeup equipment. Since the flour is not exposed to lengthy fermentation, higher levels of fungal protease are required to achieve the desired mellowing effect on the doughs.

Protease originally found little use in the continuous breadmaking processes employing liquid preferments or brews with which flour and shortening are mixed continuously with the very soft dough being deposited directly into baking pans for proofing and baking. The application of protease in this process subsequently became feasible, because modifications in the original process, such as the inclusion of up to 60% of the total flour in the preferment, have been made. It appears that use of enzymes may be effective in increasing output through the continuous mixer with lowered power requirements.[82, 267]

The use of microbial protease in cracker, biscuit, and cookie doughs has become widespread. Both bacterial (subtilisin, EC 3.4.21.14) and fungal proteases are used in these bakery products.

α-Amylase present in microbial protease preparations contributes partially to the advantages observed. When bacterial protease is employed, the thermal stability of the amylase creates no problem as it does in bread making. The internal temperature in crackers and cookies rises rapidly in the oven and reaches much higher levels than in bread. The amylase is, therefore, so quickly inactivated during the baking process that the sticky crumb problem is eliminated. According to unpublished reports, use of microbial protease in cracker doughs greatly decreases the mixing and holding time necessary to develop the gluten, promotes uniform rolling of the dough without stickiness or "bucking" at the edges, increases spread, gives exceptionally even baking with uniform and improved browning, produces more open grain with enhancement of tenderness and flavor, and reduces shortening requirements. Similar favorable results are obtained in the baking of sugar and vanilla wafers and other cookies.

Another major application for proteases is in the brewing industry. A problem which has long been recognized in brewing is the tendency for an undesirable haze to develop in beer and ale when they are cooled. The chill haze consists of a very fine precipitate which forms when beer is cooled below about 10°C. The formation of haze is accelerated by the action of light, heat, and oxygen or traces of copper or iron. Degree of haziness increases rapidly with extended storage or higher storage temperatures. The chill haze has a variable composition but generally contains protein, tannin, and carbohydrates.[159] Since protein is a necessary component of chill haze, addition of protease enzymes for its prevention was the logical approach for solving the chill haze problem. Use of proteolytic enzymes has consequently become the dominant method for chillproofing beers.

Almost any protease active at the normal pH of beer (about 4.5) — papain, pepsin, ficin, bromelain, and fungal and bacterial proteases — can be used for chillproofing. Papain, alone or in combination with other proteases, is most commonly employed. Enzymes are added during cellar operations, amounts used vary depending upon the raw materials, processing method, anticipated length of storage, and storage temperature. About 8 ppm of papain seems about an average level of addition to digest enough of the protein to prevent haze formation.[272] Foaming properties of beer are not adversely affected by nominal levels of chillproofing enzymes.

There is no good correlation between enzyme potency as determined by the conventional methods for protease assay, using hemoglobin, casein, gelatin, or other proteins as substrates, and the performance of the enzyme in chillproofing. It is, therefore, necessary to evaluate enzymes for chillproofing by testing them at various levels on untreated, fresh beer as substrate. Although quite laborious, methods have been developed and published for conducting such evaluations.[243, 279] Once an evaluation has been made on beer to determine the proper use level for a particular enzyme, this enzyme can then be used on the basis of its assay potency determined by a conventional method. For example, if 1 g of papain assaying 1000 units per gram is required per barrel, 2 g of papain assaying 500 units per gram would be necessary. However, with another enzyme, such as bacterial protease assaying 1000 units per gram by the same method, no judgment can be made from this assay value as to the required amount, which can only be determined by actual evaluation with beer.

Considerable interest has occurred in brewing circles by the possibility of using immobilized papain in chillproofing beer.[140] The hoped-for advantages include economy (reuse of enzyme) and purity (no additives). No commercialization has yet occurred.

Still another use for proteolytic enzymes is in tenderizing meat, particularly beef. There is a great deal of variation in toughness (or its opposite, tenderness) of beef depending upon the age of the animal and the part of the carcass from which it came. As discussd in a previous section of this chapter, beef is usually aged to improve tenderness by action of the natural enzymes present in the meat. Except for the choice grades of beef, additional enzymatic tenderization is usually necessary or desirable.

A great deal of work has been published on the effect of different proteolytic enzymes on different beef proteins as determined by chemical and histological methods.[67,150,273] Concentrations of enzymes employed in this work have of necessity been many times higher than those used for the practical tenderization of beef. Tenderization of meat can be recognized organoleptically, long before extensive hydrolysis can be detected by chemical analysis or before any structural changes can be shown histologically under the microscope. It is quite clear from this work that almost all proteolytic enzymes have some tenderizing action on meat and that the several available proteases have differing actions on the various meat proteins. The plant proteases, papain, bromelain, and ficin, have considerable effect on connective tissue, mainly collagen and elastin, and show some action on muscle fiber proteins. On the other hand, microbial proteases, bacterial and fungal, have considerable action on muscle fibers but only a slight effect on collagen and none on elastin fibers. Advantage has been taken of these differences in formulating meat tenderizers containing combinations of enzymes, thus taking advantage of the hydrolyzing effect of plant proteases on connective tissue and the high activity of fungal proteases in hydrolyzing muscle fiber proteins.[253]

Meat tenderizers are applied in the household, in restaurants, by distributors of retail cuts of meat in frozen form, and by meat packers. One report indicated that about one third of the papain sold in the U.S. (total import about 500,000 lb/year) was used by consumers for tenderization, and 500 million pounds of beef (about 5% of the total) were tenderized by meat packers.[10] Application of proteases in tenderizing meat presents problems. It requires uniform distribution of the enzyme in low concentration to give limited proteolysis to a particular degree of tenderness and no farther. Too high an enzyme concentration or too long a period of treatment may result in overtenderization, mushiness, and even formation of an undesirable hydrolyzate flavor.

Meat tenderizing by the ultimate consumer is usually accomplished by sprinkling a meat cut with a powdered enzyme preparation or by immersion in a solution of the enzyme. Often, the meat is pierced with a fork over its entire surface to improve penetration of the enzyme into the meat. In this way, tenderization of small steaks is easy to control since the meat is cooked immediately following enzyme treatment. Similarly, distributors dip steaks into enzyme solutions and market them as prepackaged, frozen steaks, which are broiled directly from the frozen state. Most of the tenderizing action occurs during the brief period during cooking before the temperature in the meat rises to the point where the enzyme is inactivated. A typical composition of the material used in surface treatment of beef cuts is 2% commercial papain or 5% fungal protease, 15% dextrose, 2% monosodium glutamate, and salt.[263] It is, also, possible to use interleaving sheets impregnated with papain in packaging cuts of meat, then freezing, in order to apply a predetermined amount of enzyme to each cut of meat.[9]

The most recent development in enzymatic meat tenderization is injection of proteolytic enzyme solutions into the vascular system of cattle before slaughter (ante-mortem). The vascular system effectively distributes the proteolytic enzyme throughout the tissues. Ante-mortem tenderization is now practiced on a fairly large scale by meat packers using low concentrations of specially purified papain, equivalent to 5 to 30 ppm of commercial papain based on the total weight of the animal.[27] Cows and low-grade steers require the highest concentration of enzyme, and prime heifers, the lowest.[80] The meat must be refrigerated until used to avoid overtenderization. This ante-mortem pretenderizing process permits the production and sale of a much higher percentage of tender cuts from all grades of beef, particularly the lower grades. This method also enables tenderization of large roasts, which was not possible with surface application. The only real problem encountered seems to be possible overtenderization of highly vascular organs, such as the liver, which may develop mushiness and hydrolyzate flavors.

Although the requirement for thorough cooking of chicken or pork at high temperatures usually makes enzymatic tenderization unnecessary, some attention has been given to tenderization of poultry and hams. The tenderness of chicken meat can be increased by suitable applications of proteases, the results being most dramatic with old roosters and hens. One method is ante-mortem injection into the humoral vein of birds;[27] another is injection into the peritoneal cavity 6 to 12 hr before slaughter.[97] Immersion in solutions of papain or fungal protease have improved tenderness in freeze-dehydrated chicken meat.[228] A process has also been patented for enzymatic tenderization of hams, employing bacterial protease.[205]

Several minor applications of proteases in food or feed industries, such as production of fish solubles, may be noted.[76, 85] In order to obtain drying oils from them, menhaden and similar fish are cooked and pressed, the residue, dried to fish meal, and the liquid portion, centrifuged to recover the oils. The centrifugate is concentrated to about 50% protein; the viscous solution, sold as a feed rich in proteins and vitamins. Undesirable solidification of the fish solubles is prevented by treating the liquid with a crude bacterial protease prior to concentration. The fish can also be ground and treated with proteolytic enzymes before pressing.[142]

Enzyme hydrolysis of renderer's meat scrap with fungal protease, papain, or bromelain is superior to acid hydrolysis, yielding recovered protein fractions comparable in amino acid composition with the proteins of meat and bone meal. These fractions can be used as a feed ingredient in, for example, pig starters.[46, 51]

Protein hydrolyzates are prepared most commonly by acid hydrolysis of soybean protein, wheat gluten, or milk proteins.[29] However, a number of enzyme hydrolyzates are marketed commercially as flavoring agents. Enzyme hydrolyzates can be prepared from almost all available proteins, with the advantage over acid hydrolysis that tryptophan is not destroyed. Enzyme hydrolyzates of vegetable proteins have bland and pleasant flavors. Bitter peptides are often produced by enzymatic hydrolysis of milk and other proteins. The bitter tastes of such peptides can be removed by the action of caroxypeptidases or aminopeptidases.

Industrial, nonfood uses for proteases are even more numerous and varied than the applications in food industries. These important uses are given in Table 9 as a matter of interest but will not be further discussed.

Lipases

Lipases are widely distributed in nature, in animals, plants, and microorganisms. In spite of their great natural importance in digestion and metabolism of fats, lipases have achieved only limited industrial application. The most important lipases from the standpoint of industrial uses are pancreatic lipase, animal pregastric lipases, and the lipases of certain microorganisms. Some uses are shown in Table 10.

True lipases are enzymes which hydrolyze insoluble fats and fatty acid esters occurring in a separate, nonaqueous phase.[35] Rate of lipase action is dependent upon the surface area of the emulsion upon which it acts.[63] Most lipases have considerable specificity, with regard to the type of fatty acid, chain length, and degree of saturation and with regard to the position of the fatty acids, as 1- or 3-outer chains or the 2-inner chain of glycerides. Pancreatic lipase, shows a preference for the 1- or 3-position, and its relative rates of hydrolysis of tri- di-, and monoglycerides are in that order.[44] Milk lipase has similar specificity, as do many but not all microbial lipases.[4,110]

Major application of lipase activity is made in the dairy industry. Controlled lipolysis is necessary in cheesemaking in order to develop characteristic flavors. A high degree of lipolysis is particularly important in Italian type cheeses. Traditionally, Italian cheeses were made using rennet paste as the coagulant. Such paste is made by drying the entire stomach of calves, kids, or lambs, including the milk contents. The characteristic piquant flavor of Italian cheeses made with rennet pastes cannot be obtained with rennet extracts and is due to the presence of lipase enzymes which are derived from enzyme-secreting glands at the base of the tongue of the animals.[77] These enzymes, extracted from the excised glands of calves, lambs, and kids, are now commercially available. They are variously called oral lipases, oral glandular lipases, or pregastric lipases.[196,221] The lipolytic activity of the pregastric lipases is of the same degree and kind as that of rennet pastes from the same species.[91,92,141] Pregastric lipase from calf, kid, and lamb are manufactured for and standardized according to their ability to lipolyze butterfat. They produce a specific and reproducible ratio of free fatty acids as a result of the lipolysis of butterfat. Specific enzyme action on butterfat results in a characteristic flavor for each animal species. Pregastric lipases are now widely used, along with ordinary rennet for coagulation, in the production of Italian type cheeses.

Table 9
INDUSTRIAL NONFOOD USES FOR PROTEINASES

Industry	Use	Enzymes employed
Animal feeds	Ingredient	Bacterial, fungal
Leather	Bating of hides	Bacterial, fungal
Unhairing of hides	Bacterial	
Laundering, dry cleaning	Spot removal	Bacterial
Textiles	Degumming, desizing	Bacterial
Photographic	Film stripping	Bacterial, plant
Medicine	Digestive aid	Fungal, bacterial, plant
Medicine	Wound debridement	Bacterial, plant
Medicine	Relief of inflammation, bruises, and blood clots	Bacterial, fungal, plant

Table 10
SOME USES FOR LIPASES

Flavor of dairy products — ice cream, cheeses, margarine, lipolyzed cream
Flavor in chocolate confections
Improvement of egg white whipping properties
Enzymatic drain cleaners and waste treatment
Digestive aids

Efforts to employ pancreatic, plant, or fungal lipases have been unsuccessful since they produce different ratios of free fatty acids giving atypical flavors, even undesirable, soapy or rancid flavors.

Pregastric lipase also has wide application in the food industry for treating butterfat-containing products to produce a wide variety of food flavors.[56] Butter oil, cream, half and half, condensed milk, reconstituted whole milk powder, or fresh milk are suitable substrates. There are now commercially available dairy-flavor products produced by such pregastric lipase treatment. Lipolyzed butterfat emulsions are used to give enhanced butter flavor to margarines, shortenings, popcorn oils, bakery products, vegetable oils, and confections. Lipolyzed cream and lipolyzed cultured cream are employed in candies and confections, baked products, margarines, snack coatings, and butter sauces. Modified butterfat-containing products are used to enhance the flavors of cheese dips, cheese sauces, cheese soups, cheese dressings, chocolates, confections, gravies, macaroni and noodle dishes, baked products, etc.

Of particular interest are the significant contributions to the flavor of chocolate confections by use of dairy products. Free fatty acids make appreciable contributions to the flavor of milk chocolate, butter creams, caramels, and toffee. Such flavors are obtained by use of cultured butter or more conveniently by use of pregastric lipase-modified, butterfat-containing products. Low levels of free fatty acids enhance the flavor of the confection without creation of new flavor notes. Intermediate levels give buttery flavors, and high levels, cheesy flavors.

Pancreatic lipase has a minor use in treating egg whites before drying in order to enhance whipping qualities. Minute amounts of lipids from traces of yolk have an adverse effect on whipping properties and are removed by the lipase treatment.[50]

Pancreatic lipase also has use in digestive aids and in the therapy of such diseases as cystic fibrosis in which patients have bulky, fatty stools. Lipase of pancreatic or microbial origin is an indispensible ingredient in enzymatic drain cleaner preparations, since fats, along with other food ingredients, are major contributors to stoppages in drain pipes and traps in sanitary plumbing systems.

A potential use for lipase is the manufacture of gelatin from bones which substitutes enzymatic degreasing for solvent extraction.[135] The preferred enzyme is derived from *Rhizopus arrhizus* var. *delemar.* This lipase is similar in action to pancreatic lipase. Commercial animal lipase is not suitable for this use, since it contains protease which would reduce the yield of gelatin.

Oxidoreductases

Oxidoreductases constitute a large group of enzymes catalyzing electron transfer through a series of mechanisms. These enzymes tend to be intracellular, and several are important in food processing. Many deteriorative changes are due to the action of enzymes of this group naturally present in foods. These changes include enzymatic browning (polyphenol oxidase), bleaching (lipoxidase), destruction of ascorbic acid (ascorbic acid oxidase), and oxidative flavor deterioriation (peroxidase). Also in this group are certain unique enzymes added to foods for various specific purposes, such as lipoxidase, glucose oxidase, and catalase.

Lipoxidase or Lipoxygenase

Hydrolytic lipase activity is undesirable in flour for baking since free fatty acids have a detrimental effect in doughs, but lipoxidase, or lipoxygenase (linoleate:oxygen oxidoreductase, EC 1.13.11.12), has attained considerable importance in baking.[19] Lipoxidase catalyzes oxidation of polyunsaturated fatty acids containing *cis, cis*-1,4-pentadiene groups by molecular oxygen. Soybean meal appears to be the richest source of lipoxidase.[7] Defatted soybean flours are used commercially as the source of the enzyme. About 0.5 to 1.0% of such soybean flour is used in baking to achieve the desired effect. Lipoxidase is used extensively in production of bread for bleaching the natural pigments of flour and producing a very white crumb. Lipoxidase also has an effect on dough-mixing properties of flour as well as on bread structure and flavor. These particular effects of lipoxidase are complex and little understood. Addition of lipoxidase gives stronger doughs. It has been suggested that binding occurs between lipids and wheat gluten and that the effect of lipoxidase during mixing is due to its direct action in forming peroxides and indirect effect in catalyzing oxidation of sulfhydryl groups in gluten.[49] The presence of lipoxidase in sponge doughs gives a characteristic, desirable nutty flavor. It is now being used commercially by several plants using the brew continuous mix process of breadmaking in order to obtain the desirable nutty flavor.[123]

Although useful in baking, lipoxidase has three detrimental effects in certain foods:[19,282] 1. There may be destruction of essential fatty acids, linoleic, linolenic, and arachidonic acids, 2. the free radicals produced may damage other components, including vitamins and proteins, and 3. there may be development of off-flavor and -odor.

Glucose Oxidase

A number of fungi produce glucose oxidases (β-D-glucose:oxygen 1-oxidoreductase, EC 1.1.3.4) which catalyze the reaction of glucose with molecular oxygen forming gluconic acid.

Most extensively studied have been the glucose oxidases produced by *Penicillium notatum*,[116-118] *P. amagasakiense*,[134] and *Aspergillus niger*.[79,177,235,252,255] Enzymes from each of these organisms have about the same molecular weight (about 150,000), the same isoelectric point (4.2 to 4.3), and the same pH optimum (5.5 to 5.8) and contain two flavin adenine dinucleotide units per mole of enzyme.

Glucose oxidase shows a high specificity for β-D-glucose.[1,79,177,235] Other hexoses, pentoses, or disaccharides are not oxidized or are oxidized only at negligible rates.

Table 11
SOME USES FOR LIPOXIDASE

Whitening of bread by oxidation of carotene in flour
Improvement of dough properties and flavor in baking
Preparation of peroxidized oils for flavoring

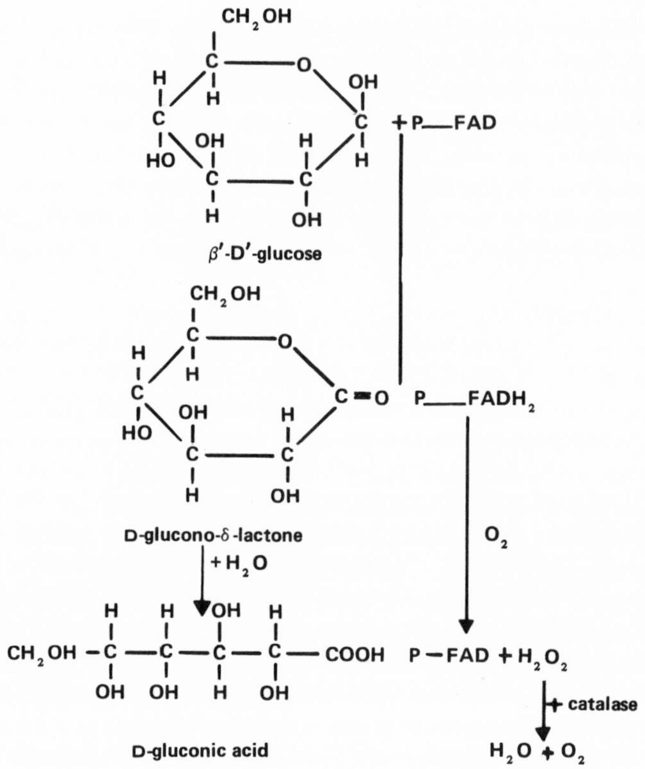

FIGURE 5. Overall reactions of the commercial glucose oxidase catalase system.

Pazur and Kleppe reported that with their purified *A. niger* preparation, mannose and galactose gave 1% and 0.5% of the rate of glucose oxidation.[177] For all practical purposes, purified glucose oxidase reacts only with glucose; thus the enzyme has become a valuable tool for analysis as well as for other purposes.

The important commercial glucose oxidase preparations are derived from *A. niger*[252,255] Usual commercial preparations also contain catalase, which is advantageous in most of the applications, and traces of various carbohydrases. Where carbohydrases are undesirable, purified preparations free of these are also available. The overall reactions of the commercial glucose oxidase-catalase system are shown in Figure 5. In industrial applications, the glucose oxidase system is used both to remove glucose and to remove oxygen.

The most important application of glucose oxidase for the removal of glucose is from egg albumin and whole eggs prior to drying. Powdered egg products are unstable and deteriorate during storage due to the nonenzymatic browning reaction between glucose and proteins. The first evidence of this change is a loss of functional properties, such as whipping properties and foam stability. On longer storage, discoloration, off-

Table 12
SOME USES FOR GLUCOSE OXIDASE

Removal of glucose from eggs before drying
Removal of oxygen from beverages — beer, soft drinks, and wines
Removal of oxygen from canned food products — dry or dehydrated foods
Removal of oxygen from mayonnaise
Preparation of gluconic acid
Analyses for glucose

flavors, and loss of protein solubility occur. The best method for stabilizing dried egg products has been found to be removal of glucose before drying; the most satisfactory method is by glucose oxidase.[21,210] In commercial processing of egg whites, pH is adjusted by addition of the required amount of citric or hydrochloric acid to about pH 7.4. For egg yolk or whole, liquid egg, no pH adjustment is necessary. Enzyme is then added, and the batch, held at about 90°F. Throughout the incubation period, gentle agitation is maintained and excess oxygen is supplied by continuously or periodically adding hydrogen peroxide. Completion of desugaring is determined by a negative test with Somogyi reagent, after which the egg product is dried. The time of desugaring can be controlled by the concentration of enzyme employed and in different operations may vary from about 2 to 7 hr, about 3 hr generally preferred.

Although presence of hydrogen peroxide during the desugaring process has a bacteriocidal effect, the requirement for low bacterial count and freedom from *Salmonella* in egg products has led to the introduction of low-temperature (7 to 15°C) desugaring by glucose oxidase followed by pasteurization for yolk and whole egg.[218]

Oxygen is responsible for a wide range of types of food deterioration, chief among which are flavor and color changes. In canned acid foods, oxygen also accelerates can corrosion. While somewhat dependent upon the constituents of the foods with which it is used, glucose oxidase has been found effective in protecting certain foods and beverages by removal of oxygen from the foods and containers. The effectiveness of the enzyme in removing oxygen will vary with each product depending upon the pH, level of glucose, and many other factors which may vary widely.

The presence of oxygen in canned and bottled beer has very deleterious effects on flavor and stability and is minimized, but not eliminated completely, during filling operations. It has been demonstrated that addition of glucose oxidase is very effective in removing residual oxygen from beer.[168,252] Commercial trial packs employing glucose oxidase in several breweries consistently indicate a beneficial effect of the enzyme on beer stability.[195] Taste panels could detect flavor differences between treated and untreated beers and preferred those which had been enzyme-treated. Although some breweries employ glucose oxidase, it is not used by a majority of brewers, due in part to a reluctance to add glucose to beer which does not always contain enough of the sugar to permit effective oxygen removal by the glucose oxidase system.

Glucose oxidase has been investigated in the treatment of wines. Removal of oxygen from unpasteurized apple wine by glucose oxidase prevented growth of microorganisms and off-flavor development.[294] Addition of 0.1% glucose along with glucose oxidase to white wines to prevent browning and flavor changes was quite effective in depleting the free oxygen content within 24 hr.[170] More recent work with numerous types of wines showed variations in the amount of oxygen removed, but in all cases, removal was substantial.[151]

Limited commercial use is made of glucose oxidase to prevent color and flavor changes in bottled or canned soft drinks. Different flavors vary greatly in their susceptibility to changes caused by traces of oxygen remaining in the sealed containers. Glucose oxidase may protect in some cases and have no effect in others. The greatest application for glucose oxidase has been in citrus drinks containing natural juices or oils which are particularly prone to oxidative flavor changes upon exposure to light. The natural cloud stability of citrus drinks seems to be related to the presence of cellulose, hence, the glucose oxidase employed must be very low in cellulase in order to retain the desirable, stable cloud and, at the same time, protect against flavor changes. Glucose oxidase is effective in protecting cans of carbonated beverages against oxidative corrosion and is used in canned soda, for example.[255]

Mayonnaise is an outstanding example of a food product which can be protected against deterioration by addition of glucose oxidase. Mayonnaise is an oil-in-water emulsion, containing much air, packed in glass containers, and exposed to light and ambient temperatures. Despite the best precautions and use of the highest quality ingredients, the product undergoes gradual deterioration, noticeable to the expert after a month and to the general public after 3 or 4 months. Removal of oxygen from mayonnaise by incorporating glucose oxidase has been found to afford significant protection to the product. No detectable change in regard to color, rancidity, or increase in peroxide values occurred in 6 months; whereas, controls without glucose oxidase had faded perceptibly and were rancid as early as the third month.[30] Glucose oxidase has not yet been added to the list of optional ingredients in the standards of identity for mayonnaise, but glucose oxidase is used in some nonstandardized salad dressings. The glucose oxidase preparation must be free of amylase for use in salad dressings which contain starch.

Oxidative deterioration of dry or dehydrated foods, such as whole milk powders, roasted coffee, active dry yeast, and white cake mixes, is frequently a serious problem. It is, of course, not feasible to add glucose oxidase directly to such food items since the enzyme is inactive in the dry state. In order to adapt the enzyme for use with dry foods, a procedure has been devised in which the enzyme, glucose, buffer, filler, and water are placed in a water-impermeable, but oxygen-permeable, packet. When this packet is placed in a hermetically sealed container along with a dry food product, the contents of the packet rapidly take up the oxygen in the container, leaving an oxygen-free atmosphere.[211,217,255] Studies in the use of the glucose oxidase scavenger packets have been reported for whole milk powder and for whole, dry milk or dry ice cream mix but never published.[133, 154]

While the enzyme scavenger packet is an intriguing and unique concept and may find specialized applications, it is not in commercial use, largely due to its relatively high cost and inconvenience in use. After manufacture, packets must be sealed into their containers immediately, since they will rapidly be exhausted if exposed to the atmosphere.

Because of its specificity, glucose oxidase is widely used as an analytical tool for detection and determination of glucose in the presence of other carbohydrates. It is of particular advantage to medical technicians and to food technologists. For example, an important application is in detecting and monitoring diabetic conditions by using it in the form of test strips to detect the presence of glucose in the urine. High-purity glucose oxidase, free of undesirable contaminating carbohydrases, is now available commercially for use in quantitative analysis procedures for glucose.[23]

Galactose Oxidase

Another hexose oxidase of interest is galactose oxidase,(D-galactose:oxygen 6-oxidoreductase, EC 1.1.3.9) a fungal enzyme isolated from *Dactylium dendoides.*[48] The enzyme requires Cu^{++} for activity and oxidizes the 6-position in galactose, forming D-galactohexodialdose and producing hydrogen peroxide in the process. It is active on a wide variety of substances other than galactose,[18] particularly in galactosides. It is not known to have any usefulness in the food industries but is important in analysis for galactose in biological fluids, as in following the disappearance of intravenously administered galactose as a liver function test.

Peroxidase

Peroxidases (donor:hydrogen peroxide oxidoreductase, EC 1.11.17) are widely distributed in plants, animals, and microorganisms. Besides the specific action of peroxidase in deteriorative changes of flavor, previously mentioned in this chapter, it is widely used as an index of pasteurization, blanching, and other heat treatments because of its resistance to thermal inactivation. If peroxidase is destroyed, then it is unlikely any other enzyme systems will survive.

A peroxidase catalyzes the reaction between a peroxide (hydrogen or organic peroxide) and hydrogen donors, such as phenols or amines. The peroxidase most studied and of greatest commercial importance is horseradish peroxidase. Large quantities of purified horseradish peroxidase are used in conjunction with glucose oxidase for detection and determination of glucose. In this use, hydrogen peroxide generated by the glucose oxidase action on glucose is used in the peroxidative reaction to produce a colored reaction product when a suitable substrate is present. The usual substrates are *o*-tolidine or *o*-dianisidine, although other substrates can be used.

Catalase

Catalases (hydrogen peroxide:hydrogen peroxide oxidoreductase, EC 1.11.1.6) convert hydrogen peroxide to water and oxygen. They are obtained for commercial uses from animal (liver), bacterial *(Micrococcus lysodeikticus*), and fungal (*Aspergillus niger*) sources.

Catalase is employed to remove the last traces of hydrogen peroxide in peroxide bleaching. It finds application in the food industry in conjunction with glucose oxidase as previously discussed. Its principal additional use in the food industry, world-wide, is in connection with cold sterilization or preservation of milk by means of hydrogen peroxide.

In many undeveloped countries of the world, cooling facilities are not available on farms, and prompt heat pasteurization is not technically possible. In these circumstances, hydrogen peroxide treatment is the only feasible method for preserving milk. For such purposes it has been suggested that 1 ml of 33% hydrogen peroxide per liter of milk be added at the farm, followed by another milliliter at the dairy plant. The milk should then be heated to 50°C for 30 min followed by cooling to 35°C and addition of catalase to destroy the residual peroxide.[199]

In the U.S., hydrogen peroxide treatment, instead of heat pasteurization of milk, is permitted for cheesemaking. Hydrogen peroxide treatment is quite effective in reducing the counts of pathogenic microorganisms. It inactivates milk catalase and peroxidase, but lipase, proteases, and phosphatases are not destroyed. Several publications have described the peroxide catalase method for cheesemaking and the quality of the cheeses produced.[200,201,240] Both flash and batch methods may be employed. In the flash method, about 0.03% of 35% hydrogen peroxide is added to the cool milk which is then passed through a plate heat exchanger where it is heated to 52°C for 25 sec. It is then cooled to the setting temperature for the cheese, and an excess of catalase, added to destroy the residual hydrogen peroxide. In the batch method, about 0.06% of 35% hydrogen peroxide is added to the vat of milk at normal setting temperature, and the milk, held for about 20 min, prior to the addition of catalase. After the catalase action, lactic starter cultures are added, and the cheese is made in the normal manner.

Table 13
SOME USES FOR CATALASES

Removal of hydrogen peroxide
- From skin (if spilled)
- In organic syntheses
- In foods, when formed due to light or added for sterilization
- After bleaching of textiles or hair so dyes may be applied

Liberating oxygen
- With glucose oxidase for desugaring eggs
- For sterilization
- For microbial growth
- For production of foams (rubber, plastics, and cement)
- For baked goods

For economic reasons, cheddar cheese is not manufactured by the hydrogen peroxide catalase treatment, but about half of the Swiss cheese made in the U.S. is produced with this treatment because of the improved body characteristics and reduction of undesirable microorganisms.

Flavor Enzymes

The use of certain enzymes in the formation of food flavors has previously been mentioned, for example, the use of pregastric lipases in dairy products. A recent important development is the availability to the food industry of the specific flavor enhancers, 5′-inosine monophosphate (IMP) and 5′-guanosine monophosphate (GMP). These compounds accentuate meaty flavors and have application in canned vegetables, sauces, gravies, soup bases, etc. They are produced in large quantities by extraction of yeast ribonucleic acid (RNA) and enzymatic hydrolysis of this RNA to the 5′-nucleotides. The process only became practical when methods for obtaining fungal phosphodiesterases from suitable strains of *Penicillium* and *Streptomyces* were developed. This work has been reviewed[167] with several good review papers providing the background for the use of enzymes in the production of 5′-nucleotide flavor enhancers.[132,220] The details are beyond the scope of the present discussion.

Considerable interest has been shown in flavor improvement of foods by direct enzymatic reactions. Many fresh foods contain volatile flavor compounds which have been investigated by gas-liquid chromatogrphic methods. During processing, for example, by dehydration, all or part of the volatile constituents may be lost with a resultant change of flavor. However, the foods may also contain nonvolatile flavor precursors. By treating flavorless, processed food with an appropriate enzyme, the flavor precursors may be converted to flavor compounds, thereby improving or restoring the flavor of the food. These discoveries are of importance in our understanding of development of flavors in foods. They may lead to practical applications for restoration of flavors in processed foods, but such enzymatic processes are not yet in commercial use. Several reviews have covered present knowledge in this interesting and promising field.[41,93,95,126,206]

Most of the reported work has been done with vegetables or fruits having strong and readily recognized flavors. For example, members of the cabbage family (*Crucifera*), cabbage, mustard, horseradish, and water cress, have been studied extensively. The flavors of these plants are derived mainly from their mustard oils (isothiocyanates), which are produced from thioglycosides, such as sinigrin (black mustard) or sinalbin (white pepper), by the action of thioglycosidases (sinigrinase or myrosinase). The hydrolytic reaction catalyzed by these enzymes produces organic isothiocyanates, sugar, and sulfuric acid salt from the thioglycosides. Allylisothiocyanate seems to be a major component, with others, such as butenyl-, *n*-butyl-, methyl-, and methylthiopropylisothiocyanates, also contributing to the flavor.[20] Dimethyl sulfide and related sulfides from *S*-methylcysteine sulfoxide also appear important in cabbage flavor.[57] Enzymes which have been studied with the cabbage family have been obtained by extracting fresh plant materials. Enzymes from cabbage and mustard are somewhat interchangeable. When the mustard enzyme is employed with dehydrated cabbage, the restored flavor is similar to, but not identical with, the flavor of fresh cabbage. The cabbage enzyme produces a pleasant, fresh cabbage flavor initially but on longer action a strong mustardlike flavor may develop.[144]

Among the flavor precursors of onions are *S*-substituted cysteine compounds. Sulfoxidase enzyme systems convert these to sulfenic acids, ammonia, and pyruvic acid.[207, 209] The unstable sulfenic acids decompose further to sulfur compounds responsible for the biting, lacrymatory effect, characteristic of onions. These enzymes have been isolated from onion and from an ornamental shrub *Albizzia lophanta*.[96, 208]

Similar flavor enzymes have been found with certain fruits. An enzyme, produced from the white center core of raspberries and acting upon a raspberry substrate freed from volatile flavors by steam distillation, produced characteristic gas chromatographic peaks and the odor of ripe raspberries.[280] Some of the peaks, but not all, were also produced by two commercial enzyme preparations designated as a cellulase and a β-glycosidase.

An enzyme preparation from fresh bananas restored much of the fresh banana odor to a heat-processed banana puree. Potential precursors, particularly pyruvic acid, valine, and oleic acid, shortened the time required for development of the fresh fruit flavor; however, such precursors not only accelerated formation of fresh aroma but also the formation of undesirable odors characteristic of overripe bananas.[98]

In most of the considerable amount of work done on enzymatic flavor improvement, enzymes have been isolated from fresh vegetables and fruits by extraction with water and precipitation by an organic liquid, such as acetone. In many instances, the natural flavor of the processed vegetables or fruits was restored by such enzyme preparations. In others, an off-flavor not characteristic of the particular food product developed. Schwimmer has tabulated the rather extensive results with cabbage, horseradish, broccoli, peas, string beans, carrots, tomatoes, and onions.[206] It has been claimed that the principle of enzymatic flavor restoration can be applied to most processed foods.[96] It may not be necessary to actually isolate the flavor enzyme in some cases. It has been claimed that the flavors of dehydrated cabbage, potatoes, tomatoes, carrots, grapefruit, or orange juice can be restored by the addition of a small portion of the fresh food after freeze-drying the dehydrated foods, in which the natural flavor-producing enzymes have been destroyed during the blanching or dehydration steps.[158]

Disadvantages which have prevented any significant commercial acceptance of flavor enzymes to date seem to be their lack of convenience in use and the problem of controlling the extent of reaction. For example, to restore the flavor to practically tasteless, dehydrated cabbage, the enzyme would have to be added by the consumer as he reconstitutes and rehydrates the product. Considerable care is necessary as to temperature and time in order to attain the desired enzyme activity and, thus, obtain the desired flavor change.

Several conditions must be met for flavor restoration to be practical. Essential flavor precursors must be present, stable, and available to enzyme action in the processed food. The necessary enzymes must be found and economically produced from the fresh foods or other sources. The enzyme actions must be simple hydrolytic or oxidative reactions; they must also be controllable and convenient to use. These requirements have been met in part in some vegetables and fruits where flavor depends on hydrolytic or oxidative production of sulfides, isocyanates, alcohols, and carbonyl compounds and where strong, readily recognized flavors are produced as in cabbage, onion, or some fruits. More subtle changes of flavor by enzymatic action still require attention. Flavor enhancement or production by enzymes is a challenging field; research is very difficult and expensive, but progress can be expected in the future.

Another, perhaps more promising, application of enzymic flavor modification is conversion of specific, undesirable flavor compounds to tasteless species. Microbial enzymes may do this without heat or drastic chemical treatment of food or drink. Debittering citrus juice, previously discussed above, is an example. Perhaps, flavor might also be modified using enzyme specificity to change some components of the flavor mix without affecting others.

NEW ENZYME APPLICATIONS AND NEW ENZYME PRODUCTS

Although major progress is being made in expanding the industrial applications of enzymes, only a limited number of them, as is apparent from previous discussion in this chapter, have, to date, attained commercial significance, even though the number of individual enzymes produced by living cells is in the thousands. The literature is replete with suggestions for new uses of known enzymes, mostly in applications to products other than foods. Removing dental plaque with dextranase (1,6-α-D-glucan 6-glucanohydrolase, EC 3.2.1.11) in toothpaste,[288] eliminating hair with keratinase (EC 3.4.99.11),[164] solubilizing cold tea solids with tannase (tannin acylhydrolase, EC 3.1.1.20),[212] and producing valuable organic compounds, such as L-aspartic acid, L-malic acid, and 6-amino-penicillanic acid, a precursor of semisynthetic penicillins,[226] are but a few.

New industrial applications for old and new enzyme products are continually being developed and put to use. In such new developments, the first requisite, of course, is that an enzyme which can bring about the desired reaction exists in nature. If a new enzyme product is desired, active search usually leads to the discovery of a microbial source for the enzyme. From the standpoint of industrial enzymology, whether such a search is justified is dictated by the potential importance of the reaction.

In the search for a new enzyme, the best source must be intelligently sought from appropriate microorganisms by suitable screening methods. Having found a source organism, extensive laboratory work is required to develop the best media and conditions for maximum enzyme yield and most suitable procedures for isolating, purifying, and stabilizing the enzyme. Finally, the fermentation and isolation methods must be scaled up through pilot plant to full manufacturing scale. While production methods are being developed, the enzyme must also be characterized as to its specificity, mode of action, pH and temperature optima, and the like; the best procedures for its use, investigated and found. After all these steps have been accomplished, it then becomes possible for the manufacturer to produce, market, and recommend the new enzyme product for appropriate applications. The books by Reed, Whitaker, and Wiseman are excellent general references treating both current and prospective applications for enzymes in food processing and other industries.[190,191,281,283,286]

REFERENCES

1. **Adams, E. C., Mast, R. I., and Free, A. H.**, Specificity of glucose oxidase, *Arch. Biochem. Biophys.*, 91, 230, 1960.
2. **Aitken, H. C.**, Apple juice, in *Fruit and Vegetable Juice Processing*, Tressler, D. K. and Joslyn, M. A., Eds., AVI Publishing, Westport, Conn., 1961, 619.
3. **Albersheim, P. and Killias, V.**, Studies relating to the purification and properties of pectin transeliminase, *Arch. Biochem. Biophys.*, 97, 107, 1962.
4. **Alford, J. A., Pierce, D. A., and Suggs, F. G.**, Activity of microbial lipases on natural fats and synthetic triglycerides, *J. Lipid Res.*, 5, 390, 1964.
5. **Allen, W. G. and Dawson, H. G.**, Technology and uses of debranching enzymes, *Food Technol.*, (Chicago), 29(5), 70, 1975.
6. **Amos, J. A.**, The use of enzymes in the baking industry, *J. Sci. Food Agric.*, 6, 489, 1955.
7. **Andre, E.**, Sur la lipoxidase des graines de soja. Etat actuel de nos connaissances, *Oleagineux*, 19, 461, 1964.
8. **Anon.**, Proteolytic activity of flour, *Cereal Laboratory Methods*, 7th ed., American Association of Cereal Chemists, St. Paul, Minn., Method 22—60, 1962.
9. **Anon.**, Enzyme interleaving sheets, *Food Process.*, 25, 84, 1964.
10. **Anon.**, Pretenderizers sizzle, *Chem. Week*, 98, 33, 1964.
11. **Anon.**, Enzyme Nomenclature, Recommendations (1972) of the International Union of Pure and Applied Chemistry and the *International Union of Biochemistry*, Elsevier, Amsterdam, 1975.
12. **Anon.**, Bright future for high-fructose corn syrup, *Chem. Eng. News*, 54(17), 13, 1976.
13. **Arima, K. and Iwasaki, S.**, Milk Coagulating Enzyme "Microbial Rennet" and Method of Preparation Thereof, U.S. Patent 3,151,039, 1964.
14. **Arima, K., Yamasaki, M., and Ysai, T.**, Studies on pectic enzymes of microorganisms. I. Isolation of microorganisms which specifically produce one of several pectic enzymes, *Agric. Biol. Chem.*, 28, 248, 1964.
15. **Armbruster, F. C.**, Enzyme Preparation, U.S. Patent 3,012,944, 1961.
16. **Ashby, J., Brooks, J., and Reid, W. W.**, Preparation of pure mono-, di-, and trigalacturonic acids, *Chem. Ind.* (London), p. 360, 1955.
17. **Aspinall, G. O.**, Pectins, plant gums, and other plant polysaccharides, in *The Carbohydrates, Chemistry and Biochemistry*, Pigman, W., Horton, D., and Herp, A., Eds., Academic Press, New York, 1970, 515.
18. **Avigad, G., Amaral, D., Bretones, C. A., and Horecker, B. L.**, The D-galactose oxidase of *Polyporus circinatus*, *J. Biol. Chem.*, 237, 2736, 1962.
19. **Axelrod, B.**, Lipoxygenases, in *Food Related Enzymes*, Whitaker, J. R., Ed., *Adv. Chem. Ser.*, 136, 324, 1974.
20. **Bailey, S. D., Bazinet, M. L., Driscoll, J. L., and McCarthy, A. I.**, The volatile sulfur compounds of cabbage, *J. Food Sci.*, 26, 163, 1961.
21. **Baldwin, R. R., Campbell, H. A., Thiessen, R., and Lorant, G. J.**, The use of glucose oxidase in the processing of foods with special emphasis on the desugaring of egg white, *Food Technol.*, (Chicago), 7, 275, 1953.
22. **Barfoed, H. C.**, Enzymes in starch processing, *Cereal Foods World*, 21, 588, 604, 1976.
23. **Barton, R. R.**, A specific method for quantitative determination of glucose, *Anal. Biochem.*, 14, 258, 1966.
24. **Beck, H., Johnson, J. A., and Miller, B. S.**, Studies on the soluble dextrin fraction and sugar content of bread baked with alpha-amylase from different sources, *Cereal Chem.*, 34, 211, 1957.
25. **Beckhorn, E. J., Labbee, M. D., and Underkofler, L. A.**, Production and use of microbial enzymes for food processing, *J. Agric. Food Chem.*, 13, 30, 1965.
26. **Bergemeyer, H. U.**, *Methods of Enzymatic Analysis*, Academic Press, New York, 1965.
27. **Beuk, J. F., Savich, A. L., and Goeser, P. A.**, Method of Tenderizing Meat, U.S. Patent 2,903,362, 1959.
28. **Blakemore, S. M.**, Puree and Method of Making Same, U.S. Patent 3,031,307, 1962.
29. **Blish, M. J.**, Protein Hydrolyzates, in *Kirk-Othmer Encyclopedia of Chemical Technology*, Vol. 11, Interscience, New York, 1953, 212.
30. **Bloom, J., Scofield, G., and Scott, D.**, Oxygen removal prevents rancidity in mayonnaise, *Food Packer*, 37(13), 16, 1956.
31. **Blouin, J. and Barthe, J. C.**, Utilization pratique d'enzymes pectolytiques en oenologie, *Ind. Aliment. Agric.*, 30, 1169, 1963.
32. **Bode, H. E.**, In Situ Dextrose Production in Crude Amylaceous Materials, U.S. Patent 3,249,512, 1966.

33. **Bode, H. E.**, Production and Use of Amyloglucosidase, U.S. Patent 3,249,514, 1966.
34. **Bolaski, W., and Gallatin, J. C.**, Enzymic Conversion of Cellulosic Fibers, U.S. Patent 3,041,246, 1962.
35. **Brockerhoff, H.**, Lipolytic enzymes, in *Food Related Enzymes*, Whitaker, J. R., Ed., *Adv. Chem. Ser.*, 136, 131, 1974.
36. **Burns, J. A.**, Food process instrumentation using immobilized enzymes, *Cereal Foods World*, 21, 594, 1976.
37. **Celmer, R. F.**, 1961, Continuous fruit juice production, in *Fruit and Vegetable Juice Processing*, Tressler, D. K. and Joslyn, M. A., Eds., AVI Publishing, Westport, Conn., 1961, 254.
38. **Chaiet, L., Kempf, A. J., Harman, R., Kaczak, E., Weston, R., Nollstadt, K., and Wolf, F. J.**, Isolation of a pure dextranase from *Penicillium funiculosum*, *Appl. Microbiol.*, 20, 421, 1970.
39. **Chance, B.**, The kinetics of the enzyme substrate compound of peroxidase, *J. Biol. Chem.*, 151, 553, 1943.
40. **Charles, R. L., Gertzman, D. P., and Melachouris, N.**, Milk-clotting Enzyme Product and Process Therefore, U.S. Patent 3,549,390, 1970.
41. **Chase, T.**, Flavor enzymes, in *Food Related Enzymes*, Whitaker, J. R. Ed., *Adv. Chem. Ser.*, 136, 241, 1974.
42. **Chibata, I., Tosa, T., Sato, T., Mori, T., and Matsuo, Y.**, Preparation and industrial application of immobilized amino-acylases, in *Fermentation Technology Today, Proc. 4th Int. Fermentation Symp.*, Society of Fermentation Technology, Japan, Osaka, 1972, 383.
43. **Cochrane, A. L.**, Invertase: its manufacture and uses, in *Production and Application of Enzyme Preparations in Food Manufacture, Soc. Chem. Ind.*, (London), Monograph 11, MacMillan, New York, 1961, 25.
44. **Coleman, M. H.**, Rapid lipase purification, *Biochim. Biophys. Acta*, 67, 146, 1963.
45. **Coles, D.**, Fungal enzymes in bread baking, in *Proc. 28th Annu. Meet. Am. Soc. Bakery Eng.*, American Society of Bakery Engineers, Chicago, 1952, 49, 1952.
46. **Connelly, J. J., Vely, V. G., Mink, W. H., Sachsel, G. F., and Litchfield, J. H.**, Studies on improved recovery of protein from rendering plant materials and products. III. Pilot plant studies on the enzyme hydrolysis process, *Food Technol.*, (Chicago), 20, 829, 1966.
47. **Cook, A. H., Ed.**, *Barley and Malt*, Academic Press, New York, 1962.
48. **Cooper, J. A. D.**, Galactose Oxidase, U.S. Patent 3,005,714, 1961.
49. **Coppock, J. B. M. and Daniels, N. W. R.**, Wheat flour lipids, their role in breadmaking and nutrition, in *Recent Advances in Processing Cereals, Soc. Chem. Ind.*, (London), Monograph 16, Macmillan, New York, 1962, 113.
50. **Cotterill, O. J.**, Effect of pH and lipase treatment on yolk-contaminated egg white, *Food Technol., (Chicago)*, 17, 103, 1963.
51. **Criswell, L. G., Litchfield, J. H., Vely, V. G., and Sachsel, G. F.**, Studies on improved recovery of protein from rendering plant materials and products. II. Acid and enzyme hydrolysis, *Food Technol.* (Chicago), 18, 1493, 1964.
52. **Croxall, W. J.**, Process for Removing Transglucosidase from Amyloglucosidase, U.S. Patent 3,254,003, 1966.
53. **Cruess, W. V., Quacchia, R., and Ericson, K.**, Pectic enzymes in winemaking, *Food Technol.*,(Chicago), 9, 601, 1955.
54. **Dalby, G.**, The role and importance of enzymes in commercial bread production, *Cereal Sci. Today*, 5, 270, 1960.
55. **Dale, J. K. and Langlois, D. P.**, Syrup and Method of Making Same, U.S. Patent 2,201,609, 1940.
56. Dairyland Food Laboratories, Inc., Bulletin No. LMMFP-1, Waukesha, Wis., 1972.
57. **Dateo, G.P., Clapp, R. C., Mackay, D. A. M., Hewitt, E. J., and Hasselstrom, T.**, Identification of the volatile sulfur components of cooked cabbage and the nature of the precursors in fresh vegetables, *Food Res.*, 22, 440, 1957.
58. **deBecze, G. I.**, Enzymes, Industrial, in *Kirk-Othmer Encyclopedia of Chemical Technology*, Vol. 8, 2nd ed., John Wiley & Sons, New York, 1965, 173.
59. **deBecze, G. I.**, Food enzymes, *CRC Crit. Rev. Food Technol.*, 2, 479, 1970.
60. **Demain, A. L. and Phaff, H. J.**, Recent advances in the enzymatic hydrolysis of pectic substances, *Wallerstein Lab. Commun.*, 20, 119, 1957.
61. **Denault, L. J., and Underkofler, L. A.**, Conversion of starch by microbial enzymes for production of syrups and sugars, *Cereal Chem.*, 40, 618, 1963.
62. **Dennis, G. E. and Quittenton, R C.**, Enzymes in Brewing, Canadian Patent 634,865, 1962.
63. **Desnuelle, P. and Savary, P.**, Specificity of lipases, *J. Lipid Res.*, 4, 369, 1963.
64. **Dixon, M. and Webb, E. C.**, *Enzymes*, 3rd ed., Academic Press, New York, 1974.
65. **Doesburg, J. J.**, *Pectic Substances, Phytochemistry*, Vol. 1, Miller, L. P., Ed., Van Nostrand Reinbold, New York, 1973, 270.

66. **Dunlap, R. B.**, Ed., *Immobilized Biochemicals and Affinity Chromatography,* Plenum Press, New York, 1975, 377 pp.
67. **El-Gharbawi, M. and Whitaker, J. R.**, Factors affecting enzymic solubilization of beef protein, *J. Food Sci.,* 28, 168, 1965.
68. **Emert, G. H., Gum, E. K., Lang, J. A., Liu, T. H., and Brown, R. D.**, Cellulases, in *Food Related Enzymes,* Whitaker, J. R., Ed., *Adv. Chem. Ser.,* 136, 79, 1974.
69. **Endo, A.**, Pectic enzymes of molds. V. and VI. The fractionation of pectolytic enzymes of *Coniothyrium diplodiella, Agric. Biol. Chem.,* 27, 741, 751, 1963.
70. **Endo, A.**, Pectic enzymes of molds. VII. Turbidity of apple juice clarification and its application to determination of enzymatic activity, *Agric. Biol. Chem.,* 28, 234, 1964.
71. **Endo, A.**, Pectic enzymes of molds. VIII., IX., and X. Purification and properties of endo-polygalacturonase I, II, III, *Agric. Biol. Chem.,* 28, 535, 543, 551, 1964.
72. **Endo, A.**, Pectic enzymes of molds. XI. Purification and properties of exo-polygalacturonase, *Agric. Biol. Chem.,* 28, 639, 1964.
72a. **Endo, A.**, Pectic enzymes of molds. XII. Purification and properties of pectin-esterase, *Agric. Biol. Chem.,* 28, 757, 1964.
73. **Endo, A.**, Pectic enzymes of Molds. XIII. Clarification of apple juice by the joint action of purified pectolytic enzymes, *Agric. Biol. Chem.,* 29, 129, 1965.
73a. **Endo, A.**, Pectic enzymes of molds. XIV. Properties of pectin in apple juice, *Agric. Biol. Chem.,* 29, 137, 1965.
73b. **Endo, A.**, Pectic enzymes of molds. XV. Effects of pH and some chemical agets on the clarification of apple juice, *Agric. Biol. Chem.,* 29, 222, 1965.
73c. **Endo, A.**, Pectic enzymes of molds. XVI. Mechanism of enzymatic clarification of apple juice, *Agric. Biol. Chem.,* 29, 229, 1965.
74. **Erenthal, I. and Block, G. J.**, High Fermentable Noncrystallizing Syrup and the Process of Making Same, U.S. Patent 3,067,066, 1962.
75. **Ernstrom, C. A.**, Rennin and other enzyme action, in *Fundamental of Dairy Chemistry,* Webb, B. H., Johnson, A., and Alford, J. A., Eds., AVI Publishing, Westport, Conn., 1974, 662.
76. **Faith, W. T., Steigerwalt, R. B., and Robbins, E. A.**, Fish Protein Solubilization Using Alkaline Bacterial Protease, U.S. Patent 3,697,285, 1972.
77. **Farnham, M. G.**, Cheese-modifying Enzyme Product, U.S. Patent 2,531,329, 1950.
78. **Fuleki, T. and Hope, G. W.**, Effect of various treatments on yield and composition of blueberry juice, *Food Technol.,* (Chicago), 18, 568, 1964.
79. **Gibson, Q. H., Swoboda, B. E. P., and Massey, V.**, Kinetics and mechanism of action of glucose oxidase, *J. Biol. Chem.,* 239, 3927, 1964.
80. **Goeser, P. A.**, Tenderized meat through ante-mortem vascular injection of proteolytic enzymes, *Proc. Res. Conf. Res. Counc. Am. Meat Inst. Found. Univ. Chicago,* 13, 55, 1961.
81. **Goldstein, L.**, Use of water-insoluble enzyme derivatives in synthesis and separation, in *Fermentation Advances,* Perlman, D., Ed., Academic Press, New York, 1969, 391.
82. **Gross, H., Bell, R. L., and Redfen, S.**, Use of fungal enzymes with flour brews, *Cereal Sci. Today,* 11, 419, 1966.
83. **Guilbault, G. G.**, *Enzymatic Methods of Analysis,* Pergamon Press, London, 1970, 347 pp.
84. **Hajny, G. J. and Reese, E. T.**, Eds., *Cellulases and Their Applications, Adv. Chem. Ser.,* 95, 470 pp., 1969.
85. **Hale, M. B.** Relative activities of commercially available enzymes in the hydrolysis of fish protein, *Food Technol.*(Chicago), 23(2), 107, 1969.
86. **Halliwell, G.**, Hydrolysis of fibrous cotton and reprecipitated cellulose by cellulolytic enzymes from soil microorganisms, *Biochem. J.,* 95, 270, 1965.
87. **Halliwell, G. and Bryant, M. P.**,The cellulolytic activity of pure strains of bacteria from the rumen of cattle, *J. Gen. Microbiol.,* 32, 441, 1963.
88. **Hansen, V.**, Pilot-versuche zur Dextroseherstellung, *Starke,* 16, 258, 1964.
89. **Hanson, A. M., Bailey, T. A., Malzahn, R. C., and Corman, J.**, Plant scale evaluation of fungal amylase process for grain alcohol, *J. Agric. Food Chem.,* 3, 866, 1965.
90. **Harada, T., Kobayashi, K., and Misaki, A.**, Formation of isoamylase by Pseudomonas, *Appl. Microbiol.,* 16, 1439, 1968.
91. **Harper, W. J.**, Lipase systems used in the manufacture of Italian cheese. II. Selective hydrolysis, *J. Dairy Sci.,* 40, 556, 1957.
92. **Harper, W. J. and Gould, I. A.**, Lipase systems used in the manufacture of Italian cheese. I. General characteristics, *J. Dairy Sci.,* 38, 87, 1955.
93. **Hasselstrom, T., Bailey, S. D., and Reese, E. T.**, Regneration of flavors through enzymatic action, *U.S. Dep. Commerc. Off. Tech. Serv. AD,* 286,640, 285, 1962.
94. **Heady, R. E. and Armbruster, F. C.**, Preparation of High Maltose Conversion Products, U.S. Patent 3,565,765, 1971.

95. **Hewitt, E. J.**, Flavor enhancement review. Enzymatic enhancement of flavor, *J. Agric. Food Chem.*, 11, 14, 1963.
96. **Hewitt, E. J., Hasselstrom, T., Mackay, D. A. M., and Konigsbacher, K. S.**, Natural Flavors of Processed Foods, U.S. Patent 2,924,521, 1960.
97. **Huffman, D. L., Palmer, A. Z., Carpenter, J. W., and Shirley, R. L.**, The effect of ante-mortem injection of papain on tenderness of chickens, *Poult. Sci.*, 40, 1627, 1961.
98. **Hultin, H. O., and Proctor, B. E.**, Banana aroma precursors, *Food Technol.* (Chicago), 16, 108, 1962.
99. **Hurst, T. L. and Turner, A. W.**, Method of Refining Amyloglucosidase, U.S. Patent 3,047,471, 1962.
100. **Hurst, T. L. and Turner, A. W.**, Method of Refining Amylogulcosidase, U.S. Patent 3,067,108, 1962.
101. **Hurst, T. L. and Turner, A. W.**, Method of Refining Amyloglucosidase, U.S. Patent 3,117,063, 1964.
102. **Ingleton, J. F.**, Use of invertase in the confectionary industry, *Confect. Prod.*, 29, 773, 776, 1963.
103. **Ishii, S. and Yokotsuka, T.**, Pectin transeliminase with fruit juice clarifying activity, *J. Agric. Food Chem.*, 19(5), 958, 1971.
104. **Ishii, S. and Yokotsuka, T.**, Clarification of fruit juice by transeliminase, *J. Agric. Food Chem.*, 20(4), 787, 1972.
105. **Ishii, S. and Yokotsuka, T.**, Clarification by pectin lyase and its relation to pectin in fruit juice, *J. Agric. Food Chem.*, 21(2), 269, 1973.
106. **Janssen, F.**, Invertase and cast cream centers, *M. Confectioner*, 40(4), 41, 56, 1960.
107. **Janssen, F.**, Factors affecting invertase action in cast cream centers, *M. Confectioner*, 41(8), 56, 60, 1961.
108. **Janssen, F.**, Composition and production of cordial cherries, *Confect. Prod.*, 2854, 56, 58, 60, 75, 1962.
109. **Janssen, F.**, Invertase. An important ingredient for cream centers, *Candy Ind. Confect. J.*, 121(6), 30, 33, 1963.
110. **Jensen, R. G., Sampugna, J., Parry, R. M., and Shahani, K. M.**, Lipolysis of laurate glycerides by pancreatic and milk lipase, *J. Dairy Sci.*, 46907, 1963.
111. **Johnson, J. A.**, Enzymes in wheat technology in retrospect, *Cereal Sci. Today*, 10, 315, 1965.
112. **Johnson, J. A. and Miller, B. S.**, Studies on the role of alpha-amylase and protease in breadmaking, *Cereal Chem.*, 26, 371, 1949.
113. **Joslyn, M. A., Mist, S., and Lambert, E.**, The clarification of apple juice by fungal pectic enzyme preparations, *Food Technol.*, (Chicago), 6, 133, 1952.
114. **Kathrein, H. R.**, Treatment and Use of Enzymes for the Hydrolysis of Starch, U.S. Patent 3,108,928, 1963.
115. **Kay, H. D. and Graham, W. R.**, The phosphatase test for pasteurized milk, *J. Dairy Res.*,6, 191, 1935.
116. **Keilin, D. and Hartree, E. F.**, Properties of glucose oxidase (notatin). *Biochem. J.*, 42, 221, 1948.
117. **Keilin, D. and Hartree, E. F.**, The use of glucose oxidase (notatin) for the determination of glucose in biological material and for the study of glucose-producing systems by manometric methods, *Biochem. J.*, 42, 230, 1948.
118. **Keilin, D. and Hartree, E. F.**, Specificity of glucose oxidase (notatin). *Biochem. J.*, 50, 331, 1952.
119. **Kertesz, Z. I.**, *The Pectic Substances*, Interscience, New York, 1951.
120. **King, K. W.**, Enzymatic attack of highly crystalline hydrocellulose, *J. Ferment. Technol.* (Japan), 43, 79, 1965.
121. **King, K. W. and Smibert, R. M.**, Distinctive properties of β-glucosidases and related enzymes derived from a commercial *Aspergillus niger* cellulase, *Appl. Microbiol.*, 11, 315, 1963.
122. **King, K. W. and Vessal, M. I.**, Enzymes of the Cellulase Complex, in *Cellulases and Their Applications*, Hajny, G. J. and Reese, E. T., Eds., *Adv. Chem. Ser.*, 95, 7, 1969.
123. **Kleinschmidt, A. W., Higashiuchi, K., Anderson, R., and Ferrari, C. G.**, Soya lipoxidase as a means of flavor improvement, *Bakers Dig.*, 37(5), 44, 1963.
124. **Klis, J. B.**, Heat stable enzyme, *Food Process.*, 23, 70, 1962.
125. **Kneen, E. and Hads, H. L.**, EFfects of variety and environment on the amylases of germinated wheat and barley, *Cereal Chem.*, 22, 407, 1945.
126. **Konigsbacher, K. S. and Donworth, M. E.**, 1965, Beverage flavors, *Advances in Chemistry*, Monograph Series, American Chemical Society, Washington, D.C.
127. **Kooi, E. R., Harjes, C. F., and Gilkison, J. S.**, Treatment and Use of Enzymes for the Hydrolysis of Starch, U.S. Patent 3,042,584, 1962.
128. **Kulp, K.**, Enzymolysis of pentosan of wheat flour, *Cereal Chem.*, 45, 379, 1968.
129. **Kulp, K.**, Debranching enzymes, in *Enzymes in Food Processing*, 2nd ed., Reed, G., Ed., Academic Press, New York, 1975, 84.

130. **Kulp, K. and Bechtel, W. G.,** Effect of water-insoluble pentosan fraction of wheat endosperm on the quality of white bread, *Cereal Chem.,* 40, 493, 1963.
131. **Kulp, K. and Bechtel, W. G.,** Effect of tailings of wheat flour and its subfractions on the quality of bread, *Cereal Chem.,* 40, 665, 1963.
132. **Kuninaka, A., Kibi, M., and Sakaguchi, K.,** History and Development of flavor nucleotides, *Food Technol.,* (Chicago), 18, 287, 1964.
133. **Kurtz, G. W. and Yonezawa, Y.,** The glucose oxidase catalase system as an oxygen scavenger for hermetically sealed containers, presented at 11th Annu. Meet. Inst. Food Technologists, Pittsburgh, *Food Technol.,* 11(4), 16, 1957, Abstr.
134. **Kusai, K., Sekuzu, I., Hagihara, B., Okunuki, K., Yamauchi, S., and Nakai, M.,** Crystallization of glucose oxidase from *Penicillium amagasakiense, Biochim. Biophys. Acta,* 40, 555, 1960.
135. **Laboureur, P. and Villalon, M.,** Process for Enzymatic Degreasing of Bones, U.S. Patent 3,634,191, 1972.
136. **Langlois, D. P.,** Application of enzymes to corn syrup production, *Food Technol.,* (Chicago), 7, 303, 1953.
137. **Lee, C. K., Hayes, L. E., and Long, M. E.,** Process of Preparing Glucose Isomerase, U.S. Patent 3,645,848, 1972.
138. **Li, L. and King, K. W.,** Fractionation of β-glucosidases and related extracellular enzymes from *Aspergillus niger, Appl. Microbiol.,* 11, 320, 1963.
139. **Li, L., Flora, R. M., and King, K. W.,** Purification of β-glucosidases from *Aspergillus niger* and initial observations on the C_1 of *Trichoderma koningii*, *J. Ferment. Technol.*(Japan), 41, 98, 1963.
140. **Lieberman, E. R.,** Enzymes in the beer industry, *Enzyme Technol. Dig.,* 4, 69, 1975.
141. **Long, J. E. and Harper, W. J.,** Italian cheese ripening. VI. Effect of different types of lipolytic enzyme preparations on the accumulation of various free fatty and free amino acids and the development of flavor in provolone and Romano cheese, *J. Dairy Sci.,* 39, 245, 1956.
142. Lumino Feed Co., Fish Meal, German Patent 1,171,248, 1964.
143. **MacAllister, R. V., Wardrip, E. K., and Schnyder, B. J.,** Corn syrups containing glucose and fructose, in *Enzymes in Food Processing,* 2nd ed., Reed, G., Ed., Academic Press, New York, 1975, 346.
144. **Mackay, D. A. M. and Hewitt, E. J.,** Application of flavor enzymes to processed foods. II. Comparison of the effect of flavor enzymes from mustard and cabbage on dehydrated cabbage, *Food Res.,* 24, 253, 1950.
145. **Mackinlay, A. G. and Wake, R. G.,** *Milk Proteins, Chemistry and Microbiology,* Vol. 2, McKenzie, H. A., Ed., Academic Press, New York, 1971, 175.
146. **Macmillan, J. D. and Sherman, M. I.,** Pectic Enzymes, in *Food Related Enzymes,* Whitaker, J. D., Ed., *Adv. Chem. Ser.,* 136, 101, 1974.
147. **Marshall, R. O.,**Enzymatic Process, U.S. Patent 2,950,228, 1960.
148. **Marshall, R. O. and Kooi, E. R.,** Enzymatic conversion of D-glucose to D-fructose, *Science,* 125, 648, 1957.
149. **McCready, R. M. and Gee, M.,** Determination of pectic substances by paper chromatography, *J. Agric. Food Chem.,* 8, 510, 1960.
150. **McIntosh, E. N. and Carlin, A. F.,** The effect of papain preparations on beef skeletal muscle proteins, *J. Food Sci.,* 28, 283, 1963.
151. **McLeod, R. and Ough, C. S.,** Recent studies with glucose oxidase in wine, *Am. J. Enol. Vitic.,* 21(2), 54, 1970.
152. **Melachouris, N. P. and Tuckey, S. L.,** Comparison of the proteolysis produced by rennet extract and pepsin preparation metroclot during ripening of cheddar cheese, *J. Dairy Sci.,* 47, 1, 1964.
153. **Mesnard, P., Devaux, G., Monnier, M., and Fraux, J. L.,** Action de la cellulase sur quelques poudres vegetales lyophilisees. *Prod. Probl. Pharm.,* 18, 628, 1963.
154. **Meyer, R. I., Jokay, L., and Sudek, R. E.,** The effect of an oxygen scavenger packet, desiccant in packet system, on the stability of dry whole milk and dry ice cream mix, *J. Dairy Sci.,* 43, 844, 1960.
155. **Michaelis, L. and Menton, M. L.,** Die Kinetik der Invertinwirkung, *Biochem. Z.,* 49, 333, 1913.
156. **Miller, B. S. and Johnson, J. A.,** Differential stability of α-amylase and proteinase, *Cereal Chem.,* 26, 359, 1949.
157. **Miller, B. S., Johnson, J. A., and Palmer, D. L.,** A comparison of cereal, fungal, and bacterial α-amylases as supplements for bread baking, *Food Technol.,* (Chicago), 7, 38, 1953.
158. **Morgan, A. I. and Schwimmer, S.,** Preparations of Dehydrated Food Products. U.S. Patent 3,170,803, 1965.
159. **Morton, B. J., Martin, E. G., Dahlstrom, R. V., and Sfat, M. R.,** Some aspects of beer colloidal instability, *Am. Soc. Brew. Chem. Proc.,* p. 30, 1962.
160. **Neubeck, C. E.,** Pectic enzymes in fruit juice technology, *J. Assoc. Off. Agric. Chem.,* 42, 374, 1959.

161. **Neubeck, C. E.,** Flavonoid-hydrolyzing enzymes in citrus technology, in *Enzymes in Food Processing,* 2nd ed., Reed, G., Ed., 1975, 431.
162. **Neuberg, C. and Roberts, I. S.,** *Invertase Monograph,* Sugar Research Foundation, New York, 1946.
163. **Neukom, H.,** Pectin-cleaving enzymes, *Z. Ernahrungswiss.,* Suppl. 8, 91, 1969.
164. **Nickerson, W. J. and Faber, M. D.,** Microbial degradation and transformation of natural and synthetic insoluble polymeric substances, *Dev. Ind. Microbiol.* , 16, 111, 1976.
165. **Nisizawa, K., Tomita, Y., Kanda, T., Suzuki, H., and Wakabayashi, K.,** Substrate specificity of C_1 and C_x cellulase components from fungi, in *Fermentation Technology Today,* Proc. 4th Int. Fermentation Symp., Society of Fermentation Technology, Japan, Osaka, 1972, 719.
166. **Notter, G. K., Brekke, J. E., and Taylor, D. H.,** Factors affecting behavior of fruit and vegetable juices during vacuum puff drying, *Food Technol. (Chicago),* 13, 341, 1959.
167. **Ogata, K., Nakao, Y., Igarasi, S., Omura, E., Sugino, Y., Yoneda, M., and Subara, I.,** Degradation of nucleic acids and their related compounds by microbial enzymes. I. On the distribution of extracellular enzymes capable of degrading ribonucleic acid into 5′-mononucleotides in microorganisms, *Agric. Biol. Chem.,* 27, 110, 1963.
168. **Ohlmeyer, D. W.,** Use of glucose oxidase to stabilize beer, *Food Technol.,* 11, 503, 1957.
169. **Olson, A. C. and Stanley, W. L.,** Lactase and other enzymes bound to a phenol-formaldehyde resin with glutaraldehyde, *J. Agric. Food Chem.,* 21, 440, 1973.
170. **Ough, C. S.,** Die Verwendung von Glucose Oxydase in trockenem Weiszwein, *Rebe Wein,* 10, 14, 1960.
171. **Pal, P. N. and Basu, S. N.,** Properies of some fungal cellulases with particular reference to their inhibition, *J. Sci. Ind. Res., Sect. C.,* 20, 336, 1961.
172. **Park, Y. K. and Lima, D. C.,** Continuous conversion of starch to glucose by an amyloglucosidase-resin complex, *J. Food Sci.,* 38(2), 358, 1973.
173. **Pettersson, L. G., Axio-Fredriksson, U. B., and Berghem, L. E. R.,** The mechanism of enzymatic cellulase degradation, in *Fermentation Technology Today,* Proc. 4th Int. Symp., Society of Fermentation Technology, Japan, Osaka, 1972, 727.
174. **Pazur, J. H. and Ando, T.,** The Action of an amyloglucosidase of *Aspergillus niger* on starch and maltooligosaccharides, *J. Biol. Chem.,* 234, 1966, 1959.
175. **Pazur, J. H. and Ando, T.,** The hydrolysis of glucosyl oligosaccharides with α-D-(1→4) and α-D-(1→6) bonds by fungal amyloglucosidase, *J. Biol. Chem.,* 235, 27, 1960.
176. **Pazur, J. H. and Kleppe, K.,** The hydrolysis of α-D-glucosides by amyloglucosidase from *Aspergillus niger, J. Biol. Chem.,* 237, 1002, 1962.
177. **Pazur, J. H. and Kleppe, K.,** Oxidation of glucose and related compounds by glucose oxidase from *Aspergillus niger, Biochemistry,* 3, 578, 1964.
178. **Pederson, C. S.,** Grape juice, in *Fruit and Vegetable Juice Processing,* Tressler, D. K. and Joslyn, M. A., Eds., AVI Publishing, Westport, Conn., 1961, 787.
179. **Pence, J. J., Nimmo, C. C., and Hepburn, F. N.,** Protein, in *Wheat Chemistry and Technology,* Hlynka, I., Ed., American Association of Cereal Chemists, St. Paul, Minn., 1964, 227.
180. **Pieper, H. J.,** Microbial Amylases for the Production of Ethanol, *Hans Ulmer, Stuttgart, W. Ger-*
181. **Pitcher, W. H. and Weetall, H. H.,** Cost of saccharification by immobilized glucoamylase, *Enzyme Technol. Dig.,* 4, 127, 1975.
182. **Platt, W. C. and Posten, A. L.,** Method for Recovering Citrus Oil, U.S. Patent 3,058,887, 1962.
183. **Pomeranz, Y.,** Lactase (β-D-galactosidase). I. Occurrence and properties, *Food Technol.*(Chicago), 18, 682, 1964.

183a. **Pomeranz, Y.,** Lactase (β-D-galactosidase). II. Possibilities in the food industries, *Food Technol.* (Chicago), 18, 690, 1964.

184. **Pomeranz, Y. and Miller, B. S.,** Evaluation of lactase preparations for use in breadmaking, *J. Agric. Food Chem.,* 11, 19, 1963.
185. **Pomeranz, Y., Miller, B. S., Miller, D., and Johnson, J. A.,** Use of lactase in breadmaking, *Cereal Chem.,* 39, 398, 1962.
186. **Preece, I. A.,** *Malting, Brewing and Allied Processes. A Literature Survey.* Heffer, Cambridge, 1960.
187. **Prescott, S. C. and Dunn, C. G.,** *Industrial Microbiology,* 3rd ed., McGraw-Hill, New York, 1959.
188. **Prins, J. and Nielsen, T. K.,** Microbial rennet, *Process Biochem.,* 5(5), 34, 1970.
189. **Ramamurti, K. and Johar, D. S.,** Enzymic digestion of fiber in coconut cake, *Nature (London),* 198, 481, 1963.
190. **Reed, G.,** *Enzymes in Food Processing,* Academic Press, New York, 1966.
191. **Reed, G., Ed.,** *Enzymes in Food Processing,* 2nd ed., Academic Press, New York, 1975, 573 pp.
192. **Reed, G. and Thorn, J. A.,** Use of fungal protease in the baking industry, *Cereal Sci. Today,* 2, 280, 1957.

193. **Reese, E. T.**, Ed., *Advances in Enzymatic Hydrolysis of Cellulose and Related Materials,* Pergamon Press, New York, 1965.
194. **Reich, I. M., Redfern, S., Lenney, J. F., and Schimmel, W. W.**, Prevention of Gel in Frozen Coffee Extract, U.S. Patent 2,801,920, 1957.
195. **Reinke, H. G., Hoag, L. E., and Kincaid, C. M.**, Effect of antioxidants and oxygen scavengers on the shelf life of canned beer, *Am. Soc. Brew. Chem. Proc.*, pp. 175, 1963.
196. **Richardson, G. H. and Nelson, J. H.**, Assay and characterization of pregastric esterase, *J. Dairy Sci.*, 50, 1061, 1967.
197. **Rohm, O.**, The Preparation of Hides for the Manufacture of Leather, U.S. Patent 886,411, 1908.
198. **Rombouts, F. M.and Pilnik, W.**, Pectic enzymes, *CRC Crit. Rev. Food Technol.*, 3(1), 1, 1972.
199. **Rosell, J. M.**, Hydrogen peroxide catalase method for treatment of milk , *Can. Dairy Ice Cream J.*, 40(8), 50, 1961.
200. **Roundy, Z. D.**, Treatment of milk for cheese with hydrogen peroxide, *J. Dairy Sci.*, 41, 1460,
201. **Roundy, Z. D.**, Peroxide catalase method of making Swiss cheese, *Milk Prod. J.*, 52(7), 12, 14, 1961.
202. **Rubenthaler, G., Finney, K. F., and Pomeranz, Y.**, Effects on loaf volume and bread characteristics of α-amylases from cereal, fungal and bacterial sources, *Food Technol.*, (Chicago), 19, 239, 1965.
203. **Sardinas, J. L.**, Milk-curdling Enzyme Elaborated by Endothia parasitica, U.S. Patent 3,275,453, 1966.
204. **Sardinas, J. L.**, New sources of rennet, *Process Biochem.*, 4(7), 13, 21, 1969.
205. **Schleich, H. and Arnold, R. S.**, Composition and Methods for Processing Meat Products, U.S. Patent 3,037,870, 1962.
206. **Schwimmer, S.**, Alteration of the flavor of processed vegetables by enzyme preparations, *J. Food Sci.*, 28, 460, 1963.
207. **Schwimmer, S. and Guadagni, D. G.**, Relation between olfactory threshold concentration and the pyruvic acid content of onion juice, *J. Food Sci.*, 27, 94, 1962.
208. **Schwimmer, S. and Kjaer, A.**, Purification and specificity of the C-S-lyase of *Albizzia lophanta, Biochim. Biophys. Acta,* 42, 316, 1960.
209. **Schwimmer, S. and Weston, W. J.**, Enzymatic development of pyruvic acid in onion as a measure of pungency, *J. Agric. Food Chem.*, 9, 301, 1961.
210. **Scott, D.**, Glucose conversion in preparation of albumen solids by glucose oxidase catalase system, *J. Agric. Food Chem.*, 1, 727, 1953.
211. **Scott, D.**, Enzymatic oxygen removal from packaged foods, *Food Technol.*, (Chicago), 12(7), 7, 1958.
212. **Scott, D.**, Solubilizing tea solids, in *Enzymes in Food Processing,* 2nd ed., Reed, G., Ed., Academic Press, New York, 1975, 494.
213. **Scott, D.**, Beet sugar processing, in *Enzymes in Food Processing,* 2nd ed., Reed, G., Ed., Academic Press, New York, 1975, 496.
214. **Scott, D.**, Lysozyme, in *Enzymes in Food Processing,* 2nd ed., Reed, G., Ed., Academic Press, New York, 1975, 502.
215. **Scott, D.**, Simultaneous gelatinization and enzymatic thinning of starch, in *Enzymes in Food Processing,* 2nd ed., Reed, G., Ed., Academic Press, New York, 1975, 505.
216. **Scott, D.**, Thinning sugar cane juice, in *Enzymes in Food Processing,* 2nd ed., Reed, G., Ed., Academic Press, New York, 1975, 509.
217. **Scott, D. and Hammer, F.**, Oxygen scavenging packet for in-package deoxygenation, *Food Technol.* (Chicago), 15, 99, 1961.
218. **Scott, D. and Klis, J. B.**, Produce salmonella-free yolk and whole eggs, *Food Process.*, 23, 76, 1962.
219. **Shahani, K. M., Harper, W. L., Jensen, R. G., Parry, R. M., and Zittle, C. A.**, Enzymes in Bovine Milk. Review, *J. Dairy Sci.*, 56, 531, 1973.
220. **Shimazono, H.**, Distribution of 5′-ribonucleotides in foods, *Food Technol.*, (Chicago), 18, 303, 1964.
221. **Siewert, K. L. and Otterby, D. E.**, Effect of fat source on relative activity of pregastric esterase and on glyceride composition of intestinal contents, *J. Dairy Sci.*, 54, 258, 1971.
222. **Silman, I. H. and Katchalski, E.**, Water-insoluble derivatives of enzymes, antigens and antibodies, *Annu. Rev. Biochem.*, 35, 873, 1966.
223. **Silverstein, O.**, Heat stable bacterial α-amylase in baking, *Bakers Dig.*, 38(4), 66, 1964.
224. **Simpson, F. J.**, Recovery of Starch, U.S. Patent 2,821,501, 1958.
225. **Sipos, T.**, Syrup Conversion with Immobilized Glucose Isomerase, U.S. Patent 3,708,397, 1973.
226. **Skinner, K. J.**, Enzyme technology, *Chem. Eng. News*,53(33), 22, 1975.
227. **Smiley, K. L.**, Continuous conversion of starch to glucose with immobilized glucoamylase, *Biotechnol. Bioeng.*, 13, 309, 1971.
228. **Sosebee, M. E., May, K. N., and Powers, J. J.**, The effects of enzyme addition on the quality of freeze-dehydrated chicken meat, *Food Technol.*(Chicago), 18, 551, 1964.

229. **Sternberg, M. Z.**, Crystalline milk-clotting protease from *Mucor miehei* and some of its properties, *J. Dairy Sci.*, 54, 159, 1971.
230. **Stimpson, E. G.**, Frozen Concentrated Milk Products, U.S. Patent 2,668,765, 1954.
231. **Stone, I. M.**, Process of Producing Baked Confections and the Products Resulting Therefrom by α-Amylase, U.S. Patent 3,026,205, 1962.
232. **Strandberg, G. W. and Smiley, K. L.**, Glucose isomerase covalently bound to porous glass beads, *Biotechnol. Bioeng.*, 14, 509, 1972.
233. **Suzuki, S.**, An overall look at the dextrose industry in Japan, *Starke*, 16, 285, 1964.
234. **Suzuki, H., Yoshida, H., Ozawa, Y., Kamibayashi, A., Sato, M., Mori, A., and Endo, M.**, Increasing Yield of Sucrose, U.S. Patent 3,767,526, 1973.
235. **Swoboda, B. E. P. and Massey, V.**, Purification and Properties of the Glucose Oxidase from *Aspergillus niger*, *J. Biol. Chem.*, 240, 2209, 1965.
236. **Takamine, J.**, Process of Making Diastatic Enzyme, U.S. Patents 525,820 and 525,823, 1894.
237. **Takasaki, Y., Kosugi, Y., and Kanbayashi, A.**, *Streptomyces* Glucose Isomerase, in *Fermentation Advances*, Perlman, D., Ed., Academic Press, New York, 1969, 561.
238. **Takasaki, Y. and Tanabe, O.**, Enzyme Method for Converting Glucose in Glucose Syrups to Fructose, U.S. Patent 3,616,221, 1971.
239. **Tappel, A. L.**, Lipoxidase, in *The Enzymes*, Vol. 8, 2nd ed., Boyer, P. D., Lardy, H., and Myrback, K., Eds., Academic Press, New York, 1963, 275.
240. **Tepley, L. J., Derse, P. H., and Price, W. V.**, Composition and nutritive value of cheese produced from milk treated with hydrogen peroxide and catalase, *J. Dairy Sci.*, 41, 593, 1958.
241. **Tilbury, R. H.**, Production of Sucrose from Sugar Cane Products, British Patent 1,290,694, 1970.
242. **Thompson, K. N., Johnson, R. A., and Lloyd, N. E.**, Process for Isomerizing Glucose to Fructose, U.S. Patent 3,788,945, 1974.
243. **Thorne, R. S. W.**, The problem of beer haze assessment, *Wallerstein Lab. Commun.*, 26, 5, 1963.
244. **Toyama, N.**, Recent advances in the production and industrial application of cellulase in Japan, *Hakko Kyokaishi*, 21, 415, 459, 1963.
245. **Toyama, N.**, A cell separating enzyme as a comlementary enzyme to cellulase and its application in processing vegetables, *Hakko Kogaku Zasshi*, 43, 683, 1965.
246. **Toyama, N.**, 1969, Applications of Cellulases in Japan. *Cellulases and Their Applications*, Hajny, G. J. and Reese, E. T., Eds., *Adv. Chem. Ser.* 95, 359.
247. **Toyama, N. and Ogawa, K.**, Utilization of cellulosic wastes by *Trichoderma viride*, in *Fermentation Technology Today*, Proc. 4th Fermentation Symp., Society of Fermentation Technology, Japan, Osaka, 1972, 743.
248. **Tsugo, T. and Yamauchi, K.**, 1960, Comparison of clotting action of various milk-coagulating enzymes. I. Comparison of factors affecting clotting time of milk, *15th Int. Dairy Congr.*, Vol. 2, Congres International de Laiterie, London, 1959, 636.
248a. **Tsugo, T. and Yamauchi, K.**, Comparison of clotting action of various milk-coagulating enzymes. II. Comparison of curds coagulated by various enzymes, 15th Int. Dairy Congr., Vol. 2, Congres International de Laiterie, London, 1959, 643.
249. **Tsumura, N. and Sato, T.**, Enzymatic conversion of D-glucose to D-fructose. II. Some properties concerning fructose accumulation activity of *Aerobacter cloacae*, Strain KN-69, *Agric. Biol. Chem.*, 25, 616, 1961.
250. **Tsumura, N., Hagi, M., and Sato, T.**, Enzymatic conversion of D-glucose to D-fructose. VIII. Propagation of *Streptomyces phaeochromogenes* in the presence of cobaltous ion, *Agric. Biol. Chem.*, 31, 902, 1967.
251. **Tumerman, L., Fram, H., and Cornely, K. W.**, The effect of lactose crystallization on protein stability in frozen concentrated milk, *J. Dairy Sci.*, 37, 830, 1954.
252. **Underkofler, L. A.**, Properties and Applications of the Fungal Enzyme Glucose Oxidases, in *Proc. Int. Symp. Enzyme Chem.*, Maruzen, Tokyo, 1958, 486.
253. **Underkofler, L. A.**, Meat Tenderizer, U.S. Patent 2,904,442, 1959.
254. **Underkofler, L. A.**, Enzyme supplementation in baking, *Bakers Dig.*, 35(5), 74, 1961.
255. **Underkofler, L. A.**, Glucose oxidase: production, properties, and present and potential applications, in *Production and Application of Enzyme Preparations in Food Manufacture, Soc. Chem. Ind. (London)*, Monograph 11, Macmillan, New York, 1961, 72.
256. **Underkofler, L. A.**, Production of commercial enzymes, in *Enzymes in Food Processing*, Reed, G., Ed., Academic Press, New York, 1966, 197.
257. **Underkofler, L. A.**, Manufacture and uses of industrial microbial enzymes, *Chem. Eng. Prog. Sym. Ser.*, 62(69), 11, 1966.
258. **Underkofler, L. A., Barton, R. R., and Aldrich, F. L.**, Methods of assay for microbial enzymes, *Dev. Ind. Microbiol.*, 2, 171, 1960.

259. **Underkofler, L. A., Barton, R. R., and Rennert, S. S.**, Production of microbial enzymes and their applications, *Appl. Microbiol.*, 6, 212, 1958.
260. **Underkofler, L. A., Denault, L. J., and Hou, E. F.**, Enzymes in the starch industry, *Starke*, 17, 179, 1965.
261. **Underkofler, L. A. and Hickey, R. J.**, Eds., *Industrial Fermentations*, Chemical Publishing, New York, 1954.
262. **Vanderpoorten, R. and Weckx, M.**, Breakdown of casein by rennet and microbial milk clotting enzymes, *Neth. Milk Dairy J.*, 26, 47, 1972.
263. **Vaupel, E. A.**, Meat Tenderizer, U.S. Patent 2,825,654, 1958.
264. **Veringa, H. A.**, Rennet substitutes, *Dairy Sci. Abstr.*, 23, 197, 1961.
265. **Vilenskaya, E. I.**, Production of enzymatically clarified juices, *Spirt. Prom.*, 29(2), 23, 1963.
266. **Vohra, P. and Kratzer, F. H.**, Improvement of guar meal by enzymes, *Poult. Sci.*, 44, 1201, 1965.
267. **Waldt, L. M.**, Fungal enzymes: their role in continuous process bread, *Cereal Sci. Today*, 10, 447, 1965.
268. **Walker, J. R. L.**, Enzymic browning in foods: a review, *Enzyme Technol. Dig.*, 4, 89, 1975.
269. **Walker, L. H., Nimmo, C. C., and Patterson, D. C.**, Frozen apple juice concentrate, *Food Technol.*(Chicago, 5, 148,1951.
270. **Walker, L. H. and Patterson, D. C.**, Preparation of fresh Italian prune juice concentrates, *Food Technol.*, (Chicago), 8, 208, 1954.
271. **Wallerstein, L.**, Beer and Method of Preparing Same, U.S. Patent 995,820, 1911.
271a. **Wallerstein, L.**, Preparation for Use in Brewing, U.S. Patent 995,823, 1911.
271b. **Wallerstein, L.**, Method of Treating Beer and Ale. U.S. Patents 995,824 , 995,825, 995,826, 1911.
272. **Wallerstein, L.**, Chillproofing and stabilization of beer, *Wallerstein Lab. Commun.*, 24, 158, 1961.
273. **Wang, H., Weier, E., Birkner, M., and Ginger, B.**, Studies on enzymatic tenderization of meat. III. Histological and panel analysis of enzyme preparations from three distinct sources, *Food Res.*, 23, 423, 1958.
274. **Westall, H. H.**, Enzymes immobilized on inorganic carriers, *Res. Dev.*, 22(12), 18, 1971.
275. **Weetall, H. H.**, Preparation, characterization and applications of enzymes immobilized on inorganic supports, in *Immobilized Biochemicals and Affinity Chromatography*, Dunlap, R. B. Ed., Plenum Press, New York, 1974, 191.
276. **Weetall, H. H.**, Immobilized enzymes. Analytical applications, *Anal. Chem.*, 46, 602A, 1974.
277. **Weetall, H. H.**, Immobilized enzyme technology, *Cereal Foods World*, 21, 581, 1976.
278. **Weetall, H. H. and Hanewala, N. B.**, Continuous production of dextrose from cornstarch. Reactor parameters necessary for commercial application, *Biotechnol. Bioeng. Symp.*, 3, 241, 1972.
279. **Weissler, H. W. and Garza, A. C.**, Some physical and chemical properties of commercial chillproofing compounds, *Am. Soc. Brew. Chem. Proc.*, 225, 1965.
280. **Weurman, C.**, Gas-liquid chromatographic studies on the enzymatic formation of volatile compounds in raspberries, *Food Technol. (Chicago)*, 15, 531, 1961.
281. **Whitaker, J. R.**, Principles of Enzymology for the Food Sciences, *Marcel Dekker, New York, 1972,*
282. **Whitaker, J. R.**, Lipoxygenase (Lipoxidase) in *Principles of Enzymology for the Food Sciences*, Marcel Dekker, New York, 1972, 607.
283. **Whitaker, J. R., Ed.**, Food related enzymes, *Adv. Chem. Ser.*, pp. 136, 365, 1974.
284. **Wieg, A. J.**, Technology of barley brewing, *Process Biochem.*, 5(8), 46, 1970.
285. **Wilson, G. D.**, Quality factors in meat and meat foods, in *The Science of Meat and Meat Products*, Freeman, San Francisco, 1960, 259.
286. **Wiseman, A.**, Ed., Handbook of Enzyme Biotechnology, *John Wiley & Sons, New York, 1975, 275*
287. **Wood, T. M.**, The C_1 component of the cellulase complex, *Fermentation Technology Today*, Proc. 4th Ferment. Symp., Society of Fermentation Technology, Japan, Osaka, 1972, 711.
288. **Woodruff, H. B., Nollstadt, K. H., Walton, R. B., and Stoudt, T. H.**, Enzymes for dental hygiene, *Dev. Ind. Microbiol.*, 17, 405, 1976.
289. **Woychik, J. H. and Wondolowski, M. V.**, Lactose hydrolysis in milk and milk products by bound fungal β-galactosidase, *J. Milk Food Technol.*, 30, 31, 1973.
290. **Wuthrich, S., Richterich, R., and Hostettler, H.**, Untersuchungen uber Milchenzyme. I. Enzyme in Kuhmilch und Frauenmilch, *Z. Lebensm. Unters. Forsch.*, 124, 336, 1964.
291. **Yagi, K. and Ozawa, T.**, Mechanism of enzyme action. I. Crystallization of Michaelis complex of D-amino acid oxidase, *Biochim. Biophys. Acta*, 81, 29, 1964.
292. **Yamamoto, Y. and Yoshitake, M.**, 1962, Formation of amino acids by the action of rennet and lactic acid bacteria during the ripening of cheese, in *Proc. 16th Int. Dairy Congr.*, Vol. B, Andelsbogtrykkeriet, Copenhagen, 1962, 395.
293. **Yamanaka, K.**, Sugar isomerases. Part II. Purification and properties of D-glucose isomerase from *Lactobacillus brevis*, *Agric. Biol. Chem.*, 27, 271, 1963.

294. **Yang, H. Y.,** Stabilizing apple wine with glucose oxidase, *Food Res.,* 20, 42, 1955.
295. **Yokobayashi, K., Misaki, A., and Harada, T.,** Specificity of *Pseudomonas Isoamylase, Agric. Biol. Chem.,* 33, 625, 1969.
296. **Zaborsky, O. R.,** *Immobilized Enzymes,* CRC Press, Cleveland, 1973.

NONNUTRITIVE SWEETENERS: SACCHARIN AND CYCLAMATE

K. M. Beck

EDITOR'S NOTE

Nearly a decade has passed since saccharin and cyclamate have come under regulatory fire as potentially caracinogenic substances. The issue remains unresolved.

In 1971, the FDA proposed removing saccharin from the GRAS list and interim regulations are proposed to permit testing the sweetener. Final interim regulations restricting the use of saccharin were issued in 1972 and the FDA asks the National Academy of Sciences to review all studies regarding saccharin's carcinogenicity. Two years later, in 1974, the NAS reports that evidence for saccharin's carcinogenicity is inconclusive. Nonetheless, FDA moved to restrict the use of saccharin in diet foods and beverages. However, in 1977 Congress prohibits FDA from banning saccharin for an 18-month period. Meanwhile, a Canadian epidermiologic study concludes that saccharin is associated with an excess risk of bladder cancer in men. But other studies do not confirm the Canadian findings and a joint FDA/ NCI group proposes a large-scale epidemiologic survey of bladder cancer patients. At the same time, saccharin has been put through a battery of short-term in vitro tests and found to be weakly positive. A second extension by the Congress is currently under consideration but the NCI study surveying some 9000 bladder cancer patients remains inconclusive. Apparently, the study, which was to provide an adequate statistical base for detecting the effects of a weak carcinogen, does not exclude the possibility that the positive associations represent chance variation. Overall, the study fails to demonstrate an association between the use of saccharin and bladder cancer. Currently, the thought is that saccharin may be a "cancer promoter" rather than an initiator but this is mostly speculations.

While the analysis of data from the NCI study are still under way, it is apparent that the Congress will have a difficult time coming to a conclusion whether to extend the FDA-proposed ban.

It is with these issues at stake that the chapter on saccharin and cyclamate has been republished and updated principally with respect to the safety literature. While the author has attempted to include the most recent data, the fast moving pace in this field precludes completeness at the time of publication.

T. E. F.

INTRODUCTION

Saccharin has been used commercially as a nonnutritive sweetening agent since 1900. Cyclamate was introduced in 1950 and widely used until its use in the U.S. was restricted in October 1969 and banned in September 1970. Beginning in 1955, there was increased use of combinations of cyclamate and saccharin in most types of food products.

Although many other nonnutritive sweeteners have been investigated and a few have been used commercially to a limited extent, none has approached saccharin and cyclamate in duration or scope of usage. Saccharin is still the most widely used nonnutritive sweetening agent in the world. Cyclamate, however, is still permitted in more than 30 countries, and there are efforts underway to get the Food and Drug Administration to approve it as a food additive in the U.S.

Replacement of sugars with nonnutritive sweeteners in foods and beverages presents a variety of challenges to the food technologist due to some basic differences between the two types of sweetening agents.

1. Nonnutritive sweeteners are not carbohydrates, and they have different chemical and physical properties than sugars have.
2. Nonnutritive sweeteners often have different taste characteristics than sugars have.
3. Nonnutritive sweeteners usually are intensely sweet so they are used in very low concentrations.

The broad usage of cyclamate and saccharin led to the development of an extensive technology that provides valuable background for the use of other nonnutritive sweeteners that may be developed in the future.

A rapid growth in the consumption of low-calorie soft drinks beginning about 1960 caused the food and beverage industries to focus attention on the market for low-calorie food and beverage products. This meant, in turn, that a great deal of new attention was directed at nonnutritive sweeteners and at the other products used in conjunction with them to formulate low-calorie and other dietetic foods. One of the major reasons for the broad acceptance of low-calorie beverages was the fact that these were good tasting products in the eyes of a large segment of the population, although in too many cases application of nonnutritive sweeteners tended to run ahead of the development of adequate technology for their proper use.

Consumer acceptance of low-calorie and other dietetic foods and beverages is dependent upon the same factors that govern the acceptance of standard products. Namely, the quality must be such that the consumer can enjoy using the product and will consume it for its own sake, not view it as a therapeutic agent. Diabetics and dieters should not feel that their special dietary foods are a form of punishment. As always, the consumer buys perceived values which are promised by the manufacturer in advertising and other promotion, but if he fails to deliver what is expected, the product will fail or its continued expansion into the market will cease. Consumers have grown to expect products manufactured with nonnutritive sweeteners to be as acceptable as those with sugars. This expectation serves to underscore the need to approach low-calorie foods and beverages in much the same manner as their standard counterparts.

Accordingly, the attempt here is to provide guidelines and general principles involved in the use of nonnutritive sweeteners in the formulation and production of low-calorie products. This is done in the context of factors including flavor, texture, color, general appearance, ease of preparation, and storage stability which are known to play significant roles in consumer acceptance. At the same time, it should be recognized that such products are governed by special regulations at both the federal and state level so that product composition must conform to them. Finally, these products will move through normal food distribution channels and be merchandised in a very similar fashion to regular foods, which means that their good qualities which can be relied upon for consumer appeal must be easily recognized. The challenge is particularly great, since there are only a limited number of materials which can be used in conjunction with nonnutritive sweeteners, but the commercial facts of life will militate against compromises in quality no matter what the reason.

PROPERTIES OF CYCLAMATE AND SACCHARIN

Although cyclamate and saccharin have quite different types of chemical structures, each exists in the form of an acid and its salts. The acid forms have only limited food and beverage uses. The sodium and calcium salts are the most widely used forms of both cyclamate and saccharin. Governmental regulations for cyclamate and saccharin generally apply to the acids and their salts.

Cyclamate

The term cyclamate refers to cyclamic acid (cyclohexanesulfamic acid), sodium cyclamate (sodium cyclohexanesulfamate), and calcium cyclamate (calcium cyclohexanesulfamate). The structural formulas for the cyclamates are shown below:

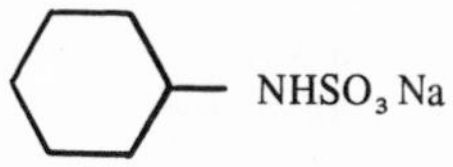

$NHSO_3Na$ — Sodium Cyclamate

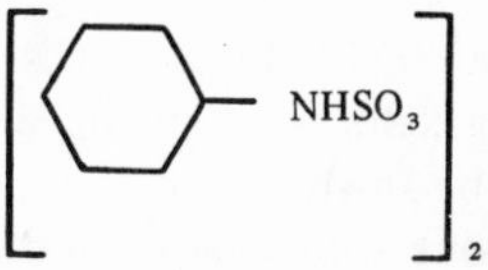

$[\,NHSO_3\,]_2$ — Calcium Cyclamate

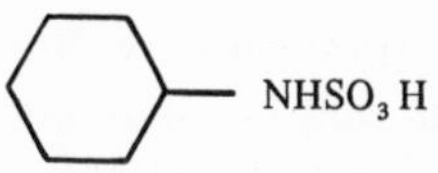

$NHSO_3H$ — Cyclamic Acid

Both sodium and calcium cyclamate occur as white crystals or as white, crystalline powders. They are odorless and in dilute solution are about 30 times as sweet as sucrose. General specifications and test methods are described in the *National Formulary*, thirteenth edition. When cyclamate is approved in the U.S. again, specifications will be published officially in the *Food Chemicals Codex.*

Typical specifications for cyclamates used in countries that permit their use are indicated below:

SODIUM CYCLAMATE

Appearance:	Odorless, white crystals or white, crystalline powder
Empirical formula:	$C_6H_{12}NNaO_3S$
Molecular weight:	201.22
Purity:	98% minimum, calculated on dried basis
Loss in drying:	1.0% maximum, when dried at 105°C for 1 hr
Solubility:	1 g dissolved in 5 ml of water and 25 ml of propylene glycol; practically insoluble in alcohol, benzene, chloroform, and ether
pH (10% aqueous solution):	Neutral to litmus
Cyclohexylamine:	Not more than 25 ppm
Dicyclohexylamine:	Not more than 0.5 ppm
Heavy metals (as Pb):	Not more than 10 ppm

CALCIUM CYCLAMATE

Appearance:	Odorless, white crystals or white, crystalline powder
Empirical formula:	$C_{12}H_{24}CaN_2O_6S_2 \cdot 2H_2O$
Molecular weight:	432.57
Purity:	98% minimum, calculated on anhydrous basis
Water:	9.0% maximum, when determined by titrimetric method
Solubility:	1 g dissolved in 4 ml of water, 60 ml of alcohol, or 1.5 ml of propylene glycol; practically insoluble in benzene, chloroform, and ether
pH (10% aqueous solution):	Neutral to litmus
Cyclohexylamine:	Not more than 25 ppm
Dicyclohexylamine:	Not more than 0.5 ppm
Heavy metals (as Pb):	Not more than 10 ppm

CYCLAMIC ACID

Appearance:	White, crystalline powder with a sweet-sour taste
Empirical formula:	$C_6H_{13}NO_3S$

Molecular weight:	179.24
Purity:	98% minimum, calculated on dried basis
Loss on drying:	1.0% maximum, when dried at 105°C for 1 hr
Solubility:	1 g dissolved in about 7.5 ml of water, 4 ml of alcohol, 4 ml of propylene glycol, or 6 ml of acetone; slightly soluble in chloroform and insoluble in hexane
Cyclohexylamine:	Not more than 25 ppm
Dicyclohexylamine:	Not more than 0.5 ppm
Heavy metals (as Pb):	Not more than 10 ppm

The titration curve for cyclamic acid (Figure 1) is that of a very strong acid. The sodium and calcium salts, being salts of a strong acid and strong base, are highly ionized in solution and tend to be fairly neutral in character. Other properties, such as vicosity and density of aqueous solutions of sodium and calcium cyclamate and temperature-solubility curves, are shown in Figures 2 through 4.

Saccharin

While the acid form of saccharin is a well-recognized article of commerce, its salts are the products actually used in the formulation of foods and beverages. The most common two are sodium and calcium, although ammonium and other salts have been prepared and used to a very limited extent. Structural formulas for saccharin and sodium and calcium saccharin are shown below.

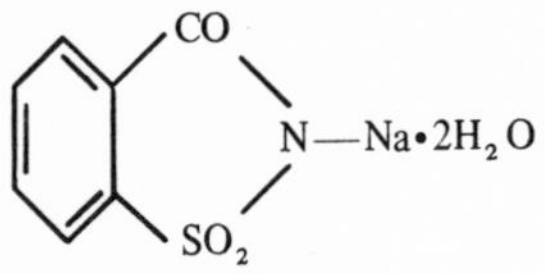

CO, N—Na•2H_2O, SO_2

Sodium Saccharin

CO, N, SO_2, 2, Ca·3½H_2O

Calcium Saccharin

CO, N—H, SO_2

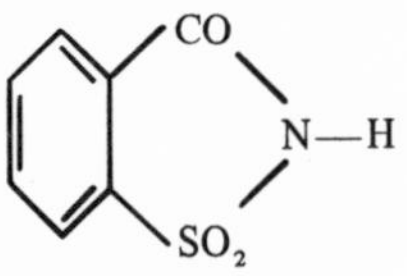

Saccharin

Saccharin (1,2-benzisothiazol-3(2H)-one 1,1-dioxide or *o*-sulfobenzimide) is described in the *United States Pharmacopeia XIX* and in *Food Chemicals Codex*, second edition. Specifications and test methods for sodium saccharin and calcium saccharin are described in the *National Formulary*, 14th edition and the *Food Chemicals Codex*, second edition.

The general properties for these three forms of saccharin are outlined below:

SODIUM SACCHARIN

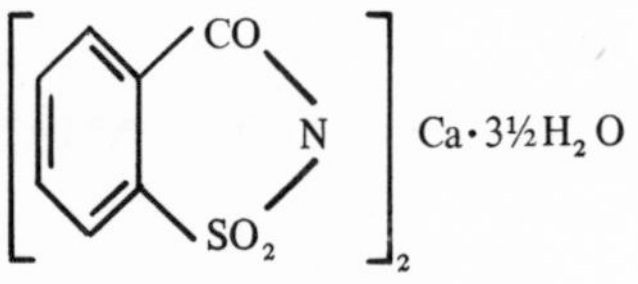

Appearance:	White crystals or white, crystalline powder
Empirical formula:	$C_7H_4NNaO_3S \cdot 2H_2O$
Molecular weight:	241.19; anhydrous, 205.16
Purity:	98% minimum of $C_7H_4NNaO_3S$, calculated on anhydrous basis
Water:	15% maximum, when determined by titrimetric method
Solubility:	1 g dissolved in 1.5 ml of water or about 50 ml of alcohol
Benzoates and salicylates:	Passes N.F. test
o-Toluenesulfonamide:	Not more than 25 ppm
Heavy metals (as Pb):	Not more than 10 ppm

CALCIUM SACCHARIN

Appearance:	White cyrstals or white, crystalline powder
Empirical formula:	$C_{14}H_8CaN_2O_6S_2 \cdot 3\frac{1}{2}H_2O$
Molecular weight:	467.48; anhydrous, 404.43
Purity:	98% minimum of $C_{14}H_8CaN_2O_6S_2$, calculated on anhydrous basis

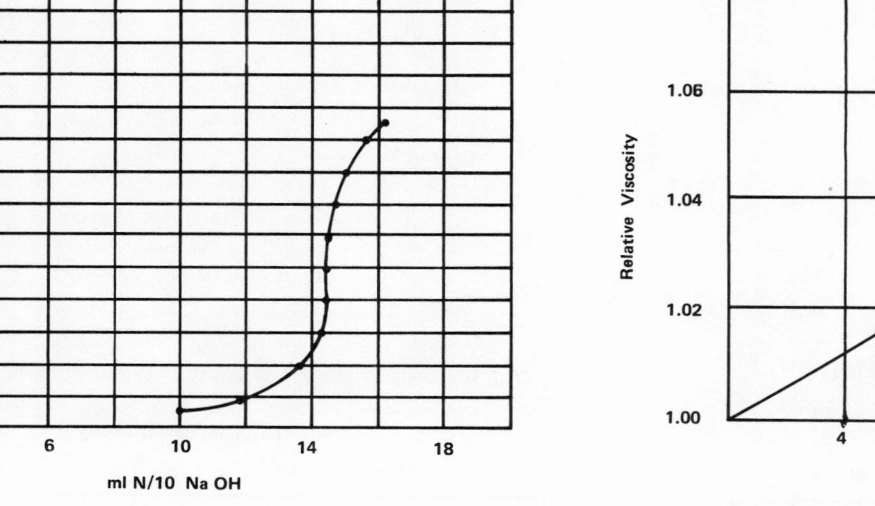

FIGURE 1. Titration of cyclamic acid with base.

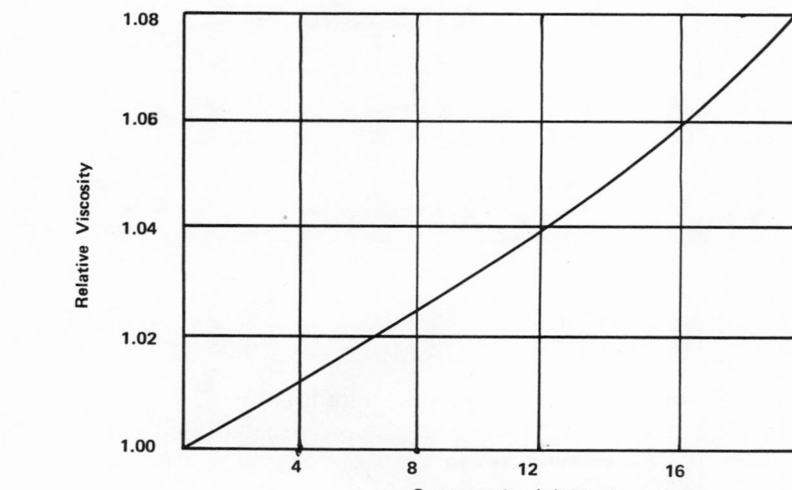

FIGURE 2. Viscosity-concentration curve: calcium and sodium cyclamate solutions at 25°C.

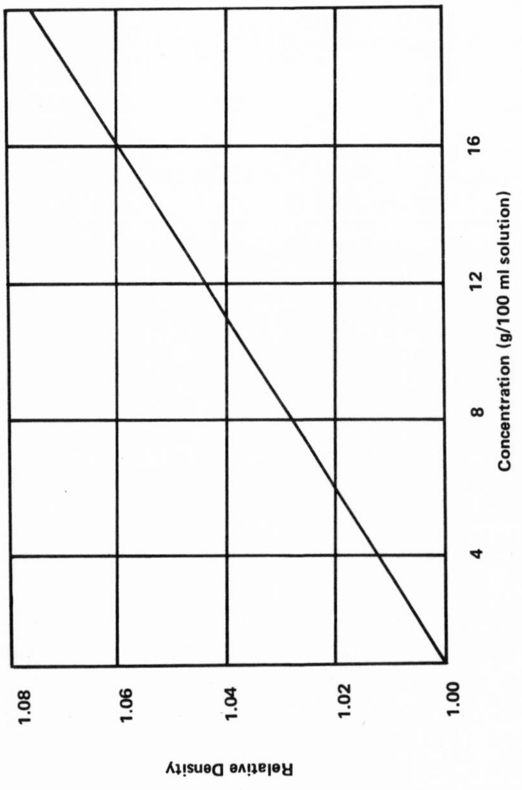

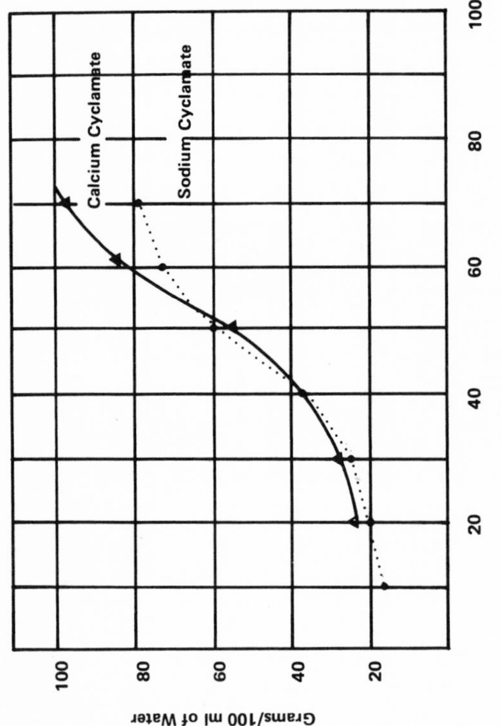

FIGURE 4. Solubility of cyclamates in water.

FIGURE 3. Density-concentration curve: calcium and sodium cyclamate at 20°C.

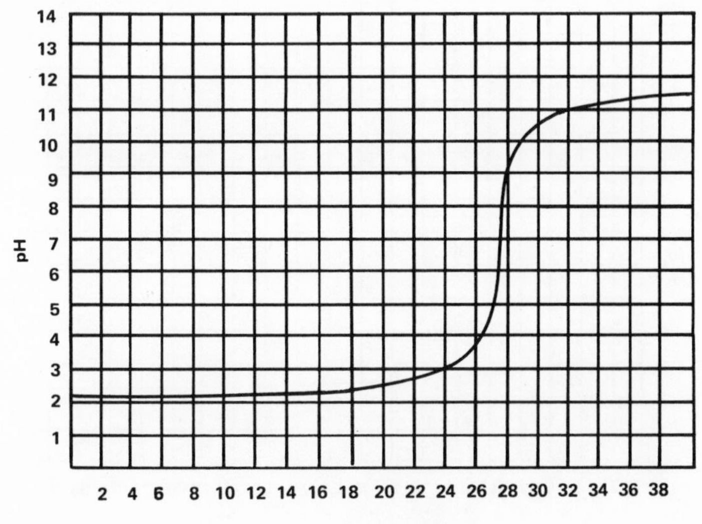

FIGURE 5. Titration of saccharin with base.

Water:	15% maximum, when determined by titrimetric method
Solubility:	1 g dissolved in 1.5 ml of water
Benzoates and salicylates:	Passes N.F. test
o-Toluenesulfonamide:	Not more than 25 ppm
Heavy metals (as Pb):	Not more than 10 ppm

SACCHARIN

Appearance:	White cyrstals or white, cyrstalline powder; odorless or a faint, aromatic odor
Empirical formula:	$C_7H_5NO_3S$
Molecular weight:	183.18
Purity:	98% minimum, calculated on dried basis
Melting range:	Between 226 and 230°C
Loss on drying:	1% maximum, when dried at 105°C for 2 hr
Solubility:	1 g dissolved in 290 ml of water at 25°C, 25 ml of boiling water, or 30 ml or alcohol; slightly soluble in chloroform and ether; dissolves readily in dilute solutions of ammonia and solutions of alkali hydroxides or alkali carbonates
Benzoic and salicylic acids:	Passes U.S.P. test
Heavy metals (as Pb):	Not more than 10 ppm
Residue on ignition:	Not more than 0.2%

Like cyclamates, the titration curves for saccharin (Figures 5 through 7) are of a strong acid. Other properties, including solubility, viscosity and density of saccharin solutions, are shown in Table 1 and in Figures 8 through 11.

SAFETY

When the Federal Food, Drug, and Cosmetic Act was amended in 1958 providing a new definition of the term, "food additive," a prior sanction approval for substances already in use and generally recognized as safe (GRAS) for their intended use was established. The first GRAS list, published in the *Federal Register* on November 20, 1959, (24 F.R. 9368, Nov. 20, 1959; CFR 121.101) included both cyclamate and saccharin.

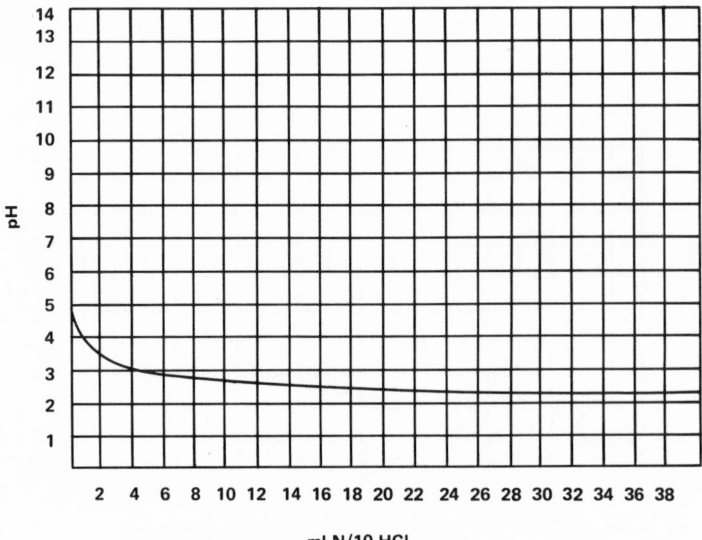

FIGURE 6. Titration of sodium saccharin with acid.

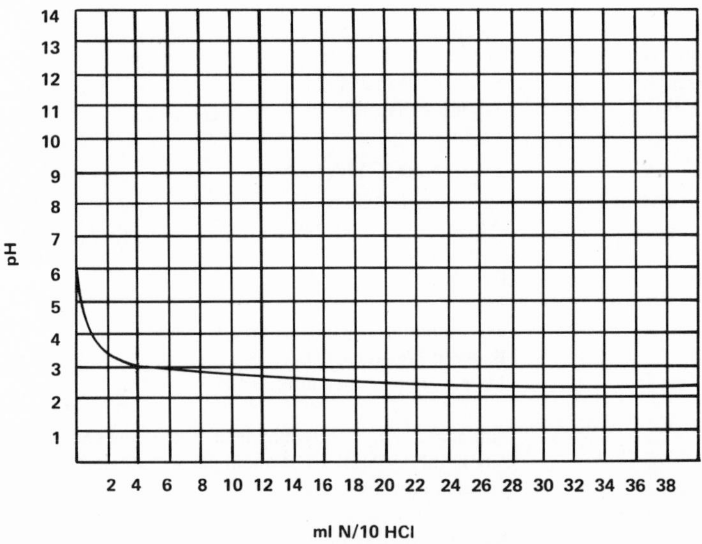

FIGURE 7. Titration of calcium saccharin with acid.

Cyclamates were removed from the GRAS list on October 18, 1969, and its further use in food and beverage production in the U.S. was banned (34 F.R. 17063, Oct. 21, 1969). Although there subsequently was some consideration of a drug classification for cyclamate to permit its use in table sweeteners, all uses of cyclamate in the U.S. were banned in August 1970 (35 F.R. 13644, Aug. 27, 1970). A food additive petition for cyclamate was submitted by the Abbott Laboratories on November 15, 1973, and was denied on October 4, 1976 (41 F.R. 43754). A request for an administrative hearing was filed and granted.

Saccharins were removed from the GRAS list on February 1, 1972, and given a provisional food additive status. The interim food additive status was renewed twice;

Table 1
SOLUBILITY OF SACCHARIN

Solubility of Saccharin in Water (g/100 g water)

Temp (°C)	Saccharin, mol 183.18	Na Saccharin · $2H_2O$, mol 241.20	Ca(Saccharin)$_2$ · 3½ H_2O, mol 467.50
10.6	—	77.3	—
20.0	—	99.8	—
25.0	0.2	—	54.8
35.0	—	138.5	—
45.0	0.6	—	115.5
50.0	—	186.8	—
65.0	1.1	220.2	194.7
75.0	—	253.5	—
85.0	—	289.1	—
90.0	—	—	262.0
95.0	—	328.3	—

Solubility of Saccharin in Ethanol-Water Mixtures at 25°C (g/100 g solvent)

% Ethanol by wt	Saccharin, mol wt 183.18	Na Saccharin · $2H_2O$, mol wt 241.20	Ca(Saccharin)$_2$ · 3½ H_2O, mol wt 467.50
20.7	0.6	87.3	58.6
43.7	1.8	60.2	60.2
69.5	4.0	23.9	55.0
92.5	4.1	2.6	30.5

Table 1 (continued)
SOLUBILITY OF SACCHARIN

Solubility of Saccharin in Glycerin and Propylene Glycol at 25°C

Solvent[a]	Saccharin, mol wt 183.18	Na Saccharin · $2H_2O$, mol wt 241.20	Ca(Saccharin)$_2$ · $3\frac{1}{2}H_2O$, mol wt 467.50
Propylene	2.6 g/100g solution	30.9 g/100g solution	26.0 g/100g solution
	2.6 g/100g solvent	44.7g/100g solvent	35.1 g/100g solvent
Glycol	2.7 g/100 ml solvent	46.2 g/100 ml solvent	36.3 g/100 ml solvent
	0.4 g/100g solution	35.8 g/100g solution	9.6 g/100g solution
Glycerin	0.4 g/100g solvent	55.8 g/100g solvent	10.6 g/100g solvent
	0.5 g/100 ml solvent	70.2 g/100 ml solvent	13.3 g/100 ml solvent

[a] Density@ 25°C is 1.033 for propylene glycol and 1.2583 for glycerin.

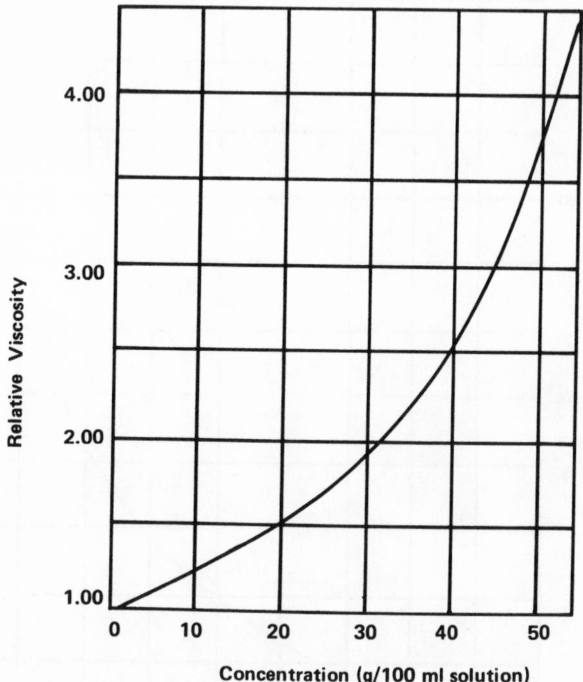

FIGURE 8. Viscosity-concentration curve: calcium and sodium saccharin solutions at 25°C.

the latest renewal was published in the *Federal Register* in January 1977 (42 F.R. 1461, Jan. 7, 1977). A proposal to revoke the interim food regulation under which saccharin and its salts may be used in prepackaged foods was issued on April 15, 1977 (42 F.R. 19996).

Cyclamate was banned as a suspect carcinogen. At the termination of a 2-year feeding study, conducted by the Food and Drug Research Laboratories on a cyclamate-saccharin mixture, papillary lesions were found in the bladders of a few rats in the highest dosage group. Although this was not designed as a carcinogenicity study and although the diet of the animals contained not only cyclamate and saccharin, but also cyclohexylamine, the Food and Drug Administration (FDA) was advised by the National Academy of Science/National Research Council (NAS/NRC) Food Protection Committee that cyclamate could no longer be classified as GRAS. Subsequent studies designed specifically as carcinogenicity tests, particularly a lifetime feeding study by D. Schmähl at the German Cancer Research Center in Heidelberg, Germany, failed to show cyclamate to be a carcinogen. At the request of the FDA, the National Cancer Institute in 1975 established a committee of experts to review the question of carcinogenicity of cyclamate. This committee concluded in the Report of the Temporary Committee for the Review of Data on the Carcinogenicity of Cyclamate, Division of Cancer Cause and Prevention, National Cancer Institute, February 1976 that "The present evidence does not establish the carcinogenicity of cyclamate or its principle metabolite, cyclohexylamine, in experimental animals."

The Federal Food and Drug Administration is still concerned about the safety of cyclohexylamine as a metabolite of cyclamate, especially the possibility that at high levels it may cause testicular atrophy or elevated blood pressure in test animals. The FDA has been unable to agree upon an acceptable daily intake (ADI) that is sufficiently high to allow for use of cyclamate in foods and beverages.

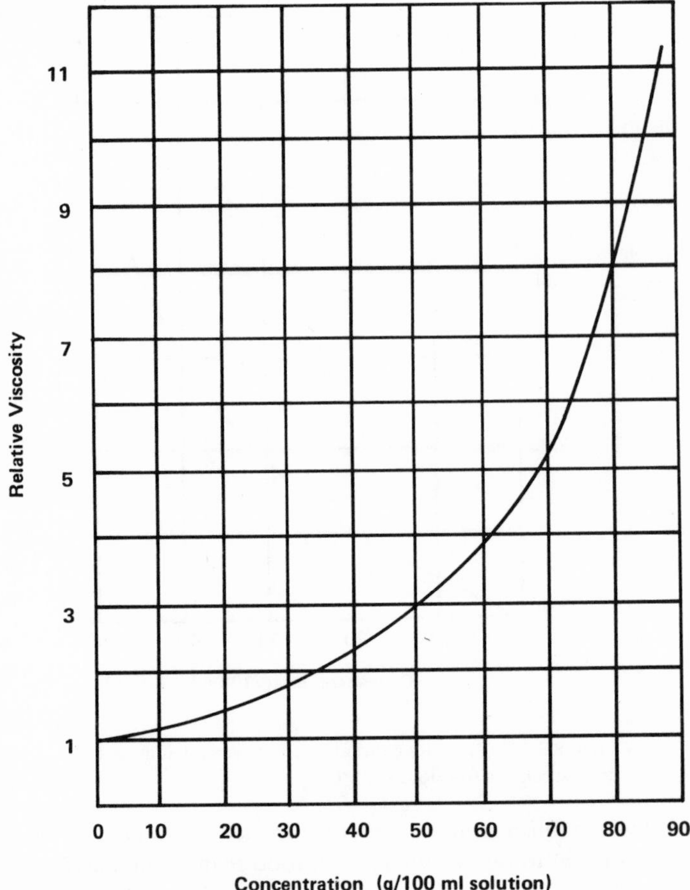

FIGURE 9. Viscosity-concentration curve: calcium and sodium saccharin solutions at 90°C.

Saccharin has also been associated with bladder tumors when fed at high levels to rats. The FDA gave a contract to the NAS/NRC in June 1972 to review all experiments on the possible carcinogenicity of saccharin. The NAS/NRC report in December 1974 concluded that the data available had not established conclusively whether saccharin is or is not a carcinogen when administered orally to test animals. Particularly questionable was the contribution to toxicity of *o*-toluenesulfonamide (OTS), an impurity in saccharin. The NAS/NRC recommended certain additional studies:

1. A carcinogenesis study of purified contaminants of commercial saccharin, especially *o*-toluenesulfonamide
2. A carcinogenesis study of pure saccharin
3. A carcinogenesis study of mixtures of known amounts of saccharin and OTS
4. A study of the interaction of stones or parasites in the bladder and saccharin in the diet
5. A study of urine composition as affected by high saccharin and OTS intake in the diet
6. A study of the significance of parenteral and in utero exposure to saccharin and OTS in carcinogenesis studies

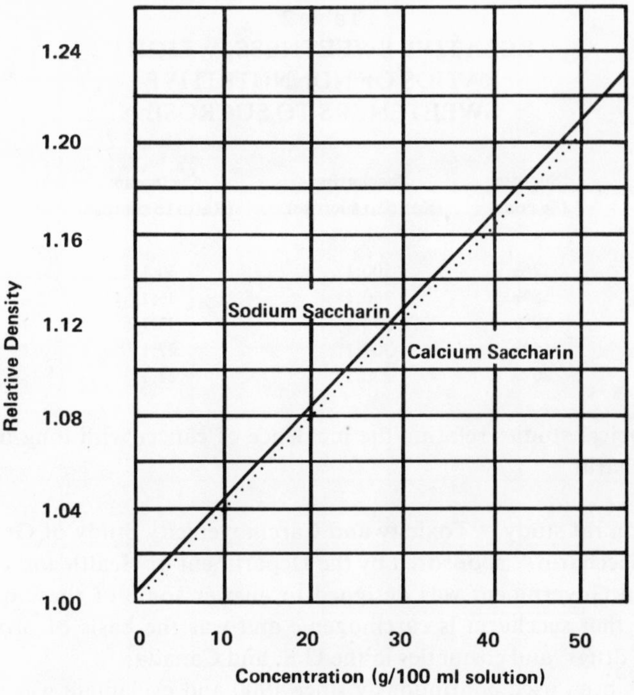

FIGURE 10. Density-concentration curve: calcium and sodium saccharin at 25°C.

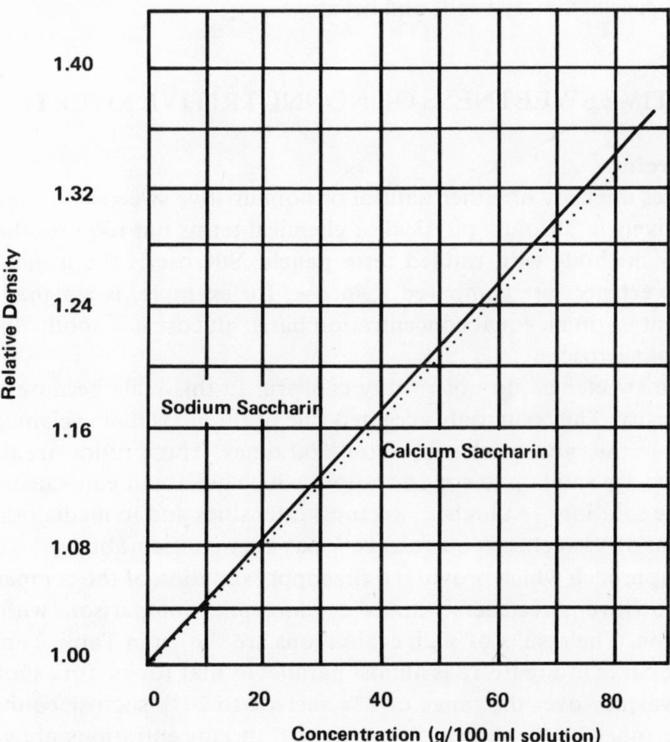

FIGURE 11. Density-concentration curve: calcium and sodium saccharin at 90°C.

Table 2
RELATIVE SWEETNESS, WEIGHT RATIOS OF NONNUTRITIVE SWEETENERS TO SUCROSE

Sucrose (% conc)	Saccharin (Ratio to sucrose)	Cyclamate (Ratio to sucrose)
2%	500:1	40:1
5%	360:1	36:1
10%	330:1	33:1
15%	300:1	27:1
20%	200:1	24:1

7. Epidemiological studies relating the incidence of cancer with long-term consumption of saccharin

A two-generation rat study, "Toxicity and Carcinogenicity Study of Orthotoluenesulfonamide and Saccharin," sponsored by the Department of Health and Public Welfare of the Canadian Government was designed to answer some of these questions. This study indicated that saccharin is carcinogenic and was the basis of proposals to ban its use in foods, drugs, and cosmetics in the U.S. and Canada.

Saccharin has been used continuously since 1900 and cyclamate was widely used in foods and beverages for 20 years. Despite questions on safety that can be raised as a result of feeding these chemicals at very high levels to experimental animals, there is no evidence that either sweetener constitutes a hazard at the levels required for the preparation of special dietary foods and beverages.

RELATIVE SWEETNESS OF NONNUTRITIVE SWEETENERS

Measuring Sweetness

The sweetness intensity of either natural or nonnutritive sweeteners cannot be measured quantitatively in absolute physical or chemical terms but requires the use of subjective sensory methods with trained taste panels. Sucrose is the usual standard, to which other sweeteners are compared. Glucose, for example, is assigned a sweetness value of 70; that is, on an equal concentration basis, glucose has about 70% the sweetness intensity of sucrose.

Nonnutritive sweeteners do not readily conform to this scale because of their high sweetness intensity. The commonly accepted rule of thumb is that cyclamate is 30 times sweeter than sucrose, and saccharin, about 300 times. These ratios are only approximate at best and are valid up to concentrations which are about equivalent in sweetness to 10% sucrose solutions. At higher sweetness intensities and in media other than pure water, the ratios of "sweetness equivalence" may vary considerably.

The simple approach which provides a first approximation of the comparative sweetness of nonnutritive sweeteners entails organoleptic comparison with sucrose in aqueous solution. The results of such evaluations are shown in Table 2 and Figure 12. The cyclamate curve in Figure 12 is almost parallel to that for sucrose showing a fairly constant equivalency over the range of 2% sucrose to 20% sucrose equivalence. Saccharin, on the other hand, shows "leveling off" at concentrations above its equivalence to 7 to 8% sucrose solutions.

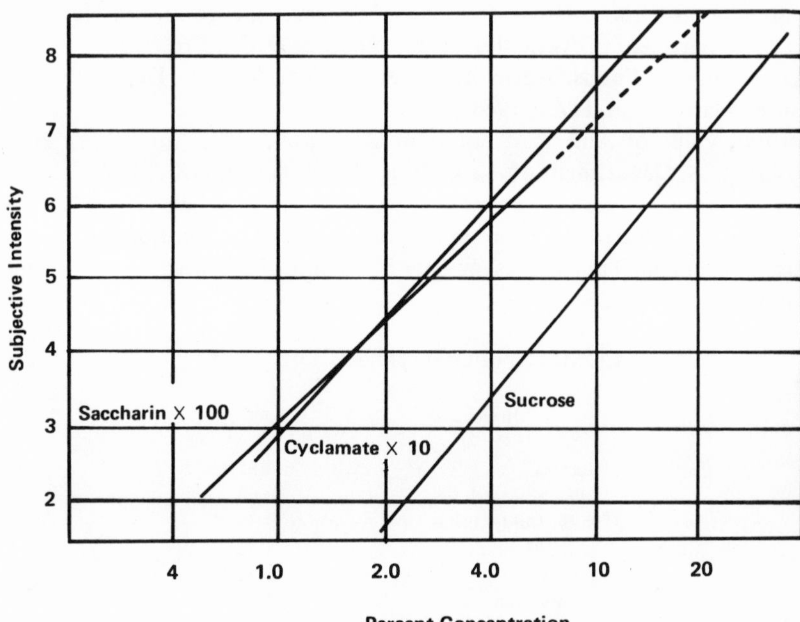

FIGURE 12. Relative sweetness of saccharin, cyclamate and sucrose. (From Schutz, H. G. and Pilgrim, F. J., *Food Res.*, 22, 206, 1957. With permission.)

At higher concentrations, sweetness intensities of nonnutritive sweeteners increase at a lower rate than natural sweeteners. Some express the opinion that this effect is due to increasing levels of bitterness and aftertaste which begin to appear at these concentrations resulting in an effect, more apparent than real. The implications of their practical application are obvious, however. At higher levels of use, there is a problem with attaining adequate sweetness intensity while attempting to avoid some of the undesirable flavor characteristics. The use of cyclamate and saccharin together as cosweeteners has provided a practical solution to this problem, since it allows each to be used at a lower concentration, where its sweetening power is more efficient and its off-taste is less noticeable. In any case, the foregoing description of relative sweetness intensities as determined in aqueous solutions can be regarded as no more than a starting point. The final and realistic determination of proper sweetness level can only be done in the actual product in which the cyclamates and/or saccharins will be used.

Although the sweetening intensity of both saccharin and cyclamate depends on the concentration, flavor, and medium in which the sweetening agent is used, generally saccharin is about ten times as sweet as cyclamate under the same conditions of use. When the two are used together, a 10 to 20% synergism often is observed. In table sweeteners, for example, 5 mg of saccharin plus 50 mg of cyclamate is as sweet as 125 mg of cyclamate or 12.5 mg of saccharin alone.

Food Energy

The approximate energy factors for calculating caloric values are as follows: proteins, 4 cal/g, carbohydrates, 4 cal/g, and fats, 9 cal/g. More specific factors are available and have been published by the Committee on Calorie Conversion Factors and Food Composition Tables, Nutrition Division of the Food and Agricultural Organization. Three important references for use in determining caloric values are the following:

1. Composition of Foods (Raw, Processed, Prepared), Agriculture Handbook No. 8, U.S. Department of Agriculture, Washington, D.C., 1963.
2. Energy Value of Foods, Agriculture Bulletin No. 74, U.S. Department of Agriculture, Washington, D.C., 1963.
3. Nutritive Value of American Foods in Common Units, Agriculture Handbook No. 456, U.S. Department of Agriculture, Washington, D.C., 1975.

Listed below are some of the more important ingredients commonly used in dessert and beverage products. They represent only a portion of the wide variety of materials covered in these references

COMMON FOOD INGREDIENTS*

Ingredient	Factor
Sucrose	3.85
Milk, fluid, whole	0.68
Milk, fluid, nonfat	0.36
Milk, dry, nonfat	3.62
Cocoa	2.93
Cream, light	2.04
Cream, heavy	3.30
Egg, whole	1.62
Egg, white	0.50
Egg, yolk	3.61
Fat, vegetable, cooking	8.80
Starch	3.62
Gelatin, plain	3.35

INGREDIENTS USED IN FOOD MANUFACTURE†

Ingredient	Factor
Citric acid	2.47
Fumaric acid	2.60
Adipic acid	4.80
Sodium citrate	1.84
Sodium phosphate	0.00
Dextrose (monohydrate)	3.62
Sorbitol	4.00
Mannitol	2.00

There are commonly used ingredients which are not listed in these references. In such cases, the manfacturer or supplier of the ingredients in question should be contacted for more precise information.

SENSORY PANEL TESTING

Provided with the foregoing objective data plus rough approximations of sweetness equivalences, the next problem, of course, is actual formulation. Before launching into this aspect of the problem, however, it would be desirable to first consider how to evaluate those products which may be developed. The basic principles of panel testing

* From Composition of Foods (Raw, Processed, Prepared), Agriculture Handbook No. 8, U.S. Department of Agriculture, Washington, D.C., 1963.

† From Energy Value of Foods, Agriculture Bulletin No. 74, U.S. Department of Agriculture, Washington, D.C., 1963.

and the types of tests used for other foods and beverages apply to low-calorie ones as well; however, they tend to be complicated by unusually severe problems of taste fatigue, carry-over, and conditioning, which may hinder the task of obtaining meaningful and reproducible sensory panel data.

Physical Conditions

Special care should be given to the test area. The need for relatively constant and comfortable combinations of temperature and humidity, as well as good isolation from outside influences is very great here. Care should be taken to provide comfortable chairs and tables. A table height of approximately 32 in. has been found suitable with a provision of approximately 32 to 36 in. of space for each judge.

Selection of Panel Members

Panelists for organoleptic tests are screened for acuteness of senses, general interest shown in previous work with flavor problems, and desirable personality characteristics. Here, one must also take into consideration the fact that the population consists of three groups: those with extremely high sensitivity to the taste characteristics of nonnutritive sweeteners, the "average" type of judge, and finally, those who are relatively insensitive to the bitterness and aftertaste of nonnutritive sweeteners. The last group is virtually useless for any type of evaluation of low-calorie products. The hypersensitive group, on the other hand, can be useful for experimental work where a high degree of discrimination is needed. Groupings are somewhat arbitrary; and therefore, sharp lines of demarcation cannot always be made.

Flavor Characteristics of Nonnutritive Sweeteners

A brief examination of their general flavor characteristics is pertinent, since they are the underlying reasons for the difficulties encountered in sensory panel evaluation of low-calorie foods and beverages. Sucrose is the usual standard for most work and will therefore be used as the basis for comparison.

Sucrose has a relatively quick sweetness impact followed by a fairly sharp, clean cutoff. Cyclamate and saccharin, on the other hand, build up to their maximum sweetness intensity at a slower rate, and the sweetness sensation persists for a much longer period. This lingering sweetness in cyclamates and saccharin is responsible for carry-over effects, taste fatigue, and the cloying effect which is encountered in some cases.

These nonnutritive sweeteners, in addition to lingering sweetness, also have other types of aftertaste, the intensity of which are a function of concentration. Generally, the severity of the aftertaste problem starts to increase at a more rapid rate at concentrations where sweetness effects start to level off. These aftertastes can be bitter, sometimes metallic, and also astringent or drying. Aside from the intrinsic unpleasantness of these particular characteristics, they can also contribute to carry-over and fatigue.

Finally, it should be noted that sucrose has a highly desirable blending and bodying effect on the overall product flavor, which neither cyclamate nor saccharin seem capable of duplicating to any significant degree. This particular characteristic, or lack thereof, is not a basic interferent in sensory panel testing, but it does complicate the problem of making direct comparisons with sugar-sweetened products. It also points up the virtual impossibility of attempting to develop products by mere substitution of nonnutritive sweeteners for sucrose. This question will be discussed at greater length and in more detail later.

Effect of Nonnutritive Sweetener Flavor Characteristics on Panel Testing

In any type of panel testing, particularly in those tests where one is attempting to differentiate among samples, it is desirable to have products where flavor cutoff is clean, relatively quick, and with minimal aftertaste. With nonnutritive sweeteners, the opposite tends to be the rule; therefore, particular care is needed in setting up test procedures.

The danger of carry-over and fatigue must be minimized insofar as possible. A common technique is to allow adequate time between tasting consecutive samples; in this case, a minute or more may be required. The use of taste refresheners, such as water, crackers, or a piece of apple is also recommended, or in some cases, a dummy sample may be used for this purpose. Tasting of two or more samples at a given sitting is, of course, unavoidable when differences in taste are being tested, but every attempt should be made to keep the number to an absolute minimum. Another precaution is to limit the number of tests, in which a judge may participate in any given day. Indeed, some cases have been found where there is evidence of carry-over from a morning test into the afternoon.

In the case of preference testing, because of the multitude of problems involved in discrimination among samples, there appears to be good reason to consider monadic testing wherever practical. Monadic testing suffers from the availability of a direct comparison for reference purposes, its relative slowness, and the limitations on the number of samples which can be evaluated. Furthermore, it has also been noted that occasionally a conditioning period is needed before judges are capable of detecting some of the less desirable flavor characteristics, particularly bitterness and aftertaste. There is no known pattern from which guidelines can be drawn, but where repeated and constant use is found to be needed before judges can discriminate properly, there may be no alternative to the monadic test.

The conclusion to be drawn from these observations is that there are no new or special sensory panel techniques that are available for use in testing of low-calorie products. Rather, there is need to be particularly alert to the problems of bias and unfair handicaps to judges in their tasting in order to ensure that meaningful and consistent data are obtained. Judges must be selected with unusual care, and an adequate number of replications employed. Testing can be one of the most troublesome and frustrating phases of low-calorie product development activity, and yet, it is obviously one of the most critical.

PRINCIPLES OF APPLICATION

In preceding sections, the general characteristics of nonnutritive sweeteners and the particular problems involved in organoleptic evaluation of products in which they are used along with some specific implications have been discussed. The importance of quality was stressed; the need to recognize limitations due to the current state of the art and to define objectives accordingly in a realistic manner should be emphasized. In essence, this means the following:

1. It is virtually impossible to duplicate the flavor, texture, and other characteristics of a sugar-sweetened product in one employing nonnutritive sweeteners. Therefore, it is important to strive for equal acceptability rather than exact duplication.
2. There are specific product areas in which it can be stated almost categorically that the present state of the art precludes the formulation and development of low-calorie products having highly acceptable organoleptic characteristics. In the baked goods area, no good solutions to the problem of textural development consistent with good flavor are known. By the same token, there is no known low-calorie system available today capable of providing the sweet glass known

> as hard candy. There are obviously products in these general areas which have achieved a certain degree of market acceptance and therefore must meet, to some extent, the needs of a segment of the consumer population. All, however, tend to have easily detectable deficiencies. It must be concluded that their low-calorie or other dietetic value is the prime reason for acceptance in the first place.

Rarely is a sugar-sweetened product not used as the target, or at least the reference standard, for the formulation of a low-calorie counterpart. It is usually satisfactory as long as the limitations discussed above are recognized and constantly kept in mind, which, in turn, leads to the next cardinal rule for the application of nonnutritive sweeteners in food and beverage product development.

Acceptable nonnutritively sweetened products cannot be developed by the simple substitution of cyclamate and/or saccharin for sugars. Rather, the new product should be completely reformulated from the beginning.

Sweeteners, whether sugars or synthetics, impart more than a taste sensation of sweetness but have characteristic flavors of their own. Cyclamate and saccharin are not identical in flavor to sucrose anymore than sucrose is identical to glucose. It also emphasizes that a finished product has flavor, texture, and other organoleptic properties, which are not the sum of the individual components but rather the result of some type of synergism. The properly formulated product, then, is a balanced whole in which the balance usually cannot be retained if one component is disturbed. These facts are fundamental and generally rather well recognized but somehow often are overlooked when it comes to low-calorie products.

The two most critical areas which must be watched in formulating low-calorie foods and beverages are flavor and texture. These two properties are closely interrelated under any circumstances, but the effect of a given flavor/texture system with nonnutritive sweeteners tends to be significantly different from that observed with sucrose or other sugars. Storage stability can be a problem in certain cases depending upon methods of preservation and packaging which are used. Color and odor generally do not offer serious difficulties, but overall appearance, quite apart from either texture or color, may be significantly affected, especially in cases where sugar has a major role to play in bulking or overall structure.

Flavor

As was noted above, there are intrinsic flavor differences between cyclamate or saccharin and sucrose; in addition, neither sweetener is capable of providing the total flavor blending function. The first conclusion, that can be drawn from these facts and which is supported by observations and actual practice, is that the same flavor will not taste the same in the two different types of systems; furthermore, there may also be a lack of body as well as a development of some of the less desirable flavors. Because of these qualities, the low-calorie product may require the use of either a modified flavor or perhaps an entirely different one. Quite often these changes mean a better flavor and one that is more expensive than sugar.

The role which flavor plays in a low-calorie product is often greater than in a sugar-sweetened one. It is important in masking undesirable sweetener aftertastes and other unwanted flavor characteristics from different sources. The overall effect of body, that quasi-flavor/tactile sensation which may have been originally derived from sugar solids or other high-calorie ingredients, may now have to be imparted or simulated by the flavor itself.

Sweetness-tartness Balance

The balance between the taste effects of the acid and sweetener components is an interplay which has been long recognized and which also, of course, plays a role in the total flavor response. With the differences in the overall sweetness character found in cyclamates and saccharin, the concentrations of acid and buffer salts probably have to be changed. If the lingering sweetness results in an apparently higher sweetness intensity, obviously some degree of compensation can be achieved by increasing the level of acidity. At the same time, however, certain products may have to be formulated with less than optimal sweetness intensity because of the rapidly increasing encroachment of undesirable aftertastes as sweetener concentrations become too high. Again, compensation may be possible to some extent by reducing the acid level. Variation of flavor type can also be used to effect the apparent sweetness-tartness intensity in either direction.

Achieving the proper level of acid and the proper ratio of acid to buffer salt usually cannot be done by pH measurement, but organoleptically in combination with at least the sweetener and flavor involved. Although this involves a potentially astronomical number of combinations, the experience of a skilled formulator can actually overcome this obstacle utilizing, if necessary, a statistically suitable design of experiments.

Texture

The problem of imparting the proper amount of feel to a low-calorie product ranges from the insignificant to the severe. The difficulty, of course, stems from the elimination of sugar solids from the product formulation; the usual approach is compensation by means of a suitable hydrocolloid system or bulking agent. The degree of success tends to vary with the type of product being developed.

In the area of carbonated beverages, a great deal of work has been done to impart improved mouth-feel by the addition of sodium carboxymethyl cellulose (CMC), pectin, and other materials, but by and large, this has proved not to be particularly necessary. A problem related to mouth-feel in low-calorie beverages is the gushing phenomenon in which there is a much greater tendency to lose carbonation in low-calorie products. Highly acceptable low-calorie beverages can be made without using a bodying agent by a proper balance of sweetness and tartness, the use of flavors particularly formulated for this type of drink, and proper carbonation levels.

Gel systems can be significantly affected in varying degrees by the absence of sugar solids; thus, in the case of gels which rely on pectin, there is need to change the type of pectin in order to achieve satisfactory set. In other types of gel systems, differences are somewhat more subtle and, therefore, not quite as severe, although syneresis can be quite troublesome in certain areas.

In the case of baked goods as was discussed previously, the texture problem is particularly acute and, indeed, presents a serious obstacle to formulation of truly high quality products. In a cake, for example, it is extremely difficult to get reasonable development of moistness, crumb structure, and overall volume. Cookies, baked with nonnutritive sweeteners and with the addition of bulking agents, tend to be hard and tough; in addition, they have texture-related flavor problems.

The problem of achieving suitable texture is in part the result of a lack of analytical methods for describing texture in a realistic sort of way and in part, stems from the absence of an ingredient or system capable of reproducing the textural functionality of sugars. These problems will be discussed in greater detail for various product areas.

Storage Stability

Storage stability can be affected by elimination of the sugar solids normally found in food products. Because of the low solid content of products sweetened with nonnutritive sweeteners, they may be more subject to microbiological spoilage. Consequently, it becomes especially important to use a preservative, such as sorbates or benzoates, or to employ a thermal process to achieve adequate shelf life.

Storage stability problems may also stem from sources other than microbial spoilage. These problems tend to be changes in appearance and texture rather than syneresis, toughening, or general drying out and are the result of an absence of sugar solids.

Governmental Regulations

The regulatory status of cyclamate and saccharin is quite different in different countries around the world. It is necessary to determine current applications and tolerance levels that are permitted in each individual country, in which one plans to market a product containing cyclamate or saccharin.

Formulations presented to illustrate the use of cyclamate and saccharin are for technical information only; they are not recommendations for approved applications or use levels. These basic formulations, however, can be valuable starting points for the use of any nonnutritive sweetener.

Carbonated Beverages

The three chief problem areas in development and production of low-calorie carbonated beverages are replacement of the sugar function, meeting flavor requirements, and achieving proper carbonation. Each of these areas will be individually discussed below.

Sugar Replacement

In a regular carbonated beverage, sugar serves to impart sweetness, to round and blend the flavor component, and to provide mouth-feel. It also helps in the retention of carbon dioxide.

The sweetness of sugar can best be imparted by a mixture of cyclamate and saccharin salts. The ratio of cyclamate to saccharin may vary from about 7:1 to 20:1. The actual ratio used would be determined by flavor, economic factors, and pertinent regulations. The best flavor level of the mixture is determined by tasting and should be done while also varying levels of acidulant. In cases where flavor considerations, particularly aftertaste problems, dictate a sweetener level which imparts inadequate sweetness intensity, reduction of acid and variation in flavor type can often compensate.

Sugar is also responsible for mouth-feel. A great deal of formulation work has been done with bodying agents, including hydrocolloids and sorbitol, but these substances generally do not appear to be necessary in carbonated beverages. The viscosity of sugar-sweetened beverages is rather low, a range of perhaps 1.1 to 1.5 cSt, compared to 0.9 cSt for a typical low-calorie beverage. The use of bodying agents can reproduce viscosity fairly readily, but achieving the same type of mouth-feel at the same time tends to be a more difficult matter. Experience indicates that perhaps the best approach toward achieving mouth-feel is by creating an illusion of more body with a flavor that tends to be "heavy in body".

In addition to sweetness, perhaps the most important function of sugars is to blend and round out flavor. Nonnutritive sweeteners do not have this property; therefore, this effect must be achieved by selection of different flavor types and adjustment of other components in the drink.

Flavor Selection

Because of the problems discussed above, when sugars are eliminated, flavors used in low-calorie carbonated beverages must be selected with greater care than usual. They should be of particularly high quality and as free as possible from undesirable notes.

They should be so formulated as to be free of harsh character, well balanced, and rounded overall; these attributes usually indicate higher quality and, as a corollary, somewhat more expensive flavor. In addition to flavor selection, the following additional adjustments can be made in order to obtain the best possible flavor, quality, and impact: 1. selection and adjustment of the acidulant or acidulant blend used, 2. in some cases, the use of bodying agents, and 3. adjustment of the carbonation level to suit the low-calorie formulation.

Typical Formulations

LEMON-FLAVORED BEVERAGE

Terpeneless lemon oil	1 fl oz
Potable alcohol	42 fl oz
Citric acid	1.5 oz
12:1 Cyclamate-saccharin mixture	11.1 oz
Water sufficient to make	1 gal

Mix the oil and alcohol; then, add water. Stir in sweetener and citric acid. Make up to 1 gal and agitate until all solids are dissolved. The resulting solution, which may be somewhat cloudy, can be clarified by filtering through a thin bed of purified, powdered asbestos. The filtrate should be clear. No addition of coloring is required.

To prepare a bottling concentrate, take the above preparation and dilute 375 cc of it to 1 gal of water. The "throw" is 1 oz of this concentration per 12-oz bottle. The bottle is filled with water carbonated at 3 to 3½ volumes. Both the carbonated water and the bottling concentrate should be refrigerated at approximately the same temperature to minimize foaming and gushing during bottling. A lemon-lime flavored beverage may be prepared in exactly the same manner except for the replacement of the lemon oil by a combination of 9 parts of terpeneless lemon oil and 1 part of terpeneless lime oil.

KOLA-FLAVORED BEVERAGE

Fivefold lime oil	2/3 fl oz
Lemon oil	2/3 fl oz
Oil, cinnamon or cassia	80 min
Solid extract kola	5/8 oz
Gum acacia	2 oz
Water	1 pt
Acid-proof caramel to make	1 gal

Mix lime, lemon, and cinnamon oils, and rub with gum acacia. Add ½ pt of water, stir well, and run mixture through a colloid mill or homogenizer to make a base emulsion. Mix the solid extract of kola with ½ pt of water; add this mixture to the base emulsion of oils. Stir while adding caramel to make 1 gal of flavor. The total flavor mix should then be put through a colloid mill or homogenizer.

To prepare a bottling concentrate, use 2½ fl oz of flavor concentrate, 1¼ fl oz of 50% citric acid, and 15 min of syrupy phosphoric acid. For sweetening, add either 2 oz of sodium cyclamate or 0.2 oz of sodium saccharin. Add water to make 1 gal of bottling concentrate. To prepare the beverage, throw 1 fl oz of bottling concentrate to a 7-oz bottle; fill with carbonated water at 3½ to 4 volumes of carbon dioxide.

Carbonation

The chief problem with carbonation in low-calorie beverages is that of gas retention, since gushing or foaming is much more common in the absence of sugar solids. Recognizing the carbonation problem, the following factors should be given careful attention during production.

1. Production water should be carefully filtered to eliminate fine particles that can serve as nucleation points for CO_2 loss.
2. Care should also be exercised in the makeup of the syrup and in equipment sanitation, again, to avoid the presence of fine particles that might serve as nucleation points.
3. If bodying agents are used in premixes or in the final formulation, they can act as foaming agents. Materials, such as methylcellulose and surfactants, should be avoided because of their tendency to enhance foaming.
4. Flavor oils should be carefully emulsified and dispersed in the beverage to prevent boiling or gushing. Emulsification procedures should be adjusted to avoid the use of the materials mentioned above, which tend to foam.
5. Temperature employed during the bottling, filling, and carbonation steps must be carefully controlled and kept at a maximum of 34°F or lower, if possible, which will again help to minimize CO_2 loss.
6. Metal contaminants in trace quantities, such as iron, tin, nickel, cobalt, and aluminum, should also be avoided.

Carbonation Requirement

Carbon dioxide has an important influence on flavor and mouth-feel, which is even greater in the case of the nonnutritively sweetened beverages. Low carbon dioxide content emphasizes the effects of sugar elimination, while increased levels of carbonation have a tendency to increase mouth-feel. Levels of carbonation used in sugar-sweetened beverages are not necessarily suitable for the dietetic product; however, they do provide convenient starting points. Volumes for some of the more common dietetic products are as follows:

Lemon	3.5—4.1 volumes
Orange	2.0—3.0 volumes
Cherry	3.0—3.5 volumes
Lemon-lime	3.5—4.0 volumes
Cola	3.5—4.0 volumes

The importance of carbonation can be quite dramatic in terms of acceptance of low-calorie beverages. A low-calorie orange beverage was prepared in which all components were kept constant except for levels of carbonation which were 2, 3, 4, and 5 volumes of carbon dioxide. When presented to a trained panel for evaluation, the preference was for the higher levels — 4 and 5 volumes — than for lower levels which were described as having a "typical dietetic" flavor. High levels were described as having more mouth-feel and less of the typical nonnutritive sweetener taste. There was some disagreement as to whether the 4-volume or 5-volume level was optimum, but in any case, higher levels were significantly preferred over lower levels.

Dry Beverage Bases

Dry, low-calorie beverage bases are powdered mixtures of cyclamate/saccharin, acidulant (fumaric, citric, or malic acid), color, flavor, and filler (optional). They provide a caloric reduction from the regular sugar-sweetened product and in addition, offer the convenience of presweetening.

Package Weight

Each packet with actual weight varying between ¼ and ½ oz is designed to yield 2 qt of finished drink. The quantity of filler determines final package weight, the higher weight being preferred for the following reasons: 1. More material ensures optimum blending and filling and helps to ensure homogeneity and uniform package weight. 2. Additional filler can help to improve body and mouth-feel, since lower weight packages tend to contain little or no filler and adequate bulk for handling by the packaging equipment is achieved by agglomeration, which also helps prevent segregation of ingredients.

Filler

Fillers, used to increase bulk also add to body or mouth-feel of the reconstituted product, include sorbitol, mannitol, lactose, and dextrose, the latter two being most popular. Economics generally dictates selection of a filler, since there is no significant difference in performance characteristics among the various alternatives.

Acidulants

Until the introduction of cold water-soluble fumaric and malic acids, cirtric acid was the only acidulant used in dry beverage bases. It may be somewhat more expensive to use than the other two, but from the view point of flavor, it is quite acceptable. When citric acid is used, the granulation recommended is 40 to 100 mesh. About 8 to 9 parts of malic acid can replace 10 parts of citric acid which provides a 10 to 20% savings in acidulant costs. Cold water-soluble fumaric acid offers even greater economic advantage. In addition, being nonhygroscopic, its use may provide improved storage stability and less expensive packaging.

Flavor

As in the case of gelatin desserts and other products, flavor type and concentration can be extremely critical. Compatibility with nonnutritive sweeteners should be evaluated; in addition, the flavor should be selected to minimize off-tastes and any artificial character.

TYPICAL FORMULATION

Fumaric acid	18.3 lb
Filler	63.0 lb
12:1 Cyclamate/saccharin mixture	13.2 lb
Flavor and color, approximately	5.5 lb

Net package weight can be varied by changing the quantity of filler used. Elimination of filler and substitution of agglomeration can, of course, provide further economies. The ingredients are mixed thoroughly with a ribbon blender.

Canned Fruits

There are Standards of Identity (21CFR27) for canned fruits which provide general and specific requirements for product quality and fill, ingredient definitions, and label statements.

Fruit Ingredients

Optional fruit ingredients are the same for both regular and dietetic products. In peaches, for example, the various types are yellow clingstone, yellow freestone, white freestone, variety groups, etc., and the various forms are peeled whole, unpeeled whole, peeled halves, peeled quarters, etc. Standards for packing media are explained in the Standards of Identity. Each medium is prepared with either water or peach juice as the liquid ingredient.

For dietetic products, the packing medium is water sweetened with saccharin, cyclamate, or a combination of these. The amount of sweetener required depends upon the natural sweetness of the fruit and the sweetness level desired by the packer. Generally, the amounts of cyclamate per gal of water as a packing liquid for various types of fruit are outlined below:

Fruits	Calcium cyclamate
Apricots, peaches, berries	1¼ oz/gal water
Plums, pears, fruit cocktail	1 oz/gal water
Cherries, figs, grapefruit	¾ oz
Grapes	½ oz

If greater viscosity is desired in the packing liquid, pectin may be used to thicken the solution. A typical formulation for this type of medium is given below:

Water	100 gal
12:1 Cyclamate/saccharin mixture	17 oz
Pectin (slow set, 150 grade)	14 oz

The density of the packing medium for dietetic proucts, measured 15 days or more after the fruit is canned, must be less than 14° Brix, which is the same as that of water.

When fruits are packed in sugar syrup, the density of the syrup is often greater than that of the fruit. The syrup thus has a higher osmotic pressure, and the flow of liquid to achieve equilibrium causes shrinkage of the fruit. Because cyclamate and saccharin are used at such low concentrations, they yield packing liquids with a density essentially like that of water. The lower osmotic pressure of the dietetic packing liquid causes a flow of fluid into the fruit, which results in a plumping effect and higher drained weight and tends to prevent a loss of flavor and texture.

Processing

The same processing requirements apply to dietetic canned fruits as to regular ones. The food is sealed in a container and heat processed to prevent spoilage. No preservative is allowed in either dietetic or regular products when water is used as the packing medium.

Gelatin Desserts

In addition to the usual problems of sweetness, tartness, and overall flavor, the role of gelatin must be given careful attention because of its importance to the total product. Also, since the final product is a dry mix, care must be given to blending and uniformity in dry packaging. Two basic formulations for dietetic gelatin desserts are given below.

FORMULA 1

Gelatin (225 Bloom)	56.2 lb
Sorbitol or mannitol	19.6 lb

Fumaric acid	10.6 lb
12:1 Cyclamate/saccharin mixture	7.1 lb
Sodium citrate	3.5 lb
Salt	3.0 lb
Flavor	As needed
Color	As needed

FORMULA 2

Gelatin (235 Bloom)	88.0 oz
Cyclamate	20.0 oz
Citric acid, anhydrous	29.0 oz
Sodium citrate	8.0 oz
Salt	2.5 oz
Flavor	As needed
Color	As needed

Gelatin

The type of gelatin selected should be one which not only gives good gels but also has a minimum of off-flavors associated with it. The preferred types are pig or calf gelatins with Bloom strengths of 200 to 250, and viscosity in the 25- to 50-cP range. Such high Bloom gelatins usually give the best clarity, are free of undesirable odors and off-flavors, and have slow melting properties as well. The gelatin should be ground so that all of it will pass through the U.S. No. 60 screen and not more than 5% will pass through a No. 325. Finely ground gelatin dissolves more rapidly, particularly at temperatures below boiling (160 to 180°F), but too fine a grind causes agglomeration and caking in storage.

The usual moisture content of gelatin is in the range of 8 to 12%, but lower levels, 5 to 8%, will result in a significant increase in storage stability if the following precautions are taken: 1. Protection against moisture pickup is provided, while the gelatin is stored in bulk; 2. protection against moisture is provided during processing; and 3. the package used for the final product is moisture-proof. Gelatin concentration is determined by running test formula lots. The complete product is prepared with various levels of gelatin, and gel strength, determined after 17 hr at 10°C. The supplier should standardize Bloom strength and the viscosity of shipments utilizing this type of standard test.

Sorbitol/Mannitol

Sorbitol or mannitol in crystalline form is used primarily as a diluent dispersant to improve solubility. It also functions as a filler to standardize net weight and as a bodying agent. As an alternative to these ingredients, a low Bloom gelatin may also be used for standardizing package weights.

Acids

The three most commonly used acids are citric, adipic, and fumaric acid. Citric tends to be more hygroscopic than the other two and, therefore, can present a problem in storage. Also, its higher usage level increases ingredient costs. Adipic acid is used at about the same level as citric acid, but at this level, pH is higher while providing the same tartness making it possible to reduce gelatin concentration without sacrificing gel strength. In addition, its hygroscopicity is significantly lower than that of citric acid. Fumaric acid provides significant economies in use because of its lower initial cost and lower usage level plus its nonhygroscopicity. However, selection of an acid should be governed by overall considerations of flavor, economics, and patents, which may control their usage in gelatin desserts.

Buffer Salts

Most commonly used are citrates and phosphates, and they function to increase pH and modify tartness. Citrates are generally preferred over phosphates for tartness modification. For low-sodium products, potassium salts may be substituted for sodium salts. Calcium salts are generally not used because of their limited solubility and incompatibility problems.

Salt

The function of sodium chloride is exclusively to flavor. Since it has no other functional purpose, it may be omitted where low-sodium content is desired. If salt is removed from the formula, however, it may be necessary to adjust the acid, buffer, and flavor balance.

Flavor

Selection of flavor is also extremely critical. The flavor should be selected according to the type which is most compatible with the nonnutritive sweetener system. In addition, the flavor level should be tested with different combinations of acid, buffer, and sweetener levels. No unusual storage stability problems should be anticipated. Flavors encapsulated by spray drying in gum arabic should be quite satisfactory for most situations. No interaction with the nonnutritive sweeteners or other components should be anticipated.

Color

Normal food colors as permitted by government regulation are applicable to low-calorie gelatin desserts. However, caution should be taken to avoid two potential problems. First, if a blend of primary colors is used, there may be "flashing" problems due to the different rates of solubilities of the various components of the mix. Second, color-plating onto the dry mix can be another source of difficulty. Tests should be conducted on large batches in order to determine what optimum conditions are. Addition of water in order to help color development is not advisable, because moisture can have a deleterious effect on storage stability.

Storage Testing

Tests on gelatin desserts as on any other product are run to determine if there are any incompatibilities due to interaction of ingredients, if the package provides adequate protectin against excessive moisture pickup and product deterioration, and if product assays are within dietary claims. Typical storage conditions are 90°F at 70 and 85% relative humidity (R.H.). Evaluation under 85% R.H. conditions are run every 2 weeks and every 4 weeks at the lower humidity. If there is no significant deterioration at 85% R.H. after 8 weeks, there should be a minimum of 1 year of shelf life under normal conditions.

Processing

The dry product is satisfactorily blended in a ribbon mixer with a 400- or 600-lb capacity being adequate for most purposes. About 100 lb of mix will make approximately 32,000 ½-oz envelopes, each sufficient for a 2-c recipe. Color, acid, buffer, flavor, and sorbitol or other filler are generally blended into a premix. It is desirable to develop at least partial coloration before the final mixing operation, since blending of the entire formulation for longer than 15 min can result in segregation. Uniformity of mixes, package weights, and uniformity between packages are particularly serious problems in low-calorie gelatin desserts because of their concentrated nature.

Cooked Puddings

Caloric reduction in cooked puddings is achieved by elimination of the sugar component. This reduction, in turn, results in textural changes, which must be compensated for by modification of the gel system and other components of the mix. Given below are two prototype formulations, one of which is fruit-flavored requiring the use of an acid:

VANILLA, BUTTERSCOTCH, OR CARAMEL

Nonfat dry milk solids	80.1 lb
Cornstarch	19.7 lb
Carrageenan (protein reactive)	0.08 lb
12:1 Cyclamate/saccharin mixture	0.08 lb
Flavor	As needed
Color	As needed

FRUIT-FLAVORED

Cornstarch	19.7 lb
Sorbitol	76.6 lb
Fumaric acid	3.5 lb
Carrageenan (protein reactive)	0.086 lb
12:1 Cyclamate/saccharin mixture	0.08 lb
Flavor	As needed
Color	As needed

Nonfat Dry Milk Solids

Use of nonfat dry milk solids is optional but extremely helpful for improved flavor and texture. Where water is used for reconstitution, the addition of nonfat dry milk solids is recommended. This is true even with lemon-flavored puddings where a pure water recipe is acceptable in the sugar-sweetened version. The above formula for lemon-flavored pudding is considerably improved when 25 to 50% of the sorbitol is replaced by nonfat dry mik solids. Care should be taken that the milk solids used are fresh and free of cooked flavors.

Cornstarch

Regular raw cornstarch may be used although textural improvements and better tolerance in recipe preparation can be achieved in many cases by means of modified starches or blends. In addition to flavor and texture, starches should also be evaluated for consistency changes upon standing, since upon storage in a refrigerator, many starches continue to set and take on a rubbery character. Syneresis is also found to be particularly critical in low-calorie products, where not enough solids are present to "bind" the water.

Carrageenan

Carrageenan can help provide tolerance to over- or under- gelatinization of the cornstarch. Other gums, such as pectin, alginates, and cellulose, may also be used for this purpose. The actual gum selected depends on the type of cornstarch used. Care should be exerted to avoid incompatibilities or other types of undesirable synergism. When incompatibility occurs, there will be low gel strength and syneresis, while synergistic effects will be manifested by heavy gelling or a pasty texture. The carrageenan or other selected gum should provide the following tolerances: 1. A ± 10% liquid variation of the recipe preparation, and 2. two extremes of cooking, one in which the mixture is removed from the heat at the first sign of a bubble and the other where it is allowed to boil for 2 full min.

Acid

Only fruit-flavored, cooked puddings require the use of an acidulant. Fumaric acid is recommended because of its low hygroscopicity and cost savings, although other acids such as citric, malic, and adipic can be used as well. It should be noted that there are patents covering the use of acidulants in certain types of pudding systems.

Instant Puddings

These formulations and products are very similar to cooked puddings except for the use of pregelatinized starch to eliminate the need for cooking of the pudding and the use of additional ingredients to ensure achievement of proper texture. Again, as in the case of cooked puddings, elimination of sugar components creates a texture problem, which must be solved in order to prepare acceptable products. Similarities and differences between cooked and instant puddings can be seen in the prototype formulas below:

IMITATION-FLAVORED PUDDING

Nonfat dry milk solids	73.0 lb
Pregelatinized starch	18.0 lb
Disodium orthophosphate	3.6 lb
Tetrasodium pyrophosphate	3.6 lb
12:1 Cyclamate/saccharin mixture	1.8 lb
Carrageenan	0.7 lb
Flavor	As needed
Color	As needed

CHOCOLATE-FLAVORED PUDDING

Nonfat dry milk solids	50.8 lb
Cocoa	30.4 lb
Pregelatinized starch	8.3 lb
Disodium orthophosphate	4.2 lb
Tetrasodium pyrophosphate	4.2 lb
12:1 Cyclamate/saccharin mixture	2.1 lb
Flavor	As needed
Color	As needed

Nonfat Milk Solids

These solids serve the same function in instant puddings as they do in cooked puddings. That is, they act as a diluent to improve dispersibility and solubility, provide bulk for easier packaging, improve flavor and texture, and also serve to standardize net weight.

Cocoa

Although there are a variety of cocoa types available for use in chocolate-flavored pudding, one should be selected that has the flavor, color, and overall compatibility with nonnutritive sweeteners. Either plain or Dutch cocoa with a fat content between 10 and 20% can be used. While high-fat cocoas are acceptable, they are more expensive and increase caloric content.

Pregelatinized Starch

This substance provides body and gives texture to the product. The best types are those made from potato and tapioca starch because of bland flavor even at high concentrations. Furthermore, pregelatinized tapioca starch provides higher viscosity (better body) and a pudding texture, closer to that of a cooked pudding.

As in the case of the raw starches, the pregelatinized starch also has inherent variation. In some cases, addition of a small amount of cold water-soluble gum helps improve uniformity of performance. Cold water-soluble gums, such as alginates and carrageenans, can also be used in the formulation of instant pudding. Their setting and thickening properties, however, differ from that of pregelatinized starch leading to different product texture.

Phosphate Salts

The principle gelling agent is tetrasodium pyrophosphate which reacts with milk protein to form a complex. The use of phosphates in instant pudding is covered by a number of patents so that any formulation should be checked to be sure it does not conflict with them. Disodium orthophosphate is used chiefly as a buffer to aid in the function of the tetrasodium salt as a thickening agent. The phosphate salts used should be very finely granulated. Generally, 100% should pass through a U.S. No. 60 screen in order to insure rapid solubility in cold water. The amount of phosphate used is determined by the concentration and type of gum employed.

Additional Ingredients

The basic instant pudding formulation may be modified by the addition of texture-improving ingredients, all of which are protected by patents. These substances include calcium salts, vegetable oils, mono- and di-glycerides or other emulsifiers.

Salad Dressings

A favorite item of people on weight-reducing diets is a tossed salad with a low-calorie salad dressing. It is nutritious and provides the bulk necessary to assuage hunger feelings with a minimal number of calories. The major source of calories in a salad dressing is the oil component, and the usual approach to formulating a low-calorie counterpart is to eliminate or reduce the oil component. Listed below are some brief, general descriptions of three of the more common types of salad dressings:

1. French dressing is a separable liquid food or an emulsified, viscous fluid food. It is prepared from edible vegetable oil, one or two acidifying ingredients, optional seasonings or flavorings, and, if desired, optional emulsifying ingredients. The oil content may not exceed 35% by weight.
2. Mayonnaise is an emulsified, semi-solid food, prepared from edible vegetable oil, one or more acidifying ingredients, and one or more egg yolk-containing ingredients. The minimum vegetable oil content in mayonnaise is 65%.
3. Salad dressing is an emulsified, semi-solid food, prepared from edible vegetable oil, one or more acidifying ingredients, one or more egg yolk ingredients, and a

cooked or partly cooked starch. Salad dressing should have a minimum of 30% by weight vegetable oil and not less than 4% by weight egg yolk.

The formulation problem in creating low-calorie counterparts of the foregoing is to maintain the flavor and texture and overall appearance with a significantly reduced oil content.

Salad Dressing Ingredients

Emulsifiers — Emulsions used in salad dressings are the oil-in-water types; thus, the emulsifiers must be suitable for these systems. In addition, various types of stabilizers are used. The following ingredients are used to stabilize salad dressings: eggs, cornstarch, flour, gelatin, xanthan gum, gum tragacanth, gum arabic, pectin, carrageenan, locust bean gum, gum karaya, propylene glycol alginate, sodium carboxymethyl cellulose, mustard, and paprika. In addition, eggs can also function as fairly efficient emulsifiers primarily due to its phospholipid component.

Egg yolks — Care should be taken that yolks are salmonella-free; consequently, frozen egg yolks are preferred. Saline solution is generally added to frozen egg yolks prior to freezing; this addition increases the viscosity of the thawed product and tends to improve its emulsifying properties. The egg yolk color will vary from a deep yellow-red to a pale yellow and can be an important contributor to the color of the finished mayonnaise.

Starch — Starch is widely used as a stabilizer. Its properties vary depending upon the particular type employed; thus, properties, such as gelatinization temperature, taste, texture, and clarity, can vary significantly depending upon whether the starch is derived from one of a number of sources including potato, corn, tapioca, amioca, arrowroot, wheat, rice, and sago. In addition, there are pregelatinized or cold water-soluble starches and chemically modified, natural, raw starches. With this wide selection, a specific type or combination of types which will yield a product with both acceptable mouth-feel and shelf life is possible.

Oil — Although the oil content in low-calorie products may be significantly reduced compared to the usual product, it can still have a profound effect on flavor, texture, and shelf life; thus, care should also be exercised in the selection of this ingredient. Purity, free fatty acid content, and the American Oil Chemists' Society (AOCS) cold test specifications are consequently critical to the low-calorie product. Use of acidulants and spices to suppress the off-flavors of poor-quality or oxidized oils is not recommended, since they may enhance rather than suppress these qualities.

Acidulants — The usual acidulants employed are vinegar and citric acid. Ideally, they should be mild in odor and taste but should enchance the flavor of the product. Metals should also be removed from the vinegar in order to avoid oxidative catalysis. Citric acid, when used, will generally serve as only a partial replacement for vinegar.

Spices and seasoning — The most popularly used seasonings in salad dressings are mustard flour, which also has some emulsifying properties, as well as pepper, paprika, onion, and garlic. The spices, ginger, mace, cloves, tarragon, and celery seed are used to a lesser extent in salad dressings. Dry spices can be contaminated by dirt and bacteria and require very strict quality control.

Processing

A variety of processes are available and used by different manufacturers, but the following is fairly typical and representative.

1. Water-vinegar and starch are combined in a large, steam-jacketed kettle fitted with an agitator. The vinegar, in some cases, is added later with the egg yolk and salad

oil. Care should be taken to add the starch slowly with constant agitation in order to minimize lumping.

2. Heat is applied to the mixture with continuous agitation until a heavy paste is formed. The product temperature at this point is somewhere between 180 and 200°F.
3. The paste is cooled to a temperature of about 140 to 150°F. The egg yolk and salad oil are then thoroughly blended until a homogeneous mixture is formed.
4. The preblended dry ingredients are then added slowly and mixed to form a fine emulsion, which should be smooth in appearance and mouth-feel.
5. The emulsion is then poured into jars and sealed tightly.

Storage Stability

The objective for product shelf life should be a minimum of 3 to 6 months at temperatures ranging between 30 and 100°F. Types of spoilage which can occur include separation of the emulsion, fermentation, and off-flavor development. The chief reason for mayonnaise spoilage is emulsion breakdown as a result of oil oxidation or improper emulsion preparation; bacterial spoilage is less frequent due to the inhibition by acid.

Emulsion breakdown can occur as a result of mechanical shock, particularly in the case of mayonnaise. Allowing the product to cool to too low a temperature may result in oil crystallization and can also cause separation. Finally, improper techniques employed in the original mixing procedures, the egg component in the case of mayonnaise and salad dressings, or improper operating temperatures may also be responsible. Fermentation problems generally are the result of failure to observe proper sanitary procedures. It can be avoided by use of thorough and regular antiseptic cleaning of plant and equipment.

Off-flavors and odors can be minimized by proper selection, care, and handling of ingredients. High quality ingredient standards should be maintained along with well-balanced formulas and plant procedures. The oil used should be free of any incipient rancidity and should not be stored for excessively long periods of time. All equipment used to store and handle ingredients should be clean and of stainless steel, whenever possible. Finally, all jar and other container closures should be air-tight.

Typical Formulas

The following formulas are intended primarily as guides in preparing low-calorie salad dressings. Considerable variation is possible in the actual ingredients used, their order of addition, speed and temperature of mixing, as well as sources of raw materials.

FRENCH-TYPE DRESSING

Water	72 lb	3.6 oz
Vinegar, cider type	12 lb	0.4 oz
Vinegar, wine type	6 lb	0.0 oz
Starch	2 lb	6.4 oz
Garlic powder	1 lb	15.4 oz
Egg yolks	1 lb	3.2 oz
Salt	1 lb	3.2 oz
Propylene glycol alginate	1 lb	0.4 oz
Tomato powder		9.6 oz
Onion powder		8.6 oz
Mustard powder		5.8 oz
12:1 Cyclamate/saccharin mixture		3.2 oz

White pepper, powdered	2.6 oz
Nutmeg powder	1.2 oz
Celery powder	1.2 oz
Paprika oleoresin	0.4 oz

MAYONNAISE-TYPE DRESSING

White vinegar	33 lb	3.2 oz
Water	29 lb	1.6 oz
Salad oil	26 lb	9.6 oz
Starch	4 lb	2.4 oz
Egg yolk	3 lb	5.6 oz
Salt	2 lb	7.8 oz
Propylene glycol alginate		13.2 oz
12:1 Cyclamate/saccharin mixture		2.4 oz
White pepper, powdered		1.2 oz
Mustard powder		1.2 oz

SALAD-TYPE DRESSING

White vinegar	39 lb	0.8 oz
Water	34 lb	3.2 oz
Salad oil	15 lb	9.6 oz
Starch	4 lb	14.4 oz
Salt	2 lb	14.8 oz
Egg yolk	1 lb	15.2 oz
Propylene glycol alginate		15.6 oz
12:1 Cyclamate/saccharin mixture		3.2 oz
White pepper		1.6 oz
Mustard powder		1.6 oz

The foregoing salad dressing formulas differ from their high-calorie counterparts as a result of the reduction or complete elimination of the oil content and the use of hydrocolloids to help impart proper texture. In the case of the French-type dressing, there is no vegetable oil at all. In the mayonnaise- and salad-type dressings, oil content has been reduced by 50% or more. The results of such changes are significant differences in flavor, mouth-feel, pourability, and appearance as compared to their high-calorie counterparts. Removal of most or all of the oil in salad dressings underlines the need for particular care in formulation to overcome what might otherwise be serious deficiencies that could destroy consumer acceptability.

Jams, Jellies, and Preserves

Gelling Ingredients

Regular products contain approximately 65% or more soluble solids; whereas, low-calorie products generally have a content range of 15 to 20%. Because of the low solid content, the more commonly used pectins with relatively high methoxyl content will not give satisfactory gels. Therefore, a special type, a low-methoxyl (LM) pectin, must be employed. This LM pectin is chemically modified to form a gel in the presence of calcium, regardless of the solid content. Additional gelling ingredients, which are occasionally used, include xanthan gum, agar-agar, locust bean gum, guar gum, gum karaya, gum tragacanth, algin, and carrageenan. These substances may be used in some cases singly, but more often in combination, to obtain the specific type of texture and gel that is desired.

Gel strength and quality are affected by variation in the natural solid content, pH, and the natural salts present. There is no firm rule for deciding in advance which particular type of system should be employed; it can only be done by modification and adjustment involving all the ingredients, not only the gelling agent, but the acids, salts, and sequestering agents. It should be noted that sequestering agents can be very important to the controlled release of calcium and other divalent salts, to which pectin is sensitive, and thus important in the control of texture. Locust bean, guar, tragacanth, and carboxymethyl cellulose are generally used to modify gel texture. They may be used to alter spreadability and texture and to impart increased resistance to gel cleavage during shipment.

Agar is either used alone or in combination. It does not require calcium or other chemicals to obtain the set as does the low-methoxyl pectin. The stiff, nonmelting gels produced with agar tend to be somewhat similar to pectin gels. Algin is very similar in behavior to pectin. It requires the presence of calcium and pH adjustments as does pectin. Carrageenan, on the other hand, requires a potassium salt in order to obtain gel formation.

Acid Adjustment

In order to obtain satisfactory gels, it is necessary to have a proper concentration of acid, i.e., proper pH and buffer salts. The natural fruit content and its characteristics will, of course, determine what levels and types are needed. Generally, for flavor reasons and also to get proper gel characteristics, a pH range of 3.0 to 3.8 is preferred. However, each batch of fruit should be checked for pH.

Syneresis

Texture and spreadability will be affected by pH. The standards (21CFR29) list the acidulants and concentrations which are permitted.

Preservatives

Because of low solid content, a preservative must be added. Normally, the products are heat processed to prevent spoilage so the added preservative serves as protection only after opening the container. The preservatives permitted include ascorbic acid, sorbic acid, sorbate salts, propionate salts, and benzoates with a limit of about 0.1% preservative by weight of the finished food.

Nonnutritive Sweeteners

The most common sweetener system used is a combination of saccharin and cyclamates. Calcium salts of the sweeteners may be employed to function at the same time as a source of calcium to aid in the setting of the pectin gel. It should be pointed out, however, that with the normal sodium content of the natural component of the product, it would be difficult to develop a product low in sodium, even with the calcium salts as sweeteners.

Processing

The gelling ingredient should be added to the full amount of water in the kettle and heated to boiling in order to disperse and hydrate it. The juice, sweeteners, and other ingredients are added, and the mixture is again brought to a boil. The mixture is poured at 180°F into jelly glasses which are then cooled as rapidly as possible in order to prevent darkening and development of cooked flavor. It should be noted that in the case of preserves and jams, where there is fruit, the mixture is cooled to about 160°F with slow stirring prior to the filling of the glasses. The lower temperature is necessary in order to allow the mixture to set rapidly enough to prevent floating of the fruit.

Typical Formulations

Listed below are some typical formulations illustrative of the products described:

DIETETIC APPLE JELLY

Apple juice	94.6 lb
Low methoxyl pectin	1.4 lb
Citric acid, anhydrous	0.3 lb
12:1 Sodium cyclamate/sodium saccharin mixture	0.2 lb
Calium chloride, 1% solution	3.5 lb
Total	100.0 lb

DIETETIC PEACH JAM

Crushed peaches	60.0 lb
Low methoxyl pectin	1.6 lb
Citric Acid, anhydrous	1.0 lb
12:1 Sodium cyclamate/sodium saccharin mixture	0.5 lb
Calcium chloride, 1% solution	5.0 lb
Water	32.0 lb
Total	100.0 lb

LOW SODIUM DIETETIC PRESERVES

Fruit	60.0 lb
Low methoxyl pectin	1.0 lb
Calcium cyclamate	6 oz
Calcium saccharin	⅔ oz
Potassium citrate	5 oz
EDTA	3⅓ oz
Water	38.0 lb
Total	100.0 lb

Frozen Desserts

The caloric values for some of the optional ingredients used in ice cream and frozen desserts are as follows:

Cream, 35% fat	330 cal/100 g
Whole milk	68 cal/100 g
Milk solids, nonfat	362 cal/100 g
Condensed unsweetened milk	138 cal/100 g
Sucrose	385 cal/100 g
Dextrose	348 cal/100 g
Dried egg yolk	693 cal/100 g
Stabilizer-emulsifiers (estimated average)	450 cal/100 g

A regular ice cream formulation with 12% fat contains about 210 cal/100 g of finished product. The objective in making a dietetic product is to achieve at least a 30% reduction in calories. The obvious route is to replace fats and carbohydrates. Nonnutritive sweeteners are used as substitutes for sugars, while fat is reduced or eliminated with the necessary formula modifications to give good texture.

Given below is a basic formula for a vanilla or fruit-flavored frozen dessert with a caloric value of 144 cal/100 g. It contains 3.5% fat and 20.5% nonfat milk solids. In order to obtain a chocolate flavor, part of the milk solids and/or sorbitol can be replaced with cocoa.

Nonfat milk solids	22.7 lb
Heavy cream, 35% fat	10.1 lb
Sorbitol	7.0 lb
12:1 Cyclamate/saccharin mixture	0.2 lb
Stabilizer[a]	0.5 lb
Water	59.5 lb
Flavor, color, nuts	As needed

[a] 50% CMC, 7 HOP® and 50% gum karaya

Fat

Fat generally can be reduced to less than 6% in a frozen dessert. The usual sources of fat are the dairy products, including milk solids, whole milk, fresh sweet cream, unsalted butter, butter oil, sweet cream buttermilk, and plain condensed milk.

Milk Solids

These substances should be selected for quality, cost, and performance. Any of several sources may be used in order to lower the cost. Milk solids should have a clean, fresh, creamy flavor and odor so that they will retain overall flavor quality during storage. Usable sources of milk solids are whole milk, sweet skim milk, nonfat dry milk, powdered whole milk, plain condensed skim milk, and plain condensed milk. As fat content is varied, formula adjustments will be needed.

Stabilizers and Emulsifiers

These substances are extremely critical in any low-solid, low-fat system, such as a dietetic frozen dessert. They are needed to bind free water and prevent large crystal formation during freezing. The stabilizers also contribute to the smooth texture that is desirable and normally found in high fat-content products. Stabilizers can be a source of undesirable flavor and texture; therefore, they must be carefully selected and tested. Some common ones used, either singly or in combination, are gelatin, agar, locust bean gum, guar gum, algins, carrageenans, and cellulose gums.

Emulsifiers increase creaminess and smoothness by increasing the effectiveness of the fat used. They may also function as stabilizers. The types of emulsifiers and their level of use are restricted. Included are such common emulsifiers as the mono- and diglycerides of edible fats, which are permitted at levels of up to 0.2%. Polyoxyethylene emulsifiers are permitted up to a level of 0.1%

Sorbitol/Mannitol

These ingredients act as conditioning or blending agents and, to some extent, may also help control crystallization. Nonfat milk solids can often be used in place of sorbitol and mannitol.

Processing

The general procedure for preparing a dietetic frozen dessert is as follows:

1. Thoroughly blend the dry ingredients.
2. Combine all liquid ingredients, and heat slowly. Agitate and add the dry ingredients before the temperature reaches 120°F.
3. Depending upon the stabilizers used, disperse and dissolve these ingredients.
4. Pasteurize the entire mix to meet the legal requirements which may include a bacterial count and pasteurization time and temperature.

5. Homogenize the mix to reduce the droplet size of the fat and thereby, increase stability, impart smoother texture, improve shipping ability, lessen the risk of churning in the freezer, and allow for a slight reduction of the stabilizer.
6. Cool the mix to approximately 32 to 40°F. Color and flavor may be added at this point.
7. Age the mixture in this temperature range for 3 to 4 hr before freezing. It will allow the fat to solidify, the stabilizer to combine completely with water, and the viscosity to build up to a maximum. Aging improves smoothness, resistance to melting, and ease of whipping.
8. Freeze the mixture. Fast freezing is essential for small ice crystal formation. Actual time and temperature for freezing are primarily determined by the type of freezer and the composition of the mix.

Baked Goods

Sugar is an extremely critical ingredient in baked goods and replacing it presents some very difficult technical problems. These problems stem from the important functions of sugar, including the following:

1. Caramelization of sugar is necessary for a desirable golden brown color.
2. Sugar is the part of the system that supports the flour protein in forming a structure or framework and is the basis for what is generally called texture. In heavy doughs, such as that for cookies, structure formation may be less important.
3. High-sugar concentration and minimum protein in cake yield the most tender crumbs. If th protein content is too high, cakes and cookies become tough.
4. Sugar is the most important source of bulk for "creaming", which is the development of an emulsion that retains the gas formed during leavening.
5. Sugar, shortening, and leavening agents act as tenderizers. Flour, milk solids, egg solids, and other proteinaceous materials generally act as tougheners.

As a general rule of thumb for high-quality baked products, the amount of sugar is slightly higher than that of flour. Sugar provides good keeping quality, moistness, tenderness, and flavor. When the sugar in a cake is replaced by nonnutritive sweeteners, its solids are reduced by 25% and the appearance, volume, and texture of the cake tend to be undesirable. The cake is tough, small, and poor in color and eye appeal. Various types of compensating ingredients are necessary.

Compensating Ingredients

For a dietetic product, compensating ingredients should have either a lower caloric value or function at lower concentrations. Since there are none readily available whose caloric value is below four, obviously, lower levels of these must be used. Some, which are employed at times, include carboxymethyl cellulose, mannitol, sorbitol, and dextrins, but by and large, they have not been completely satisfactory. The best results, thus far, have been obtained with cookies.

PROTOTYPE FORMULAS

Dietetic Chocolate Fudge Cake

Unsweetened chocolate	2 oz
Butter	1.5 oz
Cake flour	4.3 oz
Nonfat milk solids	2.4 oz
Eggs	3.8 oz
Baking powder	0.23 oz
Sodium aluminum pyrophosphate	0.12 oz
Sorbitol	3.5 oz
12:1 Cyclamate/saccharin mixture	0.035 oz
Vanilla extract	0.2 oz
Salt	0.04 oz
Water	4 oz

Dietetic Cookies

Butter	10 oz
Vanilla extract	0.1 oz
Nonfat milk solids	2.4 oz
Eggs	3.8 oz
Cake flour	7.0 oz
Salt	0.04 oz
12:1 Cyclamate/saccharin mixture	0.035 oz
Ginger	0.035 oz
Cinnamon	0.035 oz

Development of ingredients and processes for preparing high-quality, low-caloric baked goods remains one of the outstanding technical challenges that has yet to be met by the food industry. For the most part, the available compensating ingredients and formulation procedures have not been adequate for the production of a variety of high-quality products.

REFERENCES

1. **Adam, W. B.,** Control of sweetness in canned fruits and vegetables, *Campden, Ann. Rept.,* Univ. Bristol Fruit Veg. Preservation Res. St., 15, 1940; *Chem. Abstr.,* 35, 578, 1942.
2. **Aichinger, F.,** Osmotically effective compound for chaoul and plastic mass irradiation, *Strahlentherapie,* 91, 393, 1953; *Chem. Abstr.,* 47, 12753g, 1953.
3. **Akagi, M. and Tejima, S.,** Spectrophotometric determination of *p*-ethoxyphenylurea (dulcin) *Yakugaku Zasshi,* 77, 1043, 1957; *Chem. Abstr.,* 52, 985g, 1958.
4. **Alberti, C.,** Variations in the taste of dulcin, *Atti V Congr. Nazl. Chim. Pura Applicata* (Rome), pt. I, 271, 1935, *Chem. Abstr.,* 31, 3890_4, 1937.
5. **Alberti, C.,** Transformations of dulcin, *Atti X.° Congr. Intern. Chim,* 3, 21, 1939; *Chem. Abstr.,* 34, 1000_5, 1940.
6. **Alberti, C.,** The acetylation of dulcin, *Gazz. Chim. Ital.,* 65, 922, 1935; *Chem. Abstr.,* 30, 3416_8, 1936.

7. **Alberti, C.**, The pyrolysis of dulcin and of acetyldulcin, *Gazz. Chim. Ital.*, 65, 296, 1935; *Chem. Abstr.*, 30, 3417_3, 1936.
8. **Alberti, C.**, Dulcin. V. Some transformations of dulcin, *Gazz. Chim. Ital.*, 69, 150, 1939; *Chem. Abstr.*, 33, 7283_2, 1939.
9. **Aldinger, S. M., Speer, V. C., Hays, V. W., and Catron, D. V.**, Effect of saccharin on consumption of starter rations by baby pigs, *J. Anim. Sci.*, 18, 1350, 1959; *Chem. Abstr.*, 54, 13489f, 1954.
10. **Alikonis, J. J.**, Dietetic foods, their status and how to make them, *Food Eng.*, 28, 92, 1956.
11. **Althausen, T. L. and Wever, G. K.**, Effect of saccharin and of galactose on the blood sugar, *Proc. Soc. Exp. Biol. Med.*, 35, 517, 1947; *Chem. Abstr.*, 31, 4393, 1937.
12. A.M.A. Council on Foods and Nutrition, Artificial sweeteners, *JAMA*, 160, 875, 1956.
13. **Anderson, E. E., Esselen, W. B., Jr., and Fellers, C. R.**, Non-caloric sweeteners in canned fruits, *J. Am. Diet. Assoc.*, 29, 770, 1953; *Chem. Abstr.*, 47, 10147f, 1947.
14. **Annecke**, Sweeteners in the drug industry, *Dtsch. Apoth. Ztg.*, 54, 244, 1939; *Chem. Abstr.*, 33, 4373_5, 1939.
15. **Anon.**, Artificially sweetened canned pineapple; definition and standard of identity, *Fed. Regist.*, 26, 12563, 1961; *Chem. Abstr.*, 56, 7554^h.
16. **Ant-Wuorinen, O.**, Identification of saccharin and dulcin in beer, *Z. Unters. Lebensm.*, 70, 389, 1935; *Chem. Abstr.*, 30, 809_2, 1936.
17. **Arreguine, V.**, A new microchemical reaction of saccharin, *An. Asoc. Quim. Argent.*, 30, 38, 1942; *Chem. Abstr.*, 36, 5732_3, 1942.
18. **Arreguine, V.**, A new microchemical reaction for the identification of saccharin, *Rev. Univ. Nac. Cordoba*, No. 7—8, 14 pp., 1941; *Chem. Abstr.*, 36, 4060_2, 1942.
19. **Arreguine, V.**, A new microchemical reaction for the detection of saccharin, *Rev. Asoc. Bioquim. Argent.*, 8, 24, 1942; *Chem. Abstr.*, 38, 6237_i, 1944.
20. **Atanasiu, I. A.**, Electrolytic oxidation of *o*-toluenesulfonamide (to saccharin). *Bul. Inst. Natl. Cerecet. Tehnol.*, 3, 29, 1948; *Chem. Abstr.*, 43, 2521c, 1949.
21. **Audrieth, L. F. and Sveda, M.**, Preparation and properties of some *N*-substituted sulfamic acids, *J. Org. Chem.*, 9, 89, 1944; *Chem. Abstr.*, 38, 2020_7, 1944.
22. **Audrieth, L. F., Sveda, M., Sisler, H. H., and Butler, M. J.**, Sulfamic acid, sulfamide and related aquo ammonosulfuric acids, *Chem. Rev.*, 26, 49, 1940; *C.A.*, 34, 2723_3, 1940.
23. **Auerbach, F.**, Sweetening power of artificial sweetening substances, *Naturwissenschaften*, 10, 710, 1922; *Chem. Abstr.*, 16, 3982, 1922.
24. **Barnard, R. D.**, Effect of saccharin ingestion on blood coagulation and the in vitro anticoagulant effect of saccharin and ferriheme, *J. Am. Pharm. Assoc.*, 36, 225, 1947; *Chem. Abstr.*, 41, 7528c, 1947.
25. **Barnard, R. D.**, Saccharin ferrihemoglobin, *Proc. Soc. Exp. Med.*, 54, 146, 1943; *Chem. Abstr.*, 38, 557_5, 1944.
26. **Barral, F. and Ranc, A.**, The industry of sweetening agents, *Ind. Chim.*, 6, 73, 1919; *Chem. Abstr.*, 13, 1505_6, 1919.
27. **Barral, F. and Ranc, A.**, The chemistry of sweetening agents, *Rev. Sci.*, 56, 712, 1918; *Chem. Abstr.*, 13, 620_8, 1919.
28. **Baumann, A.**, Detection of saccharin and dulcin in beer, *Z. Gesamte Brauwes.*, 43, 137, 1920; *Chem. Abstr.*, 14, 3496, 1920.
29. **Bechara, E. and Huyck, C. L.**, Compatibility of sweetening agents, *Am. Prof. Pharm.*, 23, 53, 1957; *Chem. Abstr.*, 51, 7649h, 1957.
30. **Becht, F. G.**, The influence of saccharin on the catalases of the blood, *J. Pharmacol.*, 16, 155, 1920; *Chem. Abstr.*, 15, 400_3, 1921.
31. **Beck, K. M.**, Properties of the synthetic sweetening agent, cyclamate, *Food Technol.*, 11, 156, 1957; *Chem. Abstr.*, 51, 15817a, 1957.
32. **Beck, K. M.**, Sucaryl sweetened beverages improved through use of bodying agents, *Food Process.*, 5, 27, 1954.
33. **Beck, K. M.**, Non-caloric sweeteners and the dietetic beverages, Proc. 6th Annu. Meet. Soc. Soft Drink Technol., Washington, D.C., April 1959, 113.
34. **Beck, K. M.**, Basic formulation of special dietary frozen desserts, *Ice Cream Rev.*, 40, 26, 1957.
35. **Beck, K. M., Jones, R. L., and Murphy, L. W.**, New sweetener for cured meats, *Food Eng.*, 30, 114, 1958.
36. **Beintema, J., Terpstra, P., and de Vrieze, J. J.**, Crystallography of the copper-pyridine-saccharin complex $CuPy(H_2O)Sa_2$, *Pharm. Weekbl.*, 72, 1287, 1935; *Chem. Abstr.*, 30, 667_6, 1936.
37. **Belani, E.**, The Jung automatic feeder (as patented by Dallwitz-Wegner) and its use in saccharin manufacture, *Chem. Ztg.*, 49, 877, 1925; *Chem. Abstr.*, 20, 1540_5, 1926.
38. **Belani, E.**, The drying of saccharin, *Chem. Ztg.*, 51, 261, 1927; *Chem. Abstr.*, 21, 2260_2, 1927.

39. **Bell, F.**, Stevioside: a unique sweetening agent, *Chem. Ind.*, p. 897, 1954; *Chem. Abstr.*, 49, 534e, 1955.
40. **Bergmann, M., Camacho, F., and Dryer, F.**, New derivatives of *p*-phenethylurea (dulcin), *Ber. Pharm. Ges.*, 32, 249, 1922; *Chem. Abstr.*, 17, 996_2, 1923.
41. **Bertolo, P. and Bertolo, A.**, *o*-Chlorobenzoic acid from the action of chlorine on saccharin, *Gazz. Chim. Ital.*, 62, 487, 1932; *Chem. Abstr.*, 27, 73_3, 1933.
42. **Best**, The action of saccharin on gastric digestion, *Muench. Med. Wochenschr.*, 64, 1231, 1917; *Chem. Abstr.*, 12, 928, 1918.
43. **Beyer, O.**, Methods for the determination of saccharin, *Chem. Ztg.*, 43, 537, 1919; *Chem. Abstr.*, 14, 578, 1920.
44. **Beyer, O.**, Chemical changes in the composition of saccharin-bicarbonate tablets, *Chem. Ztg.*, 43, 751, 1919; *Chem. Abstr.*, 14, 590_2, 1920.
45. **Beyer, O.**, New observations in the field of saccharin analysis, *Chem. Ztg.*, 44, 437, 1920; *Chem. Abstr.*, 14, 2772_7, 1920.
46. **Beyer, O.**, Estimation of P-Acid in commercial saccharin, *Chem. Ztg.*, 47, 744, 1923; *Chem. Abstr.*, 18, 1164_7, 1924.
47. **Beyer, O.**, The Beyer formula for the titrimetric determination of saccharin, *Chem. Ztg.*, 55, 509, 1931; *Chem. Abstr.*, 25, 4818_8, 1931.
48. **Beyer, O.**, What degree of sweetness does saccharin possess?, *Schweiz., Chem. Ztg.*, p. 598, 1920; *Chem. Abstr.*, 15, 713_2, 1921.
49. **Beyer, O.**, Synthetic sweetening agents, *Z. Angew. Chem.*, 35, 271, 1922; *Chem. Abstr.*, 17, 1511_4, 1923.
50. **Beythien, H.**, (Synthetic) sweetening materials, *Chem. Ztg.*, 66, 53, 1942; *Chem. Abstr.*, 37, 5508_5, 1943.
51. **Bhargava, M. G., Dhingra, D. R., and Guptan, G. N.**, Saccharin, *Indian J. Pharm.*, 13, 83, 1951; *Chem. Abstr.*, 45, 10430_i, 1951.
52. **Bhatnagar, S. S., Mathur, K. G., and Budhiraja, K. L.**, Triboluminescence, *Z. Phys. Chem.*, A163, 8, 1932; *Chem. Abstr.*, 27, 1574_2, 1933.
53. **Bianchi, A. and DiNola, E.**, The detection of saccharin and other artificial sweeteners in foods and drinks, *Boll. Chim. Farm.*, 47, 599, 1909; *Chem. Abstr.*, 3, 1312_5, 1909.
54. **Bleyer, B., Diemair, W., and Leonhard, K.**, Influence of preservatives on enzymic processes, *Arch. Pharm.*, 271, 539, 1933; *Chem. Abstr.*, 28, 1723_5, 1934.
55. **Bleyer, B. and Fischer, F.**, The effect of saccharin on biocatalyzers and metabolic processes, I, *Biochem. Z.*, 238, 212, 1931; *Chem. Abstr.*, 25, 5919_8, 1931.
56. **Blodgett, S. H.**, Saccharin, *Med. Rec.*, 97, 521, 1920; *Chem. Abstr.*, 14, 1714_3, 1920.
57. **Boedecker, F. and Rosenbusch, R.**, Sweetening Power of *p*-hydroxyphenylurea derivatives, *Ber. Pharm. Ges.*, 30, 251, 1920; *Chem. Abstr.*, 15, 1536.
58. **Bonis, A.**, Practical methods for detection and estimation of saccharin in foodstuffs, *Ann. Fals.*, 10, 210, 1917; *Chem. Abstr.*, 12, 959, 1918.
59. **Bonis, A.**, Determination of saccharin in compressed tablets, *Ann. Fals.*, 11, 369, 1919; *Chem. Abstr.*, 13, 2959_i, 1919.
60. **Bonjean, E.**, The action on the organism of saccharin when used as a sweetener for foods, *Rev. Hyg.*, 44, 50, 1922; *Chem. Abstr.*, 16, 2554_5, 1922.
61. **Bottle, R. T.**, Synthetic sweetening agents, *Manuf. Chem.*, 35, 1, 60, 1964; *Chem. Abstr.*, 60, 15048e, 1966.
62. **Breidenbach, A. W.**, Preservatives and artificial sweeteners, *J. AOAC*, 39, 646, 1956; *Chem. Abstr.*, 50, 13315i, 1956.
63. **Breidenbach, A. W.**, Preservatives and artificial sweeteners, *J. AOAC*, 40, 782, 1957; *Chem. Abstr.*, 51, 16991f, 1957.
64. **Brooks, L. G.**, Comparative tests on sodium cyclamate as a sweetening agent, *Pharm. J.*, 189, 569, 1962; *Chem. Abstr.*, 58, 8857c, 1964.
65. **Brooks, L. G.**, Use of synthetic sweetening agents in pharmaceutical preparations and foods, *Chem. Drug.*, 183, 4445, 421, 1965; *Chem. Abstr.*, 63, 2848c, 1965.
66. **Brouwer, T.**, Identification of saccharin according to Klostermann and Scholta, *Chem. Weekbl.*, 52, 184, 1956; *Chem. Abstr.*, 50, 10613c, 1956.
67. **Bruhns, G.**, The sweetening power of saccharin and of dulcin, *Centr. Zuckerind.*, 29, 725, 1921; *Chem. Abstr.*, 15, 2131_7, 1921.
68. **Bruhns, G.**, The detection and estimation of artificial sweetening materials, *Dtsch. Zuckerind.*, 59, 646, 1934; *Chem. Abstr.*, 29, 850_4, 1935.
69. **Bucci, F. and Amormino, V.**, Detection of dulcin in food, *Rend. Ist. Super. Sanita*, 20, 530, 1957; *Chem. Abstr.*, 52, 5684d, 1958.

70. **Bucci, F. and Amormino, V.,** A new procedure for investigation of dulcin in foodstuffs, *Ann. Chim.*, 47, 770, 1957; *Chem. Abstr.*, 51, 18360f, 1957.
71. **Buhr, G.,** Poisoning with dulcin, *Med. Klin.*, 43, 105, 1948; *Chem. Abstr.*, 43, 3105h, 1949.
72. **Bunde, C. A., Lackay, R. W.,** Failure of oral saccharin in influence blood sugar, *Proc. Soc. Exp. Biol. Med.*, 68, 581, 1948; *Chem. Abstr.*, 43, 315b, 1949.
73. **Buogo, G.,** Are synthetic fats antialimentary?, *Minerva Med.*, 11, 1847, 1955; *Chem. Abstr.*, 50, 5938f, 1956.
74. **Burge, W. E.,** Substitution of saccharin for sugar, *Science*, 48, 549, 1918; *Chem. Abstr.*, 13, 748_4, 1919.
75. **Burmistrov, S. I.,** Qualitative reactions of acetophenetidin, dulcin, acetanilide, and plasmocid, *Farmatsiya*, 8, 3, 1945; *Chem. Abstr.*, 41, 3018d, 1957.
76. **Burmistrov, S. I.,** Qualitative reactions of sulfonamides and saccharin, *Farmatsiya*, 9, 2, 1946; *Chem. Abstr.*, 41, 6021a, 1947.
77. **Pectin, L. M.,** in *Fruit Products for Diabetics*, Reference Sheet No. 379, California Fruit Growers Exchange.
78. **Cameron, A. T.,** The Taste Sense and the Relative Sweetness of Sugars and Other Sweet Substances, Sugar Research Foundation Science Report Series No. 9, 1947.
79. **Camilla, S. and Pertusi, C.,** Positive and sensitive method for the detection and identification of artificial sweeteners in beverages, medicines, cosmetics, etc., *G. Farm. Chim.*, 60, 385, 1912; *Chem. Abstr.*, 6, 1323_6, 1912.
80. **Carlinfanti, E. and Marzocchi, P.,** Determination of saponin and saccharin in oil emulsions, *Boll. Chim. Farm.*, 50, 609, 1911; *Chem. Abstr.*, 6, 1378i, 1912.
81. **Carlinfanti, E. and Scelba, S.,** The most important artificial sweeteners — saccharin and dulcin, *Boll. Chim. Farm.*, 51, 505, 541, 580, 613, 1913; *Chem. Abstr.*, 7, 664_9, 1913.
82. **Carlson, A. J., Eldridge, C. J., Martin, H. P., and Foran, F. L.,** Studies on the physiological action of saccharin, *J. Metab. Res.*, 3, 451, 1923; *Chem. Abstr.*, 18, 1152_3, 1924.
83. **Castiglioni, A.,** Detection of saccharin in wine by paper chromatography, *Accad. Ital. Vite Vino, Siena, Atti*, 6, 330, 1954; *Chem. Abstr.*, 50, 7389c, 1956.
84. **Castiglioni, A.,** Paper chromatography of saccharin and dulcin mixtures, *Z. Anal. Chem.*, 145, 188, 1955; *Chem. Abstr.*, 49, 7452b, 1955.
85. **Castiglioni, A.,** Paper electrophoresis of saccharin-dulcin mixtures, *Z. Anal. Chem.*, 148, 98, 1955; *Chem. Abstr.*, 50, 3161c, 1956.
86. **Ceccherelli, F.,** Identification and estimation of saccharin in foodstuffs, *Boll. Chim. Farm.*, 54, 641, 1915; *Chem. Abstr.*, 10, 2772_7, 1916.
87. **Ceriotti, A.,** Guide to bromatological analysis. Artificial sweetening agents, *Ediciones Soc. Nac. Farm.*, 18 pp., 1932; *Chem. Abstr.*, 26, 4887_9, 1932.
88. **Chang, K. T.,** Reaction of mustard gas with saccharin sodium, *J. Chin. Chem. Soc. Ser. II*, 2, 101, 1955; *Chem. Abstr.*, 50, 1217a, 1956.
89. Changes in methods of analysis made at the Seventh Annual (AOAC) Meeting, *J. AOAC*, 40, 39, 1957; *Chem. Abstr.*, 51, 4865i, 1957.
90. **Chew, A. P.,** Shall the food law be toothless?, *Fruit Prod. J. Am. Vinegar Ind.*, 6, 2, 10, 1926; *Chem. Abstr.*, 21, 139_9, 1927.
91. **Ciaccio, C. and Racchiusa, S.,** The action of saccharin in various doses of hyperglucemia by dextrose, *Boll. Soc. Ital. Biol. Sper.*, 2, 309, 1927; *Chem. Abstr.*, 22, 2206_3, 1928.
92. **Cicconetti, E.,** Sodium saccharin for sweetening the more frequently prepared medicinal prescriptions, *Farm. Ital.*, 10, 527, 1942; *Chem. Abstr.*, 38, 3416_3, 1944.
93. **Comanducci, E.,** Note on the detection of saccharin, *Boll. Chim. Farm.*, 49, 791, 1911; *Chem. Abstr.*, 5, 2276_7, 1911.
94. **Commerford, J. D. and Donahoe, H. B.,** *N*-(W-Brom oalkyl) saccharins and *N,N*-undecamethylenedisaccharin, *J. Org. Chem.*, 21, 583, 1956; *Chem. Abstr.*, 51, 2736i, 1957.
95. Communications from the food inspection laboratories of the city of Amsterdam, *Chem. Weekbl.*, 23, 361, 1926; *Chem. Abstr.*, 20, 3317_8, 1926.
96. **Condelli, S.,** Spontaneous decomposition of saccharin, *Boll. Chim. Farm.*, 52, 639, 1914; *Chem. Abstr.*, 8, 2433_2, 1914.
97. **Condelli, S.,** Results and discussions of the analyses of various products with reference to edulcorants and questions concerning "zuccher di stato," *Staz. Sper. Agrar. Ital.*, 54, 326, 1921, *Chem. Abstr.*, 17, 3548_7, 1923.
98. **Cox, W. S.,** Report on (determination of) artificial sweeteners, *J. AOAC*, 33, 688, 1950; *Chem. Abstr.*, 45, 282d, 1951.
99. **Cox, W. S.,** Report on artificial sweeteners, *J. AOAC*, 35, 321, 1952; *Chem. Abstr.*, 46, 11473a, 1952.

100. **Cox, W. S.,** Report on artificial sweeteners — methods for the detection and determination of P-4000(2-propoxy-5-nitroaniline), *J. AOAC,* 36, 749, 1953; *Chem. Abstr.,* 48, 9569f, 1954.
101. **Cox, W. S.,** Artificial sweeteners. P-4000 and dulcin, *J. AOAC,* 37, 383 1954; *Chem. Abstr.,* 48, 9569g, 1954.
102. **Cox, W. S.,** (Determination of) artificial sweeteners (P-4000 and dulcin), *J. AOAC,* 39, 652, 1956; *Chem. Abstr.,* 50, 13315i, 1956.
103. **Cox, W. S.,** (Determination of) artificial sweeteners, *J. AOAC,* 40, 785, 1957; *Chem. Abstr.,* 51, 16991g, 1957.
104. **Cragg, L. H.,** Sour taste, *Trans. R. Soc. Can.,* 31, 3, 131, 1937; *Chem. Abstr.,* 32, 2965_8, 1938.
105. **Cremer, H. D.,** Comprehensive review. The physiological importance of chemical treatment of foods, *Z. Lebensm. Unters. Forsch.,* 96, 188, 1953; *Chem. Abstr.,* 17, 6057f, 1953.
106. **Crosby, D. G. and Niemann, C.,** Further studies on the synthesis of substituted ureas, *J. Am. Chem. Soc.,* 76, 4458, 1954; *Chem. Abstr.,* 49, 13136f, 1955.
107. **Cross, L. J. and Perlman, J. L.,** Colorimetric Determination of Saccharin in Beverages, New York State Department of Agriculture and Markets Annu. Report, 1930, 89; *Chem. Abstr.,* 26, 783_6, 1932.
108. **Cuffi-Roura, U.,** The manufacture of saccharin, *Ind. Quim.,* 5, 54, 1943; *Chem. Abstr.,* 39, 512s, 1945.
109. **Dalal, N. B. and Shah, R. C.,** Synthesis of saccharin from anthranilic acid, *Curr. Sci.,* 18, 440, 1949; *Chem. Abstr.,* 44, 4881c, 1950.
110. **De Boer, H. W. and Bosgra, O.,** The harmful effect of saccharin, *Chem. Weekbl.,* 40, 26, 1943; *Chem. Abstr.,* 38, 4042, 1944.
111. **DeGarmo, O., Ashworth, G. W., Eaker, C. M., and Munch, R. H.,** Hydrolytic stability of saccharin, *J. Am. Pharm. Assoc.,* 41, 17, 1952; *Chem. Abstr.,* 46, 3678a, 1952.
112. **Demianovski, S. T. and Hefter, R.,** Is saccharin indifferent for the animal body?, *Vrach. Delo,* pp. 16, 179, 1921; *Chem. Abstr.,* 17, 2616_8, 1923.
113. **Deniges, G.,** Microcrystalline Reactions of Saccharin, Surcamin, and "Sucrose," Bull. Soc. Pharm. Bordeaux No. 2, *1921; Chem. Abstr.,* 16, 312_1, 1922.
114. **Deniges, G. and Tourrou, R.,** Microchemical reactions of dulcin, *C. R.,* 173, 1184, *Chem. Abstr.,* 16, 594_6, 1922.
115. **De Nito, G.,** The toxic action of saccharin. Histopathological studies, *Boll. Soc. Ital. Biol. Sper.,* 11, 934, 1936; *Chem. Abstr.,* 31, 4724_4, 1937.
116. **Deshusses, J. and Desbaumes, P.,** Study and identification of dulcin in foods by paper chromatography, *Mitt. Lebensm. U. Hyg.,* 47, 264, 1956; *Chem. Abstr.,* 51, 5317e, 1957.
117. **Diemair, W. and Fischler, F.,** Effect of saccharin on the biocatalysts and on metabolic processes. II. Influence on saccharin on the blood sugar and the glycogen content of the rabbit liver, *Biochem. Z.,* 239, 232, 1931; *Chem. Abstr.,* 26, 523_3, 1932.
118. **Dietzel, R. and Taufel, K.,** Ultra-violet spectroscopy and its significance in food chemistry, *Z. Nahr. Genussm.,* 49, 65, 1925; *Chem. Abstr.,* 19, 2091_8, 1925.
119. **Dobreff, M.,** The effect of the constant use of saccharin upon the digestive juices, *Arch. Hyg.,* 95, 320, 1925; *Chem. Abstr.,* 21, 275, 1927.
120. **Dominikiewicz, M. and Kijewska, M.,** Methods of obtaining double saccharins (several new derivatives of *m,m'*-bitoly), *Arch. Chem. Farm.,* 3, 27, 1936; *Chem. Abstr.,* 32, 8390_2, 1938.
121. **Drews, H.,** Evaluation of the saccharin content of horseradish preparations, *Ind. Obst. Gemueseverwert.,* 48, 7, 8, 1963; *Chem. Abstr.,* 58, 14630_9, 1964.
122. **Dubaquie, J.,** Graphical method for calculating the compsotion of a saccharin mixture containing sucrose, dextrose and levulose, *Ann. Fals.,* 22, 352, 1929; *Chem. Abstr.,* 23, 4908_9, 1929.
123. **Durand, H.,** Experimental data on the determination of saccharin in foods with a modification of Schmidt's methods, *J. Ind. Eng. Chem.,* 5, 981, 1914; *Chem. Abstr.,* 8, 385_3, 1914.
124. **Durocher, P.,** The preparation of saccharin, *La Nature,* 53, 361, 1925; *Chem. Abstr.,* 20, 1226_1, 1926.
125. **Dutt, S.,** Dyes derived from saccharin. The sulfamphthaleins, *J. Chem. Soc.,* 121, 2389, 1922; *Chem. Abstr.,* 17, 554_5, 1923.
126. **Dyson, G. M.,** The chemistry of synthetic sweetening compounds, *Chem. Age,* 11, 572, 1924; *Chem. Abstr.,* 19, 823_3, 1925.
127. **Dyson, G. M.,** Saccharin, *Flavours,* 2, 1, 24, 46; 3, 42, 4, 33; 1939; *Chem. Abstr.,* 34, 7471_7, 1940.
128. Eighty years of sweetening agents, *Chem. Prod.,* 23, 111, 1960; *Chem. Abstr.,* 54, 11325h, 1960.
129. **Ekkert, L.,** Color Reactions of Saccharins, *Pharm. Zentralhalle,* 67, 821, 1926; *Chem. Abstr.,* 21, 875_9, 1927.
130. **Endicott, C. J. and Gross, H. M.,** Artificial sweetening of tablets, *Drug Cosmet. Ind.,* 85, 176, 1959; *Appl. Sci. Technol.,* 1259, 1959.

131. **Erb, J. H.,** Sweetening agents for use in ice cream, *South. Dairy Prod. J.*, 31, 5, 28, 32, 1942; *Chem. Abstr.*, 36, 5572_6, 1942.

132. **Evans, T. W. and Dehn, W. M.,** Constitution of salts of certain cyclic imides, *J. Am. Chem. Soc.*, 52, 1028, 1930; *Chem. Abstr.*, 24, 1844_6, 1930.

133. **Fantus, B. and Hektoen, L.,** Saccharin feeding of rats, *J. Am. Pharm. Assoc.*,12, 318, 1923: *Chem. Abstr.* 17, 2012_2, 1923

134 **Fazio, C. C.,** Arreguine's test for saccharin and its use in bromatology, *Rev. Asoc. Bioquim. Argent.*, 13, 3, 1946; *Chem. Abstr.*, 40, 5164_9, 1946.

135. **Feigl, F., Auger, V., and Frehden, O.,** Use of spot tests for the detection of organic compounds, *Mikrochemie,* 17, 29, 1935; *Chem. Abstr.*, 29, 2471_2, 1935.

136. **Feigl, F. and Bondi, A.,** Reactivity of iodine in organic solvents. II.*Monatsh.*, 53 and 54, 508, 1929; *Chem. Abstr.*, 24, 351_4, 1930.

137. **Feigl, F. and Bondi, A.,** Reactivity of iodine in organic solvents, *Monatsh.*, 49, 417, 1928; *Chem. Abstr.*, 22, 3816_9, 1928.

138. **Ferguson, L. N. and Lawrence, A. R.,** The physiochemical aspects of the sense of taste, *J. Chem. Educ.*, 35, 436, 1958.

139. **Fichter, F.,** New aspects of the electrochemical oxidation of organic and inorganic compounds. II. The electrochemical preparation of vanillin, saccharin and dyestuffs, *J. Soc. Chem. Ind.*, (London), 48, 329, *Chem. Abstr.*, 24, 2060_7, 1930.

140. **Fichter, F. and Lowe, H.,** Electrolytic oxidation of *o*-toluenesulfonamide, *Helv. Chim. Acta,* 5, 60, 1922; *Chem. Abstr.*, 16, 1571, 1922.

141. **Finberg, A. J.,** Big strides of dietetic foods, *Food Eng.*, 27, 70, 1955.

142. **Finzi, C. and Colonna, M.,** Chemical constitution and sweet taste, *Gazz. Chim. Ital.*, 68, 132, 1939; *Chem. Abstr.*, 2, 6638_8, 1938.

143. **Fischler and Schroter,** Does saccharin, orally administered, influence the blood sugar?, *Dtsch. Med. Wochenschr.*, 61, 1354, 1935.

144. **Fischer, R.,** Identification of organic preservatives and commercial sweetening substances in foodstuffs, *Z. Unters. Lebensm.*, 67, 161, 1934; *Chem. Abstr.*, 28, 3137_6, 1934.

145. **Fischer, R. and Stauder, F.,** Detection of benzoic, salicylic and cinnamic acids of saccharin and of the esters of *p*-hydroxybenzoic acid in wine, *Z. Unters. Lebensm.*, 62, 658, 1931; *Chem. Abstr.*, 26, 4409_3, 1932.

146. **Fitzhugh, O. G. and Nelson, A. A.,** Comparison of the chronic toxicity of synthetic sweetening agents, *Fed. Proc. Fed. Am. Soc. Exp. Biol.*, 9, 272, 1950.

147. **Fitzhugh, O. G., Nelson, A. A., and Frawley, J. P.,** A comparison of the chronic toxicities of synthetic sweetening agents, *J. Am. Pharm. Assoc.*, 40, 583, 1951: *Chem. Abstr.*, 46, 1165i, 1952.

148. **Flamand, J.,** Detection of saccharin in beer, *Bull. Soc. Chim. Belg.*, 26, 477, 1913; *Chem. Abstr.*, 7, 859_6, 1913.

149. **Font-Altaba, M. and Gutierrez, F. H.,** The manganous derivative of saccharin, *An. R. Soc. Esp. Fis. Quim., Ser. B.*, 44, 355, 1948; *Chem. Abstr.*, 42, 7723i, 1948.

150. **Font-Altaba, M.,** Crystalline structure of the manganese derivative of saccharin, *Publ. Dep. Cristalogr. Mineral. C.S.I.C. Spain,* 1, 133, 1954; *Chem. Abstr.*, 49, 14416e, 1955.

151. **Font-Altaba, M.,** X-Ray characteristics of saccharin: the powder diagram, *Publ. Dep. Cristalogr. Mineral. C.S.I.C. Spain,* 2, 101, 1955; *Chem. Abstr.*, 50, 16257g, 1956.

152. Food Inspection Decisions. Saccharin in Food. 1911, *F.I.D.*, p. 138, *Chem. Abstr.*, 5, 3306_1, 1911.

153. Food Inspection Decisions Issued by the United States Department of Agriculture on the Use of Saccharin in Foods., 1912, *F.I.D.*, p. 146, *Chem. Abstr.*, 6, 2473_2, 1912.

154. Food Preservatives, *J. Assoc. Off. Agric. Chem.*, 2, pt. 2, 83, 1916.

155. **Fortelli, M. and Piazza, E.,** Detection and quantitative estimation of saccharin in fatty, starchy and albuminous foods, *Z. Nahr. Genussm.*, 20, 489, 1911; *Chem. Abstr.*, 5, 539_9, 1911.

156. **Fresenius, W. and Grunhut, L.,** The chemical analysis of wine, *Z. Anal. Chem.*, 59; 9, 209, 415; 1920; 60; 168, 257, 353; 1921; *Chem. Abstr.*, 16, 459_6, 1922.

157. **Frisch, H. R.,** Synthetic compounds with a sweet taste, *Chem. Can.*, 2, 6, 22, 1950; *Chem. Abstr.*, 46, 9259c, 1952.

158. **Fuchs, P.,** Indirect volumetric analysis in organic technical chemistry, *Chem. Fabr.*, 1934, 430, 1934; *Chem. Abstr.*, 29, 1034_7, 1935.

159. **Funck, E.,** Has saccharin any effect on biocatalyst?, *Pharmazie,* 2, 543, 1947; *Chem. Abstr.*, 42, 3045a, 1948.

150. **Gandini, A.,** A rapid method for determining saccharin, *Farmaco Sci. e Tec.*, 1, 34, 1946; *Chem. Abstr.*, 40, 4317_1, 1946.

161. **Gandini, A. and Borghero, S.,** Photomettric determination of dulcin, *Ig. Mod.*, 41, 11, 1948; *Chem. Abstr.*, 45, 6971g, 1951.

162. **Gaubert, P.,** Liquid crystals of some cholesterol compounds, and their crystalline superfusion, *C. R.,* 202, 141, 1936; *Chem. Abstr.,* 30, 1625_3, 1936.
163. **Gaudiano, A. and Toffoli, F.,** Saccharin, *Atti Accad. Naz. Lincei Cl. Sci. Fis. Mat. Nat. Rend.,* 21, 109, 1956, *Chem. Abstr.,* 51, 7355e, 1957.
164. **Genth, F. A., Jr.,** Confirming the presence of saccharin in foods and beverages, *Am. J. Pharm.,* 81, 536, 1910; *Chem. Abstr.,* 4, 352_8, 1910.
165. German Legislation Concerning Artificial Sweetening Agents, *Chem. Ztg.,* 46, 361, 1922; *Chem. Abstr.,* 16, 2183_6, 1922.
166. **Gialdi, F.,** The use of 2-amino-4-nitrophenol propyl ether as a sugar substitute and its colorimetric determination, *Farm. Sci. e Tec.* (Pavia), 3, 44, 1948; *Chem. Abstr.,* 42, 4714g, 1948.
167. **Gianferrara, S.,** Determination of dulcin in food, *Ann. Chim. Appl.,* 38, 326, 1948; *Chem. Abstr.,* 43, 4391e, 1949.
168. **Gianferrara, S. and Chiorboli, R.,** Determination of saccharin and its separation from dulcin, *Chim. Ind.* (Milan), 35, 224, 1953; *Chem. Abstr.,* 47, 11082e, 1953.
169. **Gianformaggio, F.,** The reduction of benzoic sulfonimide, *Gazz. Chim. Ital.,* 50, 1,327, 1920; *Chem. Abstr.,* 15, 520_1, 1921.
170. **Gilman, H., Brown, R. E., Dickey, J. B., Hewlett, A. P., and Wright, G. F.,** Utilization of agricultural wastes, *Proc. Iowa Acad. Sci.,* 36, 265, 1929; *Chem. Abstr.,* 25, 1245_4, 1931.
171. **Gilman, H. and Hewlett, A. P.,** Some correlations in constitution with sweet taste in the furan series, *Iowa State Coll. J. Sci.,* 4, 27, 1929.
172. **Gilta, G.,** Crystallography of tryparsamide and related compounds, *Bull. Soc. Chim. Belg.,* 46, 263, 1937; *Chem. Abstr.,* 32, 851_6, 1938.
173. **Gnadinger, C. G.,** Determination of saccharin in foods, *J. Assoc. Off. Agric. Chem.,* 3, 25, 1917; *Chem. Abstr.,* 11, 2373_9, 1917.
174. **Griebel, C.,** "Ultra Suss" (sweetening agent), *Apoth. Ztg. Hanslian Ed.,* 69, 16, 1949; *Chem. Abstr.,* 44, 3629h, 1950.
175. **Griebel, C.,** Adulteration of saccharin, *Pharm. Ztg. Nachr.,* 87, 314, 1951; *Chem. Abstr.,* 46, 2753e, 1952.
176. **Grossfeld, J.,**The sweetening power of artificial sweeteners, *Z. Ges. Kohlensaure Ind.,* 15, 253, 1921; *Chem. Abstr.,* 15, 3881_7, 1921.
177. **Grossfeld, J.,** The detection and estimation of saccharin and benzoic acid in foods, *Z. Ges. Kohlensaure Ind.,* 26, 143, 159, 1920; *Chem. Abstr.,* 15, 3152_1, 1921.
178. **Gaureschi, R.,** Research on saccharin in vermuth, *G. Farm. Chim.,* 61, 400, 1912; *Chem. Abstr.,* 7, 204_7, 1913.
179. **Gutierrez, F. H.,** Acetone solubility of some sulfonamide medicinal products, *An. Fis. Quim.* (Madrid), 41, 537, 1945; *Chem. Abstr.,* 41, 6546g, 1947.
180. **Gutierrez, F. H.,** Acetone as a new solvent in the extraction of saccharin. Solubilities in acetone. V., *An. Fis. Quim.* (Madrid), 2, 1105, 1946; *Chem. Abstr.,* 41, 5120c, 1947.
181. **Gutierrez, F. H.,** The use of acetone in the gravimetric determination of saccharin in beverages and chocolate. Solubility in acetone, *An. Fis. Quim.,* 43, 393, 1947; *Chem. Abstr.,* 41, 6640d, 1947.
182. **Gutierrez, F. H.,** Solubility of saccharin and its sodium salt in water and organic solvents, *Farm. Nueva,* 11, 342, 1946; *Chem. Abstr.,* 41, 2304e, 1947.
183. **Gutierrez, F. H. and Altaba, M. F.,** New microchemical reactions of saccharin, *An. Fis. Quim.,* 43, 471, 1947; *Chem. Abstr.,* 42, 724b, 1948.
184. **Gutierrez, F. H. and Altaba, M. F.,** New microchemical reactions of saccharin, *Publ. Inst Invest. Microquim. Univ. Nac. Litoral* (Rosario, Argent.), 21, 20, 1954; *Chem. Abstr.,* 51, 3373i, 1957.
185. **Gyenes, I. and Vali, A.,** Determination of saccharin sodium by titration with perchloric acid, *Magy. Kem. Foly.,* 69, 90, 1955; *Chem. Abstr.,* 49, 15635g, 1955.
186. **Halla, F.,** The electrochemical oxidation of *o*-toluenesulfonamide to saccharin, *Z. Elektrochem.,* 36, 96, 1930; *Chem. Abstr.,* 24, 2060_5, 1930.
187. **Hamann, V.,** A new (substitute) sweetening agent, *Dtsch. Lebensm. Rundsch.,* 44, 52, 1948; *Chem. Abstr.,* 45, 3959i, 1951.
188. **Hamilton, W. F. and Turnbull, F. M.,** Substituted ammonium saccharins for nasal medication, *J. Am. Pharm. Assoc.,* 39, 378, 1950; *Chem. Abstr.,* 44, 9631c, 1950.
189. **Hamor, G. H. and Soine, T. O.,** Synthesis of some derivatives of saccharin, *J. Am. Pharm. Assoc.,* 43, 120, 1954; *Chem. Abstr.,* 4, 3892g, 1955.
190. **Hand, D. B. et al.,** The Use of Chemical Additives in Food Processing, National Academy of Science-National Research Council (U.S.) Publication No. 398, 1956, 91 pp.; *Chem. Abstr.,* 50, 10291g, 1956.
191. **Hansen, A.,** Artificial sweetening substances, *Dan. Tidsskr. Farm.,* 1, 114, 1926; *Chem. Abstr.,* 21, 285_6, 1927.

192. **Haramaki, K.**, The influence of saccharin on certain functions of the digestive tract and kidneys, *Z. Phys. Diaet. Ther.*, 26, 183, 1922; *Chem. Abstr.*, 16, 3341_9, 1922.
193. **Hardeggar, E. and Jucker, O.**, Derivatives of 3,6-anhydroglucose and glucose-6-iodohydrin, *Helv. Chim. Acta*, 32, 1158, 1949; *Chem. Abstr.*, 43, 7905h, 1949.
194. **Harvey, E. H.**, Efficiency of some common anti-ferments, *Am. J. Pharm.*, 90, 105, 1923; *Chem. Abstr.*, 17, 1675_2, 1923.
195. **Hayakawa, M. and Masuda, S.**, Preparation of dulcin from sodium *p*-nitrophenolate, *Res. Rep. Nagoya Ind. Sci. Res. Inst.*, No. 1, 25, 1949; *Chem. Abstr.*, 49, 215g, 1955.
196. **Hefelmann**, Determination of saccharin, *Pharm. Post*, p. 675, 1917; *Chem. Abstr.*, 12, 1276_1, 1918.
197. **Heiduschka, A. and Schmid, J.**, Determination of moisture in saccharin material, *Pharm. Zentralhalle*, 54, 956, 1914; *Chem. Abstr.*, 8, 2757_9, 1914.
198. **Heilmann, P.**, Saccharin intoxication, *Muench. Med. Wochenschr.*, 69, 968, 1922; *Chem. Abstr.*, 17, 827_1, 1923.
199. **Heitler, M.**, Induced spontaneous transformation of the cardiac depressant action of saccharin into cardiac stimulation, *Wien. Klin. Wochenschr.*, 35, 935, 1922; *Chem. Abstr.*, 17, 1081_8, 1923.
200. **Heitler, M.**, Sugar and saccharin, *Wien. Med. Wochenschr.*, 70, 1050, 1922; *Chem. Abstr.*, 16, 1283_2, 1922.
201. **Helch, H.**, Equivalents of saccharin solutions and sugar syrups of equal sweetening power, *Schweiz. Apoth. Ztg.*, 55, 239, 1917; *Chem. Abstr.*, 11, 2532_3, 1917.
202. **Helgren, F. J., Lynch, M. J., and Kirchmeyer, F. J.**, A taste-panel study of the saccharin "off-taste," *J. Am. Pharm. Assoc.*, 44, 353, 1955; *Chem. Abstr.*, 49, 11243c, 1955.
203. **Heller, G.**, Tautomeric phenomena in heterocyclic compounds, *J. Prakt. Chem.*, 111, 1, 1925; *Chem. Abstr.*, 20, 381_3, 1926.
204. **Henrichson, C. B.**, Produce high levels of sweetness with low calorie foods, *Food Process.*, 19, 46, 1958.
205. **Hensel, S. T.**, The use of saccharin as a sugar substitute, *J. Am. Pharm. Assoc.*, 7, 609, 1918; *Chem. Abstr.*, 13, 58_7, 1919.
206. **Hernandez, F. and Font, M.**, Microchemical reactions of saccharin, *Mon. Farm. y. Terap.*, 53, 359, 1947; *Chem. Abstr.*, 42, 1530h, 1948.
207. **Herrmann, P.**, Derivatives of dulcin, *Ann.*, 429, 163, 1922; *Chem. Abstr.*, 17, 381, 1923.
208. **Herzfeld and Reischauer**, Test for saccharin in foods, *Naturwiss. Wochenschr.*, p. 165, 1913.
209. **Herzog, W.**, Utilization of by-products of saccharin manufacture in synthesis of drugs in medical science, *Chem. Rundsch. Mitteleur. Balk.*, 7, 15, 107, 1930; *Chem. Abstr.*, 25, 2810_1, 1931.
210. **Herzog, W.**, Utilization of by-products from saccharin manufacture, *Chem. Umsch. Geb. Fette Oele Wachse Harze*, 37, 296, 1930; *Chem. Abstr.*, 25, 428_2, 1931.
211. **Herzog, W.**, Progress in the field of synthetic sweetening agents and related compounds in 1925—1926. *Fortschrittsber. Chem. Ztg.*, 51, 66, 1927; *Chem. Abstr.*, 21, 2747_2, 1927.
212. **Herzog, W.**, Fighting vermin and exterminating weeds by use of the by-products of saccharin manufacture, *Chem. Ztg.*, 54, 50, 1930; *Chem. Abstr.*, 24, 1926_7, 1930.
213. **Herzog, W.**, Progress in the field of synthetic sweet substances and related compounds, *Chem. Ztg.*, 57, 574, 1933; *Chem. Abstr.*, 27, 4598_5, 1933.
214. **Herzog, W.**, Advances in the field of synthetic sweet substances and related compounds in 1933—34, *Chem. Ztg.*, 59, 408, 1935; *Chem. Abstr.*, 29, 4848_7, 1935.
215. **Herzog, W.**, The utility of by-products from saccharin manufacture in the chemistry of synthetic tans and in the tannery, *Collegium*, p. 203, 1926; *Chem. Abstr.*, 20, 3586_8, 1926.
216. **Herzog, W.**, Advances in the field of synthetic sweetening agents and related compounds in 1927 and 1928, *Fortschrittsber. Chem. Ztg.*, 53, 99, 1928; *Chem. Abstr.*, 23, 4978_3, 1929.
217. **Herzog, W.**, The utilization of the by-products from saccharin manufacture in the industry of synthetic resins and plastic masses, *Kunststoffe*, 16, 105, 1926; *Chem. Abstr.*, 20, 2910_4, 1926.
218. **Herzog, W.**, Progress in the utilization of saccharin by-products in 1927, *Metallborse*, 18, 903, 1015, 1126, 1183, 1238, 1928; *Chem. Abstr.*, 23, 2426_7, 1929.
219. **Herzog, W.**, Waste products of the saccharin industry used in synthesizing tanning agents and in tanning, *Metalborse*, 19, 1853, 1929.
220. **Herzog, W.**, The value of by-products of saccharin manufacture in analytical chemistry, *Oster. Chem. Ztg.*, 29, 26, 1926; *Chem. Abstr.*, 20, 1612, 1926.
221. **Herzog, W.**, Progress in the field of synthetic sweetening agents and related compounds in 1935 and 1936, *Oster. Chem. Ztg.*, 40, 201, 1937; *Chem. Abstr.*, 31, 4968_8, 1937.
222. **Herzog, W.**, Medicaments derived from saccharin and its secondary products, *Pharm. Zentralhalle*, 67, 81, 1926; *Chem. Abstr.*, 20, 1301_4, 1926.
223. **Herzog, W.**, Progress in the realm of synthetic and related compounds during the years 1935—1937, *Pharm. Zentralhalle*, 78, 76, 1937; *Chem. Abstr.*, 32, 2244_5, 1938.

224. **Herzog, W.**, Advances in the chemistry of synthetic sweetening agents in 1918—21, *Z. Angew. Chem.*, 35, 133, 1922; *Chem. Abstr.*, 16, 1625_6, 1922.
225. **Herzog, W.**, Advances in the chemistry of synthetic sweeteners and related compounds in 1922, *Z. Angew. Chem.*, 36, 223, 1923; *Chem. Abstr.*, 17, 2155_5, 1923.
226. **Herzog, W.**, Advances in the field of synthetic sweetening agents and related substances in 1923 and 1924, *Z. Angew. Chem.*, 38, 641, 1925; *Chem. Abstr.*, 19, 2988_7, 1925.
227. **Herzog, W.**, The utilization of by-products from saccharin manufacture for dye synthesis, dyeing, finishing and bleaching in the period 1927 to 1929, *Z. Gesamte Textilnd.*, 32, 857, 873, 889, 1929; *Chem. Abstr.*, 24, 6021_4, 1930.
228. **Herzog, W.**, Application of by-products of saccharin manufacture in photography and photometry, *Z. Wiss. Photogr.*, 27, 177, 1929; *Chem. Abstr.*, 24, 3717_7, 1930.
229. **Herzog, W. and Kreidl, J.**, The quantitative separation of saccharin and *p*-sulfamylbenzoic acid (para acid), *Oster. Chem. Ztg.*, 24, 165, 1921; *Chem. Abstr.*, 16, 1197_9, 1922.
230. **Hist, J. F., Holmes, F., and Maclennan, G. W. G.**, Determination of dulcin (*p*-phenetylcarbamide), *Analyst*, 66, 450, 1941; *Chem. Abstr.*, 36, 722_7, 1942.
231. **Hoeke, F.**, Substitute materials. I., *Chem. Weekbl.*, 39, 390, 1942; *Chem. Abstr.*, 40, 4444_9, 1946.
232. **Hoeke, F.**, Determination of synthetic sweetening agent 1-propoxy-2-amino-4-nitrobenzene, *Chem. Weekbl.*, 43, 283, 1947.
233. **Holleman, A. F.**, Artificial sweet materials, *Recl. Trav. Chim.*, 40, 446, 1921; *Chem. Abstr.*, 15, 3821_5, 1921.
234. **Holleman, A. F.**, On some derivatives of saccharin, *Rev. Trav. Chim.*, 42, 839, 1923; *Chem. Abstr.*, 18, 239_4, 1924.
235. **Hucker, G. J. and Pederson, C. S.**, A review of the microbiology of commercial sugar and related sweetening agents, *Food Res.*, 7, 459, 1942; *Chem. Abstr.*, 37, 2206_5, 1943.
236, **Hurd, C. D. and Kharasch, N.**, Reaction of the dioxane-sulfotrioxide reagent with aniline classification of the sulfamic acids, *J. Am. Chem. Soc.*, 69, 2113, 1947; *Chem. Abstr.*, 42, 138i, 1948.
237. **Hwang, K.** et al., Comparison of cyclamate (cyclohexylsulfamate) salts in the alimentary tract of the rat, *Food Process.*, 15, 441, 1956.
238. **Hynes, L. J.**, Sugar saving substitutes, *Dairy Ind.*, 4, 348, 1939; *Chem. Abstr.*, 33, 9466_7, 1939.
239. **Illies, R.**, Sweetening agents and their use in the brewery, *Dtsch. Essigind.*, 44, 178, 184, 190, 1940; *Chem. Abstr.*, 35, 1927_5, 1941.
240. **Illing, E. T., Dalley, R. A., and Stephenson, W. H., Saccharin in soft drinks,**Food, *16, 140, 1947;*
241. **Imbesi, A. and DeAngelis, V.**, The effect of temperature on the conductivity of saccharin solutions, *Ann. Chim. Appl.*, 25, 265, 1935; *Chem. Abstr.*, 29, 6489_6, 1935.
242. International conference concerning saccharin and analogous substances, *Ann. Fals.*, 6, 165, 1913, *Chem. Abstr.*, 7, 2069_8, 1913.
243. **Ishimura, K.**, Some physical and chemical properties of dulcin, *Rep. Gov. Ind. Res. Inst. Nagoya*, 2, 87, 1953; *Chem. Abstr.*, 50, 15446e, 1956.
244. **Ivanov, A.**, Preparation of edible sorbitol, *Pisch. Prom.*, 1, 3, 30, 1941; *Chem. Abstr.*, 39, 5355_3, 1945.
245. **Jacobs, A. L. and Scott, M. L.**, Factors mediating food and liquid intake in chickens, I. Studies on the preference for sucrose or saccharin compounds, *Poult. Sci.*, 36, 8, 1957; *Chem. Abstr.*, 51, 15719i, 1957.
246. **Jacobs, M. B.**, Tobacco flavors, *Am. Perfum.*, 49, 607, 609, 1947; *Chem. Abstr.*, 41, 7059i, 1947.
247. **Jacobs, M. G.**, Artificial sweeteners, *Am. Perfum. Essent. Oil Rev.*, 57, 49, 51, 1951; *Chem. Abstr.*, 45, 3998h, 1951.
248. **Jacobs, M. B.**, The sweetening power of stevioside, *Am. Perfum. Essent. Oil Rev.*, 66, 6, 44, 46, 1955; *Chem. Abstr.*, 50, 2883i, 1956.
249. **Jaeger, F. M.**, A contribution to the theory of Barcow and Pope, *Z. Kryst.*, 44, 61, 1908; *Chem. Abstr.*, 2, 498_7, 1908.
250. **Jagoda, G.**, Fodder tests with additions of (beet) raw sugar and molasses, also saccharin, *Z. Ver. Dtsch. Zucker Ind.*, p. 243, 1927; *Chem. Abstr.*, 22, 1813_1, 1928.
251. **Jakobsen, F.**, Artificial sweetening agents, *Tidsskr. Hermetikind.*, 27, 129, 1941; *Chem. Abstr.*, 35, 7568_2, 1941.
252. **Jakobsen, F. and Jakobsen, A.**, The use of artificial sweetening agents in the preparation of berry and fruit preserves, *Tidsskr. Hermetikind.*, 28, 225, 1942; *Chem. Abstr.*, 38, 3377_4, 1944.
253. **Jamison, G. S.**, The determination of saccharin in urine, *J. Biol. Chem.*, 41, 3, 1920; *Chem. Abstr.*, 14, 954_5, 1920.
254. **Jorgensen, G.**, Detection of saccharin in beer, *Ann. Falsif.*, 2, 58, 1909; *Chem. Abstr.*, 3, 1056_8, 1909.

255. **Jorgensen, H.**, Influence of saccharin on blood sugar, *Acta Physiol. Scand.*, 20, 33, 1950; *Chem. Abstr.*, 44, 5475f, 1950.
256. **Kamen, J.**, Interaction of sucrose and calcium cyclamate on perceived intensity of sweeteners, *Food Res.*, 24, 3, 279, 1959.
257. **Karas, J.**, Determination of saccharin in food, *Z. Nahr. Genussm.*, 25, 559, 1913; *Chem. Abstr.*, 7, 3174_6, 1913.
258. **Kaufmann, H. P. and Schweitzer, D.**, Synthesis of sweetening agents, *Fette Seifen*, 55, 321, 1953; *Chem. Abstr.*, 49, 2342a, 1955.
259. **Kegler, W.**, Preparation of *p*-phenetylurea (dulcin sweetening agent), *Seifen Oele Fette Wachse*, 77, 606, 1951; *Chem. Abstr.*, 47, 7451f, 1953.
260. **Khlopin, N. Y., Litvinova, N. S., and Privalova, K. P.**, Polarographic determination of saccharin, *Gig. Sanit.*, 2, 48, 1951; *Chem. Abstr.*, 45, 10142b, 1951.
261. **Kielhofer, E. and Aumann, H.**, Two new sweet substances and their detection in wine, *Wein Wiss. Beiheft Fachzeit Dtsch. Weinbau*, 6, 1, 1955; *Chem. Abstr.*, 50, 5975e, 1956.
262. **Kiliana, H., Loeffler, P., and Matther, O.**, Derivatives of saccharin, *Ber.*, 40, 2999, 1907; *Chem. Abstr.*, 1, 2467_7, 1907.
263. **Kinugasa, Y. and Nishihara, S.**, Antiseptic power of chloramine-T, *J. Pharm. Soc. Jpn.*, 454, 1006, 1919; *Chem. Abstr.*, 14, 1410_6, 1920.
264. **Klarmann, B.**, Saccharinimine, *Chem. Ber.*, 85, 162, 1952; *Chem. Abstr.*, 46, 9539h, 1952.
265. **Klasens, H. A. and Terpstra, P.**, Crystallography of cupric saccharinate, *Recl. Trav. Chim.*, 56, 672, 1937; *Chem. Abstr.*, 31, 6076_6, 1937.
266. **Kliffmuller, R.**, Determination of sweetening agents and preservatives by paper chromatography, *Dtsch. Lebensm. Rundsch.*, 52, 182, 1956; *Chem. Abstr.*, 51, 634g, 1957.
267. **Kling, A., Bovet, D., and Ruiz, I.**, Saccharin and dulcin as substitutes for sugar, *Bull. Acad. Med.*, 124, 99, 1941; *Chem. Abstr.*, 38, 1844_6, 1944.
268. **Kling, A., Bovet, D., and Ruiz, I.**, The toxic action of the sweetening substances "dulcin," *Bull. Acad. Mex.*, 125, 69, 1941; *Chem. Abstr.*, 38, 3360_5, 1944.
269. **Klostermann, M. and Scholta, K.**, Critical consideration of the qualitative and quantitative determination of saccharin and a new method for the qualitative detection of the sweetener, *Z. Nahr. Genussm.*, 31, 67, 1916; *Chem. Abstr.*, 10, 1561_6, 1916.
270. **Kogan, I. M. and Dziomko, V. M.**, Reaction of arylsulfonamides with amines, II., *Zh. Obshch. Khim.*, 23, 1234, 1953; *Chem. Abstr.*, 12280d, 1953.
271. **Kolthoff, I. M.**, Th significance of dissociation constants in identifying acids and detecting impurites, *Pharm. Weekbl.*, 57, 514, 1920; *Chem. Abstr.*, 14, 3040_6, 1920.
272. **Kolthoff, I. M.**, The acid character of saccharin and related acids. The detection and determination of *p*-sulfamylbenzoic acid in saccharin and crystallose, *Recl. Trav. Chim.*, 44, 629, 1925; *Chem. Abstr.*, 19, 3408_6, 1925.
273. **Krantz, J. C., Jr.**, Report on (food) preservatives, *J. AOAC*, 16, 316, 1933; *Chem. Abstr.*, 27, 5118_1, 1933.
274. **Krantz, J. C., Jr.**, Report on (the determination of) preservatives, *J. AOAC*, 17, 193, 1934; *Chem. Abstr.*, 28, 4493_2, 1934.
275. **Krantz, J. C., Jr.**, Report on (the determination of) preservatives, *J. AOAC*, 18, 372, 1935; *Chem. Abstr.*, 29, 6966_5, 1935.
276. **Krantz, J. C., Jr.**, Report on (the determination of) preservatives (in foods), *J. AOAC*, 19, 205, 1936; *Chem. Abstr.*, 30, 4935_8, 1936.
277. **Kroll**, The use of artificial sweetening substances in the preparation of hot beverages, *Dtsch. Destill. Ztg.*, 63, 319, 1942; *Chem. Abstr.*, 38, 3378, 1944.
278. **Kun, E. and Horvath, I.**,Inluence of oral saccharin on blood sugar, *Proc. Soc. Exp. Biol. Med.*, 66, 175, 1947; *Chem. Abstr.*, 42, 1354g, 1947.
279. **Kurzer, F.**, Sweeter than sugar. New synthetic compounds more potent than saccharin, *Discovery*, 10, 175, 1949; *Chem. Abstr.*, 43, 5880e, 1949.
280. **LaParola, G. and Mariani, A.**, New color test for dulcin, *Ann. Chim. Appl.*, 36, 134, 1946; *Chem. Abstr.*, 40, 6367_5, 1946.
281. **Lapicque, L.**, Some remarks on the use of saccharin, *Bull. Acad. Med.*, 124, 116, 1941; *Chem. Abstr.*, 37, 5196_5, 1943.
282. **Lasheen, A. M. and Russell, T. S.**, Saccharin sodium as an artificial sweetening agent in fresh fruits, *Proc. Am. Soc. Hortic. Sci.*, 77, 140, 1961; *Chem. Abstr.*, 56, 9176d, 1962.
283. **Lavague, J.**, The colorimetry of saccharin, *Ann. Pharm. Fr.*, 3, 26, 1945; *Chem. Abstr.*, 40, 5204_8, 1946.
284. **Ledent, R.**, Saccharin in beer, *Ann. Chim. Anal.*, 18, 314, 1914; *Chem. Abstr.*, 8, 2025_2, 1914.

285. **Lehman, A. J.,** Some toxicological reasons why certain chemicals may or may not be permitted as food additives, *Assoc. Food Drug Off. U.S. Q. Bull.*, 14, 82, 1950; *Chem. Abstr.*, 45, 3517h, 1951.
286. **Lehman, A. J.,** Chemicals in foods: a report to the association of food and drug officials on current developments, *Assoc. Food Drug Off. U.S. Q. Bull.*, 15, 82, 1951; *Chem. Abstr.*, 46, 2701i, 1952.
287. **Lehmann, K. B.,** The feeding of saccharin to mice. A study of the question of minimal toxic action, *Arch. Hyg.*, 101, 39, 1929; *Chem. Abstr.*, 23, 3977_8, 1929.
288. **Lehmstedt, K.,** A new sweetening agent, *Seifen Oele Fette Wachse*, 75, 81, 1949; *Chem. Abstr.*, 44, 1207i, 1950.
289. **LeMagnen, J.,** Influence of insulin on the spontaneous consumption of sapid solutions, *C. R. Soc. Biol.*, 147, 1753, 1953; *Chem. Abstr.*, 48, 9505g, 1954.
290. **Leroy, M.,** Determination of saccharin and dulcin, *Ing. Chim.*, 26, 18, 1942; *Chem. Abstr.*, 37, 5548_8, 1943.
291. **Lerrigo, A. F. and Williams, A. L.,** A study of the determination of saccharin colorimetrically and by the ammonia process, *Analyst*, 52, 377, 1927; *Chem. Abstr.*, 21, 3859_3, 1927.
292. **Lettre, H. and Wrba, H.,** The effect of the sweetener dulcin during protracted feeding, *Naturwissenschaften*, 4, 217, 1955; *Chem. Abstr.*, 50, 5914b, 1956.
293. **Lickint, F.,** Saccharin and gastric secretion (saccharin test meal), *Muench. Med. Wochenshr.*, 90, 586, 1943; *Chem. Abstr.*, 38, 5963_6, 1944.
294. **Loginov, N. E.,** Sweetening substances, *Pishch. Prom.*, 1/2, 22, 1943; *Chem. Abstr.*, 40, 101_3, 1946.
295. **Longwell, J. and Bass, C. S.,** Colorimetric determination of dulcin (*p*-phenethylcarbamide) in composite articles, *Analyst*, 67, 14, 1942; *Chem. Abstr.*, 36, 2235_2, 1942.
296. **Lorang, H. F. J.,** The relationship between the constitution and taste of some urea derivatives, *Recl. Trav. Chim.*, 47, 179, 1928; *Chem. Abstr.*, 22, 1147_9, 1928.
297. **Lorges, A. B.,** Artificial sweeteners (saccharin, dulcin and glucin), *Rev. Chim. Ind.*, 35, 109, 126; *Chem. Abstr.*, 20, 2211_3, 1926.
298. **Luszczak, A. and Hammer, E.,** Thymol, benzene and toluene in commodities and in air. Their spectrographic determination, *Abh. Gesamtgeb. Hyg.*, 12, 82, 1933; *Chem. Abstr.*, 28, 6086_6, 1934.
299. **Lynch, M. J. and Gros, H. M.,** Artificial sweetening of liquid pharmaceuticals, *Drug Cosmet. Ind.*, 87, 324, 412, 1960; *Chem. Abstr.*, 55, 902a, 1961.
300. **Lyons, E.,** Mercury derivatives of some imides, *J. Am. Chem. Soc.*, 47, 830, 1925; *Chem. Abstr.*, 19, 1247_7, 1925.
301. **McCance, R. A. and Lawrence, R. D.,** An investigation of quebrachitol as a sweetening agent for diabetics, *Biochem. J.*, 27, 986, 1933; *Chem. Abstr.*, 28, 227_9, 1934.
302. **McKie, P. V.,** Examination of some methods of ascertaining the purity of saccharin, *J. Soc. Chem. Ind.*, 40, 150-2T, 1921; *Chem. Abstr.*, 15, 3343_9, 1921.
303. **Maass, H.,** The Sweetening Agent Problem, *Dtsch. Lebensm. Rundsch.*, 44, 50, 1948; *Chem. Abstr.*, 45, 6598b, 1951.
304. **Maes, O.,** Detection and determination of sweetening agents in beer, *Bull. Assoc. Anc. Eleves Inst. Super. Ferment. Gand*, 39, 287, 1938; *Chem. Abstr.*, 33, 1438_9, 1939.
305. **Magidson, O. Y. and Gorbachov, S. W.,** The question of the sweetness of saccharin *o*-benzoyl-sulfimide and its electrolytic dissociation, *Ber.*, 56B, 1810, 1923; *Chem. Abstr.*, 18, 227_8, 1924.
306. **Magidson, O. Y. and Zil'Berg, I. G.,** Mechanism of the oxidation of *o*-toluenesulfonamide to saccharin, *J. Gen. Chem. U.S.S.R.*, 5, 920, 1935.
307. **Mameli, E. and Mannessier-Mameli, A.,** Pyrolysis of ammonium saccharin and of ammonium thiosaccharin, saccharinimimine and pseudosaccharin amine, *Gazz. Chim. Ital.*, 70, 855, 1940; *Chem. Abstr.*, 36, 1026_8, 1942.
308. **Mannessier-Mameli, A.,** Action of hydoxylamine on saccharin, on some of its derivatives and on 3-thio-1,2-benzodithiole, *Gazz. Chim. Ital.*, 62, 1067, 1932; *Chem. Abstr.*, 27, 2682_1, 1933.
309. **Mannessier-Mameli, A.,** Action of anilines on saccharin and on thiosaccharin, *Gazz. Chim. Ital.*, 65, 51, 1935; *Chem. Abstr.*, 29, 3996_i, 1935.
310. **Mannessier-Mameli, A.,** The pyrolysis of saccharin oxime, *Gazz. Chim. Ital.*, 65, 77, 1935; *Chem. Abstr.*, 29, 3998_8, 1935.
311. **Mannessier-Mameli, A.,** Action of anilines on saccharin and on thiosaccharin, *Gazz. Chim. Ital.*, 70, 855, 1940; *Chem. Abstr.*, 36, 1917_8, 1942.
312. **Mannessier-Mameli, A.,** The action of ammonia, of ammonium carbonate, of urea, and of diurea on saccharin and on thiosaccharin, *Gazz. Chim. Ital.*, 71, 3, 1941; *Chem. Abstr.*, 36, 1028_1, 1942.
313. **Mannessier-Mameli, A.,** The action of hydrazine on saccharin and on thiosacchrin, *Gazz. Chim. Ital.*, 71, 18, 1941; *Chem. Abstr.*, 36, 1029_4, 1942.
314. **Mannessier-Mameli, A.,** The action of semicarbazide on saccharin, on thiosaccharin, and on acetylsaccharin, *Gazz. Chem. Ital.*, 71, 25, 1941; *Chem. Abstr.*, 36, 1030, 1942.

315. **Mannessier-Mameli, A.,** Action of phenylhydrazine on saccharin and on thiosaccharin, *Gazz. Chim. Ital.*, 71, 596, 1941; *Chem. Abstr.*, 37, 100_6, 1943.
316. The Manufacture of Saccharin, *Ind. Chem.*, 1, 5, 1925; *Chem. Abstr.*, 19, 1857_5, 1925.
317. **Marina, A. V.,** Study of saccharin. Its preparation in the laboratory and in industry, *Afinidad,* 21, 323, 1944; *Chem. Abstr.*, 40, 7163_7, 1946.
318. **Marina, A. V.,** A treatise on saccharin — its laboratory and industrial preparation, *Afinidad,* 23, 321, 1946; *Chem. Abstr.*, 42, 5616_3, 1948.
319. **Marina, E.,** Chromatographic detection of saccharin, *Boll. Lab. Chim. Provinciali,* 6, 80, 1955; *Chem. Abstr.*, 50, 8084a, 1956.
320. **Marty, A.,** Foodstuff substitutes from coal tar, *J. Usines Gaz,* 68, 65, 1944; *Chem. Abstr.*, 40, 7423_2, 1946.
321. **Massatsch, C.,** Interesting saccharin contaminants, *Pharm. Ztg.*, 83, 426, 1947; *Chem. Abstr.*, 44, 74601, 1950.
322. **Massatsch, C.,** Determination of the sweetening agent in soluble saccharin and similar compounds, *Pharm. Ztg.*, 85, 222, 1949; *Chem. Abstr.*, 43, 6785g, 1949.
323. **Matsui, M., Sawamura, T., and Adachi, T.,** Electrolytic reduction of saccharin. I., Electrolysis in acid and alkaline, *Mem. Coll. Sci. Kyoto Imp. Univ.*, A15, 151, 1932; *Chem. Abstr.*, 26, 5264_8, 1932.
324. **Matsumoto, K. and Matsui, T.,** Pharmaceutical analysis by polarography. V. The *m*-nitraniline derivatives, *Pharm. Soc. Jpn,* 73, 653, 1952; *Chem. Abstr.*, 47, 953f, 1953.
325. **Meadow, J. R. and Cavagnol, J. C.,** Use of saccharin derivatives for the identification of mercaptions, *J. Org. Chem.*, 17, 488, 1952; *Chem. Abstr.*, 47, 1969g, 1953.
326. **Meillere, G.,** Milk, skimmed milk, condensed milk, powdered milk. The saccharin question. Preserved meat, *J. Pharm. Chim.*, 16, 21, 1917; *Chem. Abstr.*, 11, 2832_7, 1917.
327. **Merritt, L. L., Jr., Levey, S. and Cutter, H. B.,** Sodium saccharin as a reagent for the identification of alkyl halides, *J. Am. Chem. Soc.*, 61, 15, 1939; *Chem. Abstr.*, 23, 1692_5, 1939.
328. Methods for the examination of foods, *Verh. Kais. Gesundheitsamt,* 32, 1292, 1909, *Chem. Abstr.*, 3, 554_7, 1909.
329. **Mitchell, L. C.,** Separation and identification of cyclohexylsulfamate, dulcin, and saccharin by paper chromatography, *J. AOAC,* 38, 943, 1955; *Chem. Abstr.*, 50, 13664h, 1956.
330. **Miyadera, K.,** A note on metabolism following large doses of saccharin, *Z. Phys. Diaet. Ther.*, 26, 232, 1922; *Chem. Abstr.*, 16, 3326_7, 1922.
331. **Mohrbutter, C.,** Estimation of potassium guaiacolsulfonate in sugar, juices and syrup, *Apoth. Ztg.*, 46, 533, 1931; *Chem. Abstr.*, 25, 3434_2, 1931.
332. **Momose, T.,** Analysis of medicinals by polarography. I., *J. Pharm. Soc. Jpn.*, 64, 155, 1944; *Chem. Abstr.*, 45, 816c, 1951.
333. **Moncrieff, R. W.,** Relative sweetness, *Flavours,* 11, 5, 1948; *Chem. Abstr.*, 44, 6453i, 1950.
334. **Moncrieff, R. W.,** The sweetness of synthetics, *Food Manuf.*, 24, 29, 1949; *Chem. Abstr.*, 43, 2339a, 1949.
335. **Moncrieff, R. W.,** *The Chemical Senses,* Leonard Hill, London, 1951.
336. **Moore, C. D. and Stoeger, L.,** Estimation of saccharin in tablets, *Analyst,* 70, 337, 1945; *Chem. Abstr.*, 39, 5398_5, 1945.
337. **Muller, C.,** Saccharin in (mineral water, etc.) in Egypt, *Bull. Assoc. Chim. Sucr. Distill.*, 28, 630, 1911; *Chem. Abstr.*, 5, 2671_8, 1911.
338. **Muller, C.,** Saccharin in Egypt. Investigation into its uses in food. Artificial sugars commercial varieties of saccharin, *Ann. Fals.*, 4, 278, 1911; *Chem. Abstr.*, 5, 2403_8, 1911.
339. **Muller, E. and Petersen, S.,** A new group of sweet substances, *Chem. Ber.*, 81, 31, 1948; *Chem. Abstr.*, 43, 168c, 1949.
340. National Research Council, Food, and Nutrition Board, Summary Statement on Artificial Sweeteners, December 1954.
341. **Neiman, M. B.,** Polarographic method for determining saccharin, *Zh. Anal. Khim.*, 10, 175, 1955; *Chem. Abstr.*, 49, 15635i, 1955.
342. **Neumann, R. O.,** The influence of saccharin on the utilization of proteins, *Arch. Hyg.*, 96, 264, 1925; *Chem. Abstr.*, 22, 1622_5, 1928.
343. **Neumann, R. O.,** Experiments on the sweetening of foods with saccharin and sucrose, *Z. Nahr. Genussm.*, 47, 184, 1924; *Chem. Abstr.*, 18, 2207_1, 1924.
344. **Neuss, O.,** The degree of sweetness of natural and artificial sweeteners, *Umschau,* 25, 603, 1921; *Chem. Abstr.*, 16, 296, 1922.
345. **Notzold, R. A., Becker, D. E., Terrill, S. W., and Jensen, A. H.,** Saccharin and dried cane molasses in swine rations, *J. Anim. Sci.*, 14, 1068, 1955; *Chem. Abstr.*, 50, 5862g, 19 56.

346. **Noyce, W. K., Coleman, C. H., and Barr, J. T.**, Vinology in sweetening agents. I. A vinylog of dulcin, *J. Am. Chem. Soc.*, 73, 1295, 1951; *Chem. Abstr.*, 45, 9501i, 1951.
347. **Oakley, M.**, Report on (the determination of) preservatives and artificial sweeteners, *J. AOAC*, 28, 296, 1945; *Chem. Abstr.*, 39, 3853_8, 1945.
348. **Oakley, M.**, Report on (the determination of) saccharin, *J. AOAC*, 28, 298, 1945; *Chem. Abstr.*, 39, 3853_9, 1945.
349. **Oakley, M.**, Preservatives and artificial sweeteners, *J. AOAC*, 37, 371, 1954; *Chem. Abstr.*, 48, 8983i, 1954.
350. **Oakley, M.**, Analysis of preservatives and artificial sweeteners, *J. AOAC*, 38, 552, 1955; *Chem. Abstr.*, 49, 12738f, 1955.
351. **Oba, T.**, et al., Application of infrared spectroscopy to examination of drugs and their prparations. VII. Determination of sodium cyclohexyl sulfamate in the artificial sweetener, *Bull. Nat. Hyg. Lab. Tokyo*, 77, 61, 1959.
352. **Oddo, B. and Mingoia, Q.**, Differences in the sweetening power of saccharin and some of its derivatives, *Gazz. Chim. Ital.*, 57, 465, 1927; *Chem. Abstr.*, 21, 3202_2, 1927.
353. **Oddo, B. and Mingoia, Q.**, Variations in the sweetening power of saccharin and experiments with some of its derivatives. II., *Gazz. Chim. Ital.*, 61, 435, 1931; *Chem. Abstr.*, 26, 122_5, 1932.
354. **Oddo, B. and Perotti, A.**, Variations in the sweetening power of saccharin. IV. Effects of the association of saccharin with substances containing the ureide grouping, *Gazz. Chim. Ital.*, 67, 543, 1937; *Chem. Abstr.*, 32, 1675_7, 1938.
355. **Oddo, B. and Perotti, A.**, Changes in the sweetening power of saccharin. V. Influence of the association of saccharin with substances containing the ureide group, *Gazz. Chim. Ital.*, 70, 567, 1940; *Chem. Abstr.*, 35, 1038_7, 1941.
356. **Olsen, R. W.**, Cyclamates in citrus products, *Proc. Fla. State Hortic. Soc.*, 73, 270, 1960.
357. **Olszewski, W.**, Examination of sweetening compounds, *Pharm. Zentralhalle*, 61, 583, 1920; *Chem. Abstr.*, 15, 713_1, 1921.
358. Order of the Austrian minister of the interior concerning the Austrian alimentary codex, *Veroeff. Kats. Gesundh.*, 35, 742, 1911; *Chem. Abstr.*, 5, 3477_7, 1911.
359. **Orlov, N. I.**, Toxicity of dulcin, *Gig. Sanit.*, 13, 9, 36, 1948; *Chem. Abstr.*, 43, 5499d, 1949.
360. **Palet, L. P. J.**, An artificial edulcorant called "suessoel," *An. Soc. Quim. Argent.*, 2, 47, 1914; *Chem. Abstr.*, 8, 3601_2, 1914.
361. **Parmeggiani, G.**, Investigations upon saccharin, *Z. Oester. Apoth. Ver.*, 46, 199, 1908; *Chem. Abstr.*, 2, 3129_1, 1908.
362. **Paul, A. E.**, Report on flavoring extracts, *J. AOAC*, 3, 415, 1920.
363. **Paul, T.**, The degree of sweetness of dulcin and saccharin, *Chem. Ztg.*, 45, 38, 1921; *Chem. Abstr.*, 15, 1361i, 1921.
364. **Paul, T.**, Definitions in units of measure in the chemistry of sweeteners, *Chem. Ztg.*, 45, 705, 1921; *Chem. Abstr.*, 15, 3881_8, 1921.
365. **Paul, T.**, Physical chemistry of foodstuffs. V. Degree of sweetness of sugars, *Z. Elektrochem.*, 27, 539, 1921; *Chem. Abstr.*, 16, 973_5, 1922.
366. **Paul, T.**, The degree of sweetness of sweet substances, *Z. Nahr. Genussm.*, 43, 137, 1922; *Chem. Abstr.*, 16, 2558_6, 1922.
367. **Paul, R.**, The measurement of sweetness of artificial sweeteners, *Biochem. Z.*, 125, 97, 1921; *Chem. Abstr.*, 16, 974_2, 1922.
368. **Paul, R.**, Determination of sweetness of artificial sweeteners, *Umschau*, 24, 592, 1920; *Chem. Abstr.*, 15, 275_4, 1921.
369. **Pawlowski, F.**, Detection of saccharin in beer, *Z. Gesamte Brauwes.*, 32, 281, 1910; *Chem. Abstr.*, 4, 948_8, 1910.
370. **Pazienti, U.**, Quantitative determination of saccharin and of sodium saccharinate, *Anal. Chim. Appl.*, 2, 290, 1914; *Chem. Abstr.*, 9, 939_7, 1915.
371. **Pech, J.**, Polarographic studies with the dropping mercury cathode. XXXVIII. Reduction of some aliphatic amines, quinoline and saccharin, *Collect. Czech. Chem. Commun.*, 6, 126, 1934; *Chem. Abstr.*, 28, 4293_4, 1934.
372. **Penn, W. S.**, Plastics from a saccharin by-product, *Synth. By Prod.*, 8, 401, 1946; *Chem. Abstr.*, 41, 2603c, 1947.
373. **Perez, F. A.**, A chemical and bromatological analysis of the carbonated beverages of Lima and other localities in the republic, *Farm. Peru.*, 1, 3, 1943; *Chem. Abstr.*, 38, 1039_4, 1944.
374. **Peronnet, M.**, Saccharin in soft drinks particularly those marked "no alcohol," *Ann. Chim. Anal*, 26, 108, 1944; *Chem. Abstr.*, 40, 1246_8, 1946.
375. **Pertusi, C. and DiNola, E.**, The clarification of complex materials with basic lead acetate in the presence of alkali hydroxides, *Ann. Chim. Appl.*, 23, 311, 1933; *Chem. Abstr.*, 27, 5030_3, 1933.

376. **Pesman, J. and Luten, J.**, A case of poisoning with dulcin, *Ned. Tijdschr. Geneeskd.*, 90, 261, 1946; *Chem. Abstr.*, 40, 4431_2, 1946.
377. **Petersen, S.**, New sweetening agents, *Angew. Chem.*, 60A, 58, 1948; *Chem. Abstr.*, 45, 9194i, 1951.
378. **Poppe, J**, The detection and determination of saccharin, *Ann. Chim. Anal. Chim. Appl.*, 4, 157, 1922; *Chem. Abstr.*, 16, 2282_1, 1922.
379. **Posseto, G. and Issoglio, G.**, A new and rapid method for the extraction of saccharin from food substances, *G. Farm. Chim.*, 61, 5, 1912; *Chem. Abstr.*, 6, 1039_5, 1912.
380. **Pratt, C. D.**, Would your formula problems be helped by sorbitol?, *Food Can.*, 15, 2, 11, 1955; *Chem. Abstr.*, 49, 5701i, 1955.
381. **Pritzker, J.**, Chemical change in the composition of saccharin bicarbonate tablets, *Schweiz. Apoth. Ztg.*, 58, 78, 1920; *Chem. Abstr.*, 14, 1593_9, 1920.
382. **Pritzker, J. and Basel**, Chemical changes in the composition of saccharin bicarbonate tablets, *Schweiz. Apoth. Ztg.*, 58, 29, 1920; *Chem. Abstr.*, 14, 798, 1920.
383. **Profft, E.**, The physical compatibility of the intensively sweet 2-propoxy-5-nitroaniline and some derivatives thereof, *Dtsch. Chem. Z.*, 1, 51, 1949; *Chem. Abstr.*, 44, 5839_c, 1950.
384. **Profft, E. and Jumar, A.**, Higher homologs of sweetening agents, *Chem. Ztg.*, 80, 309, 1956; *Chem. Abstr.*, 51, 1070b, 1957.
385. **Pucher, G. and Dehn, W. M.**, Solubilities in mixtures of two solvents, *J. Am. Chem. Soc.*, 43, 1753, 1921; *Chem. Abstr.*, 16, 519_8, 1922.
386. Punishment for infringement of the regulations respecting the sale, etc., of saccharin (in Spain), *Chem. Ind.*, 33, 106; *Chem. Abstr.*, 5, 144_6, 1911.
387. **Raimon, M.**, Saccharin, *Ind. Chim.*, 66, 5, 1917; *Chem. Abstr.*, 12, 1401_4, 1918.
388. **Rashkovich, S. L.**, The use of saccharin in Russia, *Bull. Assoc. Chim. Sucr. Distill.*, 26, 353, 1909; *Chem. Abstr.*, 3, 456_5, 1909.
389. **Reid, E. E., Rice, L. M., and Grogan, C. H.**, Some *N*-alkylsaccharin derivatives, *J. Am. Chem. Soc.*, 77, 5628, 1955; *Chem. Abstr.*, 50, 6434d, 1956.
390. **Reif, G.**, The detection of saccharin and dulcin in vinegar and foods containing acetic acid, *Z. Nahr. Genussm.*, 46, 217, 1923; *Chem. Abstr.*, 18, 868_7, 1924.
391. **Reif, G.**, Determination of dulcin by the use of xanthydrol, *Z. Nahr. Genussm.*, 47, 238, 1924; *Chem. Abstr.*, 18, 2565_6, 1924.
392. **Reindollar, W. F.**, Report on (the determination of) preservatives (in foods), *J. AOAC*, 20, 161, 1937; *Chem. Abstr.*, 31, 5054_5, 1937.
393. **Reindollar, W. F.**, Report on (the determination of) preservatives (in foods), *J. AOAC*, 21, 184, 1938; *Chem. Abstr.*, 32, 5935_9, 1938.
394. **Reindollar, W. F.**, Report on (the determination of) preservatives (in foods), *J. AOAC*, 23, 288, 1940; *Chem. Abstr.*, 34, 5185_9, 1940.
395. **Reindollar, W. F.**, Report on (the determination of) preservatives (in foods), *J. AOAC*, 24, 326, 1941; *Chem. Abstr.*, 35, 5576_3, 1941.
396. **Reindollar, W. F.**, (The determination of) saccharin, *J. AOAC*, 25, 369, 1942; *Chem. Abstr.*, 36, 4610_3, 1942.
397. **Reindollar, W. F.**, Report on (the determination of) food preservatives and artificial sweeteners, *J. AOAC*, 27, 256, 1944.
398. **Repetto, E.**, Investigation of saccharin in commercial sirups, specialties, etc., *Rev. Farm.*, 60, 407, 1917; *Chem. Abstr.*, 11, 3091_8, 1917.
399. Resolutions adopted at the 58th annual conference of the Association of Food and Drug Officials of the United States, 1955, *Assoc. Food Drug Off. U.S. Q. Bull.*, 19, 36, 1955.
400. **Rice, H. L. and Pettit, P.**, An improved procedure for the preparation of alkyl halide derivatives of saccharin, *J. Am. Chem. Soc.*, 76, 302, 1954; *Chem. Abstr.*, 49, 12365b, 1955.
401. **Rice, L. M., Grogan, C. H., and Reid, E. E.**, *N*-Alkylsaccharins and their reduction products, *J. Am. Chem. Soc.*, 75, 4304, 1953; *Chem. Abstr.*, 48, 12027h, 1954.
402. **Richards, R. K., Taylor, J. D., O'Brien, J. L., and Duescher, H. O.**, Studies on cyclamate sodium (sucaryl sodium) a new non-caloric sweetening agent, *J. Am. Pharm. Assoc.*, 40, 1, 1951; *Chem. Abstr.*, 45, 3527i, 1951.
403. **Richmond, H. D. and Hill, C. A.**, Analysis of commercial saccharin. The estimation of *o*-benzoylsulfonimide from ammonia produced by acid hydrolysis, *J. Soc. Chem. Ind.*, 37, 246-9T, 1918; *Chem. Abstr.*, 12, 2215_5, 1918.
404. **Richmond, H. D. and Hill, C. A.**, Analysis of commercial saccharin. II. The detection and estimation of impurities, *J. Soc. Chem. Ind.*, 38, 8-10T, 1919; *Chem. Abstr.*, 13, 979_5, 1919.
405. **Richmond, H. D., Royce, S., and Hill, C. A.**, Examination of saccharin tablets, *Analyst*, 43, 402, 1918; *Chem. Abstr.*, 12, 621_1, 1919.
406. **Roberts and Dahle**, Low-carbohydrate ice cream for diabetics, *Ice Cream Trade J.*, 50, 4, 66, 1954; *Chem. Abstr.*, 48, 7809a, 1954.

407. **Roger, H. and Garnier, M.,** Influence of saccharin on peptic digestion, *Arch. Med. Exp.,* 19, 497, 1908; *Chem. Abstr.,* 2, 147_1, 1908.
408. **Romero, J. J. L.,** Sweetening agents and sugar, *Mem. Assoc. Tec. Azucareros Cuba,* 18, 83, 1944; *Chem. Abstr.,* 39, 4704_2, 1945.
409. **Rosenblum, H. and Mildworm, L.,** Determination of saccharin, *J. Am. Pharm. Assoc.,* 35, 336, 1946; *Chem. Abstr.,* 41, 4891e, 1947.
410. **Rosenthaler, L.,** Microchemical notes, *Mikrochem. Ver. Mikrochim. Acta,* 35, 164, 1950; A., 44, 7714i, 1950.
411. **Rosenthaler, L.,** The detection of the elements in organic materials. A test for N, *Pharm. Acta Helv.,* 19, 81, 1944; *Chem. Abstr.,* 38, 4533_3, 1944.
412. **Rosenthaler, L.,** Analytical notes. II, *Pharm. Zentralhalle,* 74, 288, 1933; *Chem. Abstr.,* 27, 3893_8, 1933.
413. **Rost, E. and Braun, A.,** The pharmacology of dulcin, *Arb. Reichsgesundh.,* 57, 212, 1926; *Chem. Abstr.,* 20, 3742_2, 1926.
414. **Rothe, M.,** Definition of the terms "aroma" and "aromatic material," *Ernaehrungsforschung,* 7, 639, 1963; *Chem. Abstr.,* 59, 2107c, 1965.
415. **Roura, U. C.,** Simple and rapid method for obtaining ethoxyphenylurea (dulcin), *Ind. Quim.,* 3, 160, 1941; *Chem. Abstr.,* 36, 2845_9, 1942.
416. **Rozanova, V. A.,** Colorimetric method of the determination of saccharin and vanillin in foodstuff, *Obshchestv. Pitan.,* 9, L-19, 1941; *Chem. Abstr.,* 37, 63506, 1943.
417. **Runti, C.,** Relation between chemical constitution and sweet taste. Further isosteres and derivatives of dulcin, *Ann. Chim.,* 46, 731, 1956; *Chem. Abstr.,* 51, 6527e, 1957.
418. **Runti, C.,** Synthetic sweetening agents, *Chim. Ind.,* 39, 354, 1957; *Chem. Abstr.,* 51, 1165i, 1957.
419. **Runti, C. and Sindellari, L.,** Report on chemical constitution and sweet taste. X. Isosteres and derivatives of dulcin, *Univ. Studi Trieste Fac. Sci. Ist. Chim. Pubbl.,* article 15, 1957; *Chem. Abstr.,* 51, 16459c, 1957.
420. **Runti, C.,** Synthetic sweetening compounds, *Bull. Soc. Pharm. Bordeaux,* 101, 3, 197, 1962; *Chem. Abstr.,* 58, 14624d, 1964.
421. Saccharin again, *Ann. Fals.,* 316, 57, 1910; *Chem. Abstr.,* 4, 1702_3, 1910.
422. The Safety of Artificial Sweeteners for use in Foods, Food Protection Committee of the Food Nutrition Board Publication No. 386, National Academy of Science/National Research Council, August 1955.
423. **Sah, P. P. T. and Chang, K. S.,** Dulcin synthesis by means of the Curtius reaction, *Ber.,* 69B, 2762, 1936; *Chem. Abstr.,* 31, 2179_9, 1937.
424. **Saillard, M.,** Intenational conference for the repression of the use of saccharin and analagous sweeteners in food products and beverage, *Ann. Falsif.,* 1, 50, 109; *Chem. Abstr.,* 3, 931_6, 1909.
425. **Salac, V.,** The polarograph on analysis in the brewery, *Kvas,* 64, 383, 1936; *Chem. Abstr.,* 33, 6520_7, 1939.
426. **Samaniego, C. C.,** Stevia Rebaudiana, *Rev. Farm.,* 88, 199, 1946; *Chem. Abstr.,* 41, 501_c, 1947.
427. **Sanchez, J. A.,** Functional analytical study of dulcin, *An. Farm. Bioquim,* 2, 63, 1931; *Chem. Abstr.,* 26, 2718_9, 1932.
428. **Sandri, R. M.,** Making good sugar-free drinks *Food Eng.,* 25, 79, 196, 1956.
429. **Sands, M.,** Sugarless jellies. Special Problems Course, Purdue U., Lafayette, Ind., *J. Am. Diet. Assoc.,* 29, 677, 1953.
430. **Sasagawa, Y., Inoue, M., and Sima, T.,** Preparation of dulcin, *Annu. Rep. Takeda Res. Lab.,* 9, 47, 1950; *Chem. Abstr.,* 46, 4497i, 1952.
431. **Schecker, G.,** Detection of saccharin in beverages (beer), *Centr. Zuckerind.,* 43, 312, 1935; *Chem. Abstr.,* 30, 7276_7, 1936.
432. **Schoenberger, J. A.,** et al., Metabolic effects, toxicity, and excretion of Ca *N*-cyclohexyl sulfamate (sucaryl) in man, *A.M.A. J. Med. Sci.,* 225, 551, 1953; *Chem. Abstr.,* 47, 11548h, 1953.
433. **Schoorl, N.,** The Vlezenbeek reaction of *p*-amino phenol derivatives, *Pharm. Weekbl.,* 74, 210, 1937; *Chem. Abstr.,* 31, 2746_4, 1937.
434. **Schowalter, E.,** Separation of saccharin and benzoic acid, *Z. Nahr. Genussm.,* 38, 185, 1919; *Chem. Abstr.,* 14, 1396_5, 1920.
435. **Schudel, H., Eder, R., and Buchi, J.,** Determination of the melting point, *Pharm. Acta Helv.,* 23, 33, 1948; *Chem. Abstr.,* 43, 1147i, 1949.
436. **Schulte, M. J.,** Determination of the sodium derivative of saccharin, *Pharm. Weekbl.,* 77, 1281, 1940; *Chem. Abstr.,* 36, 7235_9, 1942.
437. **Schutz, H. G. and Pilgrim, F. J.,** Sweetness of various compounds and its measurement, *Food Res.,* 22, 206, 1957; *Chem. Abstr.,* 51, 16992a, 1957.
438. **Schwarz, C. and Buchlmann, E.,** The physiology of digestion. X. The action of crystallose, saccharin and *p*-saccharin on salivary distaste, *Fermentforschung,* 7, 229, 1924; *Chem. Abstr.,* 18, 2913i, 1924.

439. **Seeker, A. F. and Wolf, M. G.,** Saccharin, *J. AOAC,* 3, 38, 1917; *Chem. Abstr.,* 11, 2375_6, 1917.

440. **Sen Grupta, M. C., Madiwale, M. S., and Bhatt, J. G.,** Estimation of saccharin in prepared tea, *Indian J. Pharm.,* 17, 185, 1955; *Chem. Abstr.,* 50, 6704f, 1956.

441. **Serger, H.,** Artificial sweeteners, *Chem. Ztg.,* 36, 829, 1912; *Chem. Abstr.,* 6, 2960_5, 1912.

442. **Serger, H.,** Artificial sweetening (saccharin) and its use in the fruit preservation industry, *Chem. Ztg.,* 47, 98, 1923; *Chem. Abstr.,* 17, 3382_7, 1923.

443. **Serger, H. and Clarck, K.,** The use of sugar or saccharin in the canning of cucumbers, *Konserv. Ind. Allg. Dtsch. Konserv. Ztg.,* 18, 244, 193; *Chem. Abstr.,* 25, 3739, 1931.

444. **Siedler, P.,** Artificial sweeteners, especially dulcin, *Chem. Ztg.,* 40, 853, 1916; *Chem. Abstr.,* 11, 999_7, 1917.

445. **Singleton, and Gray,** Recent work on citrus sections, *Proc. Fla. State Hort. Soc.,* 72, 263, 1959; *Chem. Abstr.,* 55, 9714h, 1959.

446. **Soifer, P. A.,** Estimation of saccharin in artificially flavored non-alcoholic beverages, *Gig. Sanit.,* 11, 6, 33, 1946; *Chem. Abstr.,* 41, 3550e, 1947.

447. **Solon, K.,** Determination of saccharin, *Food,* 16, 268, 1947; *Chem. Abstr.,* 41, 7311e, 1947.

448. **Spalton, L. M.,** Use of dulcin in foods and beverages, *Food Manuf.,* 25, 371, 1950; *Chem. Abstr.,* 46, 5743a, 1952.

449. **Stanek, V. and Pavlas, P.,** Microanalytical studies of artificial sweeteners. I. Saccharin, *Listy Cukrov.,* 53, 33, 1934; *Chem. Abstr.,* 29, 1931_7, 1935.

450. **Stanek, V. and Pavlas, P.,** Microanalytical studies with respect to artificial sweeteners. I. Saccharin, *Mikrochemie,* 17, 22, 1935; *Chem. Abstr.,* 29, 2885_3, 1935.

451. **Stanek, V. and Pavlas, P.,** A simple method for detecting saccharin in beer and other beverages, *Z. Zuckerind. Czech. Rep.,* 58, 313, 1934; *Chem. Abstr.,* 28, 6931_7, 1934.

452. **Stanek, V. and Pavlas, P.,** Microchemical studies of artificial sweetening agents. I. Saccharin, *Z. Zuckerind. Czech. Rep.,* 59, 361, 1935; *Chem. Abstr.,* 29, 7231_1, 1935.

453. **Staub, H.,** Toxicity of sweetening substances, *Mitt. Lebensm. Hyg.,* 36, 7, 1954; *Chem. Abstr.,* 41, 212d, 1947.

454. **Staub, H.,** Artificial sweetening substances, *Schweiz. Med. Wochenschr.,* 72, 983, 1942; *Chem. Abstr.,* 37, 4477_4, 1943.

455. **Struve, K.,** The ability of solutions of crystalline sweetening agents to withstand storage, *Dtsch. Fisch. Rundsch.,* p. 510, 1935; *Chem. Abstr.,* 31, 5891_6, 1937.

456. **Struve, K.,** The use of dulcin in marinades, *Fischwirtschaft,* 6, 40, 1930; *Chem. Abstr.,* 25, 4945_1, 1931.

457. **Sudo, T., Shimoe, D., and Tsujii, T.,** Rapid semi-micro determination of acetyl groups and methyl groups attached to carbon. Determnation of acetyl and C-methyl groups, *Bunseki Kagaku,* 6, 498, 1957; *Chem. Abstr.,* 52, 13540i, 1958.

458. Summary of the pharmacological studies of cyclamate sodium (sucaryl sodium), *Assoc. Food Drug Off. U.S. Q. Bull.,* 18, 165, 1954.

459. **Swafford, W. B. and Nobles, W. L.,** Modifications in the formula of wild cherry sirup, *J. Am. Pharm. Assoc. Pract. Pharm. Ed.,* 15, 99, 1954; *Chem. Abstr.,* 48, 6648d, 1954.

460. Sweetener demands booming, *Chem. Eng. News,* 36, 28, 1958.

461. Sweetening agents, *Am. Soft Drink J.,* 108, 700, 1959.

462. Sweetness sans sucrose, *Chem. Week,* October 13, 1951.

463. Synthetic sweetened soft drinks, *Chem. Eng. News,* 33, 4465, 1955.

464. **Tachi, I. and Tsukamoto, T.,** Polarography of vitamin B_1. V. Saccharin, *J. Agric. Chem. Soc. Jpn.,* 25, 335, 1951; *Chem. Abstr.,* 47, 11034f, 1953.

465. **Tarugi, N. and Ceccherelli, F.,** Observations on a test for saccharin, *Rend. Soc. Chim. Ital.,* 5, 198, 1914; *Chem. Abstr.,* 9, 806, 1915.

466. **Taschenberg, E. W.,** The antipyretic action of dulcin, *Dtsch. Med. Wochenschr.,* 48, 695, 1922; *Chem. Abstr.,* 17, 1073_2, 1923.

467. **Taufel, K. and Klemm, B.,** Investigations of natural and synthetic sweet substances. I. Studies on the degree of sweetness of saccharin and dulcin, *Z. Nahr. Genussm.,* 50, 264, 1925; *Chem. Abstr.,* 20, 951_8, 1926.

468. **Taufel, K. and Naton, J.,** The hydrolysis of *o*-benzoic sulfinide (saccharin), *Z. Angew. Chem.,* 39, 224, 1926; *Chem. Abstr.,* 21, 78_1, 1927.

469. **Taufel, K. and Wagner, C.,** The constitution of aqueous solutions of *o*-benzoic acid sulamide (saccharin) and *p-phenethyl urea (dulcin), Ber.,* 58B, 909, 1925; *Chem. Abstr.,* 19, 2347_4, 1925.

470. **Taufel, K., Wagner, C., and Dunwald, H.,** The decomposition of *p*-phenethyl carbamide (dulcin) on heating in aqueous solution, *Z. Elektrochem.,* 34, 115, 1928; *Chem. Abstr.,* 22, 2306_3, 1928.

471. **Taufel, K., Wagner, C. and Preiss, W.,** The mechanism of the hydrolysis of saccharin and *o*-sulfamino-benzoic acid, *Z. Elektrochem.,* 34, 281, 1928; *Chem. Abstr.,* 22, 3336_6, 1928.

472. **Taylor, J. D., Richards, R. K., and Davin, J. C.,** Excretion and distribution of radioactive sulfur 35-cyclamate sodium (sucaryl sodium) in animals, *Proc. Soc. Exp. Biol. Med.*, 78, 530, 1951; *Chem. Abstr.*, 46, 3157c, 1952.
473. **Terlinck, E.,** Saccharin and its manufacture, *Ing. Chim.*, 8, 233, 1924; *Chem. Abstr.*, 21, 1978_4, 1927.
474. **Testoni, G.,** The estimation of saccharin in various foods, *Z. Nahr. Genussm.*, 18, 577, 1910; *Chem. Abstr.*, 4, 626_1, 1910.
475. **Thevenon, L.,** A new reaction of saccharin, *J. Pharm. Chim.*, 22, 421, 1920; *Chem. Abstr.*, 15, 999_9, 1921.
476. **Thevenon, L.,** Contra-indication to a process for the detection of saccharin, *J. Pharm. Chim.*, 23, 7, 215, 1921; *Chem. Abstr.*, 15, 2335_9, 1921.
477. **Thompson, M. M. and Mayer, J.,** Hypoglycemic effect of saccharin in experimental animals, *Am. J. Clin. Nutr.*, 7, 80, 1959.
478. **Thomas, H.,** Changes in the sweetening power of dulcin (*p*-phenethylcarbamide) by chemical modification, *Dtsch. Zuckerind.*, 49, 1056, 1924; *Chem. Abstr.*, 19, 2674_1, 1925.
479. **Thoms, H. and Netteshiim, K.,** Changes in the taste of the sweetener dulcin (*p*-phenethylcarbamide) by chemical action, *Ber. Pharm. Ges.*, 30, 227, 1920; *Chem. Abstr.*, 14, 3638_9, 1920.
480. **Tillson, A. H. and Wilson, J. B.,** Isolation and microscopic identification of cyclamate sodium (sucaryl) as cyclohexanesulfamic acid, *J. AOAC*, 35, 467, 1952; *Chem. Abstr.*, 46, 11498i, 1952.
481. **Tracy, P. H.,** A dis cussion of sweetening agents for ice cream, *Can. Dairy Ice Cream J.*, 19, 3, 58, 1940; *Chem. Abstr.*, 35, 2231_5, 1941.
482. **Tracy, P. H. and Edman, G.,** Ice cream for diabetic using sorbitol, *Ice Cream Trade J.*, 46, 7, 50, 1950; *Chem. Abstr.*, 45, 778i, 1951.
483. **Traegel, A.,** Sugar and saccharin, *Dtsch. Zuckerind.*, 50, 1175, 1925; *Chem. Abstr.*, 20, 246_9, 1926.
484. **Traegel, A.,** Sweetening and preserving power of saccharin in comparison with sugar, *Z. Ver. Dtsch. Zuckerind.*, 75, 345, 1925; *Chem. Abstr.*, 20, 1873_1, 1926.
485. **Trauth, F.,** A method of rapid identification of saccharin in wine, *Dtsch. Lebensm. Rundsch.*, Vol. 28, 1943; *Chem. Abstr.*, 38, 4749_7, 1944.
486. **Trauth, F.,** A method for the rapid qualitative detection of sweet substances in wine, *Weinbau Wiss. Beih.*, 3, 106, 1949; *Chem. Abstr.*, 46, 9247g, 1952.
487. **Triest, F. J.,** How tobacco is flavored, *Am. Perfum. Essent. Oil Rev.*, 58, 449, 451, 453, 455, 1951; *Chem. Abstr.*, 46, 5266f, 1952.
488. **Tsuzuki, Y., Kato, S., and Okazaki, H.,** Sweet flavor and resonance, *Kagaku,* (Kyoto), 24, 523, 1954; *Chem. Abstr.*, 48, 13740h, 1954.
489. **Uglow, W. A.,** Activity of saccharin on bacteria, "plankton" and digestive enzymes, *Arch. Hyg.*, 92, 331, 1924; *Chem. Abstr.*, 10, 135_5, 1925.
490. **Uglow, W. A.,** The importance of dulcitol as a sugar substitute from a hygienic standpoint, *Arch. Hyg.*, 95, 89, 1925; *Chem. Abstr.*, 19, 2712_4, 1925.
491. **Unthoff, J. and Moragas, G.,** Synthetic preparation of dulcin (*p*-phenethylurea), *Quim. Ind.*, 5, 207, 1928; *Chem. Abstr.*, 23, 1889_5, 1929.
492. **Ulrich, K.,** The sweetness of synthetic and natural products, *Zucker,* 5, 236, 1952; *Chem. Abstr.*, 47, 2807f, 1953.
493. **Van Den Driessen, W. P. H.,** Qualitative and quantitative determination of saccharin in cacao powder, *Apoth. Ztg.*, 22, 2301, 1908; *Chem. Abstr.*, 2, 1306_1, 1908.
494. **Vandrac, I.,** Fungi from concentrated sugar solutions and sugar products, *Gambrinus Z. Bier Malz U. Hopfen*, 3, 201, 1942; *Chem. Abstr.*, 38, 4041_8, 1944.
495. **Van Eweyk, C.,** The influence of saccharin on the heart and circulation, *Z. Phys. Diaet. Ther.*, 26, 276, 1922; *Chem. Abstr.*, 17, 589_3, 1923.
496. **Van Roost, H.,** Saccharin, *Bull. Trimest. Assoc. Eleves Ec. Sup. Brass. Univ. Louvain*, 28, 49, 1928; *Chem. Abstr.*, 22, 1813_4, 1928.
497. **Van Voorst, F. Th.,** The detection of dulcin and saccharin in food products, *Chem. Weekbl.*, 39, 510, 1942; *Chem. Abstr.*, 38, 2399_5, 1944.
498. **Van Zijp, C.,** Microchemical contributions, *Pharm. Weekbl.*, 67, 189, 1930; *Chem. Abstr.*, 24, 2401_9, 1930.
499. **Van Zijp, C.,** The micro copper-pyridine reaction for saccharin, *Pharm. Weekbl.*, 71, 1146, 1934; *Chem. Abstr.*, 29, 428_2, 1935.
500. **Verkade, P. E.,** A new sweetening material, *Food Manuf.*, 21, 483, 1946; *Chem. Abstr.*, 41, 1772c, 1947.
501. **Verkade, P. E. and Meerburg, W.,** Alkoxyaminonitrobenzenes. IA. Partial reduction of 1-alkoxy-2,4-dinitrobenzenes with sodium disulfide in aqueous suspension, *Recl. Trav. Chim.*, 65, 768, 1946; *Chem. Abstr.*, 41, 3065e, 1947.

502. **Verkade, P. E., Van Dijk, C. P., and Meerburg, W.,** New sweet compounds and new local anesthetics, *Proc. Ned. Akad. Wet.*, 45, 630, 1942; *Chem. Abstr.*, 38, 4093_6, 1944.
503. **Verschaffelt, E.,** The toxicity of saccharin, *Pharm. Weekbl.*, 52, 37, 1915; *Chem. Abstr.*, 9, 1071_5, 1915.
504. **Vietti-Michelina, M.,** Chromatographic determination of saccharin and dulcin in biscuits and chocolate, *Chim. Ind.* (Milan), 38, 392, 1956; *Chem. Abstr.*, 50, 11546f, 1956.
505. **Vivas, G. V.,** Intentional chemical additives to foods. Synthetic sweeteners, *Farm. Nueva,* 21, 613, 1956; *Chem. Abstr.*, 51, 6897f, 1957.
506. **Vlezenbeek, H. J.,** A new reaction for the *p*-aminophenol function and a new sensitive reaction for dulcin in the presence of saccharin, *Pharm. Weekbl.*, 74, 127, *Chem. Abstr.*, 31, 2553_1, 1937.
507. **Vnuk, K.,** Studies on Peligot's saccharin and its isolation from molasses, *Z. Zuckerind. Czech. Rep.*, 51, 460, 467, 1927; *Chem. Abstr.*, 22, 182_3, 1928.
508. **Vollhase, E.,** Detection of saccharin in caramelized beer, *Chem. Ztg.*, 37, 425, 1913; *Chem. Abstr.*, 7, 3386_8, 1913.
509. **Von Der Heide, C. and Lohman, W.,** Detection of saccharin in wine, *Z. Nahr. Genussm.*, 41, 230, 1921; *Chem. Abstr.*, 16, 1634_5, 1922.
510. **Vondrak, J.,** A rapid method for the determination of soft drinks suspected to contain saccharin, *Z. Zuckerind. Czech. Rep.*, 53, 501, 1929; *Chem. Abstr.*, 24, 171_9, 1930.
511. **Vondrak, J.,** A rapid method for detecting saccharin in lemonades, *Listy Cukrov.*, 47, 323, 1929; *Chem. Abstr.*, 23, 2508_8, 1929.
512. **Von Scheele, O.,** The necessity for new legislation regarding saccharin, *K. Lantbruksakad. Handl. Tidskr.*, 50, 273, 1912; *Chem. Abstr.*, 6, 1784_7, 1912.
513. **Votava, J.,** A microanalytical study of artificial sweetening agents, *Listy Cukrov.*, 55, 365, 1937; *Chem. Abstr.*, 31, 8736_5, 1937.
514. **Votava, J.,** Microanalyticalstudies of artificial sweeteners. III. Further studies upon the determination of saccharin in drinks and food materials, mouth washes, tooth pastes, and tooth powders, *Z. Zuckerind. Cech. Rep.*, 62, 121, 1937; *Chem. Abstr.*, 32, 2639_8, 1938.
515. **Wachsmuth, H.,** Some new addition compounds of alkaloids and imides, *J. Pharm. Chim.*, 1, 9383, 1941; *Chem. Abstr.*, 38, 2340_4, 1944.
516. **Wagenaar, G. H.,** A new specific reagent for derivatives for barbituric acid and saccharin, *Pharm. Weekbl.*, 78, 345, 1941; *Chem. Abstr.*, 38, 2454_1, 1944.
517. **Wagenaar, M.,** Microchemical reactions of saccharin, *Pharm. Weekbl.*, 69, 614, 1932; *Chem. Abstr.*, 26, 4276_6, 1932.
518. **Wagenaar, M.,** The microchemistry of saccharin, *Mikrochemie,* 5, 132, 1932.
519. **Ward, J. C., Munch, J. C., and Spencer, H. J.,** Studies on strychnine. III. Effectiveness of sucrose, saccharin and dulcin in masking the bitterness of strychnine, *J. Am. Pharm. Assoc.*, 23, 984, 1934; *Chem. Abstr.*, 29, 293, 1935.
520. **Wauters, J.,** Reactions of saccharin in beer, etc., *J. Soc. Chem. Ind.*, 28, 733, 1910; *Chem. Abstr.*, 4, 636_9, 1910.
521. **Wertheim, E.,** Derivatives of dulcin, *J. Am. Chem. Soc.*, 53, 100, 1931; *Chem. Abstr.*, 25, 936_3, 1931.
522. **White, W. B.,** Saccharin in ice cream cones, *Annu. Rep. N.Y. Dep. Agric. Mkt. 1929 Legs. Doc.*, 37, 103, 1930; *Chem. Abstr.*, 24, 3572_6, 1930.
523. **Whittle, E. G.,** Application for Reindollars method to the estimation of saccharin, *Analyst,* 69, 45, 1944; *Chem. Abstr.*, 38, 1981_9, 1944.
524. **Wilson, J. B.,** Determination of sucaryl in sugar-free beverages, *J. AOAC,* 35, 465, 1952; *Chem. Abstr.*, 46, 11498g, 1952.
525. **Wilson, J. G.,** Determination of sucaryl, *J. AOAC,* 38, 559, 1955; *Chem. Abstr.*, 49, 12738e, 1955.
526. **Wilson, J. B.,** Determination of sucaryl (cyclamates of calcium and sodium), *J. AOAC,* 43, 583, 1960.
527. **Wolf, M. G.,** Report on preservatives (saccharin), *J. AOAC,* 6, 14, 1932; *Chem. Abstr.*, 17, 1284_1, 1932.
528. **Woo, M. and Huyck, C. L.,** Diabetic Sirups, *Dull. Natl. Formulary Comm.*, 16, 140, 1948; *Chem. Abstr.*, 43, 3144i, 1949.
529. **Zabrik, M. E., Miller, G. A., and Aldrich, P. J.,** The effect of sucrose and cyclamate upon the gel strength of gelatin, carrageenan, and algin in the preparation of jellied custard, *Food Technol.*, 16, 12, 87, 1962; *Chem. Abstr.*, 59, 6903f, 1965.
530. **Zaikov, V. S. and Sokolv, P. I.,** Use of chromic mixtures for oxidation of *o*-toluenesulfamide into saccharin, *J. Chem. Ind.*, 3, 1304, 1926; *Chem. Abstr.*, 22, 2933_2, 1928.
531. **Zaviochevskii, I. N.,** Saccharin, *Recl. Soc. Chim. Russ. Brno.*, 1, 67, 1925; *Chem. Abstr.*, 21, 2128_2, 1927.

532. **Zinkeisen, E.,** The use of artificial sweetening in effervescent citrus fruit beverages, *Dtsch. Miner. Wasser Ztg.,* 44, 196, 1940; *Chem. Abstr.,* 36, 7169_8, 1942.

533. **Zlatarov, A.,** The physiology of saccharin, *Ann. Univ. Sofia II. Fac. Phys. Math.,* 30, 2, 1934; *Chem. Abstr.,* 29, 3405_1, 1935.

534. **Zwikker, J. J. L.,** Detection of saccharin and the composition of its complexes with copper and pyridine, *Pharm. Weekbl.,* 70, 551, 1933; *Chem. Abstr.,* 27, 3898_2, 1933.

SUPPLEMENTARY REFERENCES

1. **Amerine, M. A., Pangborn, R. M., and Roessler, E. B.,** Principles of Sensory Evaluation of Food, *Academic Press,* New York, 1965.
2. **Anglemeir, A. F., Crawford, D. L., and Schultz, H. W.,** Improving the stability and acceptability of precooked, freeze-dried ham, *Food Technol.,* 14, 8, 1960.
3. **Baier, W. E.,** Citrus by-products and derivatives, *Food Technol.,* 9, 78, 1955.
4. **Beck, K. M.,** Artificial sweeteners for special purpose desserts and puddings, *Food Process.,* 26, 42, 1955.
5. **Beck, K. M. and Leffler, N. E.,** How to make dietetic frozen desserts, *Ice Cream Trade J.,* 54, 25, 1958.
6. **Beck, K. M. and Nelson, A. S.,** Latest uses of synthetic sweeteners, *Food Eng.,* 35, 96, 1963.
7. **Beck, K. M.,** Diet foods; industry status report, *Food Prod. Dev.,* 1(1), 19, 1967.
8. **Beck, K. M.,** Key to low calorie candy reformulation, *Candy Ind. Confect. J.,* 129, 57, 64, 1967.
9. **Beck, K. M.,** The non-nutritive sweeteners, *Manuf. Confect.,* 48, 92, 1968.
10. **Beck, K. M.,** The non-nutritive sweetener market, *A. C. S. Div. Chem. Mkt. Econ. Pap.,* 9, 294, 1968.
11. **Beck, K. M.,** Sweeteners, non-nutritive, in *Kirk-Othmer Encyclopedia of Chemical Technology,* Vol. 19, 2nd ed., John Wiley and Sons, New York, 1969, 593.
12. **Beck, K. M. and McCormick, R. D.,** Current saccharin actions place added stress on need for non-nutritive sweetener, *Food Prod. Dev.,* 6, 34, 1972.
13. **Beck, K. M.,** Synthetic sweeteners: past, present and future, in *Symposium: Sweeteners,* Inglett, G. D., Ed., AVI Publishing, Westport, Conn., 1974, 131.
14. **Beck, K. M.,** Practical considerations for synthetic sweeteners, *Food Prod. Dev.,* 9, 47, 1975.
15. **Beck, K. M.,** Practical considerations for synthetic sweeteners, in *Low Calorie and Dietetic Foods,* Dwivedi, B. K., CRC Press, Cleveland, 1977.
16. **Benson, G. A. and Spillane, N. J.,** Structure-activity studies on sulfamate sweeteners, *J. Med. Chem.,* 19(7), 869, 1976.
17. **Birch, G. G., Green, L. F., and Coulson, C. B.,** *Sweetness and Sweeteners,* Applied Science, London, 1971.
18. **Broeg, C. B.,** Sweeteners: natural and synthetic, *Food Eng.,* 37, 66, 1965.
19. **Brooks, M. A., de Silva, J. A., and D'Arconte, L.,** Sulfamate sweeteners: a reappraisal, *J. Pharm. Sci.,* 68, 1394, 1973.
20. **Coon, J. M.,** Evaluation of cyclamate and saccharin, *Int. Congr. Pharmacol.,* 6, 117, 1975.
21. Cyclamate still to the fore, *Food Cosmet. Toxicol.,* 10, 237, 1972.
22. **Daniels, R.,** *Sugar Substitutes and Enhancers,* Noyes Data, Park Ridge, N. J., 1973.
23. **Day, P. L.,** Nutritional aspects of the use of artificial non-nutritive sweeteners in foods and beverages, *Assoc. Food and Drug Off. U.S. Q. Bull.,* 19, 128, 1955.
24. **Etheridge, F. E. and Hard, M. M.,** Non-caloric sweeteners in freezing Elberta peaches, *J. Am. Diet. Assoc.,* 44, 281, 1964.
25. **Feigl, F. Goldstein, D., and Haguenauer-Castro, D.,** Contributions to organic spot test analysis, *Z. Anal. Chem.,* 178, 419, 1961.
26. *Food Chemicals Codex,* 2nd ed., National Academy of Sciences, Washington, D.C., 1972.
27. **Gofman, J. W., Nichols, A. V., and Dobbin, E. V.,** Dietary Prevention and Treatment of Heart
28. **Grinstead, L. E., Speer, V. C., Catron, D. V., and Hays, J. W.,** Comparison of sugar and artificial sweeteners in baby pig diets, *J. Anim. Sci.,* 19, 1264, 1960.
29. **Hoo, D. and Hu, C.,** Quantitative conversion of cyclamte to *N,N*-dichlorocyclohexylamine, and ultraviolet spectrophotometric assay of cyclamate in food, *Anal. Chem.,* 44, 2111, 1972.
30. **Huyck, C. L. and Maxwell, J. L.,** Diabetic Syrups, *J. Am. Pharm. Assoc. Pract. Pharm. Ed.,* 19, 142, 1958.

31. **Ichibagase, H. and Kozima, S.**, Studies on synthetic sweetening agents, I., *J. Pharm. Soc. Jpn.*, 82, 1616, 1962.
32. **Jorysch, D.**, Flavoring of pharmaceuticals, *Drug Allied Ind.*, 140, 10, 1958.
33. **Jukes, T. H.**, Cyclamate sweeteners, *JAMA*, 236, 1987, 1976.
34. **Kozima, S. and Ichibagase, H.**, Studies on synthetic sweetening agents III, *J. Pharm. Soc. Jpn.*, 83, 1114, 1963.
35. **Kozima, S. and Ichibagase, H.**, Studies of synthetic sweetening agents II, *J. Pharm. Soc. Jpn.*, 83, 1108, 1963.
36. **Lynch, J. S.**, The diet-conscious market, *Food Bus.*, 9, 30, 1961.
37. **Matson, E. J.**, Sucaryl in dietetic beverages, *Natl. Bottlers Gaz.*, 72, 12, 1953.
38. **McCann, M. B., Trulson, M. F., and Stulb, S. C.**, Non-caloric sweeteners in weight reduction, *J. Am. Diet. Assoc.*, 32, 327, 1956.
39. **Meer, G. and Gerard, T.**, Stabilizers, bodying agents for improved low calorie foods, *Food Proc.*, 24, 170, 1963.
40. **Mills, F., Weir, C. E., and Wilson, G. D.**, The effect of sugar on the flavor and color of smoked hams, *Food Technol.*, 14, 94, 1960.
41. **Moncrieff, R. W.**, *The Chemical Senses*, 3rd ed., CRC Press, Cleveland, 1967.
42. **Murray, E. J., Wells, H., Kohn, M., and Miller, N. E.**, Sodium sucaryl: a substance which tastes sweet to human subjects but is avoided by rats, *J. Comp. Physiol. Psychol.*, 45, 134, 1957.
43. National Cancer Institute, Division of Cancer Cause and Prevention, *Report of the Temporary Committee for the Review of Data on Carcinogenicity of Cyclamate*, National Institute of Health, Bethesda, Md., 1976.
44. *National Formulary XIII*, Mack, Easton, Pa., 1970.
45. *National Formulary XIV*, Mack, Easton, Pa., 1975.
45a. **Nofre, C. and Pautet, F.**, Sodium *N*-(cyclopentylmethyl)-sulfamate: a new synthetic sweetener, *Naturwissenschaften*, 62, 97, 1975.
46. **Oser, B. L., Carson, S., Cox, G. E., Vogin, E. E., and Sternberg, S. S.**, Chronic toxicity study of cyclamate:saccharin (10:1) in rats, *Toxicology*, 4, 315, 1975.
47. **Pintauro, N. D.**, *Sweeteners and Enhancers*, Noyes Data, Park Ridge, N.J., (1977), 392 pages.
48. **Pollack, H.**, Nutritional aspects of therapy for diabetes, *Postgrad. Med.*, 30, 598, 1961.
49. **Reynolds, T. B.**, Present needs for special purpose foods in medical diets, *Food Technol.*, (Chicago), 9, 319, 1955.
50. **Rockwell, N. P.**, Special ingredients in food products: calcium cyclamate in dietetic products, *West. Canner Packer*, 50, 23, 1958.
51. **Salunkhe, D. K., McLaughlin, R. L., Day, S. L., and Merkley, M. B.**, Preparation and quality evaluation of processed fruits and fruit products with sucrose and synthetic sweeteners, *Food Technol.* (Chicago), 17, 85, 1963.
52. **Schmähl, D.**, Fehlen einer kanzerogenen Wirkung von Cyclamat, Cyclohexylamin und Saccharin bei Ratten, *Arzneim. Forsch.*, 23(10), 1466, 1973.
53. **Segreto, V. A., Harris, N. O., and Hester, W. R.**, A stannous fluoride, silex, silicone dental prophylaxis paste with anticariogenic potentialities, *J. Dent. Res.*, 40, 90, 1961.
54. **Stavric, B.**, et al, Impurities in commercial saccharin. I. Impurities soluble in organic solvents, *J. Assoc. Off. Anal. Chem.*, 59, 971, 1976.
55. **Stone, C. D.**, Sucaryl: calorie reduction in baked products. I. Yeast fermented products, *Bakers Dig.*, 36, 53, 1962.
56. **Stone, C. D.**, Sucaryl: calorie reduction in baked products. I. Chemically aerated products, *Bakers Dig.*, 36, 59, 1962.
57. **Theivagt, J. G., Helgren, P. F., and Luebke, D. R.**, Cyclohexylamine derivatives, *Encyclopedia of Industrial Chemical Analysis*, Vol. 11, John Wiley and Sons, New York, 1971, 219.
58. **Tracy, P. H.**, Honey in low calorie frozen desserts, *Ice Cream Trade J.*, 51, 52, 1955.
59. *United States Pharmacopeia XIX*, Mack, Easton, Pa., 1975.
60. **Vincent, H. C., Lynch, M. J., Pohley, F. M., Helgren, F. J., and Kirchmeyer, F. J.**, Taste panel study of cyclamate-saccharin mixtures and components, *J. Am. Pharm. Assoc. Sci. Ed.*, 44, 442, 1955.
61. **Wagner, M. W.**, Comparative rodent preferences for artificial sweeteners, *J. Comp. Physiol. Psychol.*, 75(3), 483, 1971.
62. **Weast, C. A. and Buss, C. D.**, Special purpose foods, *Food Technol.* (Chicago), 9, 53, 1955.
63. **Wing, J. M.**, Preference of calves for a concentrated feed with and without artificial flavors, *J. Dairy Sci.*, 44, 725, 1961.

PATENT REFERENCES

1. Abbott Laboratories, Metal Salts for Cyclohexanesulfamic Acid, British Patent 669,200, 1952; *Chem. Abstr.*, 47, 5437f, 1953.
2. **Altwegg, J. and Collardeau, J.**, Saccharin, U.S. Patent 1,507,565, 1924; *Chem. Abstr.*, 18, 3386_6, 1924.
3. **Audrieth, L. F. and Sveda, M.**, E.I. duPont de Nemours, *N*-Cyclohexyl Sulfamic Acid and Salts, U.S. Patent 2,275,125, 1942.
4. **Baird, W.**, Imperial Chemical Industries, Ltd., Quaternary Ammonium Salts of Mercaptoarylene-thiazoles, U.S. Patent 2,104,068, 1938.
5. **Bebie, J.**, Saccharin, U.S. Patent 1,366,349, 1921; *Chem. Abstr.*, 15, 1030_4, 1921.
6. **Beck, K. M. and Weston, A. W.**, Abbot Laboratories, *N*-(3-Methylcyclopentyl) Sulfamic Acid and its Salts, U.S. Patent 2,785,195, 1957; *Chem. Abstr.*, 15, 12969h, 1957.
7. **Bigelow, F. E.**, Lecithinated Sugar, U.S. Patent 2,430,553, 1947.
8. **Borzhim, V. S.**, Saccharin, U.S.S.R. Patent 65,065, 1945; *Chem. Abstr.*, 40, 7233_7, 1946.
9. **Brenner, G. M.**, Sugarless Beverage, U.S. Patent 2,691,591, 1954.
10. **Cattano, R. and Supparo, G.**, Sucrose-Substituting Product, Italian Patent 417,869, 1947; *Chem. Abstr.*, 42, 6472d, 1948.
11. **Chaplin, E. D.**, Crystal Sweet Co., Sweetened Food Compound and Preparing the Same, U.S. Patent 851,221, 1907.
12. **Chem. Fab.**, *Chloroacetaldehydesulfonic Acid, German Patent 362,744.*
13. **Comte, F.**, Monsanto Chemical Co., Recovery and Purification of Saccharin, U.S. Patent 2,745,840, 1956; *Chem. Abstr.*, 50, 12349g, 1956.
14. **Cummins, E. W. and Johnson, R. S**, E. I. duPont de Nemours, Process for Making Purified Cycloh-exylsulfamates, U.S. Patent 2,799,700, 1957; *Chem. Abstr.*, 52, 612b, 1958.
15. **D'Amico, J. J.**, Monsanto Chemical Co., *N*-Haloalkenyl-*O*-Benzosulfimides, U.S. Patent 2,701,799, 1955; *Chem. Abstr.*, 50, 1086d, 1956.
16. Electrolytic Preparation of Saccharin, Netherlands Patent 41,338, 1937; *Chem. Abstr.*, 31, 8399_2, 1937.
17. **Endicott, C. J. and Dalton, E. R.**, Abbott Laboratories, Calcium Cyclamate Tablets and Process for Making Same, U.S. Patent 2,784,100, 1957.
18. **Erickson, A. M. and Ryan, J. D.**, Barron-Gray Packing Co., Extraction of Sweetening Ingredients From Fruit Juices, U.S. Patent 2,466,014, 1949; *Chem. Abstr.*, 43, 6333h, 1949.
19. **Fahlberg, C.**, A. List, Manufacture of Saccharin Compounds, U.S. Patent 319,082, 1885.
20. **Fahlberg, C.**, Saccharin Compound, U.S. Patent 326,281, 1885.
21. **Ferguson, E. A., Jr.**, Sweetening Composition, U.S. Patent 2,761,783, 1956; *Chem. Abstr.*, 50, 17252a, 1956.
22. **Glassman, J. A.**, Saccharin-Aspirin Tablets, U.S. Patent 2,134,714, 1939; *Chem. Abstr.*, 33, 1104, 1939.
23. **Golding, D. R. V.**, E. I. duPont de Nemours and Co., Process for Preparing Cyclohexylsulfamic Acid, U.S. Patent 2,814,640, 1957; *Chem. Abstr.*, 52, 7351a, 1958.
24. **Gordon, J. B.**, Dietetically Sweetened Food Products and Preparing Same, U.S. Patent 2,653,105, 1953.
25. **Gordon, J. B.**, Dietetic Food Product and Method of Preparing Same, U.S. Patent 2,629,655, 1953.
26. **Griffin, Joan M.**, Chas. Pfizer & Co., Inc., Products Sweetened Without Sugar and Free From Aftertaste, U.S. Patent 3,296,079, 1967; *Chem. Abstr.*, 66, 54403q.
27. **Grogan, C. H. and Rice, L. M.**, Geschickter Fund for Medical Research, Inc., *N*- and *O*-(Tertiary-Amino) Alkyl Derivatives of Saccharin, To: U.S. Patent 2,751,392, 1956; *Chem. Abstr.*, 51, 2060F, 1957.
28. **Hamilton, W. F.**, N-Salts of Saccharin With Vasoconstrictor Amines, U.S. Patent 2,538,645, 1951; *Chem. Abstr.*, 45, 4412d, 1951.
29. **Hayes, E. A.**, Saccharin as a Preservative for Pepper, U.S. Patent 2,444,875, 1948; *Chem. Abstr.*, 42, 6962a, 1948.
30. **Helgren, F. J.**, Abbott Laboratories, Sweetening Compositions and Method of Producing the Same, U.S. Patent 2,803,551, 1957; *Chem. Abstr.*, 51, 18288h, 1957.
31. **Hopff, H. and Gassenmeier, E.**, I. G. Farb. A-G, Sulfamic Acids and Their Salts, German Patent 874,309, 1953; *Chem. Abstr.*, 48, 12171c, 1954.
32. **Inoue, H. and Takami, T.** Saccharin, Japanese Patent 178,518, 1949; *C.A.*, 46, 222d, 1952.
33. **Kahn, L. E. and Dalve, J. A.**, E. I. duPont de Nemours and Co., Process of Curing Meat and Composition Therefor, U.S. Patent 2,946,692, 1960; *Chem. Abstr.*, 54, 20007^h.

34. **Kanami, M. and Uno, S.,** Mitsubishi Chemical Industries Co., Dulcin, Japanese Patent 177,576, 1949; *Chem. Abstr.*, 45, 7592b, 1951.
35. **Kantebeen, L. J. and Appelboom, A. F. J.,** N. V. Centrale Suikermaatschappij, Purifying Alkoxy (or Alkenyloxy) Nitroanilines, Netherlands Patent 59,306, 1947; *Chem. Abstr.*, 41, 5899a, 1947.
36. **Kawamura, B.,** Saccharin, Japanese Patent 177,789, 1949; *Chem. Abstr.*, 46, 1591g, 1952.
37. **Kracauer, P.,** Cumberland Packing Co., Artificial Sweetener Composition, U.S. Patent 3,285,751, 1966.
38. **Kreidl, I.,** Sugar Substitutes, Austrian Patent 122,657, 1930; *Chem. Abstr.*, 25, 4069_9, 1931.
39. **Kreidl, I.,** Sugar Substitutes for Use as Sweetening Agents by Diabetics, British Patent 314,500, 1928; *Chem. Abstr.*, 24, 1437_8, 1930.
40. **Kreidl, I.,** Sweetening Agents, British Patent 525,031, 1940; *Chem. Abstr.*, 35, 6688_9, 1941.
41. **Kreidl, I.,** Sweetening Agents, French Patent 677,452, 1929; *Chem. Abstr.*, 24, 3065_8, 1930.
42. **Kuderman, J.,** Diabetic and Diatetic Food Product, U.S. Patent 2,311,235, 1943.
43. **Locher, F. and Mueller, P.,** Ciba Corp. Artificial Sweeting Composition, U.S. Patent 3,118,772, 1964; *Chem. Abstr.*, 60, 12591^c.
44. **Loginov, N. E. and Polyanskii, T. V.,** Dulcin, U.S.S.R. Patent 65,779, 1946; *Chem. Abstr.*, 40, 7234_2, 1946.
45. **Lowe, H.,** Electrolytic Production of Saccharin, British Patent, 174,913, 1922; *Chem. Abstr.*, 16, 1710, 1922.
46. **Lynch, M. J.,** Abbott Laboratories, Improvement in Wine Taste, U.S. Patent 3,032,417, 1962; *Chem. Abstr.*, 57, 3878^c.
47. **Lynde, W. A.,** Process of Manufacturing Saccharin, British Patent 14,122, 1906; *Chem. Abstr.*, 1, 1492_3, 1907.
48. **Macdonald, L. H.,** Upjohn Co., Sweetening Compositions for Drugs, British Patent 810,537, 1959; *Chem. Abstr.*, 53, 10673^c.
49. **McNeill, C.,** Filters for Saccharin and Other Solutions, British Patent 384,749, 1931; *Chem. Abstr.*, 27, 2605_8, 1933.
50. **McQuaid, H. S.,** E. I. duPont de Nemours and Co. Preparation of Cyclohexylsulfamic Acid and its Salts, U.S. Patent 2,804,477, 1957; *Chem. Abstr.*, 52, 8191b, 1958.
51. **Mariotti, E.,** Dulcin, Italian patent 418,948, 1947; *Chem. Abstr.*, 43, 4786b, 1949.
52. **Matveev, V. K.,** Saccharin, U.S.S.R. Patent 66,878, 1946; *Chem. Abstr.*, 41, 4812d, 1947.
53. **Mitani, M.,** Purification of Saccharin, Japanese Patent 176, 513, 1948; *Chem. Abstr.*, 45, 7148e, 1951.
54. **Mizuguchi, J., Yokota, Y., and Umeda, M.,** Saccharin by Oxidation in Acid, Japanese Patent 180,599, 1949; *Chem. Abstr.*, 46, 7590d, 1952.
55. **Monnet, P. and Cartier, P.,** Soc. Chim. des Usines du Rhone, Saccharin, British Patent 165,438, 1921; *Chem. Abstr.*, 16, 424_8, 1922.
56. **Morani, V. and Marimpietri, L.,** Sweetening Agent, Italian Patent 425,939, 1947; *Chem. Abstr.*, 43, 5881c, 1949.
57. **Nakao, K.,** (A Substitute for) Sugar-coated Tablets, Japanese Patent 174,948, 1928; *Chem. Abstr.*, 43, 6791d, 1949.
58. **Orelup, J. W.,** Saccharin, U.S. Patent 1,601,505, 1926; *Chem. Abstr.*, 20, 3696_8, 1926.
59. **Peebles, D. D. and Kempf, C. A.,** Foremost Dairies, Inc., Sweetening Product and Method of Manufacture, U.S. Patent 3,014,803, appl. 1959; *Chem. Abstr.*, 56, 7762^e.
60. Pfizer and Co., Inc., Substituted Imidazolines, British Patent 757,650, 1956; *Chem. Abstr.*, 51, 9706c, 1957.
61. **Pilcher, F. E.,** Saccharin Composition, U.S. Patent 2,570,272, 1951; *Chem. Abstr.*, 46, 1666i, 1952.
62. N.V. Polak and Schwartz's Essence-Fabrieken, 2-Alkoxy-5-nitroaniline and 2-Alkoxy-4-nitroaniline (Sweetening Agents), Netherlands Patent 57,122, 1946; *Chem. Abstr.*, 41, 4170g, 1947.
63. **N. V. Polak and Schwarz's Essence-Fabrieken,** 2-Alkyloxy- or Alkenyloxy-5-nitroanilines and Their Salts, British Patent 613,367, 1948; *Chem. Abstr.*, 4, 5799g, 1949.
64. **N. V. Polak and Schwarz's Essence-Fabrieken,** Improvements in the Preparation of Etherified Hydroxynitroanilines, British Patent 597,835, 1948; *Chem. Abstr.*, 42, 5050f, 1948.
65. **N.V. Polak and Schwarz's Essence-Fabrieken,** 2-Propoxy-5-nitroaniline Netherlands Patent 52,980, 1942; *Chem. Abstr.*, 41, 4170e, 1947.
66. **Polya, E.,** General Foods Corp., Artificial Sweetening of Foods, German Patent 1,074,985, 1960; *Chem. Abstr.*, 55, 11702^a.
67. **Rader, W. O. and Seaton, A.,** Mixture of Glucose and Saccharin for "Sugar-tolerance" Tests U.S. Patent 1,625,165, 1927; *Chem. Abstr.*, 21, 2052_4, 1927.
68. **Raecke, B.,** Henkel and Cie. G.m.b.H., Salts of Saccharin, German Patent 822,388, 1951; *Chem. Abstr.*, 48, 4238h, 1954

69. **Robinson, J. W.,** E. I. duPont de Nemours, Sodium Cyclohexylsulfamate, U.S. Patent 2,383,617, 1945; *Chem. Abstr.*, 40, 357_9, 1946.
70. **Sahyun, M.,** Glycine Ester Salts of Cyclohexylsulfamic Acid, U.S. Patent 2,789,997, 1957; *Chem. Abstr.*, 51, 15560f, 1957.
71. **Sahyun, M. and Faust, J. A.,** 1-(*m*-Hydroxyphenyl)-2-(methylamino) Ethyl Cyclohexanesulfamate, U.S. Patent 2,746,986, 1956; *Chem. Abstr.*, 51, 1270e, 1957.
72. **Seifter, E.,** Monsanto Chem. Co., Improving the Palatability of Fruits and Vegetables by Treating the Plants with a Synthetic Sweetening Agent, U.S. Patent 2,921,409, 1960; *Chem. Abstr.*, 54, 7928d, 1960.
73. **Senn, O. F.,** Maumee Development Co., O-Sulfonyl Chloride Benzoic Acid Esters, U.S. Patent 2,667,503, 1954; *Chem. Abstr.*, 49, 3257e, 1955.
74. **Shibe, W. J., Jr., Cohen, S. I., and Frant, M. S.,** Gallowhur Chem. Corp., Quaternary Ammonium "Saccharinates," U.S. Patent 2,725,326, 1955; *Chem. Abstr.*, 50, 10134d, 1956.
75. **Snelling, W. O.,** Chewing Gum, U.S. Patent 2,484,859, 1949; *Chem. Abstr.*, 44, 2790_6, 1950.
76. **Snelling, W. O.,** Chewing Gum, U.S. Patent 2,484,860, 1949; *Chem. Abstr.*, 44, 2790_6, 1950.
77. **Soder, G. and Schnell, H.,** Farb. Bayer, A-G, Cyclohexylsulfamic Acid, German Patent 950,369, 1956; *Chem. Abstr.*, 51, 12970a, 1951.
78. **Stanko, G. L.,** Richardson-Merrell, Inc., Artificial Sweetener-Arabinogalactan Composition and Its Use in Edible Foodstuff, U.S. Patent 3,294,544, 1966; *Chem. Abstr.*, 66, 45627p.
79. **Sugino, K. and Mizuguchi, J.,** Saccharin From *o*-Toluenesulfonamide, Japanese Patent 176,553, 1948; *Chem. Abstr.*, 45, 7148d, 1951.
80. **Synerholm, M. E., Jules, L. H., and Sahyun, M.,** Melville Sahyun, Trading as Sahyun Laboratories, Imidazolines, U.S. Patent 2,730,471, 1956; *Chem. Abstr.*, 51, 121a, 1957.
81. **Terao, K.,** Takeda Drug Ind. Co., Concentrated Stable Solution of Polyhydroxyalkylisoalloxazine, Japanese Patent 173,506, 1946; *Chem. Abstr.*, 46, 2242g, 1952.
82. **Thompson, W. W.,** E. I. duPont de Nemours and Co., Production of Cyclohexylsulfamtes, U.S. Patent 2,826,605, 1958; *Chem. Abstr.*, 52, 1290a, 1958.
83. **Thompson, W. W.,** E. I. duPont de Nemours and Co., Cyclohexylsulfamic Acid, U.S. Patent 2,800,501, 1957; *Chem. Abstr.*, 51, 17987e, 1957.
84. **Thompson, W. W.,** E. I. duPont de Nemours and Co., Preparation of Salts of N-Substituted Sulfamic Acids, U.S. Patent 2,805,124, 1957; *Chem. Abstr.*, 52, 9222f, 1958.
85. **Universal Industria Dolciaria Agricola Alimentara,** Sweet Material, Italian Patent 426,150, 1947; *Chem. Abstr.* 43, 5881d, 1949.
86. **Verkade, P. E.,** N. V. Polak and Schwarz's Essence-Fabrieken, 2-Alkoxy-4-nitroanilines, British Patent 594,816, 1947; *Chem. Abstr.*, 42, 2280a, 1948.
87. **Walker, H. W.,** Ditex Foods, Inc., Synergistically Sweetened Canned Fruits, U.S. Patent 2,608,489, 1952; *Chem. Abstr.*, 47, 796b, 1953.
88. **Washington, G.,** Evaporated Amorphous Saccharine Composition, U.S. Patent 1,512,730, 1945; *Chem. Abstr.*, 19, 686_1, 1925.
89. **Weast, C. A.,** Frozen Fruit Prepared With Saccharin, U.S. Patent 2,511,609, 1950; *Chem. Abstr.*, 44, 464d, 1950.
90. **Weast, C. A.,** Dietetic Canned Fruits and Making the Same, U.S. Patent 2,536,970, 1951.

SUPPLEMENTARY PATENT REFERENCES

1. Abbott Laboratories, Improvements in or Relating to the Preparation of Salts of Cyclohexylsulfamic Acid, British Patent 662,800, 1951.
2. Abbott Laboratories, Improvements in or Relating to Metal Salts of Cyclohexylsulfamic Acid, British Patent 669,200, 1952.
3. Abbott Laboratories, Procedure for Sulfonation, Belgian Patent 628,359, 1963.
4. **Birsten, O. G. and Rosin, J.,** Baldwin-Montrose, Preparation of Cyclohexylsulfamic Acid or Metal Salts Thereof, U.S. Patent 3,361,798, 1968.
5. **Birsten, O. G. and Rosin, J.,** Baldwin-Montrose, Preparation of Cyclohexylsulfamic Acid or Metal Salts Thereof, U.S. Patent 3,366,670, 1968.
6. **Farbenfabriken Bayer,** *N*-Cyclohexylamidosulfonic Acid, Sweetening Agent, Netherlands Patent 68,16560, 1969.
7. **Dickinson, H. M. N.,** Abbott Laboratories, Cyclohexylsulfamic Acid Salt, U.S. Patent 3,074,996, 1963.

8. **Donges, E., Kohlhaas, R., and Wick, D., Farbwerke Hoeschst, A. G.**, Sodium Cyclohexylamidosulfonate, German Patent 1,249,262, 1967.
9. **Donnison, G. H.**, Abbott Laboratories, Preparation of Sodium Cyclohexylsulfamate, U.S. Patent 3,192,252,1965.
10. **Donnison, G. H.**, Abbott Laboratories, Improvements in the Preparation of Sodium Cyclohexylsulfamate, British Patent 930,289, 1963.
11. **Freifelder, M.**, Abbott Laboratories, Preparation of Cyclohexylsulfamates, U.S. Patent 3,082,247, 1963.
12. **Freifelder, M. and Meltsner, B.**, Abbott Laboratories, Reduction of Phenylsulfamic Acid and Phenylsulfamates, U.S. Patent 3,194,833, 1965.
13. **Galat, A.**, Complex of Calcium *N*-Cyclohexylsulfamate with Glycine, U.S. Patent 3,345,403, 1967.
14. **Grosvenor, N. M.**, American Sugar Refining Company, Sweetening Composition and Method of Producing the Same, U.S. Patent 3,011,897, 1961.
15. **Guadagni, D. G.**, U. S. Department of Agriculture, Preparation of Frozen Fruit, U.S. Patent 3,025,169, 1962.3
16. **Helgren, F. J.**, Abbott Laboratories, Sweetening Agent and Process for Producing the Same, U.S. Patent 3,098,749, 1963.
17. **Jucaitis, P., Bliudzius, I., and Rockwell, N.**, E. I. duPont de Nemours, Solid Dietetic Food Composition, U.S. Patent 2,876,106, 1959.
18. **Kossoy, M. W. and Voegeli, M. N.**, Swift and Company, Dietetic Cured Food Product, U.S. Patent 2,966,416, 1960.
19. **Lemaire, N. A. and Peterson, R. D.**, Kellogg Company Method of Making Freeze Dried Artificially Sweetened Fruit Products, U.S. Patent 3,356,512, 1967.
20. **Leo, H. T. and Taylor, C. C.**, Dietary Gel Composition, U.S. Patent 2,865,761, 1958.
21. **Loder, D. J.**, E. I. duPont de Nemours, Production of Cyclohexylsulfamates by Amide Interchange, U.S. Patent 2,804,472, 1957.
22. Merck and Company, Medicated Candy Lozenges, British Patent 1,083,050, 1967.
23. **Merory, J.**, Shultin, Inc., Fruit Flavors, Products and Methods, U.S. Patent 2,865,756, 1958.
24. **Mueller, P. and Trefzer, R.**, Ciba Corp., Process for the Manufacture of *N*-Cyclohexyl-sulfamic Acids U.S. Patent 3,060,231, 1962.
25. **Okuda, N. and Kamba, T.**, Daiichi Pharmaceutical Company, Procedure for Preparing Salt of Cyclohexylsulfamic Acid, Japanese Patent 24,076, 1961.
26. **Okuda, N. and Suzuki, K.**, Daiichi Seiyaku Company, Process for the Production of Cyclohexylsulfamates, U.S. Patent 3,043,864, 1962.
27. **Okuda, N., Fukuda, Y., and Suzuki, K.**, Daiichi Seiyaku Company, Process for the Preparation of Cyclohexylsulfamates, U.S. Patent 3,338, 958, 1967.
28. **Riffkin, C. and Cyr, G. N.**, Olin Mathieson Chemical Corporation, Liquid Sweetening Composition and Method of Preparing the Same, U.S. Patent 2,845,353, 1958.
29. **Rutz, W. D., Donald F. Gregg**, Flavorings Containing Cyclohexanesulfamic Acid and Cyclamate Salts for Casein-containing Products, U.S. Patent 3,353,959, 1967.
30. **Shah, V.**, Abbott Laboratories, Cyclamate Process, Canadian Patent 770,637, 1967.
31. **Shah, V. D. and Bernsen, S.**, Abbott Laboratories, Cyclamic Acid from Double Salt, Water and Sulfuric Acid, U. S. Patent 3,361,799, 1968.
32. **Tribble, T. B.**, Flavored Sweetener for Livestock Feed Products, U.S. Patent 2,932,571, 1960.
33. **Villers, C. E. and Mitchell, D. D.**, Pillsbury Company, Process for the Production of Amine Salts of Sulfamic Acid, U.S. Patent 3,412,143, 1968.
34. **Yamaguchi, H.**, Nitto Chemical Industry, Procedure for Preparing Salt of *N*-Cyclohexysulfonic Acid, Japanese Patent 17,559, 1960.
35. **Yamaguchi, H. and Nakatsuchi, U.**, Nitto Chemical Industry, Process for the Manufacture of *N*-Cyclohexylsulfamates, U.S. Patent 3,090,806, 1963.

NEW SWEETENERS

G. A. Crosby and T. E. Furia

INTRODUCTION

The past 10 years have witnessed a dramatic increase in research devoted to the search for new, nontoxic nonnutritive sweeteners. A number of reasons might account for this growth, but chief among these is the tremendous market demand created when cyclamate was banned in 1969, a demand that is now only partially satisfied by the availibility of saccharin. Less than optimum taste qualities, together with a continuing threat of toxicological problems, have resulted in limitation of the consumption of saccharin. Additionally, the threatening rise in prices of natural sweeteners has prompted a search for substitute materials.

The number of new, commercially viable sweetener candidates emerging in recent years has been severely limited by the stringent requirements that must be satisfied by such materials. To be successful in the largest markets, it is generally agreed that a new sweetener must meet the following requirements:

1. It must be safe for human consumption.
2. It must possess outstanding flavor characteristics.
3. It must provide adequate solubility and stability.
4. It must be at least equal to sucrose on a cost-per-sweetness basis.

Since any new sweetener must be subjected to a lengthy and costly program of rigorous safety testing before being approved for human consumption by the U.S. Food and Drug Administraton (FDA), it is imperative that new candidates satisfy the remaining requirements. They should possess a flavor that is clean and without aftertaste; they should be competitive with sucrose on a cost basis and preferably approach a wholesale price of \$10 to \$15/lb, assuming a relative sweetness of 300 to 400 times sucrose. This price and acceptance is based on material that is sufficiently soluble for most food applications and stable under acid and heat conditions normally encountered in foods. A major application for nonnutritive sweeteners is carbonated beverages, where the solution may be as acidic as pH 2.8. It is also possible that a particular product may be exposed to relatively high storage temperatures (125°F) and direct sunlight for many hours.

Unfortunately, none of the more promising sweeteners that have emerged in past years have satisfied all requirements, so the search for new candidates continues. This review intends to provide a survey of recent sweetener research which the reader may use as a basis for speculation on future trends.

NATURAL PRODUCTS

The interest in naturally occurring sweet substances has grown in the last several years because of two potential advantages over other sources of new sweeteners. First, all new classes of compounds that are known to elicit a sweet taste response have been found by accident. Nature provides a nearly unlimited source of new compounds that can be screened (tasted) with relative ease. Second, it is generally believed that a naturally occurring substance, which has perhaps been included in the diet of a small percentage of the population, will present less of a risk in satisfying the rigorous tests required to determine human safety. Recent isolation of three new sweet proteins, monellin and the thaumatins (I and II), as well as the steroidal saponin, osladin, together with renewed studies of stevioside and phyllodulcin, gives an indication of the interest in naturally occurring sweet substances. A review has recently appeared that discusses the sweet proteins, monellin and thaumatin, and the taste-modifying glycoprotein, miraculin.[1]

Monellin

The tropical plant *Dioscoreophyllum cumminsii*[2,3] produces a light red fruit occurring in small grapelike clusters. The pulp of this fruit is intensely sweet, while the seed is extremely bitter. A sweet substance was first isolated from the fruit of this tropical plant and characterized as a carbohydrate by Inglett and May.[4] However, Morris and Cagan of the Monell Chemical Senses Center have more recently isolated a pure substance from the same fruit which they call monellin and have determined the intensely sweet material to be a protein with a molecular weight of 10,700.[5] This finding has since been independently confirmed by Bohak and Li (Weismann Institute, Rehovat, Israel) working at Dynapol (Palo Alto, Cal.),[6] Züber of Eidgenossische Technische Hochschule (Zurich),[7] and van der Wel and Loeve of Unilever (Vlaardingen, The Netherlands).[8]

In recently published work, Bohak and Li have found that monellin consists of two nonidentical, noncovalently bound chains. The amino acid sequence of these chains has been determined and reported the first half of 1976.[6] Furthermore, Bohak has chromatographically separated both chains (using polycarboxylic acid resin with 7.5*M* aqueous acetic acid) and has found that neither is sweet by itself. However, solutions of the two separated chains could be mixed and could be seen by electrophoresis to return slowly over a period of several hours to a low yield of monellin, accompanied by a return of sweet taste. A determination of the precise structure of the monellin molecule in the crystalline state by X-ray crystallography is now underway by Wlodawer and Hodgson at Stanford University.[9]

These findings would appear to present unambiguous evidence that macromolecules are capable of eliciting a sweet taste response; thus, they provide important implications relating to the mechanism of sweet taste. In this regard, monellin is an extremely important research tool.

When calculated from threshold concentrations, monellin has been determined to be approximately 3000 times sweeter than sucrose on a weight basis.[5] Present evidence indicates that monellin has little chance of satisfying the requirements of a commercially utilizable sweetener. Monellin does not meet even the minimal stability requirements for use in a carbonated beverage. Furia and Stone demonstrated that a cola beverage sweetened with monellin lost sweetness within hours after preparation.[86] The taste properties of monellin are also less than adequate. Onset of initial taste sensation does not occur for several seconds, followed by a gradual building of total intensity and a slow decline in sweet taste over a period of time ranging up to 1 hr. It has been suggested that the lagging and lingering taste of monellin results from formation of a disulfide linkage between the single sulfhydryl group of monellin and the sulfhydryl group of a membrane component.[6]

Thaumatins

The fruit of the tropical plant *Thaumatococcus daniellii Benth* has been used as a source of sweetness for many years by the inhabitants of certain sections of western Africa.[3] In 1972, van der Wel and Loeve isolated two very similar proteins (named thaumatin I and thaumatin II) from an aqueous extract of the fruit.[10] Both proteins have molecular weights of approximately 21,000 and nearly identical sensory properties. The amino acid content of both thaumatin I and II has been published elsewhere.[1] The taste potency of the thaumatins has been estimated to be approximately 1600 times sucrose on a weight basis when determined at the threshold level. The thaumatins have been reported to display persistent licoricelike aftertastes.[10] The finding that cleavage of the disulfide linkages in thaumatin results in a loss of sweet taste also suggests that the taste sensation is a result of the intact protein rather than a fragment.[10] X-ray crystallographic studies of thaumatin I have just been completed.[11] There is little likelihood that the thaumatins will find large-scale applications as commercial sweeteners.

Miraculin

Miraculin is a glycoprotein that is not sweet per se when tasted as a crude extract; however, after exposing the tongue to miraculin for a short period of time, both mineral and organic acids taste sweet. The taste sensation of lemon on a tongue "preconditioned" with this glycoprotein is much like the taste of a lemon sweetened with sugar — a unique, pleasant, sour-sweet sensation is perceived.

Miraculin occurs in the fruit of the shrub *Synsepalum dulcificum*,[2,3,12] native to west Africa. The first published account of the shrub and the interesting taste-modifying properties of its fruit, called "miraculous berries," was made by Daniell in 1852.[13] The opening lines from Daniell's paper give an interesting description of the fruit:

> Among other remarkable productions furnished by the vegetable kingdom in tropical Africa is one which, from the peculiar properties it possesses, has gained a celebrity now permanently established. European voyagers and traders who first experienced the singular effects of this fruit upon the palate were doubtless greatly astonished at what, to them, must have appeared an extraordinary power whose potency for a certain length of time could change the flavour of the most acid substances into a delicious sweetness, and on this account unanimously conferred upon it the characteristic title of the "Miraculous Berry," which it has retained and by which it has since been appropriately distinguished.

Daniell follows with more detail in a later passage:

> This tree flowers in the months of June, July, and August and in general produces a tolerable number of oblong or oval berries about two-thirds in size of, and somewhat resembling in figure, an olive; these at first are of a dull green, but gradually change as they ripen into a dusky red colour. The seeds are clothed externally by a thin, softish pulp, slightly saccharine yet imparting that extraordinary impression to the palate by which the most sour and acidulous substances become intensely sweet, so that citric or tartaric acids, lime juice, vinegar, and all immature fruit of a sourish character lose their unpleasant qualities, and taste as if they had been solely composed of saccharine matter. The duration of these effects may be stated to depend upon the amount of berries chewed and the degree of maturity they have attained, for when a sufficient quantity has been taken their influence is commonly perceptible throughout the day. Their peculiar principle, however, is soon dissipated if the fruit is suffered to remain in a ripe condition for any length of time, and their preservation in spirits, acetic acid, or syrup, does not appear to favour its retention, if I may judge from the specimens brought to England, since they have not only lost their properties, but have become extremely insipid. The purposes for which the natives of the Gold Coast usually reserve them are but few, the principle consisting in rendering their stale and acidulated kankies more palatable, and in bestowing a sweetness on sour palm wine and pitto. In Akkrah this is the more necessary from the circumstances that few or no palm trees flourish in its vicinity, and hence the wine has to be conveyed a considerable distance inland from Aquapim, and from the time occupied in the transmission, the acetous fermentation has frequently commenced before its arrival, to remove which the previous employment of these berries is considered as indispensable. In other respects they would seem to be eaten more for the novelty of the sensations they induce than for any particular object.....

Inglett and his co-workers were the first to attempt an isolation of miraculin in 1965 but were successful in realizing only a fivefold enrichment of the active materials.[14] Shortly thereafter, Kurihara and Beidler[15] and Brouwer and colleagues[12] were successful in isolating the glycoprotein using a variety of chromatographic techniques. The molecular weight of miraculin was estimated by both research teams to be in the range of 42,000 to 44,000, while the carbohydrate content, reported by Brouwer, was a mixture of only L-arabinose and D-xylose and by Beidler, a combination of glucose, ribose, arabinose, galactose, and rhamnose units. The amino acid composition of miraculin was previously reported.[12,15]

Beidler has suggested a mechanism of action for the taste-modifying protein involving the binding of the protein portion of the molecule to an appropriate location of the taste-cell membrane adjacent to a receptor site.[16] Exposure to acid is believed to result in a change in conformation of the protein, such that the carbohydrate portion(s) of the molecule is able to interact with a 'sweet'' receptor site.

Recently, one attempt was made by a commercial firm located in the U.S. to market a concentrate of miracle fuit as a dietetic aid. In September 1974, however, the FDA denied approval of the Generally Recognized As Safe (GRAS) petition, and all marketing of the product was halted.

Steviol Glycosides

The leaves of the wild shrub *Stevia rebaudiana Bertoni* have been used by the people of Paraguay to sweeten bitter drinks for at least as far back as the last century. A pure, crystalline, sweet substance was first isolated from the leaves and named stevioside by two French chemists, Bridel and Lavieille, in 1931.[17] Contributions to the structural elucidation of stevioside (Structure 1) were made by many investigators[18-22] between 1955 and 1963. Several reviews have appeared that discuss the history and properties of stevioside.[23,24]

Stevia rebaudiana Bertoni is now being extensively cultivated in Japan due to the interest in developing stevioside as a sweetener. Recently, two new, sweet steviol glucosidies, rebaudioside A and B (Figure 1), were isolated from the same plant.[87] Rebaudioside A is obtained in 1.4% yield based on the weight of dried leaves, while rebaudioside B is recovered in 0.04% yields. Rebaudioside A has been found to be 190 times more potent than sucrose and with a taste superior to stevioside.[113]

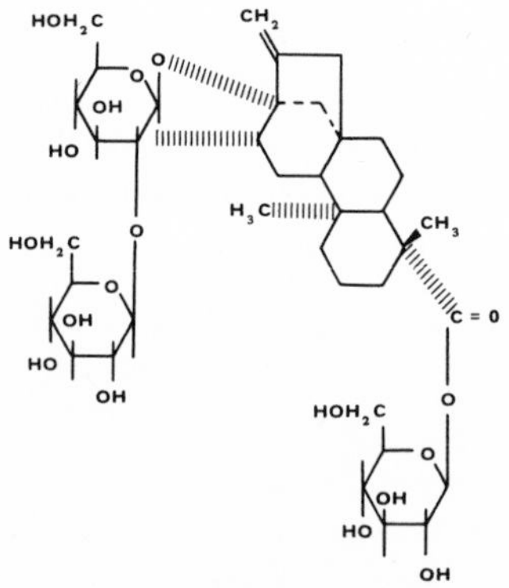

Stevioside
Structure 1

Present, limited evidence suggests stevioside is not toxic to humans.[24] However, a recent study has shown that stevioside and rebaudioside A are degraded by rat intestinal microflora, and the resulting steviol almost completely absorbed.[114] The pure substance is stated to be about 300 times sweeter than sucrose, although reports relating to the quality of sweetness are conflicting.[23,24] There are no data reported on the stability of stevioside under acidic conditions.

Rebaudioside A represents an attractive sweetener in light of its taste properties and the fact that leaves of *S. rebaudiana* contain approximately 1.4% by weight of the sweetener.

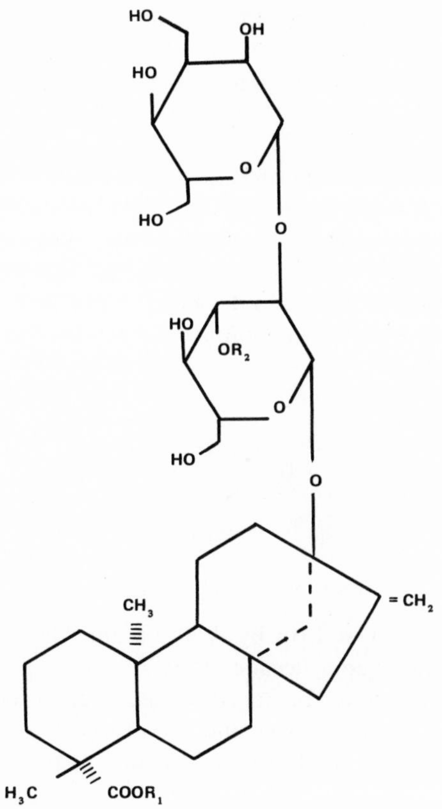

FIGURE 1. Chemical structures of rebaudosides. In rebaudoside A, $R_1 = R_2 =$ glucose; in rebaudoside B, $R_1 = H$ and $R_2 =$ glucose.

Osladin

A new sweet substance named osladin (Structure 2) has recently been isolated from the rhizomes of *Polypodium vulgare* and shown to be a steroidal saponin, the first sweet substance isolated from this structural class of compounds.[25]

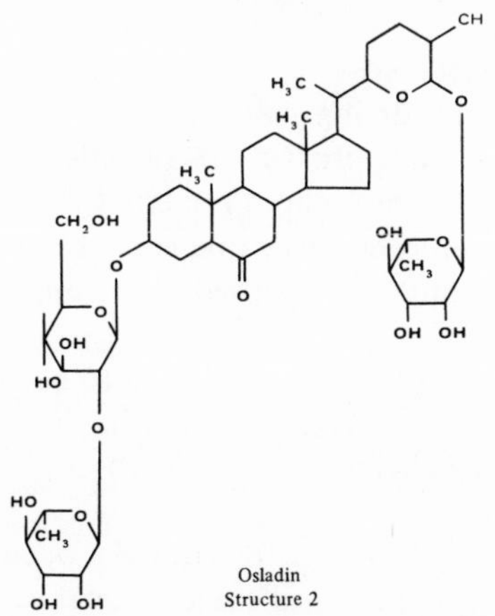

Osladin
Structure 2

It has been stated that osladin is approximately 3000 times sweeter than sucrose, although no mention has been made of the quality of sweetness. In view of the difficult isolation and potentially physiologically active structure, it is difficult to envision any significant commercial interest in this material as a sweetener.

Phyllodulcin

At the Japanese Flower Festival of Hanamatsuri (celebrating the birth of Buddha), a sweet tea is made from the dried leaves of *Hydrangea macrophylla Seringe (thubergii Makino)*. The tea can be surprisingly sweet, with a lingering taste resembling that of the dihydrochalcones. The substance responsible for the sweet taste has been identified[26] as phyllodulcin (Structure 3) and has been reported to occur in both racemic and optically active forms in *Hydrangea opuloides.*[25]

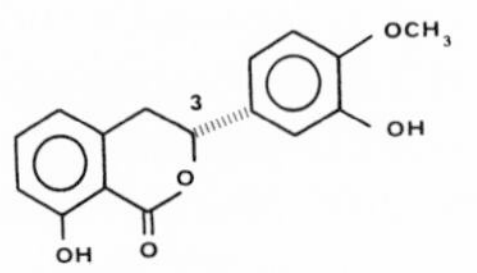

OCH_3 3 O OH OH O

Phyllodulcin
Structure 3

Phyllodulcin was isolated in 1916 by Asahina and Ueno from *Hydrangea macrophylla Seringe.*[27] A little over a decade later, Asahina and Asano demonstrated its correct chemical structure, which belonged to the 3,4-dihydroisocoumarin class;[28] however, it was not until 1959 that Arkawa and Nakazari[29] established that (+)-phyllodulcin has the 3R configuration (as shown in Structure 3) by exhaustive ozonolysis to D-malic acid.

In recent years, there has been renewed interest in phyllodulcin as a substitute sweetener.[30] Toxicological studies have not been reported. Taste-structure relationships have been studied by Yamamoto in a series of analogues and are discussed in the section on alkoxyaromatic sweeteners.

The taste of phyllodulcin is very similar to many of the dihydrochalcone sweeteners, i.e., it displays a lagging onset of sweetness and a lingering, perhaps licoricelike, aftertaste. It has been reported to be 200 to 300 times as sweet as sucrose.[31] There may be a limited market for this sweetener in products that can benefit from a lingering taste, such as hard candies, confections, chewing gum, and oral hygiene products.

Intense Sweetener from Lo Han Kuo

Most recently, a naturally occurring sweetener was reported[32] present in the dried fruit (Lo Han Kuo) of *Momordica grosvenori Swingle* found in southern China. The fruit is reported to be valued as a household remedy for colds, sore throats, and minor gastrointenstinal ailments. The sweet principle can be extracted from either the pulp or rinds by water or 50% ethanol and purified by a combination of chromatographic techniques.

The sweet taste of this new material was reported to be about 150 times more potent than sucrose, with a licoricelike aftertaste somewhat like that of stevioside and the dihydrochalcones. Preliminary chemical and spectroscopic studies indicate the sweetener to be a glycoside of a triterpenoid.

Despite a strong and growing interest in naturally occurring sweeteners, there has been little progress made in chemical and toxicological evaluation of these materials. A great deal of work remains to be done before commercialization can be realized.

Xylitol

Xylitol (Structure 4) is a sugar alcohol which can be made from a variety of natural products, including birchwood chips, berries, and leaves. It can also be obtained from mushrooms. The xylitol used in anticaries tests conducted at the University of Turku, Finland, was prepared by treating birchwood chips with acid, followed by hydrogenation of the resulting xylose. The results of the Finnish studies, in which there were two trials involving some 225 students, indicate that xylitol prevents dental caries.[88-92] Apparently, oral microorganisms that readily metabolize sucrose, glucose, fructose, etc., producing acids which cause tooth decay, do not utilize xylitol as well.

Xylitol has been on the FDA "safe products" list since 1963. It is being produced in development quantities by Hoffman®-La Roche®, and chewing gum sweetened, in part, with xylitol is being produced by Finnfoods, Inc. (Cresskill, N.J.). The National Institute of Dental Research sponsored a study on caries prevention with xylitol until the preliminary findings of a British study suggested that xylitol causes cancer in laboratory animals.[93] The Food and Drug Administration is evaluating the study.

With respect to sweetness intensity, Moscowitz determined that 4% by weight solutions of sucrose and xylitol are equisweet.[94] Little has been reported with respect to the sweetener quality of xylitol. The sweetness-intensity response of xylitol and sucrose, which are structurally dissimilar, is explained by the Shallenberger theory, which in part relates the intensity to the degree of intramolecular hydrogen bonding.[95]

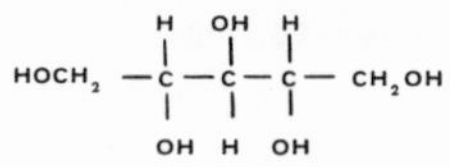

```
          H   OH  H
          |   |   |
HOCH2 —   C — C — C — CH2OH
          |   |   |
          OH  H   OH
```

Structure 4

DIPEPTIDE SWEETENERS

Aspartame

It is perhaps a sad commentary that the history of every commercially significant sweetener has been preceded by accidental discoveries, reflecting the uncertain drama of chemical research. The very term "sweetener research" is misleading, since most sweet compounds have been discovered as by-products of research directed towards other goals. Nowhere is the chemist less in control of the end product of his labors and at the same time no further away than a lick from knowing whether he has achieved success or failure. As with the accidental discovery of saccharin in 1879 and cyclamate in the 1940s, so too goes aspartame.

At least two independent groups were involved in the discovery of aspartame, but only one recognized its outstanding sweet taste. In or about 1964, Imperial® Chemical Industries (I.C.I.®) and G. D. Searle® were involved in the synthesis of gastrin and gastrin tetrapeptide analogues. During the course of studies aimed towards synthesizing the C-terminal sequence of gastrin, Davey, Laird, and Morley of I.C.I. prepared the dipeptide α-L-aspartyl-L-phenylalanine methyl ester (aspartame) (Structure 5). This was reported in 1966, but the authors did **not** comment on the taste of this product or other analogues synthesized.[33] At about the same time, the Searle group also had the occasion to synthesize the dipeptide (Structure 5) in their quest to find gastin tetrapeptide analogues for possible use as drugs for ulcer therapy; the dipeptide was prepared so that it could be employed in a bioassay. During the recrystallization of the dipeptide from ethanol, J. Schlatter spilled some on his hand. A short time later, he had the occasion to lick his fingers to facilitate lifting some weighing paper from a stack and discovered the remarkably sweet taste of the dipeptide. Mazur, Schlatter, and Goldkamp reported their findings in 1969.[34] Since that time, G. D. Searle has launched an intensive effort to commercialize the synthetic sweetener.

Aspartame is characterized as a white, crystalline powder composed of two naturally occurring amino acids. A new method has recently been developed for large-scale production of aspartame, as shown in Figure 2.[31] Some of its physical properties are shown in Table 1.

The solubility of aspartame in water under acidic conditions is appreciable, i.e., on the order of 1.3 to 1.8 wt%. As indicated in Table 2, when measured in water adjusted with buffers, aspartame is nearly twice as soluble at pH 3.72 than under neutral conditions.

The sweetening potency of aspartame has been assessed by panel tests to be 150 to 200 times that of sucrose which is quite unexpected since neither L-aspartic acid nor L-phenylalanine is sweet. In fact, while a number of small peptides are known to proudce a bitter taste, none had been reported to elicit a sweet taste response prior to the discovery of aspartame. Mazur reported that the L-aspartic acid moiety was an essential constituent for imparting the sweet taste and that the phenylalanine methyl ester portion could be altered.[34] Subsequently, a number of sweet-tasting dipeptides, each having L-aspartic acid as the common constituent, have been prepared. (See the section on structural considerations.)

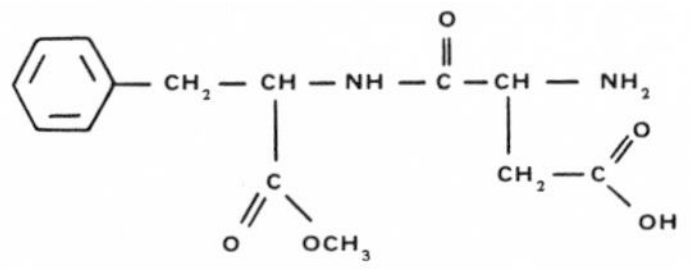

Structure 5

The potency of aspartame relative to sucrose decreases with increasing concentrations of sucrose. As indicated in Table 3, the potency of aspartame relative to saccharine at isosweet concentrations with sucrose is most apparent at recognition threshold levels. Cloninger and Baldwin highlighted the relationship of aspartame as a function of sucrose concentration.[36] As indicated in Figure 3, the sweetness of aspartame relative to sucrose was found by a panel of 20 to have almost a power function relationship to the concentration of sucrose. The difference in magnitude, however, is significant, i.e., the potency of aspartame was found to be about 180 times more than 2% sucrose solutions but only 40 times sweeter than 30% sucrose. From these data, one may expect greater replacement efficiency in the use of aspartame for products traditionally sweetened with 10 to 12% sucrose, such as carbonated beverages, than for items containing massive levels of sucrose, such as hard candy. However, data developed by G. D. Searle indicate that the potency of aspartame in various food substrates relative to the levels of sucrose employed in the same foods varies from about 130 to as much as 280 times sucrose (Table 4). Furthermore, these data suggest that aspartame may display synergism in sweetening potency as compared to sucrose when combined with other ingredients employed in foods. For example, specific reference is made by Searle to the combined effect of aspartame and sorbitol. Curiously, the potency of aspartame in chocolate or cocoa products is reported at 200 times sucrose when the sucrose level employed is 58% and 180 times sucrose in presweetened breakfast cereals containing 37% sucrose (Table 4).

Attempts have been made to determine the effect of various gums on the sweetness of aspartame.[36] This determination is important, since the viscosity of sucrose solutions increases as a function of concentration, especially at use levels employed in

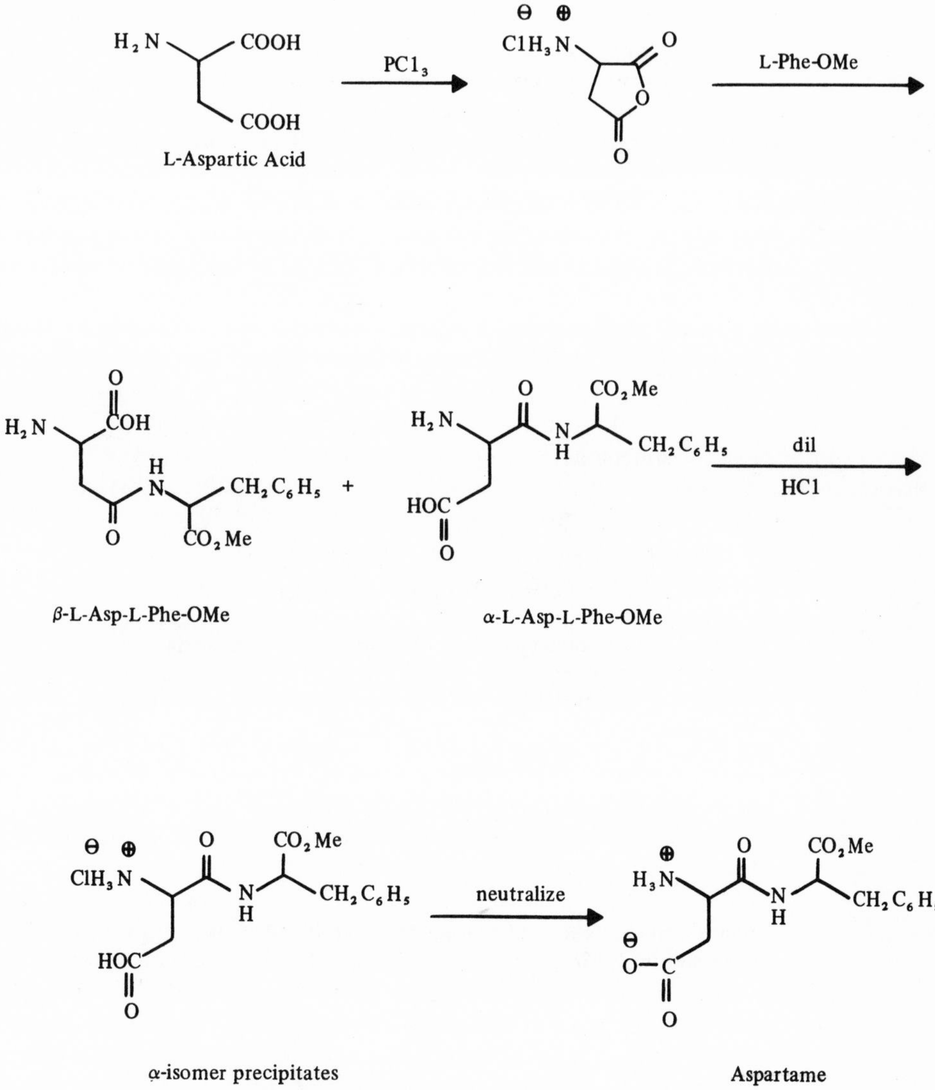

L-Aspartic Acid

PCl_3

L-Phe-OMe

β-L-Asp-L-Phe-OMe

α-L-Asp-L-Phe-OMe

dil HCl

α-isomer precipitates

neutralize

Aspartame

FIGURE 2. Method for large-scale production of aspartame. (From Crosby, G. A., *Crit. Rev. Food Sci. Nutr.*, 7(4), 297, 1976.)

Table 1
PHYSICAL PROPERTIES OF ASPARTAME[34,35]

Molecular weight	294.3
Melting point	Double m.p., about 190 and 245° C
Solubility at 25°C in:	10.20 mg/ml
Water (distilled)	8.68
Methanol	3.72
Ethanol (95%)	0.26
Chloroform	0.04
Heptane	0.001—0.007%
Optical rotation (H_2O)	(α) + 4°
Recognition threshold	

Table 2
AQUEOUS SOLUBILITY OF ASPARTAME AT VARIOUS pH LEVELS[35]

pH	Solubility (mg/ml at 25°C)
3.72	18.2
4.00	16.1
4.05	15.2
4.27	14.5
4.78	13.8
5.30	13.5
5.70	13.7
6.00	14.2
7.00 (distilled H_2O)	10.20

Table 3
THE POTENCY OF ASPARTAME AS A FUNCTION OF SWEETNESS LEVEL[35]

Isosweet concentrations (%)		Potency relative to saccharin
Sucrose	Aspartame	
0.34[a]	0.001—0.007[a]	About 400
4.3	0.02	215
10.0	0.075	133
15.0	0.15	100

[a] Recognition threshold.

From Aspartame, Technical Bulletin No. 600(060473), Searle Biochemics, div. G. D. Searle, Arlington Heights, Ill.

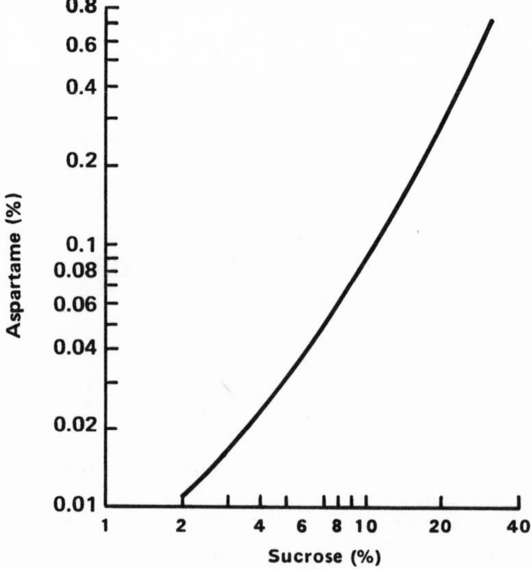

FIGURE 3. Relationship of sweetness of aspartame and sucrose. (From Cloninger, M. R. and Bladwin, R. E., *J. Food Sci.*, 39, 347, 1974. With permission.)

Table 4
POTENCY OF ASPARTAME RELATIVE TO SUCROSE IN VARIOUS FOODS[35]

Applications	Potency	Sucrose in product (%)
Still beverages		
Powdered mixes	180	11
Soft drinks	180	10
Coffee	180	4
Tea, hot	180	6
Tea, iced	180	7
Fruit drinks, cold	180	14
Carbonated beverages		
Orange	175	12.5
Cola	130	11.0
Ginger ale	175	9.0
Lemon-lime	150	9.5
Confections		
Chocolate or cocoa products	200	58
Breakfast cereals		
Presweetened	180	37 (dry)
Unsweetened	220	—
Desserts		
Gelatin	160	18
Puddings	200	17
Frozen (ice milk)	280[a]	14[b]
Raw fruits	220	—

[a] This may be due to synergism between aspartame and sorbitol in ice milk.
[b] Including other nutritive sugars.

From Aspartame, Technical Bulletin No. 600(060473), Searle Biochemics, div. G. D. Searle, Arlington Heights, Ill.

beverages, and can result in organoleptically relevant, i.e., ''mouth feel,'' differences between products sweetened with sucrose and aspartame. While gum arabic and carboxymethyl cellulose (CMC) do not affect the sweetness potency of aspartame relative to sucrose, addition of Methocel® significantly increases sweetness. In contrast, there are several reports indicating that the sweetness of sucrose and saccharine is ''masked'' (reduced) as viscosity is increased by addition of cornstarch, guar gum, and CMC.[37,38]

The stability of aspartame in aqueous media is not entirely satisfactory, especially for certain foods, such as carbonated and still beverages, which are often subjected to many months of storage prior to consumption. As a dipeptide ester, aspartame undergoes both hydrolysis and cyclization reactions (Figure 4). Under acidic conditions, hydrolysis of the ester and amide bonds is favored, resulting in formation of its constituent amino acids with a concomitant loss in sweetness. Under more neutral and alkaline environments, the dipeptide cyclizes to the corresponding diketopiperazine (Structure 6, Figure 4), which is also devoid of sweetness. Apparently, no new objectionable taste is imparted to aspartame as a result of either transformation. Mazur reports that maximum stability of aspartame in aqueous solutions occurs within the pH value range of 4 to 5.[39] The half-life under these conditions is reported to be at least 8 months at room temperature. Another report indicates that at pH 4 the half-

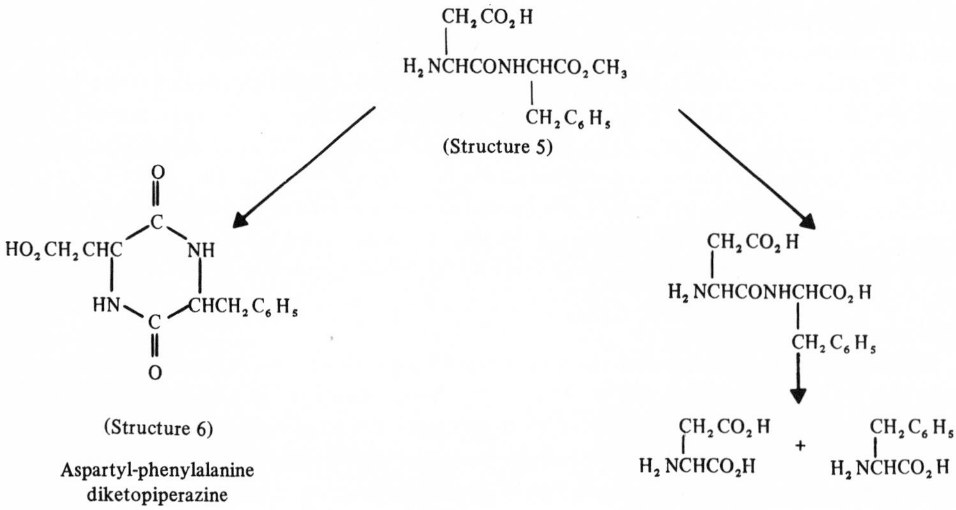

CH_2CO_2H
$H_2NCHCONHCHCO_2CH_3$
$CH_2C_6H_5$
(Structure 5)

HO_2CH_2CHC — C(=O) — NH
HN — C(=O) — $CHCH_2C_6H_5$

(Structure 6)

Aspartyl-phenylalanine diketopiperazine

CH_2CO_2H
$H_2NCHCONHCHCO_2H$
$CH_2C_6H_5$

CH_2CO_2H / H_2NCHCO_2H + $CH_2C_6H_5$ / H_2NCHCO_2H

FIGURE 4. Hydrolysis and cyclization reactions of aspartame. (From Crosby, G. A., *Crit. Rev. Food Sci. Nutr.*, 7(4), 297, 1976.)

life is in excess of 10 months at 25°C.[35] While powdered aspartame (suitably packaged in polyethylene bags contained in fiber drums) does not undergo significant conversion to the diketopiperazine after storage for 1 year at room temperature, the hydrolysis and cyclization reactions lead to consequences having relevance to: (1) its utility as a sweetener and (2) certain safety issues, as follows:

1. Since the hydrolysis and cyclization rates of powdered aspartame are relatively slow, there appears to be little cause for technical concern when it is used as a table sweetener for coffee, tea, breakfast cereals, etc. One suspects that by properly extending aspartame onto neutral, powdered carriers, premeasured packets (sachets) may be formulated. Alternatively, sugar-substitute tablets can be prepared. Other food applications identified include presweetened breakfast cereals, chewing gum, and dry bases for beverages such as instant coffee and tea, gelatins, puddings and fillings, and dairy-product analogue whipped toppings. It is, however, in prepared beverages or in cooking applications that aspartame falls short. This instability, leading to loss of sweetness, limits its value. Thus far, attempts to stabilize the hydrolysis or cyclization reactions have not proved sufficiently successful to permit its expanded use in foods.
2. While the safety of aspartame has been supported by extensive studies,[40,96-103] there is some concern that phenylalanine released during the course of normal metabolism or generated during externally induced hydrolysis reactions may affect persons with phenylketonuria (PKU).[35] About 1 in 10,000 humans is afflicted with this genetically induced metabolic defect.[35] As a result, they cannot metabolize phenylalanine and must restrict intake of this amino acid. Accordingly, in the July 26, 1974, decision by the FDA to approve the use of aspartame, labeling provisions were made in which foods not containing protein or intended for admixture to protein sweetened with aspartame should bear the statement "contains protein."[41]

In further developments, the FDA has asked for reexamination of studies in which uterine polyps were found in rats fed high levels of diketopiperazine (Structure 6, Figure 4), the cyclization product of aspartame. In the August 18, 1975, issue of *Food*

Chemical News, it was reported that "...the polyps were of a form that shows no predisposition for malignant change and that form spontaneously in aging female rats." Furthermore, the pathologists concluded that "...the possibility that uterine polyps will occur as a result of aspartame ingestion appears to be very remote." On December 5, 1975, in a surprising action, the FDA stayed for an indefinite period G. D. Searle's authority to sell aspartame under its original order of July 26, 1974.[84] The FDA has indicated "... the need for a comprehensive review of certain of the research data..."[85] This review was completed in late 1978 and validated the original Searle studies. The FDA intends to convene a public Board of Inquiry in 1980.

Structural Considerations

Since the initial discovery of the sweet taste properties of α-L-aspartyl-L-phenylalanine methyl ester (aspartame) by Mazur,[34] a large number of additional dipeptides have been synthesized in an effort to determine the relationship between chemical structure and sweet taste. Mazur and co-workers[34] have suggested that the following conditions are necessary for a dipeptide to taste sweet:

1. L-Aspartic acid (L-asp) is an indispensable component. Both the amino and β-carboxyl grop must be free.
2. The phenylalanine (Phe) component may be replaced by other amino acids in special cases; however, these replacements must be L-amino acids, and the ester group which is preferably small, is indispensable.

In a later publication these same researchers pointed out that many α-alkylamides of L-aspartic acid are also sweet.[42] According to their observations, sweetness is generated when an alkylamine, replacing the phenylalanine component, is an L-2-aminoalkane containing six or more carbon atoms. Thus, according to Mazur's observations, derivatives of the α-carboxyl group of L-aspartic acid may possibly be sweet when the structural requirements shown in Structure 7 are satisfied. R^1 is preferably a small hydrophobic group (such as the CO_2CH_3 of aspartame) and R^2 a large hydrophobic group such as the benzyl portion of aspartame. According to these conclusions, L-phenylalanine can be replaced with bulky hydrophobic esters of D-amino acids without resulting in loss of sweet taste. In support of this hypothesis, the dipeptide α-L-aspartyl-D-alanine isopropyl ester (Structure 8, Figure 5) has been reported by Mazur to be intensely sweet.[43] In this case, the methyl group of Structure 8 corresponds to the small hydrophobic group (R^1) and the isopropyl ester group corresponds to the large hydrophobic group (R^2). More recently, several exceedingly intense sweeteners have been reported by Fujino et al.[44] These researchers replaced the phenylalanine portion with diesters of aminomalonic acid, such that one ester was a small hydrophobic group and the other ester a large hydrophobic group, in keeping with the structural conclusions drawn above. Thus, α-L-aspartylaminomalonic acid fenchyl methyl ester (Structure 9) is claimed to be approximately 22,000 to 33,000 times as sweet as sucrose, without any bitter taste.[44]

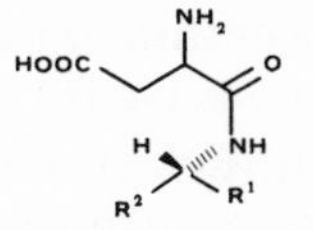

HOOC NH$_2$ O H NH R^2 R^1

Structure 7

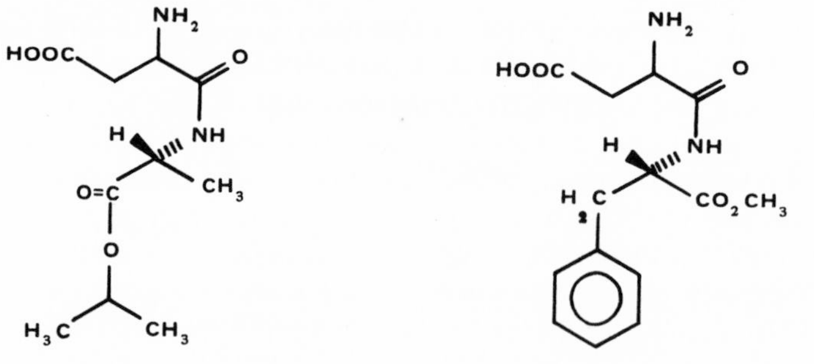

Structure 8

FIGURE 5. Comparison of a variety of derivatives of L-aspartic acid with aspartame. Structure 8 is α-L-aspartyl-D-alanine isopropyl ester where R_1 = CH_3 (small) and R_2 = $CO_2C_3H_7$ (large). In aspartame, R_1 = CO_2CH_3 (small) and R_2 = $H_2CC_6H_5$ (large).

Ariyoshi has also reported on the synthesis of a number of dipeptide sweeteners derived from L-aspartic acid and drawn additional conclusions regarding relationships between structure and taste.[31] Many of the L-aspartic acid dipeptide sweeteners reported by both Ariyoshi and Mazur are listed in Table 5.

The dipeptides listed in Table 5 are all derivatives of L-aspartic acid, which Mazur concluded was an indispensable component necessary for sweetness. Indeed, it has been shown that L-aspartic acid cannot be replaced by D-aspartic acid, a β-substituted asparaginic acid, glutaminic acid, or a β-substituted glutaminic acid.

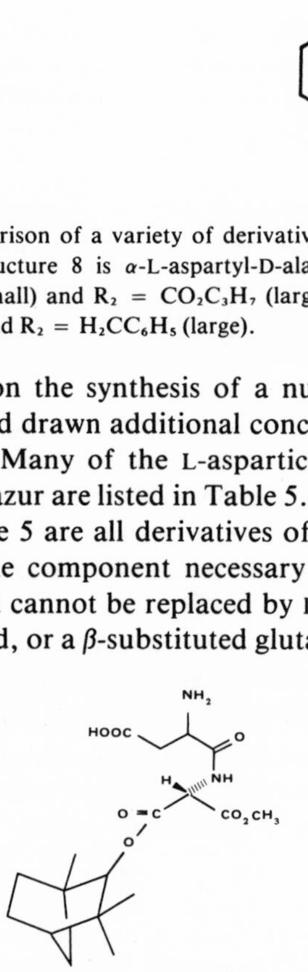

Structure 9

Briggs and Morley recently pointed out, however, that the L-aspartic acid component of aspartame can be replaced by aminomalonic acid without resulting in a loss of sweet taste.[45] Many laboratories are currently studying the taste-structure relationships of dipeptide sweeteners.

ALKOXYAROMATIC SWEETENERS

Structural Considerations

Over the years, a large number of para-substituted alkoxyaromatic compounds have been reported to be sweet. Structures shown in Figure 6 are representative examples. Despite the large differences in gross constitution of these compounds, there appears to be a unifying structural feature responsible for the stimulation of a sweet taste response. The generalized formula (Structure 14) represents what might be envisioned as the active portion of these molecules or biological probe that interacts with a common "receptor site."

Studies in both the dihydrochalcone[46] and nitroaniline[47] (Structure 10, Figure 6) se-

Table 5
L-ASPARTIC ACID DIPEPTIDE SWEETENERS (α-L-ASP-X)

X[a]	Sweetness[b]	X[a]	Sweetness[b]
Gly-OMe	8	DL-(a-Me)Glu-$(OMe)_2$	8
Gly-OPr^n	14	D-Ala-OMe	25
β-Ala-OMe	1∼2	D-Ala-OPr^n	125
β-Ala-OPr^i	7	D-Ala-Gly-OMe	3
γ-Abu-OMe	1∼2	D-Ala-Sar-OMe	Bitter
L-Abu-OMe	Tasteless	L-2-Amino-octane	10
L-Nva-OMe	4	L-1,3-Dimethylbutylamine	Bitter
L-Val-OMe	Bitter	D-Ser-OMe	45
L-Nle-OMe	45	D -Ser-OEt	115
L-Nle-OEt	5	D-Ser-OPr^n	320
L-Ile-OMe	Bitter	D-Ser-OPr^i	120
L-Leu-OMe	Bitter	D-Ser-OBu^n	70
L-Cap-OMe	47	D-Ser-OBu^i	200
L-Cap-OEt	Bitter	D-Ser-O′-cyclohex	60
L-Phe-OMe	140	D-Abu-OMe	16
L-HPhe-OMe	400	D-Abu-Opr^n	95
L-Tyr-OMe	50	D-Abu-Opr^i	70
L-Amphetamine	50	D-Thr-OMe	25
L-1-Hydroxymethyl-2-phenyl-ethylamine	1∼2	D-Thr-OEt	110
		D-Thr-OPr^n	150
L-HyNle-OMe(erythro)	18	D-Thr-Opr^i	105
L-HyNle-OMe(threo)	7	D-Thr-OBu^n	30
L-Ser-OMe	Tasteless	D-Thr-OBu^i	110
L-Ser-(Bu^t)-OMe	50	D-Thr-O-cyclohex	30
L-Ser(Bu^t)-OMe	130	D-aThr-OMe	7
L-Met-OMe	100	D-aThr-OEt	6
L-Thr-OMe	Tasteless	D-aThr-OPr^n	40
L-Thr(Bu^t)-OMe	Bitter	D-aThr-OPr^i	10
L-aThr(Bu^t)-OMe	Bitter	D-aThr-Obu^n	20
L-(p-NH_2) Phe-OMe	3	D-aThr-OBu^i	27
L-Lys-OMe	Bitter	D-aThr-O-cyclohex	22
L-Lys(Ac)-OMe	1∼2	Ama-O-2-Me-cyclohex-O-Methyl	5,450∼7,300
L-Orn(Ac)-OMe	1∼2		
DL-(a-ME)Phe-OMe	5	Ama-O-fenchyl-O-Methyl	22,200∼33,200
L-Glu-$(OMe)_2$	12		

[a] Symbols are based on *Arch. Biochem. Biophys.*, 150, 1, 1972.
[b] Intensity relative to sucrose.

From Ariyoshi, Y., *Kagaku To Seibutsu*, 12, 274, 1974. With permission.

ries suggest that the alkoxy group (OR) plays an important role in generation of a sweet taste; thus, maximum sweet taste response occurs when R = n-C_3H_7 and decreases when R is both smaller and larger. The work of Deutsch and Hansch implies that the propoxy group offers an optimum balance between the hydrophobic and hydrophilic character of the alkoxy group.[47] The ability of the ortho-substituted X-group to act as a source of hydrogen bonding (NH_2,OH) also appears to be critical for the taste response. It is interesting to note that Structure 15 (intermediate between P-4000 [Structure 10, Figure 6] and β-neohesperidin dihydrochalcone [Structure 12, Figure 6] by virtue of its hydroxyl group) has been reported to elicit a sweet taste response.[48]

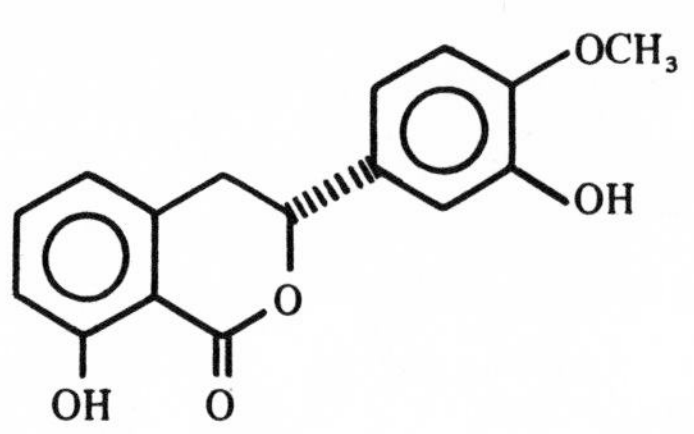

Phyllodulcin

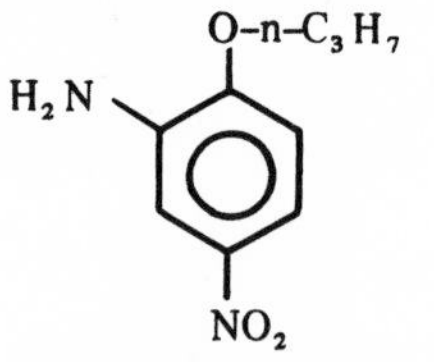

P-4000
Structure 10

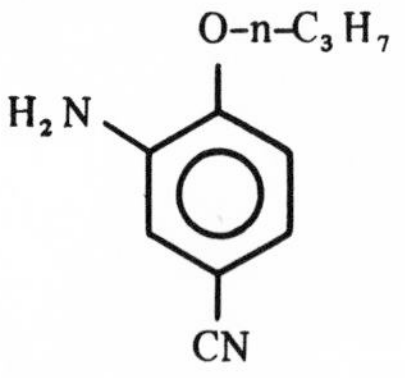

Structure 11

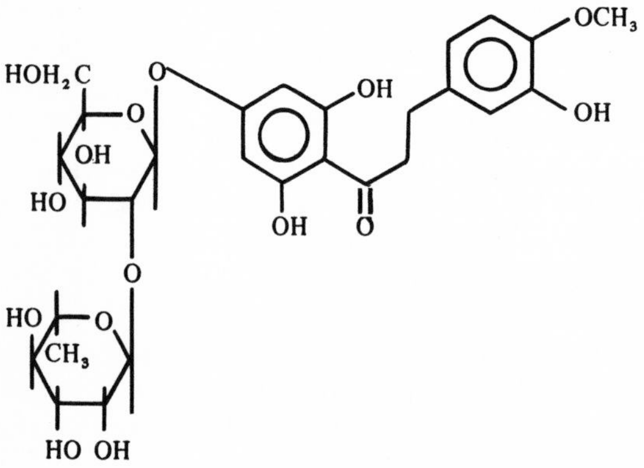

β-Neohesperidin dihydrochalcone
Structure 12

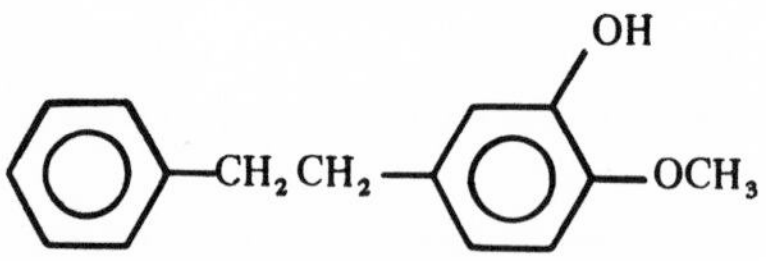

Structure 13

FIGURE 6. Alkoxyaromatic sweeteners. (From Crosby, G. A., *Crit. Rev. Food Sci. Nutr.*, 7(4), 297, 1976.)

The major variation in the sweeteners depicted in Figure 6 occurs in the portion of the molecules represented by the Y unit (see Structure 14). The most logical function of this unit would appear to be to bind the sweet molecule to the appropriate location on the taste receptor membrane such that the probe end of the molecule (Structure 14) can interact with the receptor site.

A word of caution is necessary when making structure to activity comparisons between the sweet alkoxyaromatic molecules in Figure 6: the taste properties of these substances are quite different. Thus, the nitroaniline (Structure 10) displays rapid onset of the sweet taste sensation followed by rapid loss of the clean, sweet taste. The dihydrochalcones and phyllodulcin are much slower in eliciting a taste sensation (up to several seconds) and also display a lingering taste, often described as licoricelike. These differences in taste timing might be due to the difference in the size of the molecules; however, several dipeptides having molecular weights nearly the same as some sweet dihydrochalcones were reported in the previous section to have a clean taste with no lagging or lingering sensation. It is interesting to speculate that the dihydrochalcones and related structures might undergo a slow complexation (binding?) reaction with a

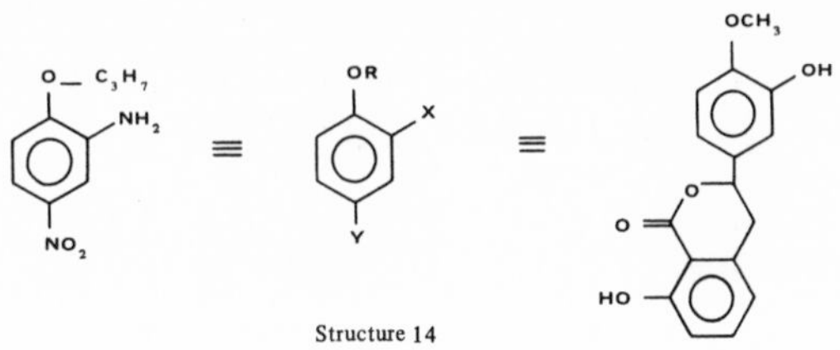

Structure 14

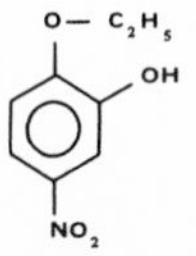

Structure 15

membrane component prior to eliciting the sweet taste response. The resulting chemical complex might then result in a slow release of sweetener over a period of time, accounting for both the slow onset of taste and the lingering aftertaste.

Whether or not there is any real mechanistic relationship between the various alkoxyaromatic sweeteners is unknown. The interest raised over the years in P-4000 and β-neohesperidin dihydrochalcone, however, suggests that the search for more understanding of the taste-structure relationships for this class of substances could lead to the discovery of new, improved sweeteners.

β-Neohesperidin Dihydrochalone

As part of a program started in 1958 to study the taste to structure relationships of bitter phenolic glycosides, Horowitz and Gentili[49] converted the bitter flavanone naringin (Structure 16) to the dihydrochalcone (Structure 17) with the expectation that this material would also be bitter.

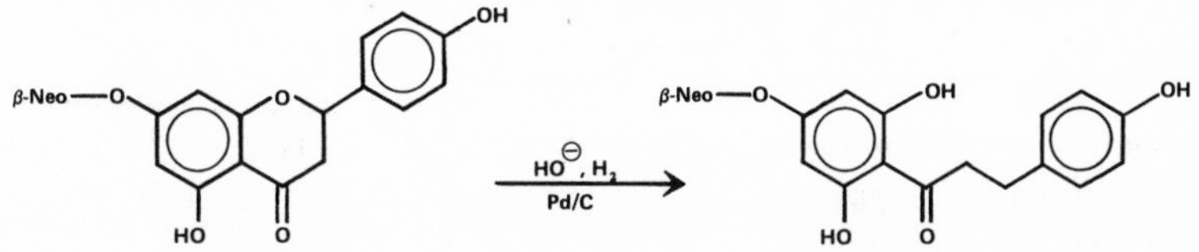

Naringin
Structure 16

Naringin dihydrochalcone
Structure 17

Surprisingly, the latter compound displayed an intensely sweet taste. This discovery led to the reported[50] synthesis (in 1963) of the intensely sweet β-neohesperidin dihydrochalcone (Structure 12, Figure 6) from the corresponding flavanone.[50] The sweetness potency of β-neohesperidin dihydrochalcone is greater than that of naringin dihydrochalcone (Structure 17) and capable of eliciting a threshold response at a concentration 1/1100 that of sucrose on a weight basis.[46] At higher concentrations, which are nearly equisweet with 10 to 12% sucrose solutions, the dihydrochalcone has a concentration of about 1/600 that of sucrose.[52] Typical of most dihydrochalcones, this material displays a somewhat slow onset of sweet taste followed by a lingering aftertaste. The taste sensation is quite clean, however, with only a minor bitter sensation. These properties

Table 6
PHYSICAL PROPERTIES OF β-NEOHESPERIDIN DIHYDROCHALCONE[54]

Empirical formula	$C_{28}H_{36}O_{15}$
Molecular weight	612.6
Solubility in:	
H_2O at 25°C	0.5 g/l
H_2O at 80°C	653.0 g/l
Ethanol at 25°C	20.4 g/l
Acetone at 25°C	0.7 g/l
Insoluble in ether, ethyl acetate, and mineral acids	
Melting point	152—154°C
Stability	No decomposition noted when held at 100°F for 1 month in air; stable in buffered solutions between pH 2 and 10 at 100°C for 8 hr

make formulation difficult in some food systems such as carbonated beverages but can offer potential advantages to products such as chewing gum, candies, and mouthwash formulations.[53] The physical properties of β-neohesperidin dihydrochalcone are indicated in Table 6.

The Toxicology and Biological Evaluation Research Unit of the U.S. Department of Agriculture (USDA) Western Regional Research Center (Berkeley, Cal.) has carried out extensive safety studies with β-neohesperidin dihydrochalcone (NDHC). In 2-year feeding studies, no indication of ill effects was observed in rats at high dose levels. Similar studies in dogs conducted by the same group indicate that "... NDHC possesses a very low order of toxicity."[104] Studies of the metabolic fate of ^{14}C-labeled NDHC in rats show that 90% of the oral dose is excreted in urine and the remainder in the feces in 24 hr, with no accumulation in the tissues.[104] Additional metabolic fate studies aimed at identifying NDHC metabolites in rats indicate that at an oral dose of 1 mg/kg of body weight, only one major metabolite appears in urine, while at 100 mg/kg a second, minor metabolite appears.[105] The major and minor metabolites are neither unchanged NDHC nor the aglycone, hesperetin dihydrochalcone; the enzyme β-glucuronidase appears to be without effect. When the urine of rats fed 1 mg/kg of ^{14}C-neohesperidin is extracted, one major and two minor metabolites are recovered. Cochromatography with neohesperidin and NDHC shows no identity, although the major metabolites from the flavone and the dihydrochalcone have close R_f values.

Currently, Nutrilite® Products, Inc. (Buena Park, Cal.) and California Aromatics and Flavors (Sun Valley, Cal.) have submitted a petition to the FDA for the use of NDHC in toothpaste, mouthwash, and chewing gum.[106]

The USDA laboratories[55] have also conducted a careful examination of the pilot plant synthesis of β-neohesperidin dihydrochalcone starting from the citrus by-product naringin (Structure 16 and Figure 7). The three-step sequence proceeds with at least a 10% overall yield to produce material of greater than 98% purity. The key intermediate, phloroacetophenone 4′-neohesperidoside (Structure 18, Figure 7), derived from naringin by a retro-Aldol reaction, is condensed with isovanillin to produce the flavanone neohesperidin (Structure 19, Figure 7). Conversion of the latter cyclic structure to dihydrochalcone is conducted in one reactor by hydrogenation of the intermediate chalcone (Structure 20, Figure 8), formed by the action of base on the flavanone, as

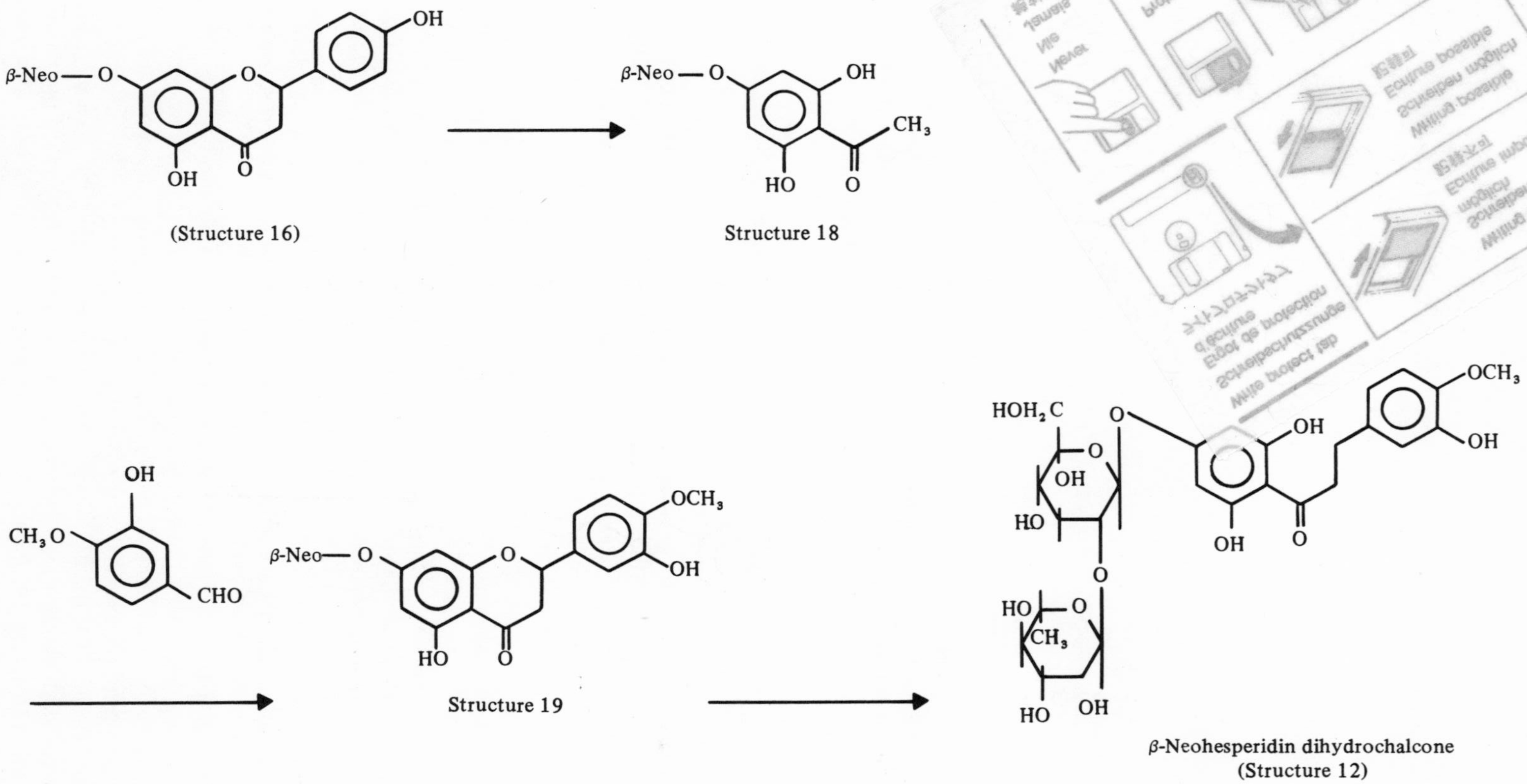

(Structure 16)

Structure 18

Structure 19

β-Neohesperidin dihydrochalcone
(Structure 12)

FIGURE 7. Synthesis of β-neohesperidin dihydrochalcone. (From Crosby, G. A., *Crit. Rev. Food Sci. Nutr.*, 7(4), 297, 1976.)

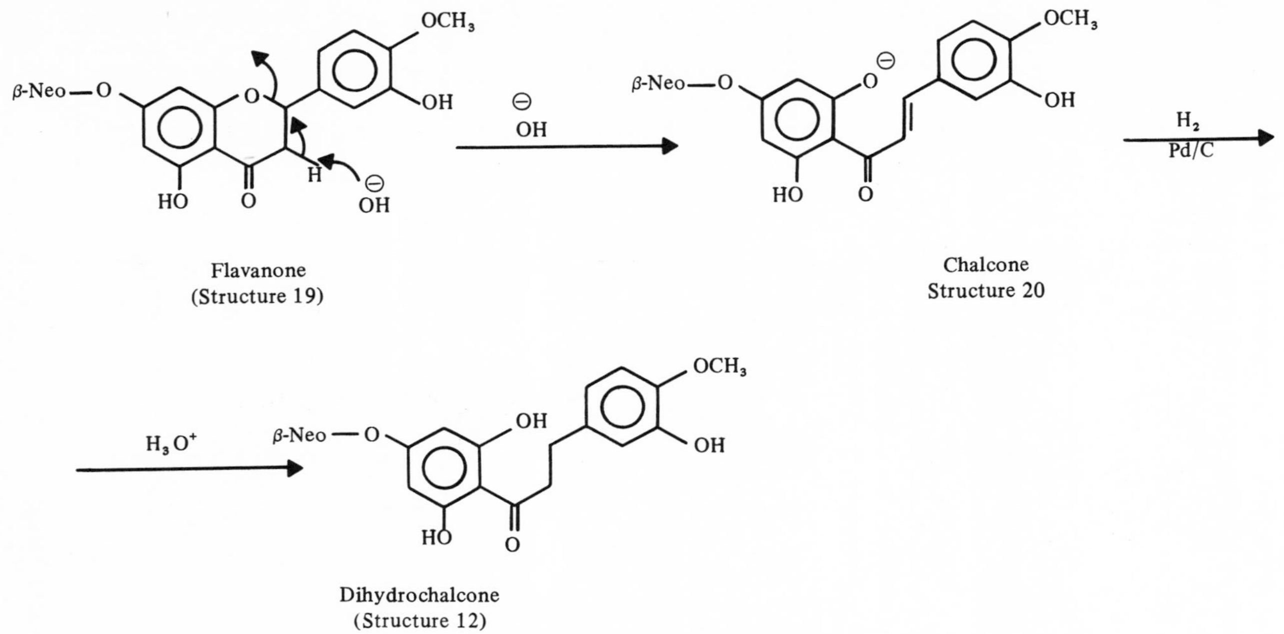

Flavanone
(Structure 19)

$\overset{\ominus}{OH}$

Chalcone
Structure 20

H_2
Pd/C

H_3O^+

Dihydrochalcone
(Structure 12)

FIGURE 8. Mechanism of reduction in base. (From Crosby, G. A., *Crit. Rev. Food Sci. Nutr.*, 7(4), 1976).

shown in Figure 8. It has been reported that one commercial firm has the capacity to produce the sweetener at a rate of 5000 lb/month;[56] however, the necessity of conducting commercial synthesis with a natural product isolated from citrus fruits places a limitation on the total annual production of β-neohesperidin dihydrochalcone.

Extensive chemical and toxicological studies indicate that dihydrochalcones may be close to commercialization, provided satisfactory applications can be found for a sweetener with lingering taste properties. Nevertheless, demonstration that a new food additive is safe for human consumption is perhaps the single greatest hurdle to overcome before commercialization can be realized. The dihydrochalcones have come a long way in this regard.

Dihydrochalcone Analogues

Since the initial disclosure of the sweet taste properties of naringin dihydrochalcone and β-neohesperidin dihydrochalcone by Horowitz and Gentili,[50] a large number of analogues have been prepared and reported, as indicated in Table 7. Several conclusions can be drawn from structural alterations of the B-ring of the dihydrochalcone nucleus:[58,59]

1. A B-ring hydroxyl group is necessary but does not guarantee sweetness.
2. The absence of a hydroxyl group assures nonsweetness or bittersweetness.
3. For sweetness in hydroxy-alkoxy di-substituted compounds, the order of the substituent groups must be R-H-OH-OR′ or R-OH-OR′, where R is the substituent at C-1 of the B-ring.
4. Taste is abolished if the order of groups in the hydroxy-alkoxy di-substituted compounds is R-H-OR′-OH.
5. Taste is abolished if three adjacent groups are present in addition to R.

A fair number of compounds varying in Ring A substitution have also been prepared, allowing the following generalizations to be made:

1. Sweetness is possible with only one hydroxyl group at either C-2 or C-6.
2. With the exception of β-rutinosyloxy, almost any hydrophilic alkoxyl group at the C-4 position seems to result in sweetness. Thus hesperetin dihydrochalcone (Table 7) as well as the carboxymethyl analogue shown in Structure 21 (Compound 20 in Table 7) are reported to be sweet.

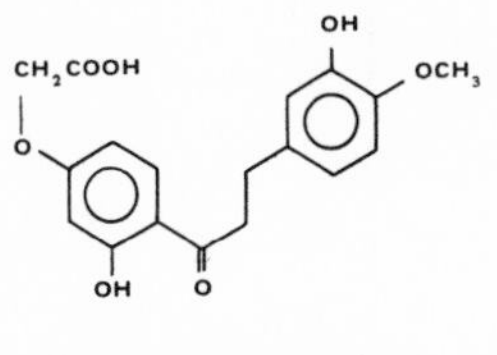

Structure 21

New, Simplified Dihydrochalcones (DHCs)

Recently, some new derivatives of hesperetin dihydrochalcone which are intensely sweet have been reported; specifically, these include the C_4-O-carboxyalkyl and sulfoalkyl analogues. The structures and sensory properties of a series of carboxyalkyl analogues are shown in Table 8.[107] Tables 9 and 10 list the structure and sensory properties, respectively, of a series of sulfoalkyl derivatives.[108] Several of these analogues, e.g., the carboxymethyl 1 in Table 8 and the sulfopropyl c and sulfomethyl a in Table

Table 7
TASTE QUALITY AND RELATIVE SWEETNESS OF DIHYDROCHALCONES REPORTED IN THE LITERATURE[a]

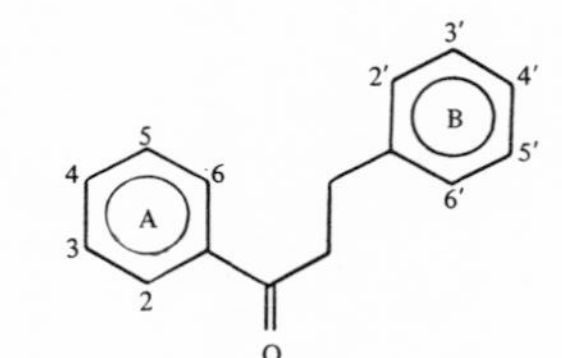

Compound (common name)	Substituents 2	4	5	6	2′	3′	4′	5′	Characteristic taste	Sweetness intensity × sucrose (weight basis)	Ref.
(Pinocembrin DHC)	OH	NH		OH					Tasteless		46
(Naringin DHC)	OH	NH		OH			OH		Sweet	100	49,50,57—59
(Neohesperidin DHC)	OH	NH		OH		OH	OMe		Sweet	100(601)[b]	49,50,57—59
(Poncirin DHC)	OH	NH		OH			OMe		Slightly bitter		49,50,57—59
(Neoeriocitrin DHC)	OH	NH		OH		OH	OH		Slightly sweet	NA	49,50,57—59
	OH	NH		OH					Sweet	100	57
	OH	NH		OH	OH				Bitter		57
	OH	NH		OH		OH	OH	OH	Tasteless		59
	OH	NH		OH		OH	OEt		Sweet	1000	57
	OH	NH		OH		OH	O-n-Pr		Sweet	2000	57
	OH	NH		OH		OH	O-i-Pr		Sweet	<1000	49
	OH	NH		OH	OH	OMe			Sweet	NA	58
	OH	NH		OH		OMe	OH		Tasteless		58,59
	OH	NH		OH		OEt	OH		Tasteless		58,59
	OH	NH		OH		OH	OMe	OH	Tasteless		58,59

	OH	NH	OH	OMe	OH	OMe		Tasteless		58,59
	OH	NH	OH		OH	OCH_2COOH		Tasteless		59
	OH	NH	OH		OMe	OMe		Bittersweet		57,58
	OH	NH	OH		Me	OMe		Tasteless		58,59
	OH	NH	OH		OH	OMe	OMe	Tasteless*		59
	OMe	NH	OH		OH	OMe		Sweet	NA	59
	OEt	NH	OH		OH	OMe		Sweet	NA	59
	OCH_2COOH	NH	OH		OH	OMe		Tasteless		59
	OH	β-D-glucosyloxy	OH		OH	OMe		Sweet	80	49, 58, 59
	OH	β-D-xylosyloxy	OH		OH	OMe		Sweet	160	59, 60
	OH	β-D-galactosyloxy	OH		OH	OMe		Sweet	140	59
	OH	6-*O*-methyl NH	OH		OH	OMe		Sweet	1000	59
	OH	6-*O*-α-L-rhamnosyl-NH	OH		OH		OMe	Slightly sweet	1	64
	OH	(4-*O*-α-D-glucosyl)$_n$, β-D-glucosyloxy-, *n*-2-4	OH		OH	OMe		Sweet	NA	65
(Hesperidin DHC)	OH	β-rutinosyloxy	OH		OH	OMe		Tasteless		58,59
	OH	NH			OH	OMe		Sweet	NA	61
	OH	NH			OH	OEt		Sweet	NA	61
Compound 20	OH	OCH_2COOH			OH	OMe		Sweet	180	62
	OH	OCH_2COOH			OH	OMe		Sweet	NA	62
	OMe	NH	OH			OH		Sweet	NA	59
	OMe	NH	OH		OMe	OMe		Bittersweet		59
	OH	2-*O*-α-L-rhamnosyl-β-D-galactosyloxy-	OH			OH		Sweet	1000	59,63
(Phloridzin)	β-D-glucosyloxy-	OH	OH			OH		Bitter		59

Table 7 (continued)
TASTE QUALITY AND RELATIVE SWEETNESS OF DIHYDROCHALCONES REPORTED IN THE LITERATURE[a]

Compound (common name)	Substituents								Characteristic taste	Sweetness intensity × sucrose (weight basis)	Ref.
	2	4	5	6	2′	3′	4′	5′			
(Glycyphyllin)	α-L-rhamnosyloxy	OH		OH			OH		Bittersweet		59
	OH	β-rutinosyloxy		OH			OH		Tasteless		58,59
	OH	β-rutinosyloxy		OH			OMe		Tasteless		58,59
(Eriocitrin DHC)	OH	β-rutinosyloxy		OH		OH	OH		Tasteless		58,59
Hesperetin DHC	OH	OH		OH		OH	OMe		Sweet		64

[a] These abbreviations will be used in the table: DHC = dihydrochalcone; NH = β-neohesperidosyl; NA = not available.
[b] The lower value (601×) has been determined by Dynapol at a concentration of sweetener equivalent to 10 to 12% sucrose (use level).

From Crosby, G. A., *Crit. Rev. Food Sci. Nutr.*, 7(4), 297, 1976.

Table 8
DIHYDROCHALCONE STRUCTURE-TASTE RELATIONSHIPS

Compound	Substituents X	Y	Z	N[a]	Concentration of tastant solution (ppm)	Intensity relative to sucrose[b]	Flavor judgment (%)[c] Sweet	Sour	Salty	Bitter
a	OH	O-β-neohesperidosyl	OH	12	250	663	88	3	2	5
b	OH	OCH_2COOH	H	12	90	76	38	6	7	7
c	OH	OCH_2COONa	H	12	95	63	42	0	8	32
d	OH	OH	H	12	90	57	1	0	1	4
e	OH	H	OH	12	90	247	21	5	9	23
f	OH	OH	OH	24	90	661	84	0	2	11
g[d]	H	OCH_2COOH	H	1	1,800	0				
h[d]	H	OCH_2COONa	H	1	1,680	0				
i[d]	OH	OCH_2COOH	OCH_2COOH	1	2,300	0				
j[d]	OH	OCH_2COONa	OCH_2COOH	1	2,200	0				
k	OH	OCH_2COOH	OH	12	90	200	66	0	0	12
l	OH	OCH_2COONa	OH	36	95	501	82	1	3	8
m	OH	$OCH(CH_3)COOH$	OH	12	90	29	19	0	0	32
n	OH	$OCH(CH_3)COONa$	OH	12	95	117	22	0	0	57
o[d,e]	OH	$OCH(COONa)_2$	OH	1	495	0				
p	OH	$O(CH_2)_3COOK$	OH	12	250	308	74	9	5	11
q[d]	OH	$OCH(COOH)CH_2CH_2COOH$	OH	1	2,400	0				
r[d]	OH	$OCH(COONa)CH_2CH_2COONa$	OH	1	2,400	0				

Table 8 (continued)
DIHYDROCHALCONE STRUCTURE-TASTE RELATIONSHIPS

Compound	Substituents			N[a]	Concentration of tastant solution (ppm)	Intensity relative to sucrose[b]	Flavor judgment (%)[c]			
	X	Y	Z				Sweet	Sour	Salty	Bitter
Sucrose				24	85,500	1	98	0	1	1
Saccharin (sodium salt)				36	60	490	70	3	5	18
Cyclamate (sodium salt)				36	1,250	29	70	8	14	4

[a] N = number of judgments.
[b] Intensities are compared to sucrose on the basis of weight.
[c] Flavor judgments totaled 100%.
[d] The saturated aqueous solution was found tasteless and therefore was not analyzed by the taste panel.
[e] The free acid form decarboxylates in aqueous soution to produce sweet DHC 11.

Reprinted with permission from DuBois, G. E., Crosby, G. A., and Saffron, P., *Science,* 195, 397, 1977. Copyright 1977 by the American Association for the Advancement of Science.

Table 9
SULFOALKYL-, SULFO-, AND SULFATODIHYDROCHALCONES

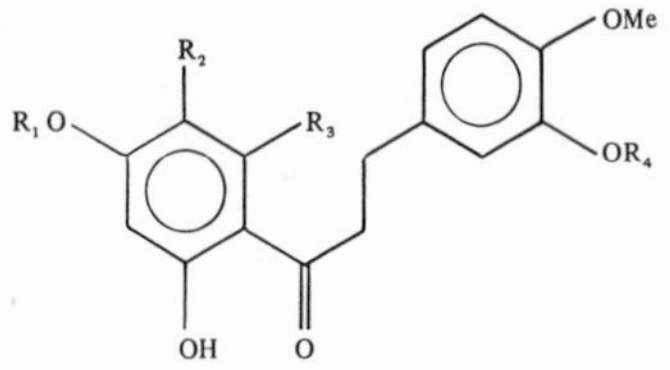

Compound	Substituents			
	R_1	R_2	R_3	R_4
a	CH_2SO_3M	H	OH	H
b	$(CH_2)_2SO_3M$	H	OH	H
c	$(CH_2)_3SO_3M$	H	OH	H
d	$(CH_2)_4SO_3M$	H	OH	H
e	$(CH_2)_2CH(COOM_4)SO_3M_2$	H	OH	H
f	$(CH_2)_2CH(CHOHPh)SO_3M$	H	OH	H
g	$(CH_2)_2CH[(CH_2)_3SO_3M]SO_3M$	H	OH	H
h	$(CH_2)_3SO_3M$	H	H	H
i	$(CH_2)_3SO_3M$	H	OCH_3	H
j	$(CH_2)_3SO_3M$	H	OCH_2Ph	H
k	$(CH_2)_3SO_3M$	H	$O(CH_2)_3SO_3M$	H
l	H	H	$O(CH_2)_3SO_3M$	H
m	H	H	OH	SO_3M
n	H	SO_3M	OH	H
o	H	CH_2SO_3M	OH	H

Reprinted with permission from DuBois, G. E., Crosby, G. A., Stephenson, R. A., and Wingard, R. E., Jr., *J. Agric. Food Chem.*, 25, 763, 1977. Copyright by the American Chemical Society.

10, have taste properties described as better than neohesperidin DHC, i.e., less lingering menthol-like aftertaste.

Synthesis of the new analogue starts with the readily available flavone aglycone hesperetin and is completed in two steps (as shown below): (1) alkylation and (2) reduction.[109]

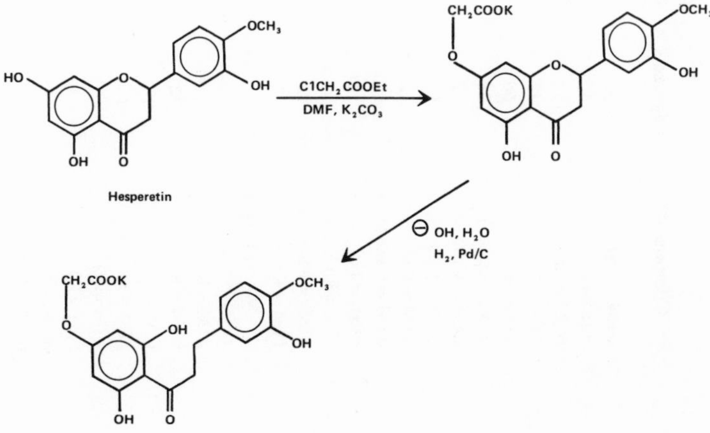

Hesperetin

$ClCH_2COOEt$, DMF, K_2CO_3

$^{\ominus}OH$, H_2O; H_2, Pd/C

Table 10
SENSORY EVALUATION OF SIMPLIFIED DIHYDROCHALCONES

Compound	Judgments	Conch (ppm)	Perceived intensity	Calcd intensity Wt basis	Calcd intensity Molar basis	Flavor judgment Sweet	Sour	Salty	Bitter	Other[b]	Aftertaste[c]
a (M = Na)	32	250	1.27	432	534	84	1	3	10	2	28
b (M = K)	12	228	1.88	705	929	83	2	3	7	5	75[d]
b (M = Ca 0.5)	12	250	1.63	557	703	84	0	2	7	7	58[d]
c (M = Na)	12	250	1.45	496	650	80	0	3	9	8	42
c (M = K)	60	250	1.13	386	525	77	4	5	9	5	65
c (M = Ca 0.5)	108	250	1.46	499	650	83	0	3	8	6	78
c (M = Mg 0.5)	12	250	1.52	520	665	82	1	3	10	4	50
c (M = Zn 0.5)	12	250	1.42	486	650	80	0	1	10	9	50
d (M = K)	12	250	0.66	226	316	50	0	8	19	23	50[e]
d (M = Ca 0.5)	12	250	0.63	215	289	68	0	8	11	13	25[e]
e (M_1 = H; M_2 = K)	12	250	0.45	154	229	72	9	6	13	0	42[e]
e (M_1 = M_2 = K)	12	250	0.36	123	197	50	7	8	35	0	67[e]
f (M = K)	2[f]	1,130				~50			~50		Strong lingering sweetness
g (M = K)	1[g]	1,000		0	0						
h (M = K)	32	250	0.52	178	233	81	2	2	15	0	31[e]
i (M = K)	12	255	0.73	245	342	57	0	0	25	18	67
j (M = K)	12	235	0.66	240	389	28	2	0	38	32	92
k (M = K)	1[g]	2,000		0	0						
l (M = K)	1[g]	1,000		0	0						
m (M = Na)	1[g]	1,000		0	0						
n (M = Na)	6	400	0.62	133	157	7	2	2	70	19	100 (bitter)
o (M = K)	1[g]	10,500							100		
p	42	250	1.79	612	1,097	80	4	4	12	0	76[d]
q	10	90	0.70	665	592	84	0	2	11	3	60
r	27	90	0.53	504	566	82	0	4	7	7	30[e]
s	12	250	0.90	308	386	74	9	5	11	1	67

Sodium cyclamate	24	1,250	0.53	36	21	75	5	13	4	3	17
Sodium saccharin	17	60	0.32	456	274	75	3	6	12	4	77 metallic and bitter[e]
Sucrose	221	85,500	1.00	1.0	1.0	97	1	1	1	0	4

[a] Times the 0.25 M (8.55%) sucrose reference.
[b] The balance of flavor was made up by tastes described as medicinal, phenoloic, and most commonly menthol- and licoricelike.
[c] Percentage of judgments where presence of unpleasant lingering aftertaste was reported; aftertastes were sweet except where noted.
[d] Percentage somewhat high due to a high perceived intensity.
[e] Percentage somewhat low due to a low perceived intensity.
[f] Determined by two judges to be intensely and unpleasantly bittersweet and not subjected to detailed panel analysis.
[g] Compounds judged tasteless in a preliminary analysis were not analyzed by the sensory panel.

Reprinted with permission from DuBois, G. E., Crosby, G. A., Stephenson, R. A., and Wingard, R. E., Jr., *J. Agric. Food Chem.*, 25, 763, 1977. Copyright by the American Chemical Society.

Chlorogenic Acid

Chlorogenic acid (Structure 22) is structurally not unlike the dihydrochalcones and might be considered as belonging to the class of alkoxyaromatic sweeteners except that it has hydroxyl groups rather than an alkoxyl group bound to the aromatic ring. Chlorogenic acid is not sweet but has been

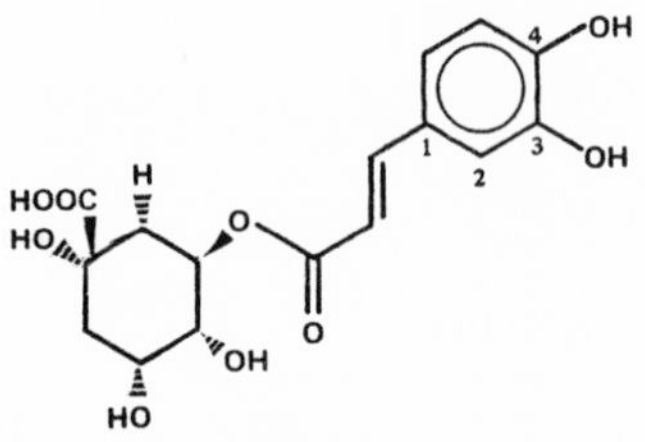

Chlorogenic acid
Structure 22

reported to be one of the naturally occurring components in artichokes which is responsible for causing water to taste sweet for several minutes after the tongue is exposed to these chemical substances.[66] Apparently, not all test subjects are sensitive to the taste-modifying effect of chlorogenic acid.

Phyllodulcin Analogues

Yamato and colleagues have recently studied the taste properties of a number of phyllodulcin analogues (listed in Table 11) and have also come to the conclusion that the alkoxy aromatic ring is important in eliciting a sweet taste response.[67,68]. On the basis of these studies, Yamato et al. prepared several new alkoxy aromatic compounds, indicated in Table 12.[69-71] A few were found to be sweet, although no indication is given of the taste intensity or quality of these materials. The low solubility of these compounds places a limit on the perceived taste intensity.

Nitroanilines

In the late 1930's Blanksma and van der Weyden prepared 1-*n*-propoxy-2-amino-4-nitrobenzene (Structure 23) by partial reducton of the corresponding 2,4-dinitro compound with sodium disulfide.[72] They reported the nitroaniline derivative to be no less than 3300 times sweeter than sucrose.

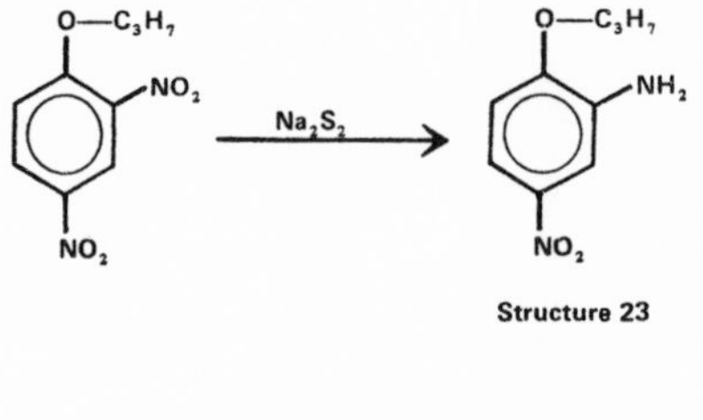

Structure 23

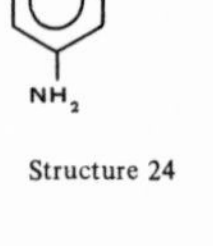

Structure 24

TABLE 11
PHYLLODULCIN ANALOGUES[67,68]

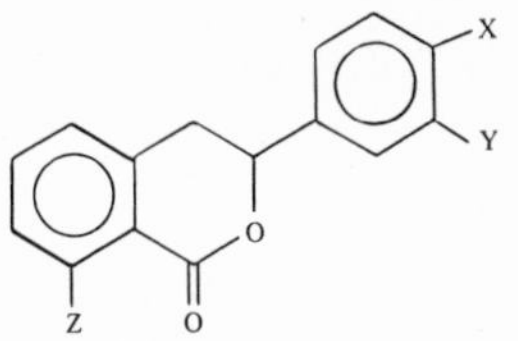

Substituents			
X	Y	Z	Taste
OCH_3	OH	OH	Very sweet
OCH_3	OCH_3	OH	Bitter
OCH_3	OCH_3	OCH_3	Tasteless
OCH_3	$OCOCH_3$	$OCOCH_3$	Slightly sweet
OH	OH	OH	Tasteless
OH	H	OH	Tasteless
OH	OH	H	Tasteless
OCH_3	OH	H	Very sweet
OH	OCH_3	H	Tasteless

From Crosby, G. A., *Crit. Rev. Food Sci. Nutr.*, 7(4), 297, 1976.

Table 12
ALKOXYAROMATIC SWEETENERS[69-71]

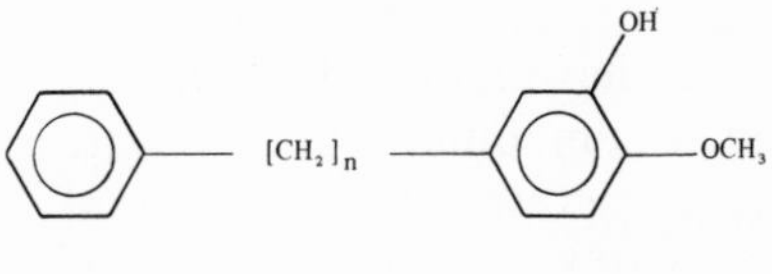

n	Taste
1	Tasteless
2	Sweet
3	Slightly sweet
4	Tasteless

From Crosby, G. A., *Crit. Rev. Food Sci. Nutr.*, 7(4), 297, 1976.

A few years later, Verkade and co-workers repeated this reaction and found the desired sweet product was contaminated with approximately 40% of the isomeric 1-n-propoxy-2-nitro-4-aminobenzene (Structure 24).[73] Upon separation of the mixture, it was found that the pure 2-amino-4-nitro isomer (Structure 23) was at least 4000 times sweeter than sucrose, while the positional isomer (Structure 24) was tasteless.

The original sweetener of this series, 1-n-propoxy-2-amino-4-nitroaniline (Structure

Table 13
ALKOXYNITROANILINES[74]

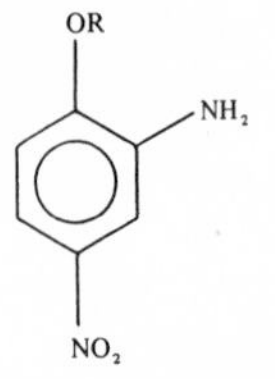

R	Sweetness (× sucrose)
H	120
CH_3	220
CH_2CH_3	950
$CH_2CH_2CH_3$	4100
$CH_2CH_2CH_2CH_3$	1000
$CH{=}CHCH_3$	2000
1-Propyl	600
n-Pentyl	Insoluble

From Crosby, G. A., *Crit. Rev. Food Sci. Nutr.*, 7(4), 297, 1976.

23), was actually used as a commercial sweetening agent in the Netherlands during the German occupation period of World War II and later in Berlin during the blockade of that city.[74] A major drawback of this material, called P-4000, is its powerful local anesthetic effect on the tongue and mouth.[75] Approval for use as a sweetener in this country is not possible because of the finding that P-4000 is toxic in rats at dosage levels of 0.1%.[76]

In the years that followed the initial discovery of P-4000, Verkade examined the synthesis and taste properties of a large number of alkoxynitroaniline isomers. An extensive review of this work appeared in 1968.[74] A number of the sweet nitroanilines reported during these years are listed in Table 13. It is interesting to note that none of the positional isomers, i.e., 3-amino-5-nitro- or 2-amino-6-nitroalkoxybenzenes, are sweet.

More recently, Crosby et al. have reported several new sweet nitroanilines (Structures 25 and 26).[77,78] The latter compound is prepared from commercially available 3-nitro-4-hydroxybenzoic acid (Structure 27). The potential chemical reactivity of the 3-bromopropoxy analogue (Structure 25) suggests that this material might be a useful tool for the study of the mechanism of taste.

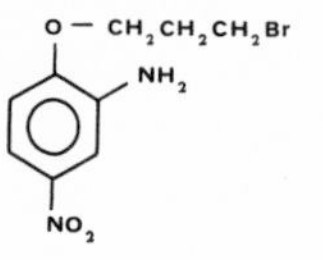

O — $CH_2CH_2CH_2Br$, NH_2, NO_2

Structure 25

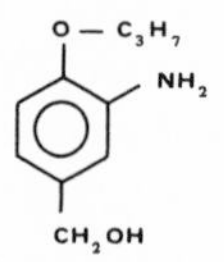

O — C_3H_7, NH_2, CH_2OH

Structure 26

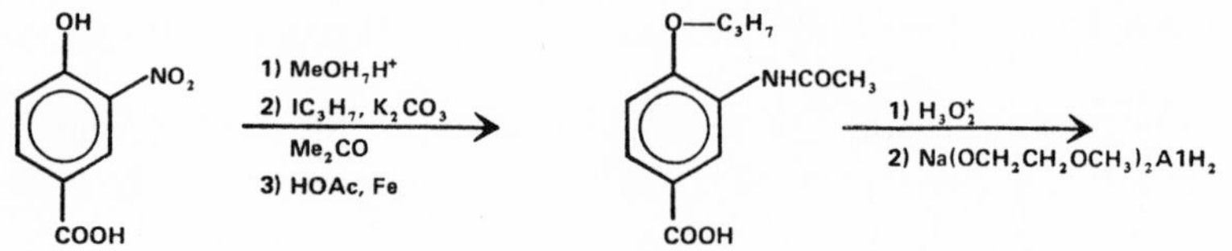

OH, NO_2, COOH → 1) MeOH,H^+ 2) IC_3H_7, K_2CO_3, Me_2CO 3) HOAc, Fe → O—C_3H_7, $NHCOCH_3$, COOH → 1) H_3O^+ 2) $Na(OCH_2CH_2OCH_3)_2AlH_2$

Structure 27

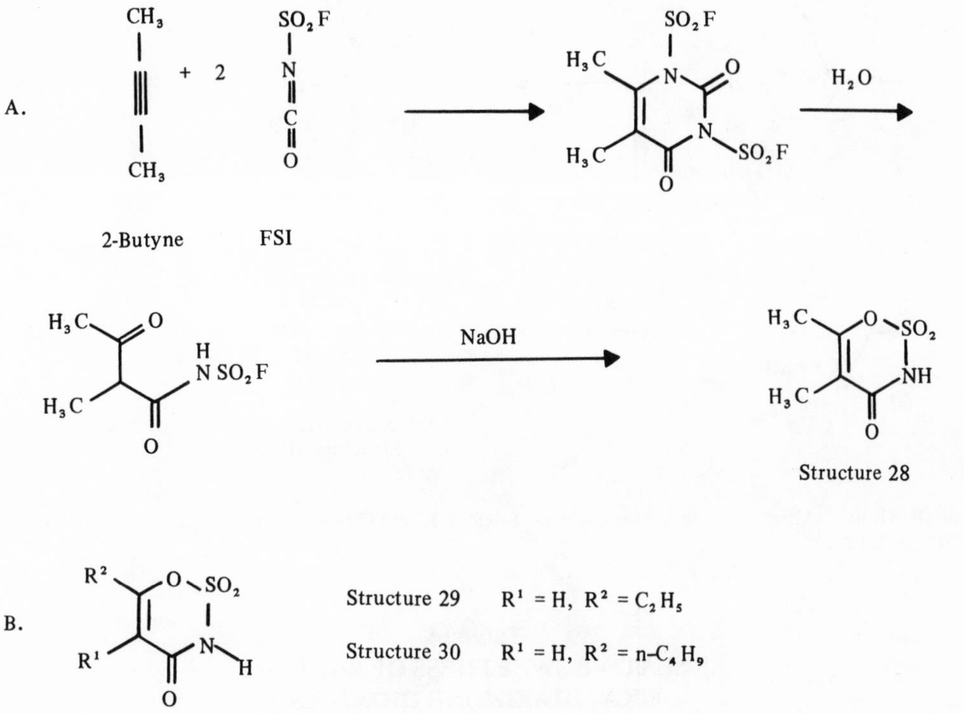

FIGURE 9. Synthesis of oxathiazinone dioxides. A. Synthesis. B. Analogues prepared by synthetic route in A. (From Crosby, G. A., *Crit. Rev. Food Sci. Nutr.*, 7(4), 297, 1976.)

MISCELLANEOUS SWEETENERS

Oxathiazinone Dioxides (Acesulfames)

In November 1973, a comprehensive report appeared describing the discovery of a new class of sweeteners, the oxathiazinone dioxides.[79] The first substance (Structure 28, Figure 9) in the series was prepared by reaction of 2-butyne with fluorosulfonyl isocyanate (FSI), as shown in Figure 9. The reaction was also found to work with 1-butyne and 1-hexyne, leading to compounds (Structures 29 and 30, respectively, in Figure 9) with a sweet taste. Further research demonstrated that additional oxathiazinone dioxides, i.e., acesulfames (Structure 31, Figure 10) could be synthesized from FSI and a variety of active methylene compounds, including α-unsubstituted ketones, β-diketones, and β-keto acids and -esters, as shown for *t*-butyl acetoacetate in Figure 10. Furthermore, chlorosulfonyl isocyanate (CSI) was found to react with the more reactive β-dicarbonyl derivatives, such as *t*-butyl acetoacetate, in the same manner as FSI.

A variety of oxathiazinone dioxides were prepared by these methods and found to be relatively intense sweeteners, as indicated in Table 14. Because of the strong acidity

Acesulfam, acid form
Structure 31

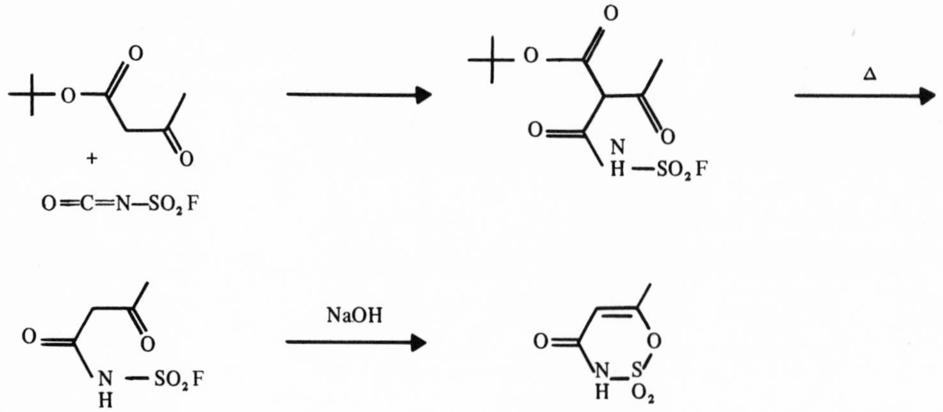

FIGURE 10. Alternate route to oxathiazinone dioxides. (From Crosby, G. A., *Crit. Rev. Food Sci. Nutr.*, 7(4), 297, 1976.)

Table 14
RELATIVE SWEETNESS OF SALTS OF OXATHIAZINONE DIOXIDES

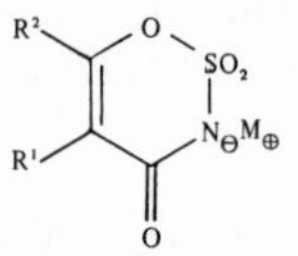

Substituents			
R^1	R^2	M	Sweetness[a]
H	H	Na	10
H	CH_3	Na	130
H	CH_3	K	130
H	CH_3	Ca	130
CH_3	H	Na	20
CH_3	CH_3	Na	130
H	C_2H_5	Na	150
C_2H_5	H	Na	20
CH_3	C_2H_5	Na	130
C_2H_5	CH_3	Na	250
H	n-C_4H_9	Na	30
n-C_3H_7	CH_3	Na	30
i-C_3H_7	CH_3	Na	50
CH=CH–CH=CH		Na	About 50
C_2H_5	n-C_3H_7	Na	70

[a] Relative to 4% sucrose.

From Crosby, G. A., *Crit. Rev. Food Sci. Nutr.*, 7(4), 297, 1976.

Table 15
SOME CHEMICAL AND PHYSICAL CHARACTERISTICS OF ACESULFAME[110,111]

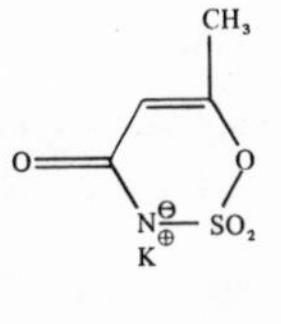

Empirical formula	$C_4H_4O_4NSK$
Molecular weight	201.2
Melting point of:	
Potassium salt	225°C
Free acid	123.5°C
Absorption maximum	227 mm
Extinction coefficient (ε)	10,900
Density	1.83 g/cm³
Solubility at 20°C in:	
Water, 20°C	30 g/100 ml
Water, 40°C	40
Water, 80°C	90
Ethanol	0.4
Ethanol, 80% vol	5
Ethanol, 60% vol	10
Ethanol, 40% vol	15
Ethanol, 20% vol	23
Methanol	1.0
Acetone	0.08
Acetic acid	ca. 13
DMF	ca. 15
DMSO	>30
Formic acid	>60

of the NH proton, these materials are best utilized in the form of highly water-soluble alkali or alkaline earth metal salts.

The potassium salt of the 6-methyl derivative, also called simply acesulfame-K (Structure 31, Figure 10), has been reported to display the optimum purity of sweet taste.[79] In terms of sweetness, the intensity of acesulfame-K is greater than cyclamate but less than saccharin.[110] The synthesis and chemical and physical characteristics of acesulfame-K have recently been reported.[110,111] As summarized in Table 15, acesulfame-K is quite soluble in water and sufficiently stable to hydrolysis for practical purposes. Although safety studies are in progress (Hoechst®, Frankfurt, West Germany),[115] the acute toxicity in rats has been reported to be approximately 7.4 g/kg body weight.[110] Also, no ill effects resulted from feeding the calcium salt of acesulfame to rats for 95 days at dosage levels as high as 5% of the diet.[79]

Kynurenine Derivatives

In 1973, members of the Western Regional Research Laboratory of the USDA reported that both the *N′*-formyl (Structure 32) and *N′*-acetyl (Structure 33) derivatives of kynurenine, a metabolite of tryptophan, are approximately 35 times sweeter than sucrose. It was also reported that both derivatives elicit a sweet taste immediately upon contact with the tongue. No safety tasting of the sweeteners has been conducted as of the time of the report.

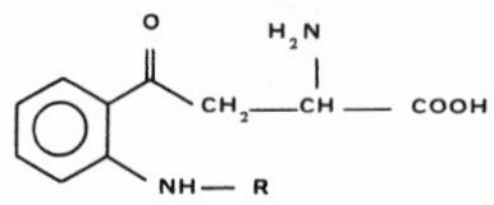

Structure 32 where R = CHO

Structure 33 where R = $COCH_3$

Oximes

Furukawa et al. reported in 1920 that the syn-oxime of perillartine (Structure 34), the purified oil component of *Perilla namkinensis,* was 2000 times sweeter than sugar.[81] More recently, Acton and co-workers have investigated the relationship between taste and chemical structure for the oxime class of sweeteners.[82] According to the results of their investigation, the syn form of perillartine is strongly sweet. By varying the terpene portion of the molecule and leaving the oxime group intact, these workers established that a double bond conjugated with the oxime is essential for sweet taste. In addition, the 1,3-diene isomer (Strucure 35) is only half as sweet as perillartine oxime (Structure 34), while the 1,4-isomer (Structure 36) is nearly of the same sweetness. The aromatic version (Structure 37) is mostly bitter.

A major factor that limits commercial utilization of the above sweeteners is their low water solubility. As a solution to this problem, Acton[56] has devised the synthesis of a new sweet oxime, SRI Oxime V. The new compound (Structure 38) is 450 times sweeter than sucrose, with a clean taste and sufficient water solubility for most uses. The material is reported to be stable in carbonated beverages if acidity is kept above pH 3.

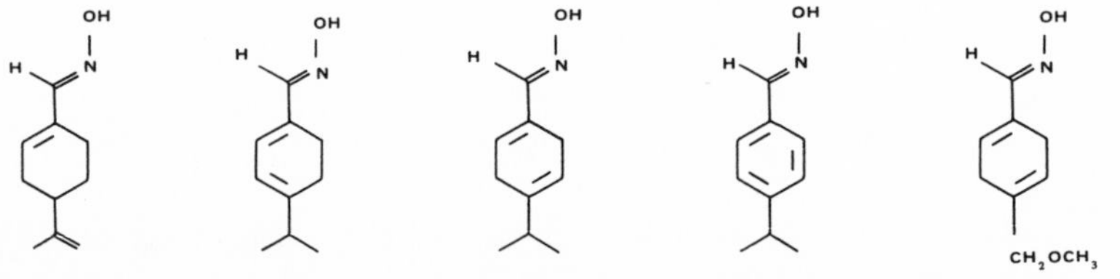

Structure 34 Structure 35 Structure 36 Structure 37 SRI Oxime V Structure 38

Chlorosucrose

Recently, two new, intensely sweet derivatives of sucrose (see B1, Figure 11) have been reported.[112] These compounds represent the first examples of simple sugars, i.e., disaccharides. Specifically, 1′,4,6,6′-tetrachloro-galactosucrose (see A2, Figure 11) is comparable in sweetness intensity to saccharin but reportedly void of unpleasant after-taste. Similarly, 1′,6′-dichlorosucrose (see B2, Figure 11) is also reported to be intensely sweet, much like the tetrachloro-derivative, while the 1′,6,6′-trichloro- (see B3, Figure 11) and 4,6,6′-trichloro- (see A3, Figure 11) derivatives are more than ten times sweeter than sucrose. Apparently, the positions chlorinated are important with respect to sweetness, e.g., 6,6′-dichlorosucrose (see B4, Figure 11) is less sweet than sucrose.

Interestingly, sucrose derivatives in which the 6-hydroxyl of the fructofuranosyl unit has been replaced by a chloro substituent are resistant to the action of invertase.[112]

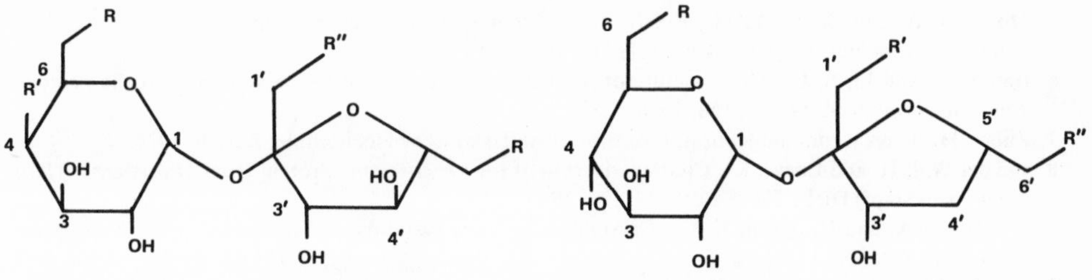

FIGURE 11. Intensely sweet derivatives of sucrose and galacto-sucrose. In A1 galacto-sucrose, R = R′ = OH; A2, R = R′ = R″ = Cl; and A3, R = R′ = Cl, R″ = H. In B1, sucrose, R = R′ = R″ = OH; B2, R = H, R′ = R″ = Cl; B3, R = R′ = R″ = Cl; and B4, R = R″ = Cl, R′ = H.

CONCLUSION

The removal of cyclamate from the GRAS list in 1969 accounts for much of the increased research effort devoted to the development of new sweeteners. Unfortunately, the present extent of knowledge regarding the mechanism of taste and taste-structure relationships of sweet substances is far from sufficient to enable one to predict structures for totally new classes of sweeteners. As a result, the recent increased activity continues to center around the study of analogues of accidentally discovered sweet substances in the hope that an improved version will be found that is free from the limitations of stability, toxicity, or taste quality. The major sweeteners presently available suffer from one or more of these problems. The approval of aspartame has recently been stayed by the FDA; β-neohesperidin dihydrochalcone appears to be safe, but suffers from problems of taste quality, while the oxathiazinone dioxides require additional toxicology studies before a conclusion can be drawn. Clearly, there is a tremendous need for more research into the mechanism of taste at all levels as well as a greater understanding of taste-structure relationships, so that major new advances can be made in the field of synthetic sweeteners.

It is unfortunate, but the tremendous costs required to bring a new sweetener to market may severely limit the number of sweeteners available to the public in the future, despite the increasing ability of research to produce better products. The development, however, of a nonabsorbable sweetener, as currently under investigation at Dynapol,[56,83] may hold real promise by reducing the risk of expensive toxicology studies as well as the risk to consumers. There is the additional possibility that cyclamate may again be approved for use at restricted levels in the U.S. In addition, public and commercial interest in safe sweeteners is very strong at this time and should continue to provide ample incentive for increased research in both new sweeteners and the field of taste mechanism for many years.

REFERENCES

1. **Cagan, R. H.,** Chemostimulatory protein: a new type of taste stimulus, *Science,* 181, 32, 1973.
2. **Irvine, F. R.,** *Woody Plants of Ghana,* Oxford University Press, London, 1961, 32, 596.
3. **Inglett, G. E. and May, J. F.,** Tropical plants with unusual taste properties, *Econ. Bot.,* 22, 326, 1968.
4. **Inglett, G. E. and May, J. F.,** Serendipity berries — source of a new intense sweetener, *J. Food Sci.,* 34, 408, 1969.

5. **Morris, J. A. and Cagan, R. H.,** Purification of monellin, the sweet principle of *Dioscoreophyllum cumminsii, Biochim. Biophys. Acta,* 261, 114, 1972.
6. **Bohak, Z. and Li, S. L.,** The structure of monellin and its relation to the sweetness of the protein, *Biochim. Biophys. Acta,* 427, 153, 1976.
7. **Züber, H.,** Private communication, Eidgenossische Technische Hochschule, Zurich, 1974.
8. **van der Wel, H. and Loeve, K.,** Characterization of the sweet-tasting protein from *Dioscoreophyllum cumminsii* (Stapf) Diels, *FEBS Lett.,* 29, 181, 1973.
9. **Wlodawer, A. and Hodgson, K. O.,** Crystallization and crystal data of monellin, *Proc. Natl. Acad. Sci. U.S.A.,* 72, 398, 1975.
10. **van der Wel, H. and Loeve, K.,** Isolation and characterization of thaumatin I and II, the sweet-tasting proteins from *Thaumatococcus daniellii Benth, Eur. J. Biochem.,* 31, 221, 1972.
11. **Lyenger, R. B., et al.,** The complete amino-acid sequence of the sweet protein Thaumatin I, *Eur. J. Biochem.,* 96, 193, 1979.
12. **Brouwer, J. N., van der Wel, H., Francke, A., and Henning, G. J.,** Miraculin, the sweetness-inducing protein from miracle fruit, *Nature,* 220, 373, 1968.
13. **Daniell, W. F.,** On the *Synsepalum dulcificum, De Cand.* or, miraculous berry of western Africa, *Pharm. J.,* 11, 445, 1852.
14. **Inglett, G. E., Dowling, B. , Albrecht, J. J., and Hoglan, F. A.,** Taste-modifying properties of miracle fruit *(Synsepalum dulcificum), J. Agric. Food Chem.,* 13, 284, 1965.
15. **Kurihara, K. and Beidler, L. M.,** Taste-modifying protein from miracle fruit, *Science,* 161, 1241, 1968.
16. **Kurihara, K. and Beidler, L. M.,** Mechanism of the action of taste-modifying protein, *Nature,* 222, 1176, 1969.
17. **Bridel, M. and Lavieille, R.,** Le principe a saveur sucree du Kaa-he-e *(Stevia rebaudiana Bertoni), J. Pharm. Chim.,* 14, 99, 1931.
18. **Wood, H. B., Jr., Allerton, R., Diehl, H. W., and Fletcher, H. G.,** Stevioside. I. Structure of the glucose moieties, *J. Org. Chem.,* 20, 875, 1955.
19. **Mosettig, E. and Nes, W. R.,** Stevioside. II. Structure of the aglycone, *J. Org. Chem.,* 20, 884, 1955.
20. **Vis, E. and Fletcher, H. G., Jr.,** Stevioside. IV. Evidence that stevioside is a sophoroside, *J. Am. Chem. Soc.,* 78, 4709, 1956.
21. **Djerassi, C., Quitt, P., Mosettig, E., Cambie, R. C., Rutledge, P. S., and Briggs, L. H.,** Optical rotatory dispersion studies. LVIII. The complete and absolute configurations of steviol, kaurene and the diterpene alkaloids of the garryfoline and atisine groups, *J. Am. Chem. Soc.,* 83, 3720, 1961.
22. **Mosettig, E., Reglinger, U., Dolder, F., Lichti, H., Quitt, P., and Waters, J. A.,** Absolute configuration of steviol and isosteviol, *J. Am. Chem. Soc.,* 85, 2305, 1963.
23. **Inglett, G. E.,** *A Potential Saccharin Replacement: Stevioside,* Botanicals, Peoria, Ill., 1972.
24. **Farnsworth, N. R.,** Current status of sugar substitutes, *Cosmet. Perfum.,* p. 27, July 1973.
25. **Jizba, J., Dolejs, L., Herout, V., and Sorm, F.,** Structure of osladin, the sweet principle of the rhizomes of *Polypodium vulgare L., Tetrahedron Lett.,* p. 1329, 1971.
26. **Dean, F. M.,** *Naturally Occurring Oxygen Ring Compounds,* Butterworth, London, 1963, 458.
27. **Asahina, Y. and Ueno, S.,** Phyllodulcin, a chemical constituent of amacha, *(Hydrangea thunbergii Sieb), J. Pharm. Soc. Jpn.,* p. 146, 1916.
28. **Asahina, Y. and Asano, J.,** Constitution of hydrangenol and phyllodulcin, *Chem. Ber.,* 62, 171, 1929.
29. **Arakawa, H. and Hakazaki, M.,** Absolute configuration of phyllodulcin, *Chem. Ind.,* p. 671, 1959.
30. Ajinomoto Co., Phyllodulcin: Sweet Component from Hydrangea Leaves, Japanese Patent 49-41565, 1974.
31. **Ariyoshi, Y.,** Taste and chemical structure (2), *Kagaku To Seibutsu,* 12, 274, 1974.
32. **Lee, C. H.,** Intense sweetener from Lo Han Kuo, *(Momordica grosvenori), Experientia,* 31, 533, 1975.
33. **Davey, J. M., Laird, A. H., and Morley, J. S.,** Polypeptides. III. The synthesis of the C-terminal tetrapeptide sequence of gastrin, its optical isomers and acylated derivatives, *J. Chem. Soc. C.,* p. 555, 1966.
34. **Mazur, R. H., Schlatter, J. M., and Goldkamp, A. H.,** Structure-taste relationships of some dipeptides, *J. Am. Chem. Soc.,* 91, 2684, 1969.
35. Aspartame, Technical Bulletin No. 600(060473), Searle Biochemics, div. G. D. Searle, Arlington Heights, Ill.
36. **Cloninger, M. R. and Baldwin, R. E.,** L-Aspartyl-L-phenylalanine methyl ester (Aspartame) as a sweetener, *J. Food Sci.,* 39, 347, 1974.
37. **Vaisey, M., Brunion, R., and Cooper, J.,** Some sensory effects of hydrocolloid sols on sweetness, *J. Food Sci.,* 34, 397, 1969.
38. **Arabie, P. and Moskowitz, H. R.,** The effects of viscosity upon perceived sweetness, *Percept. Psychophys.,* 9, 410, 1971.

39. **Mazur, R. H.,** Aspartame — a sweet surprise, in Am. College of Nutrition: First Annual Interim Meeting, Part II, Nov. 16—18, 1974, 5.
40. **Halpern, S. L.,** Chairman, Scientific review of a new sweetener, Am. College of Nutrition, First Annual Interim Meeting, II, Nov. 16—18, 1974, pp. 1—63.
41. *Fed. Regist.*, 39(145), 27317, 1974.
42. **Mazur, R. H., Goldkamp, A. H., James, P. A., and Schlatter, J. M.,** Structure-taste relationships of aspartic acid amides, *J. Med. Chem.*, 13, 1217, 1970.
43. **Mazur, R. H., Reuter, J. A., Swiatek, K. A., and Schlatter, J. M.,** Synthetic sweeteners. 3. Aspartyl dipeptide esters from L- and D-alkylglycines, *J. Med. Chem.*, 16, 1284, 1973.
44. **Fujino, M., Wakimasu, M., Tanaka, K., Aoki, H., and Nakajima, N.,** L-Aspartyl-aminomalonic acid diesters, *Naturwissenschaften*, 60, 351, 1973.
45. **Briggs, M. T. and Morley, J. S.,** British Patent 1,299,265, 1972.
46. **Inglett, G. E., Krbechek, L., Dowling, B., and Wagner, R.,** Dihydrochalcone sweeteners — sensory and stability evaluation, *J. Food Sci.*, 34, 101, 1969.
47. **Deutsch, E. W. and Hansch, C.,** Dependence of relative sweetness on hydrophobic bonding, *Nature,* 211, 75, 1966.
48. **Ferguson, L. N.,** *Organic Chemistry: A Science and An Art,* Willard Grant Press, Boston, 1972, p. 68.
49. **Horowitz, R. M. and Gentili, B.,** Taste and structure in phenolic glycosides, *J. Agric. Food Chem.*, 17, 696, 1969.
50. **Horowitz, R. M. and Gentili, B.,** Dihydrochalcone Derivatives, and Their Use as Sweetening Agents, U.S. Patent 3,087,821, 1963.
51. Flav-O-Last, Technical Bulletin, Nutralite Products, Inc. Buena Park, Cal.
52. Dynapol, unpublished data, Palo Alto, Cal., 1974.
53. **Westall, E. B., Scanlan, J. J., and Sahaydak, M.,** Flavor Preservative in Sugarless Chewing Gum Compositions and Candy Products, U.S. Patent 3,857,962, 1974.
54. **Booth, A. N., Robbins, D. J., Gumbmann, M. R., and Gould, D. H.,** Dihydrochalcone Safety Evaluation in Rats, paper presented at the Citrus Research Conference, Pasadena, Cal., Dec. 4, 1973.
55. **Robertson, G. H., Clark, J. P., and Lundin, R.,** Dihydrochalcone sweeteners: preparation of neohesperidin dihydrochalcone, *Ind. Eng. Chem. Prod. Res., Dev.*, 13, 125, 1974.
56. **Seltzer, R. J.,** Work on new synthetic sweeteners advances, *Chem. Eng. News,* p. 27, Aug. 25, 1975.
57. **Krbechek, L., Inglett, G., Holik, M., Dowling, B., Wagner, R., and Ritter, R.,** Dihydrochalcones: synthesis of potential sweetening agents, *J. Agric. Food Chem.*,16, 108, 1968.
58. **Horowitz, R. and Gentili, B.,**in *Sweetness and Sweeteners,* Birch, G. G., Green, L. F., and Coulson, C. B., Eds, Applied Science, London, 1971.
59. **Horowitz, R. and Gentili, B.,**in *Symposium: Sweeteners,* Inglett, G. E., Ed., AVI Publishing, Westport, Conn., 1974, chap. 16.
60. **Horowitz, R. M. and Gentili, B.,** Dihydrochalcone Xylosides and Their Use as Sweetening Agents, U. S. Patent 3, 826, 856, 1974.
61. **Farkas, L. and Nogradi, M.,** Dihydrochalcone sweetening agents, *Hung, Teljes,* 4026, April 28, 1972, *C. Ab.*, 77, 60321e, 1972.
62. **Farkas, L., Nogradi, M., Gottsegen, A., and Antus, S.,** New 1,3-Diphenylpropanone-1-Derivatives and The Salts Thereof, and Sweetening Agents Containing the Same, German Patent 2,258,304, 1973.
63. **Van Niekerk, D. M. and Koeppen, B. H.,** Synthesis of 2-O-α-L-rhamno pyranosyl-β-D-galacto-pyranosides, *Experientia,* 28, 123, 1972.
64. **Horowitz, R. M.,** private communication, USDA Laboratories, Pasadena, Cal., 1973.
65. **Okada, S.,** Artificial sweetening — dihydrochalcone (DC), *Kagaku To Seibutsu,* 11, 712, 1974.
66. **Bartoshuk, L. M., Lee, C. H., and Scarpellino, R.,** Sweet taste of water induced by artichoke *(Cynara scolymus), Science,* 178, 988, 1972.
67. **Yamato, M., Kitamura, T., Hashigaki, K., Kuwano, Y., Yoshida, N., and Koyama, T.,** Synthesis of biologically active isocoumarins. I. Chemical structure and sweet taste of 3,4-dihydroisocoumarins, *Yakugaku Zasshi,* 92, 367, 1972.
68. **Yamato, M., Hashigaki, K., Kuwano, Y., and Koyama, T.,** Syntheses of biologically active isocoumarins. II., *Yakugaku Zasshi,* 92, 535, 1972.
69. **Yamato, M., Kitamura, T., Hashigaki, K., Kuwano, Y., Murakami, S., and Koyama, T.,** Syntheses of biologically active isocoumarins. III. Chemical structure and sweet taste of 3,4-dihydroisocoumarins, *Yakugaku Zasshi,* 92, 850, 1972.
70. **Yamato, M., Sato, K., Hashigaki, K., Ishikawa, T., and Koyama, T.,** Chemical structure and sweet taste of isocoumarin and its derivatives. IV., *Yakugaku Zasshi,* 93, 1639, 1973.
71. **Yamato, M., Sato, K., Hashigaki, K., Oki, M., and Koyama, T.,** Chemical structure and sweet taste of isocoumarins and related compounds, *Chem. Pharm. Bull.*, 22, 475, 1974.
72. **Blanksma, J. J. and van der Weyden, P. W. M.,** Relationship between taste and structure in some derivatives of metanitraniline, *Recl. Trav. Chim.*, 59, 629, 1940.

73. **Verkade, P. E., Van Dijk, C. P., and Meerburg, W.,** Researches on the alkoxy-amino-nitrobenzenes. I. Partial reduction of 1-alkoxy-2,4-dinitrobenzenes with sodium disulphide. The taste of the alkoxy-amino-nitrobenzenes thus obtained, *Recl. Trav. Chim.,* 65, 346, 1946.
74. **Verkade, P. E.,** On organic compounds with a sweet and/or a bitter taste, *Farmaco Ed. Sci.,* 23, 248, 1968.
75. **Verkade, P. E., Van Dijk, C. P., and Witjens, P. H.,** Alkoxyaminonitrobenzenes. VI. A survey of their anesthetic properties, *Recl. Trav. Chim.,* 68, 639, 1949.
76. **Fitzhugh, O. G., Nelson, A. A., and Frawley, J. P.,** A comparison of the chronic toxicities of synthetic sweetening agents, *J. Am. Pharm. Assoc.,* 40, 583, 1951.
77. **Crosby, G. A. and Peters, G. C.,** Halogenated Aromatic Compound Having Sweetening Properties, U.S. Patent 3,845,225, 1974.
78. **Crosby, G. A. and Saffron, P. M.,** Edible Substances with 3-Amino-4-n-propoxybenzyl Alcohol as a Sweetener, U.S. Patent 3,876,814, 1975.
79. **Clauss, K. and Jensen, H.,** Oxathiazinone dioxides — a new group of sweetening agents, *Angew. Chem. Int. Ed. Engl.,* 12, 869, 1973.
80. **Finley, J. W. and Friedman, M.,** New sweetening agents, N′-Formyl-and N′-acetylkynurenine, *J. Agric. Food Chem.,* 21, 33, 1973.
81. **Furukawa, S. and Tomizawa, Z.,** Essential oil of *Perylla nankinensis* Dene, *Kogyo Kagaku Zasshi,* 23, 342, 1920.
82. **Acton, E. M., Leaffer, M. A., Oliver, S. M., and Stone, H.,** Structure-taste relationships in oximes related to perillartine, *J. Agric. Food Chem.,* 18, 1061, 1970.
83. **Zaffaroni, A.,** Novel, Non-absorbable, Non-nutritive Sweeteners, U.S. Patent 3,876,816, 1975.
84. Aspartame stay of effectiveness of food additive regulation, *Fed. Regist.,* 40(235), 56907, 1975.
85. **Spivak, J.,** *Wall Street J.,* p. 6, Dec 5, 1975.
86. **Furia, T. E. and Stone, H.,** private communication, Dynapol, Palo Alto, Cal., 1974.
87. **Kohda, H., Kasai, R., Yamasaki, K., Murakami, K., and Tanaka, O.,** New sweet diterpene glucoside from *Stevia rebaudiana, Phytochemistry,* 15, 981, 1976.
88. **Scheinin, A., Mäkinen, K. K., and Ylitalo, K.,** Turku sugar studies. I. An intermediate report on the effect of sucrose, fructose and xylitol diets on the caries incidence in man, *Acta Odontol. Scand.,* 32, 383, 1974.
89. **Mäkinen, K. K. and Scheinin, A.,** Turku sugar studies. II. Preliminary biochemical and general findings, *Acta Odontol. Scand.,* 32, 413, 1974.
90. **Larmas, M., Mäkinen, K. K., and Scheinin, A.,** Turku sugar studies. III. An intermediate report on the effect of sucrose, fructose and xylitol diets on the numbers of salivary *lactobacilli, candida* and *streptococci, Acta Odontol. Scand.,* 32, 423, 1974.
91. **Gehring, F., Mäkinen, K. K., Larmas, M., and Scheinin, A.,** Turku sugar studies. IV. An intermediate report on the differentiation of polysaccharide-forming streptococci *(S. mutans), Acta Odontol. Scand.,* 32, 435, 1974.
92. **Moore, K. K.,** Xylitol: uncut gem among sweeteners, *Food Prod. Dev.,* p. 66, May 1977.
93. FDA scientists reviewing carcinogenicity of xylitol, *Food Chem. News,* p. 33, September 4, 1978.
94. **Moskowitz, H. R.,** The sweetness and pleasantness of sugars, *Am. J. Psychol.,* 84, 387, 1971.
95. **Lindley, M. G., Birch, G. G., and Khan, R.,** Sweetness of sucrose and xylitol. Structural considerations, *J. Sci. Food Agric.,* 27, 140, 1976.
96. **Reynolds, W. A., Butler, V., and Lernkey-Johnston, N.,** Hypothalamic morphology following ingestion of aspartame or MSG in the neonatal rodent and primate: a preliminary report, *J. Toxicol. Environ. Health,* 2, 471, 1976.
97. **Koch, R., Shaw, K. N. F., Williamson, M., and Haber, M.,** Use of Aspartame in Phenylketonuric Heterozygous Adults. *J. Toxicol. Environ. Health,* 2, 453, 1976.
98. **Ranney, R. E., Oppermann, J. A., and Muldoon, E.,** Comparative metabolism of aspartame in experimental animals and humans, *J. Toxicol. Environ. Health,* 2, 441, 1976.
99. **Stern, S. B.,** Administration of aspartame in non-insulin-dependent diabetics, *J. Toxicol. Environ. Health,* 2, 429, 1976.
100. **Knopp, R. H.,** Effects of aspartame in young persons during weight reduction, *J. Toxicol. Environ. Health,* 2, 417, 1976.
101. **Frey, G. H.,** Use of aspartame by apparently healthy children and adolescents, *J. Toxicol. Environ. Health,* 2, 401, 1976.
102. **Stegink, L. D.,** Absorption, utilization, and safety of aspartic acid, *J. Toxicol. Environ. Health,* 2, 215, 1976.
103. **Munro, H. N.,** Absorption and metabolism of amino acids with special emphasis of phenylalanine, *J. Toxicol. Environ. Health,* 2, 189, 1976.
104. **Gumbruann, M. R., Gould, D. H., Robbins, D. J., and Booth, A. N.,** Safety Evaluation of the Sweetener Neohesperidin Dihydrochalcone: 2-Year Study in Dogs, paper presented at the Citrus Research Conference, Pasadena, Cal., Dec. 8, 1976.

105. **Horowitz, R. M. and Gentili, B.,** Studies of the Metabolic Rate of Neohesperidin Dihydrochalcone, paper presented at the Citrus Research Conference, Pasadena, Cal., Dec. 8, 1976.
106. Hunting a safe sweetener, *Newsweek,* p. 95, April 4, 1977.
107. **DuBois, G. E., Crosby, G. A., and Saffron, P.,** Non-nutritive sweeteners: taste — structure relationships for some simple dihydrochalcones,. *Science,* 195, 397, 1977.
108. **DuBois, G. E., Crosby, G. A., Stephenson, R. A., and Wingard, R. E., Jr.,** Dihydrochalcone Sweeteners. Synthesis and sensory evaluation of sulfonate derivatives, *J. Agric. Food Chem.,* 25, 763, 1977.
109. **DuBois, G. E., Crosby, G. A., and Saffron, P.,** Regioselective flavanone alkylation: a facile method for preparation of some simple, sweet dihydrochalcones, *Synth. Commun.,* 7, 49, 1977.
110. **Clauss, K., Luck, E., and von Rymon Lipinski, G. W.,** Acetosulfam, ein neuer Susstaff. I. Herstellung und Eigenschaften, *Z. Lebensm. Unters. Forsch.,* 162, 37, 1976.
111. **Paulus, E. F.,** 6-Methyl-1,2,3-Oxathiazin-4(3K)-on-2,2-Dioxid, *Acta Crystallogr. Sect. B,* 31, 1191, 1975.
112. **Hough, L. and Phadnis, S. P.,** Enhancement in the sweetness of sucrose, *Nature,* 263, 800, 1976.
113. **Crosby, G. A. and Wingard, R. E.,** in *Developments in Sweeteners,* Hough, C. A. M., Parker, K. J., and Vitos, A. J., Eds., Applied Science, London, 1979.
114. **Enderlin, F. E., Brown, J. P., Wingard, R. E., Dale, J. A., Hale, R. L., and Seitz, C. T.,** Degradation of diterpene glycosidic sweeteners by gut microflora and absorption and metabolism of steviol-17-[^{14}C] in rats, *Fed. Proc. Fed. Am. Soc. Exp. Biol.,* 38, 539, 1979.
115. **Arpe, H.-J.,** Acesulfame-K, a new noncaloritic sweetener, in *Health and Sugar Substitutes,* Guggenheim, B., Ed., S. Karger, Basel, 1979.

NATURAL AND ARTIFICIAL FLAVORS

Frank Fischetti, Jr.

EDITOR'S NOTE

Perhaps more so than with any other subject reviewed in these volumes, flavors constitute the most complex topic to discuss in a meaningful manner. The sheer number of individual flavor ingredients of natural and synthetic origin available to the creative flavorist is enormous and the permutations and combinations with which they may be employed in a finished flavor composition increases the complexity of the subject beyond reasonable chances for thorough coverage in a single section. Also synthetic and natural products chemists are particularly active since many flavor compounds are structurally unique and represent new frontiers of organic chemistry. A considerable force of chemists are constantly identifying new ingredients to be added to the already ample body of knowledge available to the flavor industry. Consequently, it is important to recognize that this section is intended only as a general guide and introduction. It is not geared to the long-practicing creative flavorist or flavor chemist although they, too, may derive some benefit from the author's viewpoint. Nonetheless, the very complexity that precludes thorough review dictates that a substantial level of technical detail be included for the reader to derive an appreciation for both the subjective character of creative flavor technology together with the objective nature of flavor chemistry. Those wishing to gain further insight into various aspects of flavor technology, chemistry and its application to the food industry are urged to consult, and perhaps acquire, key texts and journals devoted to the subjects. The references included at the end of this section have been selected as a guide to further readings. Some of the titles include catalogs and directories listing flavor ingredients, compositions and their regulatory status. In our opinion, these are often as valuable as more traditional texts.

Finally, the reader is cautioned to pay strict attention to those regulatory actions governing the use of all ingredients going into our food supply. The past few years have seen changes in the regulations governing flavors and their use in foods. Some of the changes in U.S. regulations are minor such as the title section numbers in the U.S. Code of *Federal Regulations* detailing the natural and synthetic flavors and flavor adjuvants (see Part II). Other changes are more substantive. Regulatory actions currently being promulgated do not always adhere to uniform codes, and there is considerable risk associated with reaching general conclusions especially on an international scale. Again, the reader wishing further information on the current regulatory status of flavor ingredients is advised to seek expert guidance.

T.E.F.

INTRODUCTION

What is flavor? Flavoring? Flavor (or flavoring) is the property of a substance which causes a seemingly simultaneous sensation of taste in the mouth and odor in the back of the nose. Flavor is the summation of all the sensations that take place in the mouth. While the sensations are predominantly those recognized by the senses of smell and taste, they also include the mouthfeel sensations received and recognized by the brain by way of tactile and pain receptors. As described by Marcel Perret, ''flavor is a memory''.

It is important to recognize that flavor per se is not a substance, but rather the very special property of a substance. Because it is a property, it can be moved or transferred to other substrates, such as foods and beverages. This characteristic is what the flavorist (and the flavor industry) relies upon when creating a flavor composition destined for a specific application.

Flavors perform several functions in a food system. Foremost is their ability to render food more acceptable and enjoyable. It is often used to create the impression of flavor where very little or none existed. More important, flavors impart food products with recognizable character. The character of flavors is sufficiently implanted in man's mind that most societies associate the name of the flavor with the finished food product as a means of generic identification: orange, lemon, lime, cream, cola, chocolate, vanilla, etc. Even colors have adopted flavor titles to convey more precise descriptions, although in the food industry itself, it is rare to use color descriptions — red, yellow, green — to describe food products.

Flavoring agents are often employed to change the flavor of a product (e.g., flavoring milk, cottage cheese, or yogurt); modify, supplement, or enhance an existing flavor (e.g., the chicken flavor in a chicken soup, the beef flavor in bouillon, and the butter flavor in margarine); and mask or eliminate an objectionable taste characteristic that may be present (e.g., the "beany" taste in some soya products, the aftertaste of artificially sweetened beverages, and the unpleasant aftertaste of certain pharmaceuticals).

Flavoring is rather ubiquitous. Cream can be added to coffee for its own flavor value or to mask the inherent bitterness. Vanilla can be added to a product to create a sweet taste or to enhance the overall aroma. Methyl cyclopenteneolone can be added to a butterscotch flavor to modify inherent "scotch-notes" or to contribute a maple character. 2-Ethyl-3-hydroxy-4H-pyran-4-one can be added to an astringent beverage to smooth it out.

It should always be kept foremost in mind that a flavor is designed to enhance the acceptability and appreciation of a product. It is not primarily intended to correct man-made errors nor to improve a poorly designed product, but it can, within certain limitations, correct natural mistakes as in the case of the beany taste of soy. A good flavor, properly formulated for its intended use and properly used is generally beneficial to the enhancement of product acceptability. However, a good flavor composition used either in a poorly designed product or incorporated in the wrong way results in a bad flavor. It can not, under these conditions, enhance the acceptability and enjoyment of a food product.

COMPOSITION OF FLAVORS

Natural Compounds

Spices and Herbs

There can be little doubt that the first natural flavoring materials to be utilized by man would have been spices. Spices, in addition to their flavors, were employed for their preserving effect on meat products. Whether they did or not is really not important; man used them because he thought they preserved food. Spices certainly helped to cover up or mask the objectionable flavor of spoilage that must have been present. Spices and herbs had other uses in addition to flavoring or preservation. For example, they were used for embalming, as ointments, medicines, and incense. Probably the smallest use for spices and herbs was in the kitchen. Many ancient civilizations sought spices to flavor what is, by modern standards, a rather drab or monotonous fare. The food, due to lack of refrigeration, was at least partially decomposed and must have exhibited quite a range of odors.

Spices can be defined as "Any of various, often pungent, vegetable products such as pepper, cinnamon, nutmeg, mace, allspice, ginger, and clove — used in cooking to season food and flavor beverages.'' The many kinds of spices represent various portions of their respective plants. For example, ginger is a root stock, clove is a flower bud, nutmeg is a seed, cinnamon is the inner bark of a tree, and black pepper corns are whole fruits. Another part of the definition for spices one usually finds is ''Spices grow only in tropical areas and are available commercially either in whole or ground form and are always dried.''

Herbs can be defined as a seed plant which does not develop woody persistent tissue, as in a shrub or tree, but is more or less soft and succulent — a plant of economic value; specifically, one used for medicinal purposes or for its sweet scent or flavor. Herbs are plants which usually grow in temperate zones and, like spices, are adaptable for flavoring, seasoning, or coloring foods. They may be available fresh or dried, whole or ground. Some examples of herbs include basil, marjoram, bay, rosemary, oregano, and tarragon.

As can be seen from the above, it is difficult to clearly differentiate between spices and herbs. However, there is general agreement that aromatic, fragrant vegetable products of tropical origin used to season foods and flavor beverages are spices. When the plant stems from temperate regions and is used to season foods, it is called an herb.

From the flavorist's point of view, crude spices and herbs are more useful in other forms, such as essential oils and oleoresins. Crude, ground spices or herbs have relatively poor flavor strength and are traditionally used in the 0.5 to 1% range in finished food. They are usually insoluble and in many instances highly colored. Nonetheless, they effectively contribute flavor, color, and sometimes texture to a food product and some impart antibacterial (clove) and antioxidant properties (sage and rosemary) to the finished food.

In the flavoring industry, spices and herbs find extensive use in processed meat, pickling, convenience foods, and baking. These applications are traditional, but there are several disadvantages. There can be considerable variation in the flavoring and coloring ability of spices and herbs, from batch to batch and season to season. Large spice and herb suppliers guard against such variations by imposing strict quality control measures. When ground spices are used in a product, often greater flavor concentrations are experienced in areas of close proximity to the spices. This situation occurs because it takes time for the ground spice to reach flavor equilibrium with its surrounding. The speed with which the medium extracts flavor from the spice and how soon it migrates away from the spice particle determines this equilibrium. Flavor ''hot spots'' resulting from the use of ground spices can constitute a serious problem for manufactured food products, but is generally accepted in home cooking.

Crude spices and herbs also result in problems of storage and stability. For example, the equivalent of 100 lb of dry cloves can be represented by 6 lb of Oleoresin Clove. Similarly, 9 lb of Oleoresin Nutmeg will replace 100 lb of nutmeg and, in the case of black pepper, 4.5 lb of the oleoresin will replace 100 lb of the dry spice (Table 1). Therefore, the use of oleoresins saves considerable shelf and storage space. It should also be recalled that spices and herbs contain volatile oils. Even under the best storage conditions, the oils tend to volatilize, causing flavor losses. Also, dry, porous spices and herbs are usually packed in burlap bags and can adsorb or absorb extraneous odors from the environment.

Finally, spices and herbs are often contaminated with bacteria and mold spores. This is due to conditions prevailing during the growth, harvest, and shipment of the spices. Although fumigation with gaseous agents such as ethylene oxide can be used to sterilize spices, such treatment is troublesome, requires considerable capital equipment, is not entirely effective in reducing the microorganism count to zero, and be-

Table 1
OLEORESINS-DRY SPICE EQUIVALENTS

Oleoresin	Volatile oil content m*l*/100 g	Spice equivalent, lb[a]
Allspice (*Pimenta officinalis L.*)	40—50	4—6
Anise (*Pimpinella anisum L.*)	15—18	7.5—9
Capsicum (*Capsicum frutescens* and other related capsicum varieties)	500,000—1,000,000 Scoville heat units	5—7
Cardamom (*Cardamom elettaria*)	50	5
Celery (*Apium graveolens L.*)	9—11	3.5—6
Cinnamon (*Cinnamomum zeylanicum N.* and other cinnamomum var.)		2.5
Clove (*Caryophyllus aromaticus L.*)	40	5
Coriander (*Coriandrum savitum L.*)	6	6.5
Cubeb (*Piper Cubeba L.*)	80	15—18
Dill seed (*Anethum graveolens L.*)	10	4—6
Garlic (*Allium sativum L.*)	5	2 (fresh), 8 (dehydrated)
Ginger (*Zingiber officinale R.*)	30—40	4—6
Laurel leaf (*Laurus nobilis L.*)	25—30	12—15
Mace (*Myristica fragrans H.*)	40—45	6—8
Marjorma (*Majorana hortensis M.*)	8—10	7—9
Mustard seed (*Brassica nigra* and related species)	5	4—5
Nutmeg (*Myristica fragrans H.*)	55—60	8—10
Onion (*Allium cepa L.*)	5	0.25 (fresh), 1 (dehydrated)
Origanum (*Coridothymus capitalus R.*)	27—30	8—10
Paprika (*Capsicum annum L.*)	40,000—100,000 color value	6—10
Parsley (*Petroselinum sativum H.*)	12—15	0.33 (fresh), 3 (dehydrated)
Black pepper (*Piper nigrum L.*)	20—30, 53—57% piperine	4.5
Rosemary (*Rosmarinus officinalis L.*)	10—15	4—7

Table 1 (continued)
OLEORESINS-DRY SPICE EQUIVALENTS

Oleoresin	Volatile oil content m*l*/100 g	Spice equivalent, lb[a]
Sage Dalmatian		
(*Salvia officinalis L.*)	25—30	6—9
Tarragon		
(*Artemisia dracunculus L.*)	12—15	2—3
Thyme		
(*Thymus vulgaris L.*)	10	5
Turmeric		
(*Curcuma longa L.*)	1000 color value	15—20

[a] The pounds of oleoresin needed to replace 100 lb of dry commercial quality spice or herb.

cause it employs high reactive gases, it can adversely affect the quality of the spice. The problem of microbiological contamination is best solved by the use of a combination of volatile oils and oleoresins. Essential oils and oleoresins are free of microorganisms. Steam distillation destroys the bacteria and mold spores in the case of essential oils, while the volatile solvents act in a similar fashion in the case of oleoresins. Further, the mechanical assemblies involved in the manufacture of essential oils and oleoresins places the raw materials containing microorganisms at some distance to the finished product. This also assists in rendering oils and oleoresins essentially free of microorganisms. In view of the many disadvantages encountered in using spices and herbs, their corresponding essential oils and oleoresins offer the best means of standardization to the flavor user.

Essential Oils and Derivatives

Essential oils can be defined as aromatic, or odorous, oily liquids (sometimes semiliquid or solid) obtained from plant material. The oils volatilize, that is, evaporate from the botanical upon heating. They are usually soluble in alcohol or ether, but are only slightly soluble in water. It is this very volatility that distinguishes essential oils from fatty oils.

Since ground spices and herbs are used in foods principally for flavor, it seems logical that the most flavorful part of the spice, its essential oils, should be of higher value — and indeed they are. The aroma and characteristic organoleptic qualities of the spice are contained and concentrated within the essential oil fraction. Among the natural products used in flavor compositions, essential oils are probably the most important, being the largest single category of flavoring substances currently available to the flavorist. The strength of some essential oils in terms of their corresponding dry spice equivalent is shown in Table 2.

The essential oils may be prepared or derived from specific parts of plants, such as flowers (Oil Neroli), buds (clove), seeds (coriander and anise), fruit (lime, lemon, and orange), leaves (spearmint), twigs (petitgrain), bark (cinnamon), and roots (ginger).

Essential oils and derivatives may be obtained from the plant material by any of the methods shown in Figure 1. These include:

Expression

In the case of the citrus oils (lemon, lime, orange, grapefruit, and tangerine), the rind of the fruit contains the oil. The cells containing the oils are either pierced or ruptured when the skin is squeezed, broken, or rasped. The original method employed

Table 2
ESSENTIAL OIL-DRY SPICE EQUIVALENTS

Spice	Spice equivalent, lb[a]
Allspice	2.5
Almonds, bitter	0.5
Angelica root	0.75
Angelica seed	1.0
Anise seed (Russian)	2.5
Anise star (Chinese)	3.0
Basil, sweet	0.15
Bay leaves	2.0
Calamus root	2.5
Caraway seed	2.5
Cardamom seed	3.0
Cassia (cinnamon)	1.5
Celery seed	2.0
Cinnamon ceylon	0.5
Cloves	15.0
Coriander seed	0.75
Cumin seed	3.0
Dill seed	4.0
Dill weed	1.0
Estragon	0.13
Fennel seed	5.0
Ginger	0.25
Horseradish	1.0
Laurel leaves (bay)	1.0
Lovage root	0.5
Mace	5.0
Marjoram, sweet	0.5
Mustard seed	0.25
Nutmeg	5.0
Parsley seed	3.0
Pepper	1.0
Sage	1.25
Thyme	2.0
Valerian root	1.0
Black pepper	5.0 Oleoresin
Capsicum N.F.	5.0 Oleoresin
Ginger	5.0 Oleoresin
Celery	5.0 Oleoresin

[a] The pounds of essential oil needed to replace 100 lb of dry commercial quality spice or herb.

for collecting citrus oils was to squeeze the rind and take up the oil with a sponge (sponge process). This was the original technique of cold pressing oils. Today, the citrus fruits are machine pressed, the juice and the oils are collected simultaneously, and allowed to separate. Since the juice and oil differ significantly in specific gravity, the oil floats on top of the juice. If any of the oil remains either suspended or emulsified in the juice after separation, it is recovered either by centrifugation or distillation. Cold-pressed or expressed oils have an advantage over distilled oils in that they are not exposed to high temperatures and, therefore, suffer very little heat degradation. However, in machine pressing, the oil is in contact with acidic, aqueous juice solutions. Since the oils contain some slightly water soluble components, these can be removed

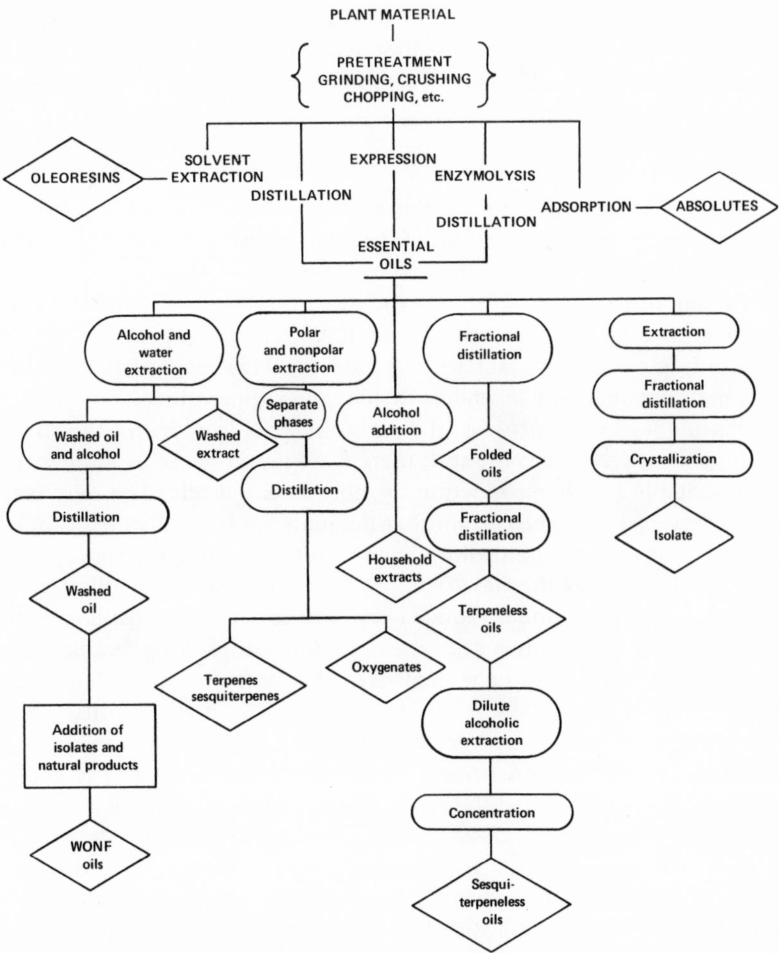

FIGURE 1. Flow chart for essential oils and derivatives obtained from plant material.

from the oil, lowering its value. In addition, the process undoubtedly fosters the hydrolysis of some esters and other acid catalyzed reactions can take place. Offered a choice between an expressed oil and a distilled one (except in the case of lime), most flavorists would probably choose the cold-pressed oils. To minimize the deleterious effects of the acidic juice, some processes call for the separation of the peel from the fruit. The peel is then processed to recover the oil in an operation which segregates it from contact with juice. However, distilled citrus oils are offered in the marketplace. These oils are distilled from the juice or from the exhausted rinds. Again, in all cases except lime, the distilled citrus oils are definitely not as acceptable, from a flavor or stability point of view, as the expressed oils.

Because clarified lime juice is a commercially valuable product, it gives rise to another valuable product of commerce, namely, distilled oil of lime. In making the juice, the whole fruit is crushed and the mixture of juice and oil is allowed to stand for several weeks. The juice is racked, drawn-off, and filtered. This results in the production of clarified lime juice. The residue remaining after filtration is steam distilled to recover the oil. However, because the oil was in contact with the juice, it is a totally different product than cold-pressed lime oil. Cold-pressed lime oil is high in terpenes, especially beta pinene, while the distilled oil is higher in oxygenates. Also, the cold-pressed oil is higher in citral, while the distilled oil is higher in alpha-Terpineol. Although the expressed oil is more like the lime fruit, most flavor chemists in the U.S. would choose the distilled oil. However, with all the other citrus oils, the product of choice would be the expressed oils.

Distillation

Water distillation — In this method, the material to be distilled is completely immersed in boiling water so that it circulates freely during the distillation. This particular method is used for floral materials such as Neroli and Ylang Ylang.

Water and steam distillation — In this type of distillation, the botaniial material is supported above the boiling water, usually on metal screens or grids. In this fashion, the botanical only comes in contact with the steam. This method is used for many types of plant material including lavender, thyme, and peppermint.

Steam distillation — In this process, the plant material is added to the still pot and live steam is injected directly into the still; there is no liquid phase as in the two above. The method is suitable for the production of most essential oils. The only precaution that must be observed is to make certain that the material to be distilled is not ground too fine since this may cause ''channelling'' and results in poor distillation (yields).

Most essential oils used by the flavorist are prepared by steam distillation. The distillation is usually conducted under reduced pressure to lower the distillation temperature. High distillation temperatures can lead to decomposition of valuable essential oil components. However, it should be remembered that with steam distillation there are definite changes that occur in the terpenoid constituents of essential oils. The terpene alcohols and their simple esters undergo rearrangement and elimination reactions. For example, if linalyl acetate (a constituent of bergamot and lavender) is subjected to steam distillation, the resulting distillate has been found to contain no less than twleve products.*

Another type of distillation used is the so-called dry or destructive distillation. In this little-used method, open-flame heat is directly applied to a still pot containing the bark or wood and the essential oil distillate is condensed and collected. Usually, the oils must be subsequently rectified to have commercial value. Cade oil is prepared in this fashion, although it has been banned in the U.S. and its toxicity is being questioned in Europe. Birch tar is also prepared in this fashion. Although not an essential oil, the so-called pyroligneous acid extract is prepared in this manner. The destructive distillate of wood is also worth mentioning because this is the initial procedure used to make rum ether and as originally employed, involved the estrification of distillates prepared in this fashion.

Distilled oils are classified as concentrated sources of flavoring. They are similar to the product from which they are obtained, but because the oils do come in contact with water or steam during the distillation, some water soluble components are lost. For example, phenylethyl alcohol, a component of rose oil, is slightly water soluble and during steam distillation is partially lost in the aqueous phase. Hydrolysis and other chemical reactions also occur during distillation. For example, caryophylene is formed from caryophylene oxide in clove. Heat also takes its toll. Some of the delicate top notes are destroyed and many of the naturally occurring antioxidants in the botanical do not distill over with the oil. During distillation, large amounts of water are condensed, and this water is always saturated with essential oil. It cannot simply be discarded as it would mean a loss of oil. The condensed water from the water and steam distillation can be used again as the water supply for the next charge of the same plant material in the still, or it can be returned to the still and itself be redistilled. This

* Picket, J. A. et al., *Chem. Ind.,* (London) No. 13, pg. 571—72 (1975)

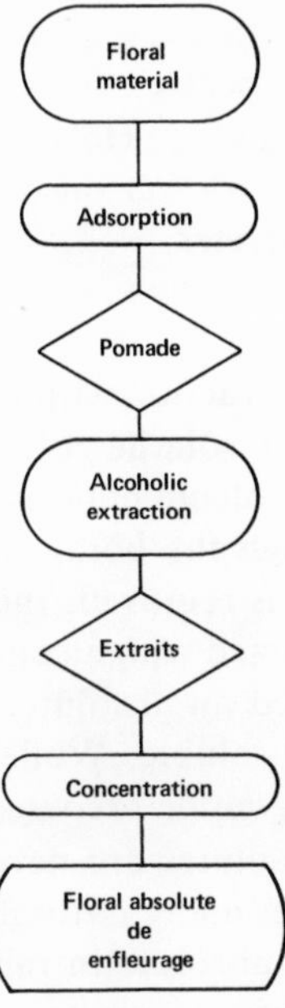

FIGURE 2. Flow chart for enfleurage process (adsorption).

process of recovering the oil from the water of distillation is commonly called cohobation.

Fermentation or enzymolysis

Through the action of fermentation or enzymolysis and subsequent distillation with water, essential oils of mustard, bitter almond, wintergreen, and sweet birch may be obtained. These products do not contain readily extractable essential oils per se. To recover essential oils, the botanicals are crushed, left in contact with water, and due to the action of enzymes, the essential oil is formed. The oil is then recovered by distillation with water. In the case of bitter almond, the oil is present as the glucoside amygdalin. When the enzyme emulsin comes in contact with amygdalin, it splits the glucose into the oil and glucose.

Enfleurage

This is a rather ancient process in which the volatile fragrances of various flowers are extracted or adsorbed by fixed oils. The process is outlined in Figure 2. If flowers are brought into contact with or immersed in deodorized fixed oils such as lard, olive oil, or beef suet, the essential oil diffuses from the flowers into the fat. In this instance, the essential oil is collected in the fat. The essential oil is subsequently recovered by extraction with ethyl alcohol or it may be distilled. In the process, the flowers are laid on frames containing the fat and as the flowers respire, they give off volatile products which are solubilized or adsorbed by the fat. When the fat becomes saturated with volatiles, it is called a pomade. When a pomade is extracted with a solvent such as ethyl alcohol, the resulting solution is known as an extrait. This may be used as such (more so in perfumes than in flavors) or the alcohol can be distilled off under vacuum to obtain the so-called absolutes (absolute of enfleurage). When this is prepared carefully, it has the advantage of good solubility in ethyl alcohol. The advantage of the enfleurage process is that for certain flowers, such as jasmine and violet, the yields of oil are greater than by extraction processes using volatile solvents. Enfleurage is an old and well-known method of extraction. However, it is currently an uneconomical process due to high labor costs. Most flower oils are currently obtained by extraction with volatile solvents to yield the concretes.

Extraction

Essential oils can be obtained by extraction with low boiling solvents. This is affected by percolation or maceration with a volatile solvent and subsequently stripping the solvent to yield the essential oil. Petroleum ether is usually used due to its low boiling point. The solvent percolates through the botanical or animal material and extracts the essential oils. When the solvent is removed, the resulting product is called a concrete. The concrete can then be extracted with alcohol, which upon removal by vaccum distillation results in a product called an absolute. Absolutes are valuable for flavor formulations since they are alcohol soluble. While absolutes are expensive, they are extremely powerful. Absolutes can even be prepared from the exhausted flowers used in the enfleurage process. While absolutes are never used in great quantities due to their cost, small quantities find use in many natural flavors. For example, jasmin absolute is used in fruit flavors, genet absolute in raisin, date, and plum, and mimosa absolute for its woody-orris-ionone taste.

Terpeneless Oils

Essential oils are composed basically of just two fractions:

1. Hydrocarbons of the general formula $(C_5H_8)n$, and
2. Oxygenated components of these hydrocarbons (e.g., citral, geraniol, and linalool)

The hydrocarbons can be further subdivided, depending on the value of the letter n. For example:

- n = 1, (C_5H_8) is isoprene (2-Methyl-1,3-butadiene). This is a hemi-terpene and is not a constituent, as such, of essential oils. It was thought for many years that isoprene was the precursor for all terpenes.
- n = 2, $(C_{10}H_{16})$ are called the terpenes. Terpenes can be divided into acyclic (aliphatic) and cyclic terpenes. Examples of monocyclic terpenes include limonene, terpinene, and phellandrene. Examples of bicyclic terpenes include pinene and camphene.
- n = 3 $(C_{15}H_{24})$ are the sesquiterpenes. These can also be subdivided into the acyclic and cyclic compounds. For example, farnescene is acyclic, bisabolene and zingerberene are monocyclic, cadinene and caryophylene are bicyclic, cedrene and copaene are tricyclic.

Under the oxygenated derivatives we find all the functional groups represented:

- Alcohols—linalool (oil bois de rose) and menthol (oil peppermint)
- Aldehydes—citral (oil lemongrass), citronellal (oil citronella), cinnamic aldehyde (oil cinnamon), and benzaldehyde (oil bitter almond)
- Esters—linalyl acetate (oil bergamot) and geranyl tiglate (oil geranium)
- Ethers—anethol (oil anise)
- Ketones—d-carvone (oil caraway) and l-carvone (oil peppermint)
- Phenols—eugenol (oil clove) and thymol (oil thyme)

The outstanding difference between the hydrocarbons and their oxygenated derivatives is that the terpenes and sesquiterpenes possess two very characteristic properties: they have very poor solubility in water and dilute alcoholic solutions, and they have a propensity to oxidize readily. Being unsaturated, they tend to polymerize (resinify) and when this occurs, a concomitant deterioration in the taste and odor quality of the essential oils is noted. It is important to note that the hydrocarbons have little flavor value per se (it is not true to say that they have none). However, when compared to the oxygenated components, their flavor contribution is indeed small. Some flavor chemists like to think of the terpenes as the diluent or solvent for the oxygenated derivatives in essential oils. Still others feel the hydrocarbons are the naturally occurring fixations (stabilizers) for the oxygenated compounds. Certainly, many show structural similarities to compounds which have antioxidant activity.

As can be seen by the compounds comprising the oxygenated fraction, it is quite obvious that its function is to impart flavor and aroma to the essential oils.

With this as a background, it is perhaps appropriate to look at an essential oil in skeleton form. It is composed of four principle fractions:

1. Terpenes (general formula $C_{10}H_{16}$)
2. The oxygenated derivatives
3. Sesquiterpenes (general formula $C_{15}H_{24}$)
4. Nonvolatile residue*

If the terpene and nonvolatile residue fractions are removed, the remaining fraction is called a terpeneless (terpene free) oil. This material still contains the oxygenated fraction as well as the sesquiterpene fraction. If this oil is now treated to also remove the sesquiterpene fraction, it is termed a sesquiterpeneless (sesquiterpene free) oil. With terpenes, sesquiterpenes, and the nonvolatile residue removed, the remaining fraction containing only the oxygenated fraction is now called a sesquiterpeneless oil. In specifying an essential oil, the flavorist recognizes:

- Terpeneless oils—terpenes and nonvolatile residues are removed.
- Sesquiterpeneless oils—terpenes, nonvolatile residues, and sesquiterpene fractions are removed.

It is quite important in terms of quality and value to recognize that sesquiterpeneless oil means none of the terpenes or sesquiterpenes remain, while the term terpeneless does not. Consequently, the term sesquiterpeneless usually implies a higher degree of concentration; the only time it does not is for oils which are relatively low in sesquiterpenes, as in the case of oil of peppermint, where the highest concentration is called simply terpeneless (see Figure 3).

How are terpeneless oils prepared? There are basically three methods of production:

Distillation under Vacuum — (fractional distillation)

Any of the four fractions can be separated from an essential oil utilizing the differences in their boiling point ranges. The lowest fraction are the terpenes (boiling point range 55 to 180°C), followed by the oxygenates (boiling point range approximately 190 to 250°C), the sesquiterpene fraction (boiling range 260 to 300°), and lastly, the nonvolatile residue fraction consisting of waxes and very high boiling sesquiterpenes and oxygenated hydrocarbons.

* It is composed of waxes and higher boiling point sesquiterpenes and oxygenates.

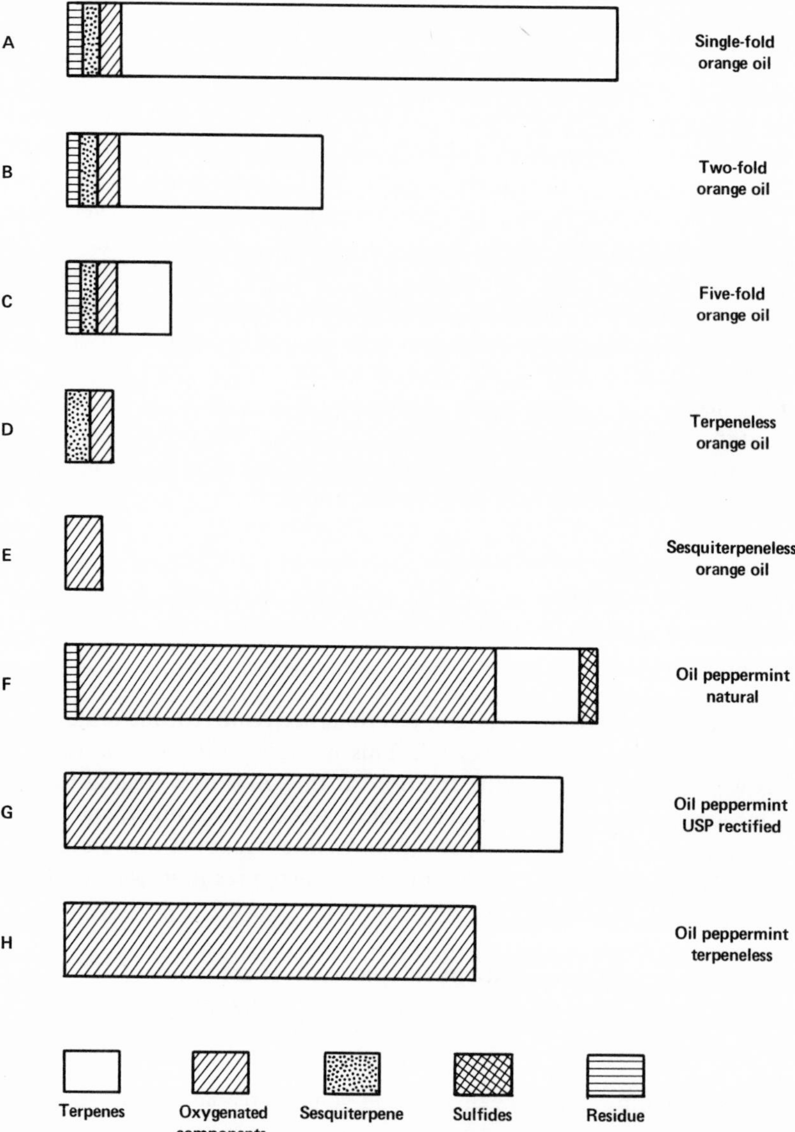

FIGURE 3. The various differences in composition of various concentrated and terpeneless oils.

Even though lower temperatures are possible under vacuum distillation, the oils are exposed to a considerable quantity of heat. Even at 5 mm Hg, the oxygenated components are exposed to fairly high temperatures, e.g., alpha-terpineol bp 92°C, phenylethyl alcohol bp 90°C, and eugenol bp 110°C. There is also the chance of localized over-heating, which imparts to the oil a still or burnt note. The separations are in no way sharp or complete. Within the terpene fractions are found some oxygenates and within the oxygenated fraction some sesquiterpenes are recovered. This can be minimized by redistilling each fraction or otherwise chemically removing the oxygenates from the terpenes. In the distillation of lemon oil, the terpene fraction comes over first and is composed mostly of limonene bp 175°C at 760 mm Hg (37°C at 2 mm Hg). Citral, a component of the oxygenated fraction, boils at 229°C at 760 mm Hg (110°C at 2 mm Hg). One would suspect a rather clean fractionation; yet in practice, the terpene fraction shows an aldehyde content. Also, at times the oil contains a considerable quantity of nonvolatile waxes which make the complete removal of the terpenes impossible by distillation alone.

Due to the problems inherent in fractional distillation, quite often steam distillation is used to prepare terpeneless oils. The advantages of steam distillation are that the boiling point of the aromatics never goes higher than 100°C and localized heating is avoided since water conducts the heat away from the still walls. The disadvantages of this method are that it takes considerably longer to complete the distillation (some four times longer) and, as mentioned previously, the problem of losses with partially water soluble constituents such as phenylethyl alcohol, geraniol, and eugenol. Such constituents can be recovered, but as always, by using extra processing steps at added cost. The hydrolysis of esters is still an inherent problem with this process. The esters are hydrolyzed to their corresponding alcohols and acids. Steam distillation under vacuum is also employed; it has the advantage of being faster than steam distillation under atmospheric conditions.

Alcohol Solvent Process

In this method, dilute alcoholic solutions of the essential oils are used. As the oils distill, the various fractions co-distill with the alcohol/water vapors. As the vapors condense, the terpenes separate since they are immiscible in dilute alcoholic solutions. The oxygenates remain in solution. After the two phases are separated, the alcohol phase is redistilled and the terpeneless oil recovered. The only disadvantages that this method has is that the repeated distillations are time consuming and costly. Also, some alcohol and water soluble components are lost.

Solvent Extraction

Dissolving the essential oil in 95% ethyl alcohol and then adding water to reduce the alcohol concentration to 60% by volume will cause the terpenes to separate. By separating the layers, chilling, and filtering, a terpeneless oil is obtained. This may be recognized by some as the washed citrus oil process. While the process has proven operable, it is rather expensive.

Another solvent extraction method depends on the fact that the hydrocarbons (terpenes) are nonpolar and tend to dissolve in nonpolar solvents, while polar oxygenated compounds show appreciable solubility in polar solvents. By treating an essential oil with a mixture of polar and nonpolar solvents (e.g., methyl alcohol and hexane), the polar solvent selectively extracts the oxygenates, while the nonpolar will extract the terpenes. The solutions are then distilled, permitting the recovery of terpene and oxygenate fractions. The most significant advantage of this method is that the temperatures employed are quite low (30 to 40°C). One difficulty encountered with this method is the formation of emulsions, which at times prove difficult to break.

What are the advantages of terpeneless oils?

1. They are more stable toward oxidation. By removing the oxygen-sensitive terpenes, the resulting oils prove generally more stable to storage and in some uses.
2. Terpeneless oils have good alcohol/water solubility important in formulations. For example, single-fold lemon oil is soluble 1:150 in 70% alcohol while the terpeneless oil shows a solubility of 1:35-50 parts in 70% alcohol and 1:2 parts in 80% alcohol. The sesquiterpeneless oil shows a solubility of 1:2-2 parts in 70% alcohol.
3. Clear beverages and gelatin desserts can be flavored by using terpeneless oils. The

cloudiness of pulp-free citrus oils is caused by the presence of terpenes. If the terpenes are removed, the oil is clear.

4. Finally, terpeneless oils are more concentrated flavor products than the corresponding oil with terpenes. This can also be a disadvantage since care must be exercised in their use in foods.

What are the disadvantages of terpeneless oils?

1. Some heat processing must be used in the manufacture of terpeneless oils and the characteristic freshness of single-fold oils is reduced.
2. The fixatives usually found in the nonvolatile fraction do not co-distill and are not present in the terpeneless oils.

Perhaps the largest use of terpeneless oils is in carbonated and still beverages. When the FDA severely restricted the use of brominated vegetable oils (BVO) in flavors, it presented a severe problem to the beverage industry. BVO was used to regulate the density of oils with that of the finished beverage. When its use levels were severely restricted, the only way in which stable beverages could be flavored was to utilize the more soluble, concentrated (folded) terpeneless oils.

Terpeneless and Concentrated Essential Oils

Often, differences in the gross chemical composition of various concentrated and terpeneless oils, particularly citrus oils, become difficult for the nonflavorist to commit to memory. Figure 3 is offered as a guide and illustrates the various differences in composition. The constitutents represented in Figure 3 are not drawn to scale relative to their actual percentage.

- "A" represents a whole single-fold orange oil. Note that almost all of the oil consists of terpenes, with small percentages of oxygenated constituents, sesquiterpenes, and nonvolatile residual materials such as waxes and substituted Coumarins.
- "B" represent a twofold oil. To prepare a twofold oil, 50% of the total oil, which consists of terpenes, is removed as a unit, leaving behind 50%, which is the twofold product. Notice that relative to the total composition, the oxygenated components and the other materials now amount to twice that of a single-fold oil.
- "C" represents a fivefold oil, in which 80% of the whole oil has been removed, leaving 20% as the product. In this case, the relative composition now contains five times the oxygenated components and other materials, as compared to the single-fold oil. In removing the terpenes, a small amount of oxygenated components tend to co-distill. Hence, a fivefold citrus oil will not necessarily have exactly five times the aldehyde content of a single-fold oil.
- "D" represents a terpeneless oil in which case all the terpenes have been removed. This is produced in a fractionation to a specific boiling point and not by removing a given percentage of the oil. In addition, the terpeneless fraction is further fractionated as a unit and results in the elimination of a nonvolatile residue as a by-product.
- "E" represents sesquiterpeneless oil. It is more highly concentrated than a terpeneless product. Since the sesquiterpenes and oxygenated components have very close boiling points, these must be separated very carefully using more sophisticated processes.

Examples "A" through "E" in Figure 3 can be applied to any citrus oil.

- "F" represents a whole peppermint oil. Note that it is almost entirely composed of oxygenated components. Consequently, the production of a "folded" peppermint oil should not be attempted since, for example, in preparing a twofold by removing 50% of the oil, it would also remove a significant portion of oxygenated components and result in the loss of valuable constituents.
- "G" represents a rectified peppermint oil. In this instance, rectification removes the sulfide fraction and the nonvolatile residue. The sulfides as a fore-run are removed in the distillation while the residue is the remnant after the main body of the oil has been distilled.
- "H" represents a terpeneless oil. All of the terpenes have been removed and the oxygenated portion of the oil is distilled, leaving the nonvolatile residue as a by-product. Example "H" can be applied to most other types of essential oils including clove, spearmint, petitgrain, and bay.

Proper Storage of Essential Oils

All essential oils keep best in dark-brown glass containers when long storage histories are contemplated. If the glass container is filled to eliminate excessive head space, tightly stoppered, and kept in a cool place (68°F) protected from light, the essential oil can be maintained indefinitely without detrimental effects. For example, a well-packaged 50-year-old oil lime expressed and oil coriander were examined and found to be completely satisfactory.

Another universally accepted type of packaging for essential oils is in stainless steel containers. Essential oils packed in full, tightly closed stainless steel containers and stored in a cool place (68°F) can be maintained indefinitely.

Aluminum containers are suitable for essential oils provided long storage is not contemplated. Aluminum containers usually acquire a thin oxide coating which prevents reaction of the oil or its components with aluminum metal. If the oxide coating has not been damaged, the oils can be favorably maintained for a considerable period of time. However, because of the softness of ordinary aluminum containers, the continuity of oxide coating is often disrupted and can result in metallic interaction with attendant side effects.

Plastic containers are generally not suitable for either packing or storage of essential oils and related materials. In most instances, the essential oil will bleed through the plastic or act as a solvent to extract the plasticizing material. This will result in softening of the container and eventual collapse. In addition, most plastic materials are not impervious to the transfer of air and the essential oil could be subject to oxidation. As a rule, polyethylene or polyvinyl chloride containers are not satisfactory even for short storage periods.

The most suitable containers for essential oils and related materials are described above, but for reasons of economy and shipping, these need not necessarily be required. The following packaging information was developed in our laboratories* over a 2-year period and the polymeric packaging materials described were found suitable for all essential oils and related materials, provided excessive storage is not contemplated.

With the availability of new types of coating materials, the effects on the quality of essential oils packed in containers lined with phenolic or epoxy-type coatings were studied over a 2-year period. The studies made use of steel drums which had a double coating to avoid the possibility of pin holes, scratches, etc. Periodically, the containers were observed to determine whether the linings were affected, that is, loosened, dissolved or discolored by the essential oil. The essential oils were compared to samples

* Fritzsche, Dodge and Olcott, Inc.

stored in glass to determine whether they become contaminated due to elution of the constituents from the lining in which case the oils would show a color change and/or a degradation of odor and flavor.

As a result of this thorough study, we have found that all essential oils including the citrus, mint, phenolic, and other types are satisfactorily packaged in phenolic-coated drums. Epoxy coatings also proved acceptable, but since these are less flexible than phenolic coatings, it was decided to use the latter since the brittleness of the epoxy liner may result in chips accumulating if the drum is subject to impact. In most cases, phenolic-lined drums are suitable for storing oils a minimum of 9 months without any evident detrimental effects. This is usually a sufficient period of time to permit shipment and use of the material without requiring transfer to another container by the user. If storage periods much greater than 9 months are contemplated, a tin-lined or galvanized drum should be specified. However, storage periods in excess of 9 months can often be permitted without concern. This is especially true of citrus, with which we have had the longest experience. Citrus oils stored in phenolic and epoxy-lined containers under prescribed conditions have been found suitable for periods of approximately 2 years. With oils other than citrus, it is recommended that if storage periods longer than 9 months are desired, that appropriate tests be conducted.

In all cases, the containers should be full and kept in a cool place (68°F); these conditions are standard for any type of packing. Full containers are a necessity since the following effects could otherwise occur:

Phenolic oils—Clove, origanum, and thyme oils will darken due to oxidation, which in some cases can require redistillation if an absolutely colorless oil is required.

Mint oils—Spearmint oil is somewhat more prone to oxidation than peppermint oil because laevo carvone (the major constituent) tends to oxidize and form carvacrol. Carvacrol has a detrimental effect on the flavor character even in low concentration. Peppermint oil will tend to mellow with age, but unfortunately it will discolor to a bluish or greenish shade due to the oxidation of menthofuran, which may be present at levels of 3 to 15% depending on the type of peppermint oil.

Aldehydic oils—Oils which contain aldehydes as the principal constituent (e.g., oil bitter almond FFPA and oil cassia) can oxidize and result in high acid values. In such cases, crystalline material may separate from the oil and a reduction in the concentration of the main flavoring constituent will take place. For example, benzaldehyde may be converted to benzoic acid.

Although all essential oils are suitably packaged in phenolic-coated containers, black-iron drums can still be used for some oils, such as eucalyptus. Isolates from essential oils, such as linalool or eugenol, various aromatic chemicals and mixtures, such as imitation essential oils, perfume compounds and most flavoring compositions, are also suitably packaged in phenolic-coated drums provided the proper storage conditions referred to previously are met.

Oleoresins

Oleoresins are prepared by percolating a volatile solvent through a ground spice or herb. Solvents commonly employed include ethyl, methyl and isopropyl alcohol, chlorinated hydrocarbons such as methylene and ethylene chloride, hexane and acetone. The spice or herb is prepared for extraction by grinding; this ruptures flavor-bearing cells to facilitate extraction. As the solvent percolates through a bed of ground spice or herb, it removes not only the flavor constituents (mainly the essential oil), but also resins, gums, and other materials that are miscible in the solvent. The solvent is then stripped for reuse, leaving behind a somewhat viscous material now called an oleoresin.

In the U.S., the amount of residual solvent remaining in the oleoresin is regulated by law. The regulations covering solvent residues are presented in Appendix 1, since the issue is currently of general concern to the food industry for safety reasons.

Oleoresins differ from essential oils in that oleoresins contain many of high boiling and nonvolatile constituents native to the spice or herb. These have been removed from the corresponding essential oil. Certain spices are extracted to produce oleoresins, which are not primarily intended for their flavor value, but rather for the intense coloring matter they contain. Such is the case with the oleoresins paprika and turmeric. These are used in the manufacture of salad dressings and for coloring pickle products. Although oleoresins are normally thick, viscous, and sometimes highly colored products, they generally impart less color to the finished product than the corresponding spices because they are used in such small quantities. The solubility characteristics of oleoresins renders them quite useful in various food applications (Table 3).

The flavor of oleoresins may not be equivalent to their corresponding essential oils. For example, the essential oils of ginger and black pepper have none of the bite of their corresponding crude spices or oleoresins, but do have a higher concentration of the aromatic properties. In general, oleoresins offer the flavorist an opportunity to use the advantages of spices without some of their disadvantages. The oleoresins are more uniform, more applicable and more potent than the dry spice (Table 1). The normal use range in food of oleoresin is 1/5 to 1/20 of their corresponding dry spice. Also, oleoresins are especially suited to high-temperature use applications such as baking and frying. The resins and fatty oils contained within oleoresins act as natural fixatives for volatile essential oils. Solvent extraction removes from spices and herbs not only the essential oils which are primarily responsible for its flavor, but other nonvolatile constituents. These will be retained in the food product undergoing heat processing and at the same time tend to fix or stabilize the more volatile essential oil component. This fixation prevents or retards evaporation or steam distillation of the volatile components during heat processing.

The following are some key advantages of using oleoresins:

- Uniformity of flavor—The flavor quality, strength, and color of spices and herbs tend to vary from year to year. The use of oleoresins minimize such variations.
- Stability—Spices and herbs tend to lose their volatile oils through polymerization, oxidation, and evaporation. With oleoresins, higher content of nonvolatile components, including antioxidants, retards aroma losses.
- Storage—The equivalent flavor value of an oleoresin to the dry spice is small, i.e., a much smaller quantity of oleoresin is needed to replace an equivalent quantity of dry spice. Smaller quantities lessens the storage space needed. Also, oleoresins are stored in metal containers, thereby lessening the chances of contamination compared to stored burlap bags.
- Microbiological—Though elaborate sterilization techniques can make spices and herbs essentially bacteria free, it cannot eliminate microbes completely. Oleoresins, because of the nature of their production, are free from bacterial molds and fungi. Also, oleoresins do not support microbial growth.
- Economy—Spices and herbs, no matter how finely ground, always leave some flavor components entrapped within the cells. This flavor is unavailable and as such is lost. Oleoresins give greater flavor value per unit of spice compared to dry spices.

Since every advantage noted for oleoresins as compared to dry spice would be the same for essential oils vs. dry spice, what advantages do oleoresins have over essential

Table 3
SUBJECT: SPICE OLEORESIN SOLUBILITY CHART

Oleoresins	Alcohol 190 proof	Propylene glycol USP	Vegetable oil	Oil dill	Oil cassia	Oil clove	Approximate % volatile oil	Color
Capsicum NF	Not complete	Not soluble	Soluble	Soluble	Soluble	Soluble		Dark red, heavy liquid
Capsicum NF high pungency	Not complete	Not soluble	Soluble	Soluble	Soluble	Soluble		Reddish, heavy liquid
Celery fivefold	Partly soluble	Slightly soluble	Soluble	Soluble	Soluble	Soluble	14.5	Light green, liquid
Cubeb NF	Soluble, hazy	Not soluble	Very hazy	Soluble	Soluble, slightly hazy	Soluble	77.0	Green liquid
Ginger NF acetone extraction	Partly soluble	Partly soluble	Insoluble	Soluble	Soluble	Soluble	37.0	Brown liquid
Ginger NF Alcoholic extraction	Partly soluble	Partly soluble	Insoluble	Soluble	Soluble	Soluble	37.0	Brown liquid
Ginger NF alcoholic extraction	Partly soluble	Partly soluble	Insoluble	Soluble	Soluble	Soluble	33.0	Brown liquid
Jamaica Mace	Partly sediment	Partly soluble	Very hazy	Soluble	Soluble, slight sediment	Soluble	31.0	Brownish, viscous liquid, green
Paprika	Not complete	Not soluble	Soluble	Soluble	Soluble	Soluble		Dark red
Parsley seed	Not soluble	Not soluble	Insoluble	Insoluble	Insoluble	Insoluble	6.2	Bright green paste

Black pepper decolorized	Not complete, Considerable sediment	Soluble, small sediment	Not complete	Soluble	Soluble	Soluble	29.0	Yellow semisolid
Pimenta berries	Partly soluble	Slightly soluble	Soluble, small sediment	Soluble, slight sediment	Soluble	Soluble	45.0	Dark green
Sage	Not complete, Considerable sediment	Partly soluble, high sediment	Soluble, slight sediment	Soluble, slight sediment	Soluble	Soluble	34.0	Brownish green
Turmeric	Soluble, slight haze	Partly soluble	Insoluble	Insoluble	Partly soluble	Soluble		Dark red

[a] The above dilutions are 1 part Oleoresin in 9 parts diluent by volume.

oil? Oleoresins display greater stability toward high-heat applications and have flavor characteristics more like the natural dry spice than for the corresponding essential oil.

Isolates

Isolates, as the name suggests, are specific materials fractionated from natural flavor substances. In one sense, essential oils, oleoresins, tinctures, and solid extracts obtained from botanical materials are isolates. Similarly, starter distillate being distilled from a butter starter culture can also be considered an isolate. However, isolates usually refer to aromatic chemicals derived from natural products. Generally, the isolates are obtained from the essential oils since the essential oil is a concentrated product compared to the botanical and would be rich in the material. The most frequently used methods for obtaining isolates from essential oils are distillation, crystallization, and extraction.

When a naturally occurring flavor material is not prepared synthetically, it is usually because it is easier and more economical to isolate it from the natural source. Some synthetic routes are just too costly. Compounds such as phellandrene, santalol, and vetiverol are not produced commercially via synthesis, but are isolated from their naturally occurring sources. However, some distilled isolates, such as citral, geraniol, and linalool, are now produced via chemical synthesis on an extensive scale. Most isolates are fractionally distilled from the essential oils and, as such, always have small amounts of impurities associated with them. This can sometimes be a problem.

Chemical isolation is another method of obtaining isolates. For example, citral can be obtained by distillation or chemical isolation. Citral is found in oil lemongrass (70 to 80%) and can be chemically removed as the solid sodium bisulfite addition complex. After the solid addition complex is removed, it is decomposed with alkali, and the citral recovered by distillation. This material, although chemically identical to the synthetic material, does not have the same odor. This is attributed to that small amount of naturally occurring impurities that is carried through with the isolate (e.g., methyl heptenone). Citronellal can also be isolated by a process similar to citral.

Many isolates are obtained by freezing or crystallization. The largest and most common compound isolated in this fashion is l-menthol, formed in oil peppermint (*Mentha piperita*) and in Brazilian or Japanese mint oil (*Mentha arvensis*). By simply cooling these oils to a low temperature, the menthol crystallizes. Until recently, the production of menthol was dependent on the availability of the naturally occurring raw material. However, l-menthol has been made synthetically and will soon be available commercially from a synthetic process. Here again is another case where the natural isolate, when compared to the chemically identical synthetic material, has been characterized as being too clean. It is the minute impurities that come along with the natural that account for the differences. Other isolates that were obtained by freezing are camphor, thymol, and 1,8-cineol; the latter is still produced in this manner.

Isolates, aside from their use directly as perfume or flavor materials, are used to manufacture other more valuable materials, e.g., vanillin was first commercially prepared from eugenol isolated from clove oil. Now vanillin is produced more economically from the waste liquors of the sulfite process employed in the paper industry (the so-called vanillin from lignin). In another example, the production of ionone, once isolated from a naturally occurring source (boronia absolute) is now produced from the condensation of citral and acetone.

Since the production of isolates is strictly dependent on the availability of the parent essential oil or other natural raw materials, industry has moved more towards their production via synthesis. However, during the height of the petrochemical shortage in 1973, many aromatic chemicals (e.g., synthetic citral and linalool) were in short supply.

At that point, producers began isolating linalool from bois de rose oil to meet their needs. As soon as the petrochemical shortage ended, they immediately returned to production of the synthetic materials. However, aside from economic factors, there is one outstanding reason why isolates will continue to survive as viable sources of aromatic ingredients. The March 15, 1977 issue of the *Federal Register* defined a natural flavor as:

The term "natural flavor" or "natural flavoring" means the essential oil, oleoresin, essence or extractive, protein hydrolysate, distillate. Natural flavors include the natural essences or extractives obtained from plants listed in Par. 182.10, 182.20, 182.30, of this section and the substances listed in Par. 172.510 (see Appendices 2 and 3).

From this definition, materials which are extracted, distilled, or crystallized from natural products are considered natural. They are natural in that the flavoring substances are isolated directly from natural sources and are not chemically synthesized. As such, they will find use in natural flavor formulations.

Fruit Juices and Derivatives

Fruit juices have been available for use by the flavorist for many years, but these are generally unsatisfactory to be used as such in many food products. The largest usage of fruit juices is for the citrus varieties (orange, lemon, and lime) by the beverage industry; other juices are used for preserves and alcoholic beverages. Juices are relatively weak in flavor value as compared to most other flavor materials. They also suffer from significantly reduced shelf life. However, they serve as starting materials for a large and important number of derivatives (Figure 4).

Fruit juices contain a rather large percentage of water (80 to 90%). Dissolved or dispersed within the juices are characteristic flavor substances, e.g., the aroma or flavor, acids, color, and other solids such as sugar, pectic substances, and minerals. The actual flavor principles in the juice are found in very small amounts. To produce fruit juice concentrates, some of the water must be evaporated to reduce the volume of the juice and thereby concentrating the flavor extractives. The usual way of concentrating juice is to distill under a vacuum. Although the temperature of the juice during distillation does not exceed 40°C, this is sufficient to affect flavor quality. As the juice is concentrated, the volatile flavor components tend to distill off with the water. The first 10% of the distillate appears to contain most of the volatiles. While evaporation of the water from the juice yields concentrated juice containing soluble substances and extractive matter, most of the volatile substances are lost with the distillate. However, the distillate is further condensed by vacuum distillation, a great deal of the volatiles are recovered, and can then be added back to the concentrated juice. This notion of condensing the distillate from juice concentration operations is known as "essence recovery". This method is employed by most large fruit juice producers in the U.S.

"True fruit extracts", as the name implies, are prepared by extracting and concentrating fresh, frozen, or dried fruits. The flavor contribution of true fruit extracts is directly dependent on the percentage of fruit per gallon, the flavor quality of the starting fruit, and the extent and success of aroma or essence recovery. A method of preparing true fruit extracts is by macerating the fruit in alcohol, stripping off the alcohol and water, and condensing the distillate. After the distillation, the stripped juice is pressed to yield the fruit juice. The juice is depectinized using pectolytic enzymes, filtered, and concentrated. Depectinization is a very important aspect of the process in order to produce extracts slated for use in the spirit industry. (Concentrated fruit juices containing high levels of pectin are unsuitable for alcoholic fruit-flavored beverages since this will give a cloudy appearance to the beverage.) This method works

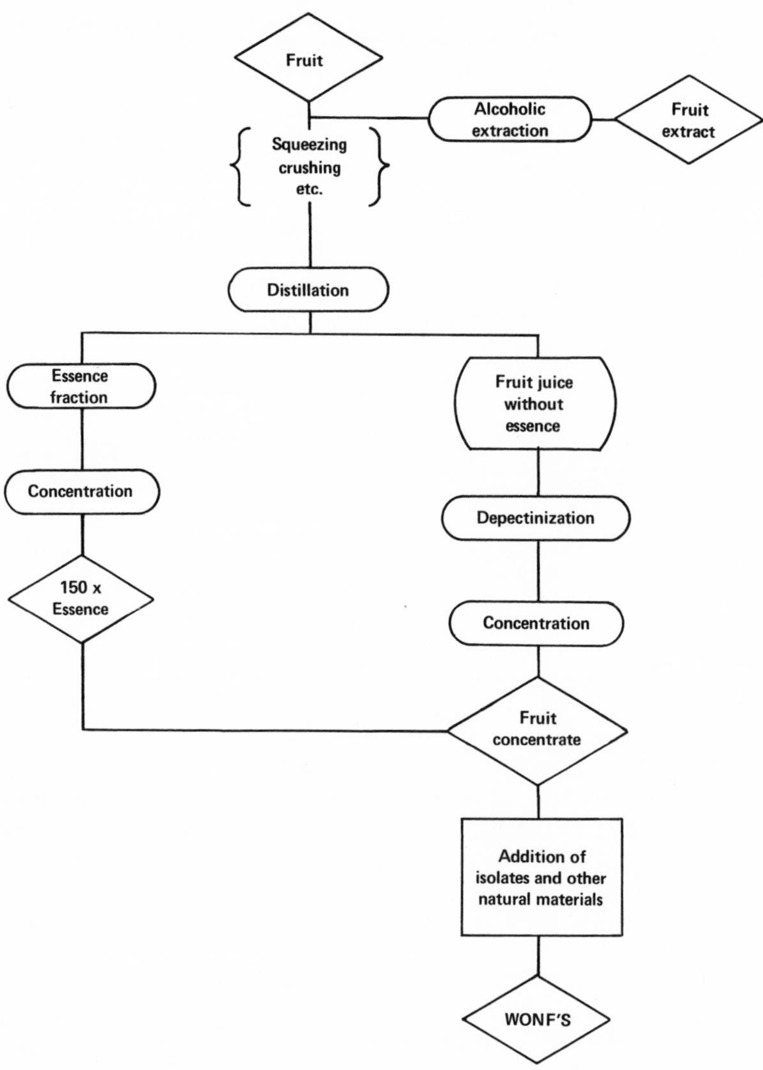

FIGURE 4. Flow chart for fruit juice and derivatives.

well and if the distillation is carried out with alcohol, results in essences of excellent quality. Finished fruit extracts, then, are blends of the aroma distillate and concentrated fruit juices.

Essence recovery systems are based on the principle that the flavor of fruit juices can easily be stripped by flashing off a certain percentage of the juice. This flashing process is accomplished rapidly by injection of high pressure steam directly into the juice in flash chambers; the vapors obtained after flashing are separated from the juice. The juice is then rapidly cooled to room temperature, depectinized, filtered, and concentrated under a high vacuum. The flavor-bearing vapors are diverted to a fractionating column where the aroma or flavor portion is separated from the water portion. The concentrated flavor vapors rise to the top of the column while the water vapors remain at the bottom of the column. The aroma vapors are collected and are called fruit essences. They are usually labeled with the name of the fruit and the fold (multiple) concentrate; e.g., "Grape Essence 100 Fold". The 100- or 150-fold means that each volume of the essence carries the flavor of 100 or 150 volumes of the original juice. Thus, one volume of 100-fold grape essence contains the flavor from 100 volumes of grape juice. The concentrated essences are then added back to the concentrated juice to produce finished extracts.

Generally fruit extracts and concentrations cannot be used in hard-candy flavor formulations, or any other application involving high-temperature processing. High temperatures tend to caramelize the sugars in the juice, causing undesirable color and off-flavors. In addition, acids in juice can invert the sugar used in hard-candy production. In general, fruit extracts can be used where low temperatures are involved and in those flavor applications where large quantities of extract can be tolerated (e.g., beverage).

To overcome the weak flavor strength of fruit extracts and fruit concentrates, the flavor industry has developed the so-called WONF*-reinforced extracts. A WONF extract is a fruit extract or other natural flavor containing both a characterizing flavor from the product whose flavor is being simulated (e.g., lemon, lime, etc.) and natural flavors from other sources which simulate, resemble, or reinforce the characterizing flavor.** It should be noted that in the U.S., these "other natural flavors must conform to 21 CFR Sec. 101.22 Food Labeling; spices, flavoring, colorings and chemical preservatives, (a)......(3) which defines natural flavors explicitly (see Appendices 2 and 3)." A raspberry WONF (a natural flavor with other characterizing natural flavors) could contain concentrated or single-fold raspberry juice or raspberry extract and any other juice that either simulates, resembles, or reinforces the raspberry flavor. In this case, it could contain some strawberry juice, tincture of orris root, or absolute jasmine. The orris root would contribute and strengthen the ionone note also found in the raspberry juice while jasmine contributes a reinforcing fruity effect. The starting point for most WONFs is a fruit concentrate which is then reinforced with various essential oils, oleoresins, botanical extracts, and other fruit extracts, juices, or juice concentrates. The overall effect is that WONFs are several times stronger than their corresponding fruit extracts and concentrates. Although the flavor strength is improved, heat sensitivity is still present. Another problem often encountered, i.e., as attempts are made to increase the flavor strength, care must be expressed not to lessen the natural quality of the flavor.

Fruit extracts, concentrated juices, and WONF flavors are extensively used in non-carbonated beverages, in specialty wines, cordials, ice cream, and dairy products.

Botanicals and Animal Extracts

There are many botanical (and animal) extracts other than herbs and spices which flavorists put to good use. The flavorist has many occasions to use materials such as extracts, tinctures, fluid and solid extracts of such diverse items as coffee, cocoa, aloe, gentian, guassia, castoreum, civet, licorice, vanilla, fenugreek, horehound, arnica, and Saint John's Bread, and these certainly do not complete the list. Botanical and animal extracts appear on the market in several forms. For example, coffee and cocoa extracts are supplied as water extracts, alcohol/water extracts, or in special cases, as the alcoholic distillates. The usual forms, however, are either tinctures, fluid extracts, or solid extracts. A flow chart detailing the various extraction methods is shown in Figure 5.

Tinctures are alcoholic or hydro/alcoholic solutions prepared from animal or vegetable materials, and sometimes from chemical substances. Most tinctures have N.F. or U.S.P. specifications to which they must conform. "Official tinctures" should contain the extractive matter of 10 to 20 g of the material per 100 mℓ of the tincture. Tinctures are rather easy to prepare. For example, most official tinctures are made by

* WONF is an acronym for With Other Natural Flavors.

** *Federal Register*, Vol. 42, No. 50, pg. 14315—17, March 15, 1977.

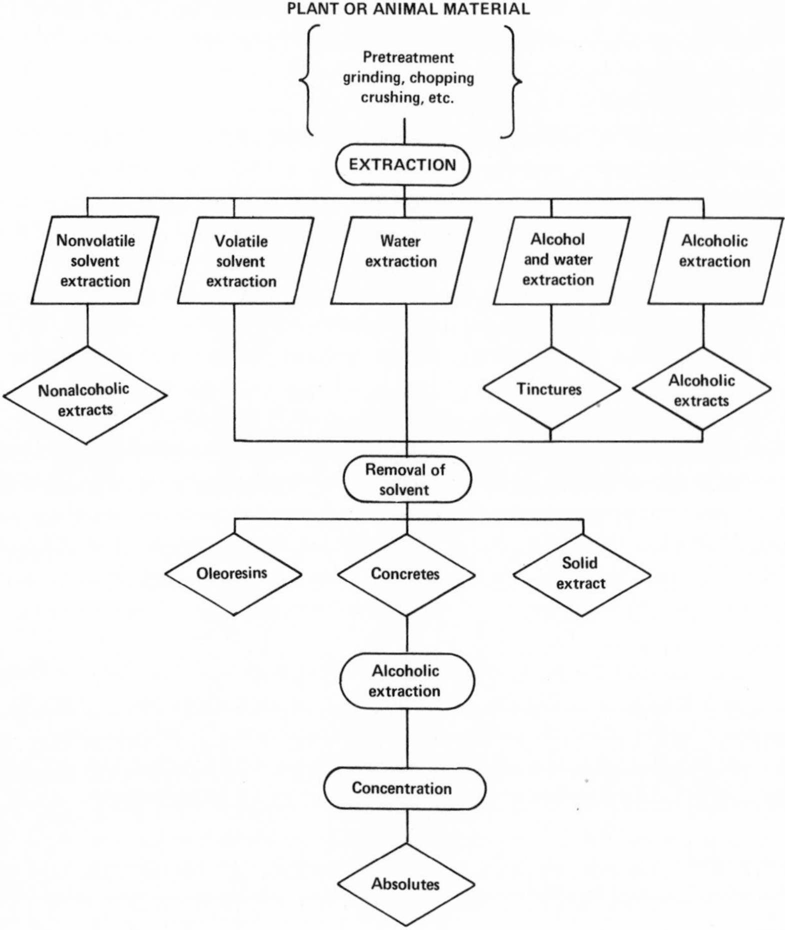

FIGURE 5. Flow chart for extraction methods.

simple percolation. However, many flavor supply houses make their own via maceration, decoction, or infusion. Also, they often use solvents other than alcohol.

Fluid or solid extracts are concentrated preparations of animal or vegetable material obtained by extracting the active constituents with a hydro/alcoholic solvent, evaporating all (or nearly all) of the solvents, and adjusting the residual solids to the prescribed standards. Most solid extracts, as in the case of tinctures, are extracted by percolation and are concentrated by distillation under reduced pressure. As readily seen, the process for making a solid extract is very similar to that for preparing an oleoresin with the only difference being that for solid extracts, alcohol and water are employed. To the best of our knowledge, oleoresins are prepared using more volatile solvents. Since solid extracts are nothing more than concentrated tinctures, it follows that tinctures and fluid extracts can be made from solid extracts by diluting with alcohol and water to definite volumes. However, whenever this approach is taken, the tinctures and fluid extracts do not always retain the same aroma top notes as when they were prepared directly from the fresh botanical or animal tissue. Tinctures, fluid extracts, and solid extracts find many uses today in the area of natural flavors, especially with natural flavors that contain alcohol and water, such as the WONFs. For example, the tinctures of castoreum, civet, and ambergris, although originally used for perfumes, are used in flavors. Castoreum adds unusual notes to raspberry and strawberry while civet adds an interesting effect to cheese and alcoholic beverage flavors. Both civet and ambergris have an ability to meld (blend) various flavor ingredients in a manner unmatched by other materials.

No discussion of botanical flavor extracts would be complete without detailing the most important one of all, vanilla. Vanilla is one extract for which the U.S. Federal Government has issued standards of identity.* These specifications give the requirements for vanilla extract, concentrated vanilla extract, vanilla flavor, concentrated vanilla flavoring, vanilla-vanillin extract, vanilla-vanillin flavor, vanilla powder, and vanilla-vanillin powder (see Appendix 4). By definition, "a vanilla extract is the solution in aqueous ethyl alcohol of the sapid and odorous principles extracted from vanilla beans." It must contain no less than 35% by volume alcohol and must contain 13.35 oz vanilla beans per fold per gallon of extract. These regulations specify the labeling, preparation, and optional ingredients allowable in each product.

Vanilla extracts and oleoresins find broad usage in flavors. Major uses include ice cream, baking, and chocolate products. Most flavorists are aware that vanillin is the main flavor component (compound) in vanilla. While the vanillin content is the bean varies from about 1 to 3%, the flavor of the vanilla extract is due to much more than just vanillin. The vanilla bean contains resins, which not only contribute taste without concomitant aroma, but also gives the extract thermal stability. When the use of vanilla extract is compared to vanillin in baked goods, more of the extract appears to remain. Vanilla has the unusual property of fitting well into almost every flavor formulation. It has wide acceptance and blends so well that it is used extensively. Vanilla extract appears in the market place in various concentrated forms, twofold, tenfold, etc. Although the overall strength of vanilla extract increases with folding, the exquisite bouquet of the single-fold extract is never quite matched by folded products.

Synthetic Flavors

Unfortunately, while the consumer views the term "artificial flavor" as synonymous with synthetic flavor compounds, it does not imply that compounds or ingredients derived from naturally occurring sources are not equally artificial in the context of their addition to food. The flavorist, by tradition, has aided in the confusion and this has been spurred by various regulatory decrees in the U.S. and many other nations. The following guidelines are offered to the reader with the view that particular legal definitions require appropriate legal advice:

- In the strictest legal terms, any food product to which flavor has been added is in fact artifically flavored. It is of no consequence that a portion or all of the added flavor is of natural origin. In effect, this legal definition supported by appropriate labelling regulations implies that food is either processed without added flavor, in which case it is indeed flavored by nature, or if flavor is added, then the food product is artifically flavored. There are no halfway measures, but the food industry is quite adept at creating less than black-or-white measures. Those members of industry wishing to make a point of complete naturalness will often make label statements declaring that it does not contain artificial flavor. However, those food processors who, because of regulatory constraints or for pure marketing attractiveness, use naturally derived flavors have complied with labelling requirements via statements such as "artificially flavored with natural flavors." Those using mixtures of naturally derived flavors and synthetics go a step further and it is not

* 21 CFR 169.175-169.182

uncommon to see food products labelled as "artifically flavored with natural and synthetic flavors." The shades of gray employed in complying with labelling of food products with added flavors is becoming increasingly complex and it is more appropriate for a discussion in food law than food technology. The reader is, therefore, cautioned and advised to seek expert guidance in such matters.

- There are five general categories of flavors employed by flavorists to produce finished flavor formulations, two of which are composed of synthetic components:

1. Natural flavors and their extracts (e.g., oil lemon, oleoresin ginger, concentrated grape juice, etc.).
2. Natural flavors WONF; a natural flavor with other characterizing naturals added (e.g., oil lemon with added citral).
3. Natural flavors with noncharacterizing naturals added (e.g., natural orange flavor with pineapple juice added.)
4. Natural flavors with noncharacterizing synthetic flavors added (e.g., natural apple flavor with gamma-undecalactone added).
5. Synthetic (artificial) flavors; that is, flavors totally dependent on synthetic materials.

Therefore, artificial flavors are composed of individual flavor materials, both natural and artificial, used singly or in combination.

The term artificial flavor is a labelling requirement for all food products to which flavor has been added regardless of its origin. This is, of course, in the strictest sense of definitions. The flavor materials used in the formulations of artificial flavors fall into four categories:

1. Those naturally occurring materials such as oil clove, oil cinnamon, and oil orange.
2. Substances that are isolated from naturally occurring flavor materials, such as benzaldehyde from bitter almond oil, cinnamic aldehyde from cassia oil, eugenol from clove, citral from lemongrass, and linalyl acetate from oil bergamot.
3. Substances that are prepared synthetically, but that are chemically identical to those found in nature, i.e., the so-called nature identical aromatics.
4. Synthetic substances which as yet have not been found to occur in nature. Two common examples are allyl caproate (allyl hexanoate) and ethyl methyl phenyl glycidate (aldehyde C-16).

All of the above materials, except the naturally occurring essential oils, are single aromatic chemicals. Being aromatic chemicals, almost all of the functional groups are represented (e.g., esters, acids, alcohols, aldehydes, ketones, lactones, mercaptans, etc.). Therefore, we can further define an artificial flavor as a blend of natural and synthetic materials or a blend of synthetic materials, compounded in such a way as to simulate known flavors.

A perusal of the many publications that contain analytical data on components of natural products show that its compositions are extremely complex. Often, one character impact item predominates, as in the case of vanillin in vanilla and cinnamic aldehyde in cassia. However, there are always other components tending to modify, and in many cases enhance, the character of the major component. For example, the use of acetaldehyde in an orange flavor acts as an enhancer.

Artificial flavors generally begin with the attempt to simulate the natural flavors.

Consequently, within artificial flavors there is also seen the use of character impact items, modified by other components to give a more accurate representation of the flavor. For example, in an artificial concord grape flavor, the use of ethyl butyrate and/or ethyl 3-hydroxy butyrate significantly enhances and synergizes the concord note. In this instance the use of ethyl 3-hydroxy butyrate in concord flavor is the subject of a U.S. patent.* From perfume work, it is known that blending of primary ingredients tends to give more character to a final perfume and adds a subtlety to the finished product not evident when making perfume based on just a few aromatics. So too with flavors: the addition of small quantities of flavor materials (contributory and differential items) gives the flavor a distinctive quality.

Each portion of a flavor, as well as each component, has a part to play in the overall rendition. It is only by understanding the role played by each ingredient and the environment in which it must perform that the creation of accurate artifical flavors is made possible. Table 4 lists the vast number of synthetic flavor compounds employed in creating flavor formulations (see also Part 11). The number is enormous and consists of both nature-identical materials and wholly synthetic compositions.

Artificial flavors are composed of a flavor portion and a diluent portion; these must be balances in creating an acceptable flavor. What is meant by flavor balance? Simply the proper combination of flavor and diluent (solvent). The aroma of solvent should never predominate. The flavor (aroma) should remain characteristic over a period of time. Its composition should be dilutable and still maintain its character.

Within the flavor portion of a flavor composition, compounds having characteristic functional groups are often employed. In most artificial flavors, especially the fruits, an ester component is usually incorporated. The esters give the flavor its characteristic aroma of fruitiness. They also serve to lift the overall sensory impression. As a class, esters have good consumer acceptance. The flavor portion of a composition always contains some character impact items, that is, discrete components which in and by themselves approximate the basic flavor being simulated. These are often small amounts of powerful aromatics and give the flavor a character and a name. These materials are effective in compositions at the 1 to 10 ppm range. An example of such a material that would be included in a typical peach flavor composition would be gamma-undecalactone (Aldehyde C-14). After deciding on the character items, one then looks at the contributory notes which, in the case of peach, might be ethylmethyl phenylglycidate (Aldehyde C-16). This would contribute a fruity note that blends well with C-14. After having established the character and contributory items, one can then devote attention to the so-called differential items. These give the flavor the creator's mark. They are the notes that the flavor chemist likes, feels most comfortable with, and has had success using.

Lastly, a flavorist will attempt fixation. Fixatives aid in the overall retention of the desired flavor, holding constant the desired flavor type. Fixatives function by causing the components to evaporate from the surface at a more uniform rate and reducing chromatographic effects so readily detected by olfactory and taste receptors. A fixative need not be a very high boiling substance, but it should be a material boiling higher than the main note and definitely similar and compatible with the character items. Methyl dihydrojasmonate is a good example of a fixative useful in raspberry flavor formulations.

With most flavors, there is one aromatic or group of aromatics that helps create and intensify the taste and odor; the so-called character impact item(s). Some examples of character impact items include: ethyl-2-methyl butyrate (apple), furfuryl mercaptan

* U.S. Patent 3,427,167.

Table 4
SYNTHETIC FLAVOR INGREDIENTS

Name	Organoleptic characteristics	Regulation	Flavors useful in
Acetal (acetaldehye diethyl acetal)	Pungent, green, woody solvent odor, whisky taste	172.515	Fruit flavors, rum; whisky
Acetaldehye (ethanal)	Sharp, penetrating, choking fruity aroma, leafy green taste	182.60	All fruits for lift; especially orange, apple, butter
Acetaldehyde benzyl methoxyethyl acetal	A floral, milk, or cream odor with a creamy taste	2148	Fruit flavors; vanilla
Acetaldehyde butyl phenethyl acetal	A vegetable-green odor with a green pepper, pea taste	3125	Vegetable and fruit flavors
Acetaldehyde phenethyl prupyl acetal (Acetal R)	A pleasant clean, green odor with a green pepper taste	172.515	Vegetable and fruit flavors
Acetanisol (*p*-methoxyacetophenone)	A sweet, floral odor with a sweet taste	172.515	Vanilla, nut, tobacco, butter flavors
Acetic acid (ethanoic acid)	Pungent vinegar odor with sour taste	182.1005	Butter, cheese, grape, fruit flavors
Acetoin (acetyl methyl carbinol)	Bland, woody, yogurt odor with fatty creamy "tub" butter taste	182.60	Butter, milk, yogurt, strawberry
Acetone (dimethyl ketone)	Solvent for extractions	3326	—
Acetophenone (methyl phenyl ketone)	A strong medicinal odor with a cherry branch taste	172.515	Fruit flavors, grape, cherry, tobacco
3-Acetyl-2,5 dimethylfuran (2,5-dimethyl-3-acetylfuran)	An almond, nut, smoke odor with walnut taste	3391	Nut, cherry smoke
2-Acetyl-3,5(and 6) dimethyl-pyrazine	A hazelnut, caramel odor and taste	3327	Roasted, hazelnut caramel
2-Acetyl-3-ethylpyrazine (2-acetyl-3-ethyl-1,4 diazine)	A characteristic potato chip odor and taste	3250	Potato chips
2-Acetyl pyrazine (methyl pyrazinyl ketone)	A popcorn odor; roasted peanut, popcorn taste	3126	Peanut, popcorn roasted flavors
2-Acetylpyridine (methyl 2-pyridyl ketone)	A stale popcorn odor, peanut popcorn, roasted taste	3251	Lingering popcorn taste peanut, meat, tobacco flavors
3-Acetylpyridine (methyl 3-pyridyl ketone)	Stale popcorn, peanut, tobacco odor with popcorn taste	3424	Lingering popcorn taste, peanut, meat, tobacco flavors
2-Acetyl thiazole	A pleasant peanut, chocolate odor with nut taste	3328	Nut flavors, popcorn
Aconitic acid (citridic acid)	Very low odor if any; a winey-sour taste	182.60	Fruit flavors, rum, brandy
Adipic acid (hexadioic acid)	Odorless with a pleasant sour taste	172.515	Acidulant
beta-Alanine (2-amino propionic acid)	Odorless white solid with mildly sweet taste	182.5118	—
Allyl anthranilate	A distinct Concord grape odor with long lasting grape taste	172.515	Grape, pineapple, citrus

Allyl butyrate	Fruity apple, pineapple odor with apple taste	172.515	Apple, pineapple, peach, apricot
Allyl cinnamate	A sweet spicy odor with a fruity-spicy taste	172.515	Berry flavors, grape, peach
Allyl cyclohexaneacetate	A pleasant fruity odor with a punch-like taste	172.515	Pineapple, apricot, apple, peach, mixed fruit
Allyl cyclohexanebutyrate	A fruity odor with a distinct pineapple taste	172.515	Fruit flavors, pineapple
Allyl cyclohexanehexanoate	A fatty fruity odor with a fruity, peach taste	172.515	Peach, pineapple, apricot
Allyl cyclohexanepropionate (allyl 3-cyclohexylpropionate)	A sweet, fruity pineapple odor and taste	172.515	Pineapple, fruit blends
Allyl cyclohexanevalerate	A woody, waxy, apple, pineapple odor and taste	172.515	Pineapple, apple, banana, pear
Allyl disulfide (diallyl disulfide)	Powerful, pungent garlic odor and taste	172.515	Garlic, onion, spice
Allyl 2-ethylbutyrate	Pungent, fruity odor; sweet fruity, bubble gum taste	172.515	Chewing gum flavors; fruit
Allyl 2-furoate	A fruity, pineapple, coffee odor with a garlic taste	2030	Pineapple, caramel, low levels in meat flavors, coffee
Allyl heptanoate	A sweet, fruity, apple, banana odor and taste	2031	Apple, banana, pineapple; brandy
Allyl hexanoate (allyl caproate)	A pungent, fatty,fruity odor with pineapple taste	172.515	Pineapple, apple, fruit blends, tutti-frutti
Allyl alpha-ionone	A woody, fruity, pineapple taste and odor	172.515	Pineapple
Allyl iso-thiocyanate (artificial mustard oil)	Overpowering mustard-like odor with stinging taste	172.515	Spice blends; mustard vegetable flavors, meat-horseradish
Allyl iso-valcrate	Pleasant over-ripe fruit odor; apple taste	172.515	Apple, cherry fruit blends
Allyl mercaptan (2 propene-1-thiol) (allyl sulfhydrate)	Pungent onion, leek odor and taste	172.515	Spice, onion, garlic
Allyl methyl disulfide (methyl allyl disulfide)		3127	
Allyl methyl trisulfide (methyl allyl trisulfide)	Strong garlic, onion odor and taste	3253	Garlic, onion imitations
Allyl nonanoate	A fatty, pineapple coconut odor with coconut, cognac taste	172.515	Coconut, cognac, pineapple
Allyl octanoate (allyl caprylate)	A fatty, fruity, pineapple, apple odor and taste	172.515	Pineapple, apple, banana fruit blends
Allyl phenoxyacetate	Pleasant fruity pineapple odor and taste	172.515	Pineapple, honey
Allyl phenylacetate	A fruity honey, pineapple odor and at low levels honey, and higher levels, pineapple taste	172.515	Honey, pineapple
Allyl propionate	Fruity, pineapple, garlic odor with garlic, mustard taste	172.515	Pineapple, spice blends
Allyl sorbate (allyl 2,4-hexadienoate)	A pungent, pineapple odor with a sweet fruity taste	172.515	Pineapple
Allyl sulfide (diallyl sulfide) (allyl thiopropionate)	Pungent garlic, horseradish-like odor and taste	172.515 3329	Garlic, horseradish

Table 4 (continued)
SYNTHETIC FLAVOR INGREDIENTS

Name	Organoleptic characteristics	Regulation	Flavors useful in
Allyl tiglate (allyl T-2-methyl-2-butenoate)	An earthy, rooty odor with an earthy, sweet taste	172.515	Fruit flavors
Allyl 10-undecenoate (allyl undecylenate)	A sweet anise, fennel odor with sassafras, root-beer taste	172.515	Root beer, pineapple, coconut
Ammonium iso valerate	A pleasant, cheesy odor and sweet cheesy taste	172.515	Cheese, nut, butterscotch butter
Ammonium sulfide	A strong rotten egg odor and at great dilution, an egg taste	172.515	Egg flavor, spice blends
Amyl acetate	A pleasant banana, pear odor with pronounced banana taste	2055	Banana, pear, fruit blends, apple
Amyl alcohol (1-pentanol)	Strong apple, banana odor and taste	172.515	Apple, banana, fruit blends, pineapple
Amyl butyrate (pentyl butyrate)	Strong estery mixed fruit odor with apple, banana taste	172.515	Apple, banana, pineapple, apricot
alpha-amyl cinnamaldehyde	Floral, jasmine odor with perfumy taste	172.515	Traces in fruit flavors, spice flavors
alpha-amyl cinnamaldehyde dimethyl acetal	Floral, green odor with perfumy taste	172.515	Traces in fruit flavors
alpha amyl cinnamyl acetate	A perfumy, fruity odor with a taste suggestive of apricot, peach	172.515	Peach, apricot, punch spice, chocolate
alpha-amyl cinnamyl alcohol	A peach, banana, pear odor with jasmine taste	172.515	Traces in fruit flavors
alpha-amyl cinnamyl formate	A sweet floral, butter-like odor with perfumy taste	172.515	Traces in maple, nut flavors
alpha-amyl cinnamyl isovalerale	A floral, fruity odor with spicy, fruity taste	172.515	Traces in nut, chocolate, grape
Amyl formate (pentyl formate)	A fresh fruity odor and taste	172.515	Apricot, peach, strawberry, cherry, plum
Amyl-2-furoate (pentyl-2-furoate)	A fruity, caramel odor and fermented taste	2030	Apricot, peach, rum, maple
Amyl heptanoate (pentyl heptanoate)	Fruity, banana-like odor with fruity taste	172.515	Fruit flavors, banana coconut
Amyl hexanoate (amyl caproate)	A green, fruity odor with apple, pineapple taste	172.515	Apple, pineapple fruits, grape
2-Amyl-5 or 6-keto-1,4 dioxane	A caramel-like fruity odor with fruity-lactone taste	2076	Peach, margarine flavors, fruits
Amyl octanoate (amyl caprylate)	A floral, fruity odor with fruity taste	172.515	Mixed fruit flavors, apple
Anethole (1-methoxy-4-propenyl benzene)	A sweet anise odor with harsh anise taste	182.60	Anise imitations, spice flavors
Anisole (methoxy benzene)	A sweet anise-like odor and sweet taste	172.515	Vanilla, anise, rootbeer
Anisyl acetate (*p*-methoxyenzyl acetate)	A fruity, sweet floral odor and taste	172.515	Cherry, strawberry, vanilla
Anisyl alcohol (*p*-methoxybenzyl alcohol)	A floral resinous vanilla odor and taste	172.515	Vanilla, chocolate, cocoa, peach

Anisyl butyrate (*p*-methoxybenzyl butyrate)	Sweet fruity oxanone-like odor — helistropine taste	172.515	Vanilla, maple, nut
Anisyl formate (*p*-methoxybenzyl formate)	Sweet floral odor with berry-like taste	172.515	Raspberry, strawberry, vanilla
Anisyl phenylacetate	A floral, honey odor with anise-like taste	172.515	Honey flavors
Anisyl propionate (*p*-methoxybenzyl propionate)	A sweet, fruity vanilla odor with sweet fruity taste	172.515	Vanilla, apricot, peach, strawberry
Ascorbic acid (vitamin C)	Acid taste	182.3013	Acidulant, fortification
Benzaldehyde	A powerful almond odor and taste	182-60	Almond, cherry, peach, apricot, nut
Benzaldehyde dimethyl acetal	A weedy almond odor with almond taste	172.515	Almond cherry, peach, apricot, nut
Benzaldehyde glyceryl acetal (2 phenyl-m-dioxan-5-ol)	A sweet cherry, almond odor with mild cherry taste	172.515	Almond, cherry, peach, apricot, nut
Benzaldehyde propyleneglycol acetal (4-methyl-2-phenyl-m-dioxolane)	Sweet, solventy odor characteristic of all p.g. acetals — taste of propyleneglycol evident	172.515	Cherry, peach, nut
Benzenethiol (thiophenol)	An unpleasant burnt odor which upon dilution tastes like meat	172.515	Meat, coffee
2-Benzofurancarboxaldehyde (2-formyl benzofuran)	A sweet coumarin-like odor with a bitter taste	3128	Nut, tobacco flavors
Benzoic acid	Bitter taste	2131	Preservative
Benzoin (2 hydroxy-2-phenyl acetophenone)	A sweet non-descript odor and taste	172.515	Sweet flavors, vanilla, butterscotch
Benzophenane (diphenyl ketone)	A mild floral odor with little character in taste	172.515	Vanilla, butter
Benzothiazole	A mild peanut odor with an oily, peanut taste	3256	Peanut, cocoa, potato
Benzyl acetate	Fresh floral, fruity odor and taste	172.515	Cherry, peach, apricot, raspberry, strawberry
Benzyl acetoacetate	Mild fruity odor suggestive of berry-mouth filling, but low flavor value — hint of berry	172.515	Strawberry, raspberry
Benzyl alcohol	A weak sweet odor with fatty taste; upon ageing, cherry-like	172.515	Cherry, nut, oil soluble solvent
Benzyl benzoate	Very little odor, prune-like taste	172.515	Prune, berry, cherry
Benzyl buthylether	A harsh floral odor with metallic taste	172.515	Fruit flavors
Benzyl butyrate	A fruity odor with apricot, pear taste	172.515	Pear, apricot, berry, plum
Benzyl cinnamate	A floral resinous odor with honey taste	172.515	Honey, peach, apricot
Benzyl 2,3 dimethyl crotanate (benzyl methyltiglate)	Powerful fruity, spicy odor and taste	172.515	Traces in spice flavors
Benzyl disulfide (dibenzyl disulfide)	A burnt-sugar odor and taste	172.515	Coffee
Benzyl ethyl ether	A penetrating mushroom odor with a solventy taste	172.515	Spice flavors, mushroom

Table 4 (continued)
SYNTHETIC FLAVOR INGREDIENTS

Name	Organoleptic characteristics	Regulation	Flavors useful in
Benzyl formate	A fruity odor with a sweet taste	172.515	Apricot, peach, pineapple
3-Benzyl-4-heptanone (benzyl dipropyl ketone)	Sweet raisin, prune odor with prune taste	172.515	Prune, raisin, apricot
Benzyl iso butyrate	A floral, fruity odor with fruity berry taste	172.515	Raspberry, strawberry, fruit blends
Benzyl iso valerate	Fruity blue-cheese odor and taste	172.515	Cheese, fruit flavors, tobacco-apple
Benzyl mercaptan (benzylthiol)	A strong onion odor and taste	172.515	Traces in coffee, meat flavors
Benzyl T-2-methyl-2-butenoate (benzyl tiglate)	An earthy, mushroom odor and taste	3330	Mushroom
Benzyl phenylacetate	A weak floral odor with a honey taste	172.515	Honey, caramel, butter
Benzyl propionate	A weak odor with fruity apricot, peach taste	172.515	Apricot, peach, cherry, raspberry, banana
Benzyl salicylate (benzyl o-hydroxybenzoate)	A sweet floral odor and taste	172.515	Apricot, peach, banana, raspberry
Biphenyl (phenylbenzene)	A pungent odor, nondescript taste	3129	Fungus growth inhibitor
Birch tar oil	A strong burnt wood, smoke odor with sour, burnt taste	172.515	Smoke flavors
Bisabolene	A resinous, spicy odor with a citrus taste	3331	Essential oil imitations, citrus flavors — lime — fixative, blender
Bis-(2,5-dimethyl-3-furyl disulfide	—	3476	—
Bis (2-furfuryl) disulfide (difurfuryl disulfide)	—	3257	—
Bis(2-furfuryl)sulfide (difurfuryl sulfide)	A coffee, meaty odor with mushroom taste	3258	Coffee, mushroom, meat
Bis(2 methyl-3-furyl)-disulfide (2 methyl-3-furyl disulfide)	—	3259	—
Bis(2-methyl-3-furyl) tetrasulfide (2-methyl-3-furyl tetrasulfide)	—	3260	—
Borneol (bornyl alcohol)	A camphor-like minty odor and taste	172.515	Mint flavors, traces in spice flavors, lime, nut
Bornyl acetate	A pine-like minty odor and taste	172.515	Mint flavors, traces in spice flavors, pineapple
Bornyl formate	Clean, pine-like odor with sweet green, woody taste	172.515	Traces in fruit flavors
Bornyl iso valerate	A minty valeric acid odor with minty taste	172.515	Traces in fruit flavors

Bornyl valerate	A minty pine-like odor with minty taste	172.515	Traces in fruit flavors
B-Bourbonene	—	172.515	Essential oil imitators
Brominated vegetable oil	A fatty oily odor and oily, bitter taste	2168	Weighing agent for citrus oils
2,3 Butandithiol	—	3477	—
1-Butanethiol (*N*-butyl mercaptan)	Cabbage skunk-like odor with bitter taste at low levels	3477	Vegetable flavors
2-Butanone (methyl ethyl ketone)	A solventy ketone odor and taste	172.515	Cheese, banana flavors, coffee
Butan-3-one-zyl butanoate	An excellent yogurt odor with woody, buttery taste	3332	Yogurt, butter, butterscotch
Butter acids	A butter, cheese odor and taste	172.515	Cheese, butter, milk flavors
Butter esters	A butter, cheese odor and taste	172.515	Cheese, butter, milk flavors, chocolate
Butter starter distillate	Pungent butter aroma and taste	172.515	Culture flavors, yogurt
Butyl acetate	Pleasant, solventy banana-like odor and taste	172.515	Banana, pear, pineapple, berry
Butyl acetoacetate	A sweet wine, brandy odor with sweet fruity taste	172.515	Fruit flavors, berry
Butyl alcohol (1-butanol)	A harsh fusel odor with a banana, fusel taste	172.515	Banana, butter, whisky flavors, cheese
Butylamine	An ammonia odor and taste	3130	Seafood flavors, chocolate
Butyl anthranilate	A sweet fruity-citrus, grape odor and taste	172.515	Grape, tangerine, grapefruit
Butylated hydroxy anisole (BHA)	—	182.3169	Antioxidant - preservative
Butylated hydroxy toluene (BHT)	—	182.3173	Antioxidant - preservative
2-Butyl-2-butenal	—	3392	—
Butyl butyrate	Choking, woody sweet, rum odor with creamy taste	172.515	Apple, pear, banana, butter
Butyl butyrllactate	A fruity, prune, bread-crust odor with creamy taste	172.515	Vanilla, butter, butterscotch
alpha-Butylcinnamaldehyde	A green floral, spice odor with perfumy, fruity, spice taste	172.515	Cinnamon type spice flavors
Butyl cinnamate	Sweet resinous, fruity odor with sweet taste	172.515	Vanilla,chocolate, berry flavors, cocoa
2-sec-Butylcyclohexanone	A clean Brazilian mint odor and taste without weediness	3261	Mint flavors, toothpaste, mouthwash flavors
Butyl 2-decenoate (butyl decylenate)	Fatty coconut odor with coconut, walnut taste	172.515	Coconut, walnut, apricot, peach
1,3,Butylene glycol	Odorless with sweet-bitter taste	121.1176	Solvent for flavors; forms solid acetals with aldehyde
sec-butyl ethyl ether	Ether solventy, odor with fruity taste	3131	Juice flavors
Butyl ethyl malonate	Green fruity odor and taste	172.515	Fruitiness to fruit flavors
Butyl formate	Volatile rum odor with rum, brandy taste	172.515	Rum, brandy fruit blends
Butyl heptanoate	Fruity, green-apple odor with fruity taste	172.515	Apple, fruit flavors
Butyl hexanoate (butyl caproate)	Fruity, nondescript odor with pleasant fruity taste	172.515	Pineapple, rum, butter

Table 4 (continued) SYNTHETIC FLAVOR INGREDIENTS

Name	Organoleptic characteristics	Regulation	Flavors useful in
Butyl *p*-hydroxybenzoate (butyl parasept)		172.515	Preservative
3-Butylidene pathalide	A characteristic fresh celery-like odor with celery, butterscotch taste	3333	Spice flavors, soup flavors, butterscotch
Butylisobutyrate	Strong fruit odor with pineapple taste	172.515	Pineapple, apple, cherry
Butyl iso valerate	Solventy "ketoney" banana odor with blue-cheese taste	172.515	Milk, cheese, banana flavors
2-Butyl-5 or 6 keto -1,4 dioxane	A fatty, green, creamy odor with peach, cream taste	2204	Peach, milk flavors
Butyl lactate	Sweet buttery odor with milk taste	172.515	Dairy flavors, milk cheese, butterscotch
Butyl laurate	A mild fruity nut-like odor with fatty oily taste	172.515	Nut, brandy, peach, apricot
Butyl levulinate	Fruity caramel odor with astringent woody fruity taste	172.515	Fruit flavors, rum-butterscotch
n-Butyl-2-methylbutyrate	A fruity, sweet, chocolate odor with an apple, grape taste	3393	Fruit flavors
Butyl phenylacetate	A sweet floral honey-like odor and taste	172.515	Honey, nut, mixed fruit
3-*n*-Butylphthalide	A characteristic celery, lovage odor and soup taste	3334	Spice flavors, soup flavors, maple
Butyl propionate	Fruity estery, rum-like odor and taste	172.515	Rum, fruit flavors, apricot
Butyl stearate (butyl octadecanoate)	Fatty fruity odor and taste	172.515	Solvent for flavors, butter, banana
Butyl sulfide (pi butyl sulfide)	Floral green-leaf odor and taste	172.515	Traces in fruit flavor to impart green notes; violet
Butyl 10-undecenoate	A fatty brandy-like odor with buttery taste	172.515	Butter, wine, brandy, nut
Butyl valerate	A choking fruity odor with fruity taste	172.515	Apple, strawberry, butter
Butyraldehyde	An irritating, fruity odor upon great dilution fruity, banana taste	172.515	Fruit flavors, banana, caramel
Butyric acid	Strong butter, cheese-like odor, butter-fat taste	182.60	Dairy flavors, butter, cheese, caramel
Cadinene	Woody, resinous, burnt odor with little taste	172.515	Essential oil imitations, fixative blender
Caffeine (methyl theobromine)	Bitter taste	182.1180	Cola beverages 0.02%
Calcium acetate	—	2228	Sequestrant
Camphene (2,2-dimethyl-3-methylene-norbornane)	Earthy camphoraceous odor and taste	172.515	Essential oil imitations, mint flavors, spice flavors
d-Camphor	A minty odor with slight minty cool taste	172.515	Essential oil imitations, mint flavors

Carboxymethylcellulose		2239	Stabilizer
Caruacrol (2 hydroxy *p*-cymene)	Medicinal smoke odor, excellent smoke taste, a 1—5 ppm	172.515	Smoke flavors, spice, meat flavors
Caruacryl ethyl ether (2 ethoxy-*p*-cymene)	Earthy, spicy odor with sweet spicy taste	172.515	Spice flavors
Carveol (*p*-mentha-6,8-dien-2-ol)	A phenol, indole aroma with animal, civet-like taste	172.515	Essential oil imitations, spearmint, caraway
4-Carromenthenol (1-*p*-menthen-4-ol) (4-terpinenol)	Woody, earthy, spicy odor with spicy, pepper taste	172.515	Citrus, spice
d, l-Carvone (6,8,(9)-*p*-menthadien-2-one)	*d*- A caraway, dillseed, rye bread odor and taste	2249	Kummel liquor, pickle, spice flavors
	l- A spearmint odor and taste		Spearmint flavors
cis-carvone oxide	—	172.515	—
Carvyl acetate	A green spearmint odor with spicy, minty taste	172.515	Art. mint oils, spice blends, meat
Carvyl propionate	A minty spearmint odor with fruity spearmint taste	172.515	Spice blends, mint
beta-caryophyllene	Woody terpene-like odor, weak woody taste	172.515	Essential oil imitations, fixative blender for flavors
Caryophyllene alcohol	Earthy, spicy odor with minty, woody taste	172.515	Mushroom flavor
Caryophyllene alcohol acetate	A fruity, woody odor and taste	172.515	Mushroom flavor
beta-caryophyllene oxide	—	172.515	—
Cedarwood oil alcohols	A cedarwood odor and taste	172.515	Cedarwood imit. — tobacco, mint oils
Cedarwood oil terpenes	Green camphoraceous odor and taste	172.515	Mint oils, cedarwood imitations, tobacco
1,4 Cineole	A minty lime oil odor and taste	172.515	Essential oil imitations, mint, toothpaste flavors
Cinnamaldehyde	A spicy cinnamon cassin odor with spicy sweet taste	182.60	Cassia, cinnamon, cola spice blends, cream soda
Cinnamaldehyde ethylene glycol acetal	A sweet spicy odor with a cinnamon, pimento taste	172.515	Spice blends
Cinnamic acid	A mild resinous odor with weak, spicy taste	172.515	Spice blends, cherry, honey, apricot
Cinnamyl acetate	A floral fruity odor with fruity spicy taste	172.515	Cinnamon, apple, cherry, pineapple
Cinnamyl alcohol	A sweet floral odor and taste	172.515	Apricot, peach, raspberry, plum
Cinnamyl anthranilate	Floral, spicy, fruity grape odor and taste	172.515	Grape fixative, cherry
Cinnamyl benzoate	Spicy resinous odor with spicy taste	172.515	Fixative for fruit flavors, grape
Cinnamyl butyrate	Fruity resinous odor with sweet taste	172.515	Grape, strawberry, honey, raspberry, mixed fruit
Cinnamyl cinnamate	Sweet resinous odor and taste	172.515	Fruit flavors — fixative
Cinnamyl formate	Fruity resinous odor with bitter taste	172.515	Strawberry, raspberry, peach, banana, spice

Table 4 (continued)
SYNTHETIC FLAVOR INGREDIENTS

Name	Organoleptic characteristics	Regulation	Flavors useful in
Cinnamyl isobutyrate	Sweet, resinous, fruity odor with sweet taste	172.515	Fruit flavors, apple, strawberry, citrus
Cinnamyl isovalerate	Green floral odor with fruity apple taste	172.515	Apple, strawberry
Cinnamyl phenylacetate	A floral, honey odor with a creamy taste	172.515	Honey, chocolate, spice flavors
Cinnamyl propionate	Spicy floral odor with fruity, spicy taste	172.515	Grape, raspberry, currant, strawberry
Citral (3,7 dimethyl-2,6-octadienal)	A lemon aroma and taste	182.60	Lemon citrus flavors, mixed fruit
Citral diethyl acetal	Green citrus-like odor and taste	172.515	Lemon, citrus flavors, mixed fruit
Citral dimethyl acetal	Green citrus, vegetable odor and citrus taste	172.515	Citrus, lemon, vegetable flavors
Citral propylene glycol acetal	Distinct citral-lemony odor and taste	172.515	Lemon, citrus, mixed fruit
Citric acid	Sour taste	182.1033	Acidulant
Citronellal (3,7,dimethyl-6-octenal)	A citrus, oil citronella odor and taste	172.515	Traces in beverages, citrus flavors, cherry
DL-citronellol (7 dimethyl-6-octen-1-ol)	A perfumy odor and a bitter taste	172.515	Traces in fruit flavors, peach, cola
Citronelloxyacetaldehyde (6,10-dimethyl-3-oxa-9-undecenal)	A perfumy odor and taste	172.515	Traces in fruit flavors
Citronellyl acetate	A fruity, floral odor and a fruity, pungent taste	172.515	Fruit flavors, apricot, citrus
Citronellyl butyrate	Sweet fruity, floral odor with apple taste	172.515	Apple, honey, banana, plum, cola
Citronellyl formate	Leafy, fruity floral odor and taste	172.515	Cucumber, vegetable flavors, plum, orange
Citronellyl isobutyrate	A sweet floral citrus odor with citrus taste	172.515	Traces in fruit flavors, raspberry, grape
Citronellyl phenylacetate	A floral, honey odor and taste	172.515	Honey, caramel
Citronellyl propionate	A floral fruity odor with fruity taste	172.515	Fruit flavors, citrus, berry flavors
Citronellyl valerate	Floral fruity, dried leaves odor, fruity, honey taste	172.515	Honey, apricot
p-Cresol (4 hydroxytoluene)	Medicinal, phenolic odor and taste	172.515	Smoke, nut flavors, vanilla
o-Cresol (*o*-hydroxytoluene)	Medicinal, phenolic odor and taste	172.515	Smoke, nut flavors
Cuminaldehyde (*p*-iso propylbenzaldehyde)	A spicy, cinnamon-like odor with a distinct spicy, cumin taste	172.515	Spice flavors, berry
Cycloheptadeca-9-en-1-one	An animal fecal odor, pleasant at great dilution	3425	Traces in flavors
Cyclohexanacetic acid	A "tub" butter aroma with a tongue coating taste	172.515	Milk, butter flavors
Cyclohexamethyl acetate	A fruity, raspberry, banana odor with banana taste	172.515	Banana, raspberry
Cyclohexyl acetate	Fruity "solventy" odor with fruity apple taste	172.515	Apple, banana, blackberry, raspberry
Cyclohexyl anthranilate	A fatty, orange flower, grape odor with an amyl-	172.515	Apple, banana, grape

	cinnamic aldehyde-like taste		
Cyclohexyl butyrate	A fatty, fruity, apple, banana odor and taste	172.515	Apple, banana, pineapple, grape
Cyclohexyl cinnamate	A fatty spicy odor with a berry-like taste	172.515	Apple, peach, strawberry, prune
Cyclohexyl formate	A choking odor, butyacetate-like with a fruity, banana taste	172.515	Banana, cherry
Cyclohexyl iso valerate	A good berry odor with a noncharacteristic fruity taste	172.515	Apple, strawberry, banana
Cyclohexyl propionate	A banana odor and a banana-rum taste	172.515	Banana, rum
Cyclopentanthiol cyclopentyl mercaptan	—	3262	—
p-Cymene (1-methyl-4-isopropyl benzene)	A characteristic terpene odor with citrus taste	172.515	Citrus and spice flavors
L-Cysteine (2 amino-3-mercaptopropionic acid)	Odorless solid with sour taste	3263	Nutritional fortification
2-*trans*,4-*trans* Decadienal	A characteristic odor of chicken with oily chicken taste	3135	Chicken
gamma-decalactone	Coconut odor, creamy milk, peach, coconut taste	172.515	Milk, cream, peach, citrus, coconut
delta decalactone	Coconut odor, creamy milk, peach, coconut taste	172.515	Peach, milk, coconut, cream
Decanal (aldehyde C-10)	A strong fatty citrus odor and citrus peel taste	182.60	Citrus flavors
Decanal dimethyl acetal	A waxy, citrus odor with a fatty orange taste	172.515	Citrus flavors
Decanoic acid (capric acid)	Fatty rancid odor and taste	2364	Dairy flavors, rum, coconut, brandy
1-Decanol (alcohol C-10)	Floral, waxy odor and taste	172.515	Orange, lemon, fruit blends, coconut
2-Decenal	A chicken, poultry, orange odor and taste	172.515	Citrus, poultry flavors
4-Decenal	—	3264	—
3-Decen-2-one (heptylidene acetone)	A floral, fruity, green odor with raspberry taste	172.515	Berry flavors
Decyl acetate (acetate C-10)	A fatty, fruity odor with an orange, pineapple taste	172.515	Orange, pineapple, tropical fruit
Decyl butyrate	A fruity, waxy odor with a peach, apricot, citrus taste	172.515	Peach, apricot, citrus, brandy
Decyl propionate	Pleasant, citrusy odor with a fatty, citrus taste	172.515	Citrus flavors
Dehydrodihydroionol	—	3446	—
Dehydrodihydroionone	—	3447	—
Diacetyl (2,3 butandione)	Rancid butter odor with butter taste at 1 ppm	182.60	Culture flavors mild cheese, certain fruits, coffee
Diallyl sulfide (allyl sulfide)	Strong garlic-leek odor and taste	172.515	Garlic, leek, meat, soup flavors
Diallyl disulfide (allyl disulfide)	Powerful garlic, mustard odor and taste	172.515	Garlic, onion imitations, spice blends, leek
Diallyl trisulfide (allyl trisulfide)	Powerful garlic, onion odor and taste	3265	Garlic, onion imitations

Table 4 (continued)
SYNTHETIC FLAVOR INGREDIENTS

Name	Organoleptic characteristics	Regulation	Flavors useful in
Dibenzyl ether	A pleasant almond odor with earthy, mushroom taste	172.515	Mushroom, cherry
Di-(Butan-3-one-1-yl) sulfide	A sulfury, meaty-like odor with a mild meaty taste	3335	Meat, coffee flavors, egg
4,4-Dibutyl-gamma-butyrolactone	A butter, nut odor with a peanut, chestnut taste	172.515	Nut flavors, butter, coconut
Dibutyl sebacate (butyl sebacate)	Mild fruity odor with fruity, fatty taste	172.515	Fruit flavors
Dicyclohexyl disulfide	—	3448	—
Diethyl malate (ethyl malate)	Pleasant cherry, apple odor and taste	172.515	Apple, cherry, banana
Diethyl malonate (ethyl malonate)	Exhibits many different fruit characters in odor and taste grape, apple, cherry, pear	172.515	Fruit flavors, apple, pear
2,3-Diethyl-5-methyl-pyrazine	A powerful chocolate odor and taste	3336	Chocolate, nut, roasted flavors
2,3-Diethyl pyrazine	Earthy, green odor and taste	3136	Potato, nuts
Diethyl sebacate (ethyl sebacate)	Fruity, pear, grape odor and taste	172.515	Grape, pear, solvent for fruit flavors
Diethyl succinate (ethyl succinate)	Fruity, green grape odor and taste	172.515	Berry flavors, solvent for fruit flavors, grape
Diethyl tartrate (ethyl tartrate)	Odor similar to hydrogen peroxide — no taste	172.515	Up to 100 ppm this material has no flavor
2,5 Diethyl tetrahydrofuran	A characteristic minty odor and taste	172.515	Spearmint, peppermint imitations
Difurfuryl ether (furfuryl ether)	—	3337	—
Dihydrocarveol (6-methyl-3-iso-propenyl-cyclohexanol)	A woody, floral odor with spicy taste	172.515	Imitation spice flavors, mint
Dihydrocarvone (1-methyl-4-iso-propenyl-cyclohexane-2-one)	A sweet spearmint-like odor and taste	172.515	Spearmint imitations, mint flavors
Dihydrocarvyl acetate	A sweet spearmint odor and taste	172.515	Spearmint imitations, mint flavors
Dihydrocoumarin (benzodihydropyrone)	A coumarin odor and a sweet then bitter taste	2381	Vanilla, butter, cream soda, tobacco flavors
4,5-Dihydro-3(2H) thiophene	A solventy, roasted, meaty odor with a bitter taste	3266	Meat, coffee flavors
5,7-Dihydro-2-methyl-thieno(3,4-D) pyrimidine	—	3338	—
m-Dimethoxybenzene (dimethyl resorcinol)	An earthy, nut-like odor and taste at low levels	172.515	Nut flavors, walnut, vanilla
p-Dimethoxy benzene (dimethyl hydroquinone)	A sweet nut-like aroma with an oily, phenolic taste	172.515	Nut flavors

1,1-Dimethoxyethane (acetaldehyde dimethyl acetal)	Vegetable, green odor with fruity taste	3426	Fruit flavors, apple, whisky, orange
2,6 Dimethoxyphenol	An excellent smoke, bacon-like odor and taste	3137	Smoke, bacon flavors
3,4-Dimethoxy-1-vinyl-benzene	Resinous, sweet, earthy phenolic odor with sweet, phenolic taste	3138	Modifier for vanilla flavors, meat
2,4-Dimethyl acetophenone	A pungent, sweet, floral, woody odor and taste	172.515	Almond, cherry, grape
1,4-Dimethyl-4-acetyl-1-cyclohexane	—	3449	—
2,4-Dimethyl-5-acetyl-thiazole	A meaty, nut aroma with an oil, nutty taste	3267	Roasted flavors, meat, chocolate, nut
2,4 Dimethyl benzaldehyde	A mild almond odor with an almond, cherry taste	3427	Fruit flavors, cherry, peach
p-alpha-dimethylbenzyl alcohol	An odor made up of nut, cherry acetophenone with cherry taste	3139	Nonbenzaldehyde, cherry
alpha,alpha-dimethylbenzyl isobutyrate	A sweet, fruity odor and taste	172.515	Banana, peach, apricot
3,4-Dimethyl-1,2-cyclopentadione	—	3268	—
3,5-Dimethyl-1,2-cyclopentadione	—	3269	—
2,5-Dimethyl-2,5,dihydroxy-1,4-dithione	—	3450	—
2,5 Dimethyl-3-furanthiol	—	3451	—
2,6 Dimethyl-4-heptanol	—	3140	—
2,6 Dimethyl-5-heptenal (melonal)	Powerful green odor with a taste like melon-rind	172.515	Melon, cucumber, tropical fruit
2,6-Dimethyl-10-methylene-2,6,11-dodecatrienal (alpha-sinensal)	A terpene citrus odor with citrus taste	3141	Artificial citrus oils
3,7-Dimethylocta-2,6-dienyl-2-ethylbutanoate (geranyl 2-ethlbutyrate)	—	3339	—
2,6 Dimethyloctanal	Sweet fruity, vegetable odor with melon taste	172.515	Melon
3,7-Dimethyl-1-octanol (tetrahydro geraniol)	Sweet floral odor with bitter taste	172.515	Citrus, fruit flavors
3,7-Dimethyl-6-octenoic acid (citronellic acid)	A green, weedy odor with very little taste	3142	Essential oil imitations
2,4 Dimethyl-2-pentenoic acid	—	3143	—
alpha-alpha-Dimethylphenethyl acetate (benzyl dimethyl carbinyl acetate)	A floral fruity odor with a fruity, pear taste	172.515	Pear, fruit flavors, cherry
alpha-alpha-dimethyl-phenethyl alcohol (dimethyl benzyl carbinol)	A sweet vanilla odor with flowery taste	172.515	Vanilla, fruit flavors
alpha,alpha-Dimethyl-phenethyl butyrate (benzyl dimethyl carbinyl butyrate)	Excellent prune odor and taste	172.515	Prune, apricot, dry fruit flavors
alpha-alpha-Dimethyl-phenethyl formate (benzyl dimethyl carbinyl formate)	A green, floral odor with spicy taste	172.515	Prune, apricot, dry fruit flavors

Table 4 (continued)
SYNTHETIC FLAVOR INGREDIENTS

Name	Organoleptic characteristics	Regulation	Flavors useful in
2,6-Dimethylphenol (2,6 xylenol)	A phenolic-like top-note with a thymol, carvacrol taste	3249	Excellent in smoke flavors
2,3-Dimethylpyrazine	Roasted, buttery, meat aroma, burnt protein taste	3271	Meat, nut, chocolate, coffee, butter, tobacco
2,5-Dimethylpyrazine	A pungent roasted peanut odor with a chocolate, butter taste	3272	Cocoa, coffee, meat, nut potato
2,6 Dimethylpyrazine	A roasted peanut odor with chocolate taste	3273	Chocolate, coffee, meat, nut
p-alpha Dimethylstyrene		3144	
Dimethyl succinate	Fruity mild citrus odor and taste — > 30 ppm extracted lemon character	172.515	Solvent for fruit flavors
4,5-Dimethyl thiazole	Possesses a fish, amine-like odor and seafood taste	3274	Seafood flavors
2,5-Dimethyl-3-thiofuroyl-furan	—	3481	—
2,5 Dimethyl-3-thioiso-valeryl furan	—	3482	
Dimethyl trisulfide (methyl trisulfide)	A strong onion, garlic odor with a cooked garlic taste	3275	Onion, garlic, chicken, vegetable flavors
2,4-Dimethyl-5-vinyl-thiazole	—	3145	—
1,3-Diphenyl-2-propanone (dibenzyl ketone)	Sweet, weak, almond odor and taste	172.515	Cherry, traces in fruit flavors
Dipropyl trisulfide (propyl trisulfide)	Powerful garlic odor and taste	3276	Meat, poultry, onion, garlic, spice flavors
Disodium phosphate	—	2398	Sesquestrant, emulsifier
Disodium succinate	Very little odor with salty, meaty taste	3277	Meat, soup flavors
Spiro-(2,4-dithia-1-methyl-8-oxabicyclo(3,3,0)octane-3,3′-	Reported to have a meat aroma	3270	—
2,8-Dithianon-4-en-4-carboxaldehyde	—	3483	—
2,2′(dithiodimethylene)-difuran	—	3146	—
gamma-Dodecalactone	Bland fatty, fruity odor, fatty, milk, taste	172.515	Margarine flavors, peach, coconut, maple
delta-Dodecalactone	Bland fatty fruity odor, creamy taste	172.515	Margarine flavors, peach, coconut, pear
2-Dodecenal	A fatty green odor with a fatty oily, citrus taste	172.515	Citrus flavors, tangerine
Dodecyl isobutyrate	Oily, fatty, rancid odor with fruity oily taste	3452	Coconut, brandy
Erythrobic acid	—	182.3041	Color preservative for meat
Estragole (methyl chavicol) (*p*-allylanisole)	A sweet anise, fennel odor and taste	172.515	Spice flavors, rootbeer, seasonings

1,2 Ethanedithiol	—	3484	—
p-Ethoxybenzaldehyde	Floral anise odor with sweet taste	172.515	Vanilla, use as a sweetener
o-(Ethoxymethyl)phenol	—	3485	—
2-Ethoxythiazole	—	3340	—
Ethyl acetate (acetic ether)	Vinegar, nail polish odor with little taste character	182.60	Lift for flavors, imparts a vinous, fruity note to flavors
Ethyl acetoacetate (ethyl 3-oxobutanoate)	Pleasant green, fruity, rum odor with ripe fruit taste	172.515	Strawberry, fruit flavors
Ethyl-2-acetyl-3-phenyl-propionate (ethyl benzylacetoacetate)	Floral, woody, fruity odor with fruity jam taste	172.515	Traces in fruit flavors
1-Ethyl-2-acetylpyrrole	—	3147	—
Ethyl aconitate	A characteristic rum odor, a rather weak rum taste	172.515	Rum, fruit flavors
Ethyl acrylate (ethyl propenoate)	Irritating, acrid odor with fruity, rum taste at low levels	172.515	Rum, pineapple, fruit blends
Ethyl alcohol (ethanol)	Extraction solvent, preservative, flavor carrier	2419	All flavors — solvent
Ethyl-*p*-anisate (ethyl-*p*-methoxybenzoate)	A mild, floral anise odor with a woody anise taste	172.515	Anise, fennel, licorice
Ethyl anthranilate	Sweet fruity, grape odor and taste	172.515	Concord grape wine flavors, mandarin
Ethyl benzoate	A fruity odor and taste less harsh than methylester	172.515	Cherry, black currant, strawberry
Ethyl benzoylacetate	Fruity, woody, acetophenone odor, nutty woody taste	172.515	Brandy, whisky
alpha-ethyl benzylbutyrate	Sweet, green fruity odor with sweet, fruity taste	172.515	Fruit flavors
Ethyl brassylate	A musky odor and taste	172.515	Pear, traces in flavors
Ethyl *trans*-2-butenoate (ethyl crotonate)	A sharp, fruity, burnt sugar odor and taste	172.515	Pineapple, strawberry
2 Ethyl butyl acetate	Fruity, oily, banana odor with sweet fruity taste	172.515	Banana, strawberry, pear
2 Ethyl butyraldehyde	A choking, chocolate odor with chocolate-cocoa taste	172.515	Chocolate, cocoa
Ethyl butyrate	A fatty, fruity buttery odor with a juicy-fruit flavor	182.60	Pineapple, grape, rum, strawberry
2-Ethyl butyric acid	Pleasant cheese odor, earthy, sour taste	172.515	Strawberry, cheese, nut
Ethyl cellulose	—	6057	Thickener, stabilizer, binder, filler
Ethyl cinnamate	A sweet resinous honey odor with sweet fruity taste	172.515	Strawberry, raspberry, plum, cherry
Ethyl cyclohexane-propionate	A sweet fruity odor on great dilution, sweet fruity taste	172.515	Pineapple, pear, apple, banana

Table 4 (continued)
SYNTHETIC FLAVOR INGREDIENTS

Name	Organoleptic characteristics	Regulation	Flavors useful in
Ethyl *trans*-2, cis-4-decadienoate	A fatty, fruity odor with a distinct bartlett pear taste	3148	Bartlett pear
Ethyl decanoate (ethyl caprate)	A fatty nut odor and taste	172.515	Nut, cheese, wine, cognac, rum
2-ethyl-3,(5 or 6)-dimethyl-pyrazine	Burnt almond, chocolate odor and taste	3149	Roasted nut flavors, chocolate, potato chips
3-Ethyl-2,6-dimethylpyrazine	A strong chocolate, peanut, moldy odor with excellent chocolate taste	3150	Chocolate, nut, smoke, cocoa
Ethyl 2,4 dioxyhexanoate	—	3278	—
Ethylene oxide	—	2433	Ripening agent for fruits, fungistat
Ethyl 2-ethyl-3-phenyl-propionate	—	3341	—
Ethyl formate	Estery, fruity, rum-like odor and taste	182.1295	Rum, apricot, peach, pineapple, mixed fruit, sherry
2-Ethyl furan	Solvent-like, burnt odor and taste	172.515	Coffee flavors
Ethyl 2-furanpropionate	A woody, pineapple odor and taste	172.515	Pineapple, brandy, apple, raspberry
4-Ethylguaiacol (4-ethyl-2-methoxyphenol)	Phenolic, bacon-like odor, mild smokey bacon taste	172.515	Smoke flavors, bacon, coffee
Ethyl heptanoate	Fruity, brandy odor with sharp, fruity taste	172.515	Brandy, cognac, berry flavors, cheese, blueberry
2-Ethyl-2-heptenal	Fatty nut-like odor with a green, woody, fruity taste	172.515	Nut flavors, coconut, pineapple
Ethyl hexanoate (ethyl caproate)	A fruity, apple odor and estery fruity taste	172.515	Apple, banana, pineapple, rum
2-Ethyl-1-hexanol	A mushroom odor with perfumy sweet taste	3151	Traces in fruit flavors
Ethyl-3-hexenoate	A fruity, pineapple odor with cognac, pineapple taste	3342	Pineapple, apple, cognac
Ethyl-3-hydroxybutyrate	A fresh, fruity, fatty odor with fruity taste	3428	Grape, pineapple, blueberry
3-Ethyl-2-hydroxy-2-cyclopenten-1-one	A caramel, butterscotch odor with milk chocolate taste	3152	Chocolate, maple, coconut
2-Ethyl-2-hydroxy-4- methylcyclopent-2-en-1-one	—	3453	—
5-Ethyl-2-hydroxy-3- methylcyclopent-2-en-1-one	A caramel-like odor with sweet caramel taste	3454	Maple, meat, fruit flavors
5-Ethyl-3-hydroxy-4- methyl-2(5H)-furanone	A caramel-like odor with sweet berry-like taste	3153	Caramel, pineapple, strawberry, meat

Ethyl isobutyrate	Pungent, fruity, buttery odor with pleasant fruity taste	172.515	All fruits, strawberry, cherry, butter
N-Ethyl-2-isopropyl-5- methylcyclohexane carboxamide	—	3455	—
Ethyl iso valerate (ethyl 3-methylbutyrate)	An excellent apple odor and taste	172.515	Apple, adds lift to all fruit flavors, pineapple
Ethyl lactate (ethyl 2 hydroxypropionate)	A rum odor with a milk cream taste	172.515	Rum, milk, cream, grape, wine, coconut, rum
Ethyl laurate (ethyl dodecanoate)	A fatty, oily odor with a fatty, fruity taste	172.515	Cheese, coconut, cognac, nut
Ethyl levulinate	Pleasant apple, strawberry odor with mild fruity taste	172.515	Excellent solvent for fruit flavors, apple
Ethyl maltol (2 ethyl-3-hydroxy-4H-pyran-4-one)	Sweet cotton candy odor with sweet taste	172.515	Strawberry, grape, pineapple
Ethyl 2-mercaptopropionate	A meaty, green onion odor with onion, cooked garlic taste	3279	Spice flavors, meat, egg flavors
2-Ethyl(or methyl) — (3,5, or 6) methoxypyrazine	—	3280	—
Ethyl 2-methylbutyrate	A sweet, fruity, strawberry-like odor with apple, strawberry taste	172.515	Apple, strawberry, lift for fruit flavors
Ethyl 2 methylpentanoate	—	3488	—
Ethyl 2-methyl-3-pentenoate	A fruity, green banana odor with sweet berry taste	3456	Strawberry, banana
Ethyl 2 methyl-4-pentenoate	A fruity, green odor with berry, banana taste	3489	Strawberry, banana
Ethyl methylphenylglycidate (aldehyde C-16 so-called)	A fruity, berry aroma with an apple, berry taste	182.60	Strawberry, apple, blends well with aldehyde C-14
2-Ethyl-5-methylpyrazine	A weedy, tobacco, smoke, whisky odor, smokey peanut taste	3154	Peanut, chocolate, roasted flavors, nut
3-Ethyl-2-methylpyrazine	Peanut, chocolate odor and taste	3155	Peanut, chocolate, flavor
Ethyl-3-methylthiopropionate	A fruity, cooked pineapple odor with a pineapple taste	3343	Pineapple
Ethyl myristate (ethyl tetradecanoate)	A meaty coconut odor with coconut taste	172.515	Coconut, solvent for flavors, cognac
Ethyl nitrite	A sweet, rum-like odor and fruity taste	172.515	Rum, brandy, fruit flavors
Ethyl nonanoate (ethyl pelargonate)	A nutty, brandy odor with coconut taste	172.515	Brandy, rum, coconut
Ethyl 2-nonynoate (ethyl octyne carbonate)	Powerful, green, cucumber odor and taste	172.515	Melon flavors, strawberry
Ethyl octadecanoate (ethyl stearate)	Very mild odor with waxy taste	3490	Creates mouthfeel for flavors
Ethyl octanoate (ethyl caprylate)	A fruity, winey odor with a fruity, brandy taste	172.515	Apple, pineapple, wine, brandy, rum

Table 4 (continued)
SYNTHETIC FLAVOR INGREDIENTS

Name	Organoleptic characteristics	Regulation	Flavors useful in
Ethyl *cis*-4-octenoate	A pleasant non-characteristic fruity odor with fatty cognac taste	3344	Cognac, melon
Ethyl oleate	A mild fatty odor and taste	172.515	Butter, mouthfeel
Ethyl palmitate (ethyl hexadecanoate)	A waxy, odor and taste	2451	Coconut, cognac, butter
p-Ethylphenol	Medicinal, phenolic odor with a sweet smokey taste	3156	Smoke flavors
Ethyl phenylacetate	A sweet, honey-like odor with fruity, honey taste	172.515	Honey, peach, tobacco flavors
Ethyl 4-phenylbutyrate	Sweet fruity plum like odor with prune taste	172.515	Peach, apricot, prune
Ethyl 3-phenylglycidate	A fruity, berry-like odor with berry-jam taste	172.515	Berry flavors, cherry
Ethyl 3-phenylpropionate (ethyl hydrocinnamate)	A fruity berry-like odor and fruity melon taste	172.515	Berry flavors, melon
Ethyl propionate	A strong, fruity, rum odor and taste	172.515	Rum, apple, pineapple
2-Ethylpyrazine	A peanut aroma with a peanut, chocolate taste	3281	Peanut, chocolate, roasted flavors, potato
3-Ethylpyridine	A roasted odor with a smoke, roasted flavor	3394	Smoke, meat flavors
Ethyl pyruvate	A sweet, fatty odor with a mouth-coating taste	172.515	Milk, coconut, butterscotch, chocolate
Ethyl salicylate	Sweet floral, fruity odor with fruity taste	172.515	Berry flavors, root beer, toothpaste, mouthwash flavors
Ethyl sorbate (ethyl 2,4 hexadienoate)	A fruity, caramel, pineapple odor and taste	172.515	Pineapple, apple, tropical fruit
Ethyl thioacetate	A cooked garlic, meat odor with garlic taste	3282	Spice flavors, poultry flavors
2-Ethylthiophenol (2 ethyl phenylmercaptan)	A smoke, meat odor with phenolic, smoke taste	3345	Smoke, meat, nut
Ethyl tiglate (ethyl trans-2-methyl-2-butenoate)	Fruity, pear, apple odor and taste	172.515	Pear, apple, rum, berry
Ethyl (p-tolyoxy) acetate	An anise "sweaty-horse" odor with honey-anise task	3157	Honey, anise
2-Ethyl-1,3,3-trimethyl-2-norbornanol	—	3491	—
Ethyl undecenoate	A fatty, nutty odor and taste	172.515	Nut, brandy, coconut
Ethyl 10-undecenoate	A fruity, fatty odor with wine-like taste	172.515	Nut, wine, coconut
Ethyl valerate	A fruity, apple odor and taste	172.515	Apple, pineapple, peach, apricot
Ethyl vanillin (3 ethoxy 4 hydroxy benzaldehyde)	A sweet chocolate, vanilla odor and taste	182.60	Vanilla, chocolate, cream soda
Eucalyptol (1-8-cineol)	A cool camphoraceous odor and taste	172.515	Mouthwash, toothpaste, cough-drops,

			mint imitations
Eugenol (4-allyl guaiacol) (4-allyl-2-methoxyphenol)	Spicy, smokey, bacon-like odor and taste	182.60	Spice flavors, smoke, bacon, nut
iso-Eugenol (1-hydroxy-2-methoxy-4-propenylbenzene)	Similar to normal compound except stronger in taste	2468	Spice flavors, bacon, smoke
Eugenyl acetate (acetyl Eugenol)	A clove-like odor with sharp, clove taste	172.515	Spice blends, vanilla, berry
iso-Eugenyl acetate (acetyl iso-Eugenol)	A clove-like odor with a sweet burning taste	2470	Spice blends, berry
Eugenyl benzoate (4-allyl-2-methoxy phenyl)	A resinous clove odor and taste	172.515	Spice blends, berry
Eugenyl formate (4 allyl-2-methoxy-phenyl formate)	Sweet, woody, floral, clove-like odor and taste	172.515	Spice blends
Eugenyl methyl ether (4 allyl veratrole) (1,2 dimethoxy-4-allylbenzene)	A spicy, clove-like odor and taste	172.515	Spice blends
Farnesol (3,7,11-trimethyl-2,6,10-dodecatrien-1-ol)	A weak citrus-lime odor and taste	172.515	Excellent blender
D-fenchone (*d*-1,3,3-trimethyl-2-norbornanone)	Sweet resinous, camphoraceous odor and cool medicinal taste	172.515	Essential oil imitations, traces in fruit flavors
Fenchyl alcohol	Sweet, camphoraceous, citrus-like odor and citrus taste	172.515	Citrus (lime) spice blends, strawberry
Formic acid	Pungent odor and sour taste	172.515	Pineapple, rum, smoke
2-Formyl-6,6-dimethyl bicyclo(3,1,1)hept-2-ene (myrtenal)	A spicy cinnamon-like odor and taste	3395	Essential oil imitation, traces in fruit flavors
Fumaric acid	Sour taste	2488	Acidulant
2-Furanmethanethiol formate	A coffee-like odor and taste	3158	Coffee, meat, nut
Furfural (2-furaldehyde)	A cereal-like, spicy odor with distinct caramel taste	2489	All heat processed type flavors, bread, butterscotch, coffee
Furfuryl acetate	An estery, floral odor with mild, fruity taste	2490	Fruit blends, coffee
Furfuryl alcohol	Very low odor with a cooked sugar taste	2491	Caramel, butterscotch, coffee
2-Furfurylidene butynaldehyde	A sweet, spicy cinnamon odor and taste	2492	Spice blends, nut, rum
Furfuryl isopropyl sulfide	A fried garlic odor and taste	3161	Garlic, onion, coffee
Furfuryl mercaptan (2 furanmethanethiol)	Egg, meat, coffee aroma with coffee taste at low levels	2493	Coffee, meat flavors, chocolate
Furfuryl 2-methyl-butyrate	A general fruity odor with tastes ranging from licorice, grape, plum	3283	Fruit flavors, molasses
Furfuryl methyl ether	Solventy odor with hint of nut taste	3159	Peanut, coffee, nut
Furfuryl methyl sulfide	At low levels, a cooked garlic odor and taste	3160	Spice flavors, coffee
alpha-furfuryl octanoate	A sweet coconut-like odor with a cheesy, creamy, sweet taste	3396	Coconut, cream, cheese

Table 4 (continued) SYNTHETIC FLAVOR INGREDIENTS

Name	Organoleptic characteristics	Regulation	Flavors useful in
alpha-furfuryl pentanoate	A woody, fruity pineapple odor with fruity taste	3397	Pineapple, grape
Furfuryl propionate	—	3346	—
N-Furfurylpyrrole (1-(2-furfuryl)pyrzole	A pungent, earthy odor and taste	3284	Garnish flavors for soups and gravies
Furfuryl thioacetate	A cheese, garlic-like odor and at low levels, a cheese taste	3162	Cheese, malt flavors
Furfuryl thiopropionate	A coffee-like aroma and taste	3347	Coffee, meat flavors
Furyl acrolein (2 furanacrolein)	A powerful, woody, cinnamon odor and taste	2494	Spice flavors, coffee
4-(2-Furyl)-3-butene-2-one (furfuryl acetone)	A woody, cinnamon odor and taste	2495	Spice flavors
2-Furyl methyl ketone (acetyl furan)	A furfural, cereal-like odor with an oily, peanut taste	3163	Nut, cereal flavors, tobacco
(2-Furyl)-2-propanone	A fruity, spicy caramelic odor, sweet, fruity, spicy taste	172.515	Nut, fruit blends
1-Furyl-2-propanone	—	172.515	—
Fusel oil, refined	A whisky odor and taste	172.515	Wine, brandy, rum, fruit flavors
Geraniol (*trans*-3,7,dimethyl-2,6-octadien-1-ol)	Sweet, floral odor with fruity apple taste	182.60	Apple, peach, strawberry, apricot, citrus
Geranyl acetate	A floral, apple-like odor and at low levels, an apple taste	2509	Apple, pear, berry
Geranyl acetoacetate	An over-ripe fruit odor with fruity, apple taste	172.515	Apple, fruit blends
Geranyl acetone	A mild berry-like aroma with a black raspberry taste	172.515	Fruit flavors, tomato
Geranyl benzoate	A mild floral odor and in great dilution, an apple taste	172.515	Apple, fruit flavors
Geranyl butyrate	Sweet, heavy, fruity odor with fruity, apple tase	172.515	Apple, pear, pineapple
Geranyl formate	Fruity, leafy odor with fruity, woody taste	172.515	Apple, pear, black currant
Geranyl hexanoate	Fruity, floral, pineapple odor at low levels a sweet taste	172.515	Pineapple, apple, fruit blends
Geranyl iso butyrate	Sweet, fruity odor and fruity, floral taste	172.515	Strawberry, peach, pear, pineapple
Geranyl iso valerate	A floral, fruity, apple odor and taste	172.515	Apple, peach, pear, tobacco
Geranyl phenylacetate	Floral honey odor with fruity, honey taste	172.515	Honey, apricot, peach
Geranyl propionate	Sweet, floral, grape-like odor, fruity, floral taste	172.515	Pear, apple, berry flavors, pineapple
Glucose pentaacetate	Extremely bitter	172.515	A bittering agent

l-Glutamic acid (2 aminoglutaric acid)	Odorless powder with acid taste	182.1045	Meat, soup, poultry flavors
Glycerol (glycerine)	Mild odor, sweet taste	182.1320	Solvent for flavors
Glycerol ester of wood rusin	Extremely bitter tasting solid	6072	Weighing agent for citrus oils, chewing gum plasticizer
Glyceryl monooleate	Sweet odor, fatty taste	172.515	Emulsifier
Glyceryl monostearate	Mild odor, fatty taste	182.1324	Emulsifier
Glyceryl triacetate (tri-acetin)	Fruity acetic odor with sweet taste	2007	Butter, butterscotch, nut solvent for flavors
Glyceryl tribenzoate	—	3398	Weighing agent for citrus beverages
Glyceryl tributyrate (tributyrin)	Fruity, buttery odor with bitter taste	182.60	Butter, margarine, solvent for flavors
Glyceryl tripropanate (tripropionin)	Fruity sour odor with sweet taste	3286	Butter, cheese, margarine, solvent for flavors
Glycine (amino acetic acid)	Odorless powder with sweet taste	3287	Taste modifier
Guiacol (*o*-methoxyphenol)	Odor reminiscent of cough syrup, hickory smoke taste at 5 ppm	172.515	Smoke, coffee, vanilla, tobacco
Guaiacyl acetate (*o*-methoxyphenyl acetate)	Subdued smoke aroma sweet, smokey, vanilla, maple taste	172.515	Smoke, maple, currant, coffee
Guaicyl phenylacetate	Phenolic resinous vanilla with sweet phenolic taste	172.515	Vanilla, honey, smoke, tobacco
Guaiene	A spicy, resinous odor and peppery taste	172.515	Artificial essential oils, spice blends, fixative, blender
Guaiol acetate	Sweet, earthy odor and fruity, peppery taste	172.515	Traces in fruit flavors, tobacco
2,4 Heptadienal	Green, pungent, fruity, spicy odor with green, spicy taste	3164	Fruit blends, blueberry, raspberry
gamma-Heptalactone	A fatty, sweet, coconut, coumarin odor and taste	172.515	Nut, coconut, fruit flavors
Heptanal (aldehyde C-7)	Fatty, pungent odor, fatty taste	172.515	Citrus, vegetable flavors, melon
Heptanal glyceryl acetal	Earthy, dirty odor with earthy, mushroom taste	172.515	Mushroom
2,3-Heptandione (acetyl valeryl)	Fatty, cheese, waxy odor "tub" butter, mozzarella taste	172.515	Butter, cheese flavors, berry
Heptanoic acid	Sweaty, fatty, rancid odor and taste	3348	Cheese, dairy flavors
2-Heptanol	A green, citrusy aroma and taste	3288	Citrus flavors
3-Heptanol	Powerful, pungent, resinous odor with bitter taste	172.515	Traces in fruit flavor, spice blends
2-Heptanone (methyl amyl ketone)	A fruity, cheese aroma with blue-roguefort taste	172.515	Cheese, banana, butter, coconut
3-Heptanone (ethyl butyl ketone)	Fruity, fatty odor with fruity "ketone" taste	172.515	Cheese, banana, melon
4-Heptanone (dipropyl ketone)	Powerful solvent odor with fruity berry taste	172.515	Berry flavors
2-Heptenal	Fatty, meaty, poultry odor with similar taste	3165	Meat, poultry, green vegetable flavors

Table 4 (continued)
SYNTHETIC FLAVOR INGREDIENTS

Name	Organoleptic characteristics	Regulation	Flavors useful in
4-Heptenal	A green, fatty aroma with a creamy taste at low levels	172.515	Buttery flavors
4-Heptenal diethyl acetal	A fruity, green odor with creamy taste at low levels	3349	Butter, milk, cheese
2-Hepten-4-one	—	3399	—
3-Hepten-2-one	—	3400	—
trans-3-Heptenyl acetate	—	3493	
trans-3-Heptenyl-2-methyl-propionate	—	3494	—
Heptyl acetate	A fatty, fruity odor with a milk, coconut, apricot taste	172.515	Apricot, coconut, pear, pineapple
Heptyl alcohol (alcohol C-7)	Green, fatty odor with sweet, fruity nut-like taste	172.515	Coconut, nut flavors
Heptyl butyrate	A fruity, fatty odor with a strawberry taste	172.515	Strawberry, plum, fruit flavors
Heptyl cinnamate	Green, earthy, woody odor with floral taste	172.515	Almond, cherry, grape, berry
Heptyl formate	Fatty aliphatic alcohol odor with a fatty, coconut taste	172.515	Coconut, plum, apricot, peach
2-Heptyl furan	—	3401	—
Heptyl iso butyrate	A fatty, fruity odor with a black pepper spicy taste	172.515	Fruit flavors, spice blends
3-Heptyl-5-methyl-2-(3H)-furanone	—	3350	—
Heptyl octanoate	A fatty, green, fruity odor with fatty fruity taste	172.515	Fruit flavors
1-Hexadecanol (cetyl alcohol)	Faint sweet odor, bland taste	172.515	Traces in flavors, chocolate
omega-6-hexadecenlactone (ambrettolide)	Musk aroma and taste	172.515	Traces in flavors, pear
trans,trans-2,4-hexadienal	Sweet, green odor with citrus taste	3429	Citrus flavors
gamma-hexalactone	A coumarin-like odor and taste	172.515	Tobacco, coconut, fruit flavors
delta-hexalactone	A coumarin-like odor with creamy, fruity, coconut taste	172.515	Coconut, tobacco, fruit flavors
Hexanal (aldehyde C-6)	Strong, green, grass odor with apple taste	172.515	Apple, tomato
2,3-Hexandione (acetyl butyryl)	Butterscotch-like odor with creamy butter-fat-like taste	172.515	Butter, butterscotch, fruit flavors, berry
3,4-Hexandione (diproprionyl)	Pungent, buttery odor and taste	3168	Butter, margarine, fruit flavors
1,6-Hexane dithiol	—	3495	—
Hexanoic acid (caproic acid)	Powerful rancid cheese odor and taste	172.515	Cheese, butter, rum, butterscotch
3-Hexanol	Solventy odor with an oily coconut taste	3315	Coconut, milk, cream

3-Hexanone (ethyl propyl ketone)	—	3290	—
2-Hexenal	A green, rhubarb odor with a fatty, green, chicken taste	172.515	Raspberry, fruit flavors, apple, strawberry
cis-3-Hexenal	A fresh green, fruity odor with fruity, green taste	3496	Fruit flavors
4-Hexene-3-one	—	3352	—
trans-2-Hexenoic acid	A pleasant, musty, cheese odor with a cheese, cream cheese taste	3169	Cheese, butter, berry flavors
3-Hexenoic acid	Musty cheese odor with mild cheese taste	3170	Cheese, berry flavors
2-Hexen-1-ol	A green vegetable odor with apple taste	172.515	Apple, fruit flavors
3-Hexen-1-ol (leaf alcohol)	A powerful green leafy odor with a long lasting green-leaf taste	172.515	Fruit flavors, mint
4-Hexen-1-ol	Fatty, fruity, green odor, fatty green taste	3430	Fruit flavors
2-Hexen-1-yl acetate	An apple, pear odor with a pear taste	172.515	Apple, pear, banana
cis-3-Hexen-1-yl acetate	A green apple, banana odor with a ''starchy'' banana taste	3171	Apple, banana, fruit flavors
cis-3-Hexen-1-yl butyrate	A fruity pear odor with a pear, apple taste	3402	Apple, pear, brandy
3-Hexenyl formate	A pungent, green fruity odor with fruity taste	3353	Fruit flavors, vegetable flavors
cis-3-Hexenyl formate	A sharp, green formic acid odor with green cucumber, melon taste	3431	Fruits, cucumber, melon
cis-3-Hexenyl hexanoate	A green fruity odor with a fatty, buttery taste	3403	Pineapple, fruit flavors
3-Hexenyl isovalerate	A green, fruity odor with a green, melon, honey taste	172.515	Strawberry, melon
3-Hexenyl-2-methyl-butanoate	A fruity, buttery odor with an apple taste	3497	Apple
3-Hexenyl 3-methyl-butanoate	A fruity, buttery odor with a green, fruit taste	3498	Strawberry, raspberry
3-Hexenyl 2-methyl-butyrate	A green, fruity odor with a black pepper taste	172.515	Pear, apple, banana, spice blends
3-Hexenyl phenyl acetate	A fruity, floral odor with an excellent spicy taste	172.515	Spice blends
Hexyl acetate	A fruity, apple, pear odor with fruity, apple taste	172.515	Apple, pear, fruit flavors
2-Hexyl-4-acetoxy-tetrahydrofuran	Sweet floral, fruity odor with sweet, fruity taste	172.515	Peach, apricot, fruit flavors
Hexyl alcohol (alcohol C-6)	Fatty, fruity odor and taste	172.515	Berry flavors, coconut
N-Hexyl-2-butenoate (hexyl crotonate)	Wood, fruity odor with a fruity, pineapple taste	3354	Pineapple flavor
Hexyl butyrate	Strong mixed fruit odor with sweet fruity, pineapple taste	172.515	Pineapple, pear, strawberry
alpha-hexylcinnamaldehyde	Oily, floral odor, sweet spicy taste at low levels	172.515	Spice blends, honey, fruit
Hexyl formate	Strong, unripe fruit odor with green fruity taste	172.515	Apple, pineapple, mixed fruit
Hexyl 2-furoate	A caramel, fruity odor with fruity taste	2571	Coffee, maple, nut, fruit flavors, mushroom
Hexyl hexanoate	Green, ''snap-beans'' odor with green, fruity taste	172.515	Strawberry, vegetable flavors

Table 4 (continued)
SYNTHETIC FLAVOR INGREDIENTS

Name	Organoleptic characteristics	Regulation	Flavors useful in
2-Hexylidene cyclo-pentanone	Green, fruity, spicy odor with floral, vegetable taste	172.515	Strawberry, peach, fruit flavors, toothpaste flavors
Hexyl isobutyrate	A pungent fruity odor and taste	3172	Strawberry, fruit flavors
Hexyl isovalerate	Sweet, pungent, green, fruity odor with green, fruity taste	172.515	Fruit flavors, tobacco flavors
2-Hexyl-5 or 6-keto-1,4 dioxane	A fatty, peach, lactone-like odor with creamy oily, nutty taste	2574	Peach flavors, milk and cream flavors
Hexyl-2-methylbutanoate	Fruity, spicy odor with spicy black pepper taste	3499	Spice blends, fruit flavors
Hexyl-3-methylbutanoate	A green, fruity odor and fruity, apple taste	3500	Apple, fruit flavors
Hexyl-2 methylbutyrate	A spicy green, peppery odor with a spicy black pepper taste	172.515	Spice flavors, black pepper without bite
Hexyl octanoate	Fruity, apple pulp aroma with fatty, cognac taste	172.515	Apple, coconut, brandy
Hexyl phenylacetate	A green fruity, winey odor with fruity, honey, taste	172.515	Honey, fruit flavors
Hexyl propionate	Earthy, sour, fruity odor with sweet, fermented, fruity taste	172.515	Fruit flavors
4-Hydroxybutanoic acid lactone	Fatty, meat-like odor with creamy milk taste	3291	Fruit flavors, meat
1-Hydroxy-2-butanone	—	3173	—
Hydroxycitrovellal (3,7-dimethyl-7-hydroxoctanal)	Pleasant, floral odor and sweet, floral taste	172.515	Traces in fruit flavors, citrus
Hydroxycitronellal dimethyl acetal	Pleasant, floral melon-like odor and taste	172.515	Melon-like at low levels, citrus, cherry
Hydroxycitronellal diethyl acetal	Pleasant, mild floral odor and taste	172.515	Traces in fruit flavors, citrus (lime)
Hydroxycitronellol (3,7-dimethyl-1,7 octanediol)	Weak, floral odor and taste	2586	Artificial citrus flavors, cherry
2-Hydroxy-2-cyclohexen-1-one	—	3458	—
4-Hydroxy-2,5 dimethyl-3(2H) furanone	Sweet, fruity, caramel odor with a fruity taste	3174	Pineapple, strawberry, meat flavors
6-Hydroxy-3,7 dimethyl octanoic acid lactone	—	3355	—
N-(4-hydroxy-3-methoxybenzyl)-8-methyl-6-nonenamide	Mild spicy odor with biting taste	172.515	Spice blends for pungency, bite
N-(4-hydroxy-3 methoxybenzyl) nonanamide	Odorless with pungent biting taste	2787	Spice blends, synthetic capsicum

2-Hydroxymethyl-6,6 dimethyl bicyclo(3.1.1)hept-2-enyl formate	Sweet, spicy odor and taste	3405	Artificial essential oils
3-(Hydroxymethyl)-2-octanone	Musty, earthy odor and taste	3292	Mushroom, nut flavors
Hydroxynonanoic acid delta-lactone (delta nonalactone)	A sweet nut-like odor with fatty, milk, cream taste	3356	Peach, coconut, milk, cream, nut
5-Hydroxy-4-octanone (butyroin)	Sweet, sharp, buttery odor with oily, buttery taste	172.515	Butter, cheese, nut
4-Hydroxy-3-pentenoic acid lactone	Pleasant, bread, molasses-like odor with raw chestnut taste		Chestnut, walnut, nut flavors
4-(*p*-hydroxy phenyl)-2-butanone (oxanone)	A pleasant, sweet odor with sweet, fruity taste	172.515	Raspberry
2-Hydroxy-3,5,5 trimethyl-2-cyclohexenone	—	3459	—
5-Hydroxyundecanoic acid lactone (delta undecalactone)	A creamy, fatty odor with creamy taste	3294	Peach, coconut, milk, nut
Indole (2,3 benzopyrrole)	A mild, fecal odor and at low levels, a floral taste	172.515	Cheese, citrus, mixed fruit, artificial essential oils
alpha-ionone 4-[2,6,6-trimethyl-2-cyclohexen-lyl]-3-buten-2-one	A woody, perfumy odor with fruity, raspberry taste	172.515	Raspberry, blackberry
beta-ionone (4-[2,6,6-trimethyl-1-cyclohexen-1yl]-3 buten-2 one)	A woody, perfumy odor with fruity, raspberry taste	172.515	Raspberry, blackberry
gamma-ionone	A wood, perfumy odor with fruity, raspberry taste	3175	Raspberry, blackberry
alpha-irone	A sweet, perfumy, orris odor with sweet, berry taste at low levels	172.515	Raspberry, strawberry, mixed fruit
isoamyl acetate	Strong, pear, banana odor and taste	172.515	Apple, pear, banana
isoamyl acetoacetate	Estery, green, fruity odor with green-apple taste	172.515	Apple, strawberry, fruit flavors
isoamyl alcohol	Pleasant, whisky, apple-like odor, mild apple-banana taste	172.515	Apple, banana
isoamyl benzoate	A floral, chocolate, banana odor with sweet talcum-powder taste	172.515	Vanilla, fruit flavors
isoamyl butyrate	A fruity, mixed fruit odor, sweet and fruity taste	172.515	Banana, pineapple, apricot, cherry, mixed fruit
isoamyl cinnamate	Sweet, heavy, chocolate-like odor and taste	172.515	Vanilla, chocolate
isoamyl formate	Volatile, sweet odor with fruity taste	172.515	Apple, fruit flavors
isoamyl-2-furanbutyrate	Sweet, caramel, buttery odor and sweet taste	172.515	Coffee, chocolate
isoamyl-2-furanpropionate	Floral, fruity, caramel-like odor and taste	172.515	Coffee, chocolate

Table 4 (continued)
SYNTHETIC FLAVOR INGREDIENTS

Name	Organoleptic characteristics	Regulation	Flavors useful in
isoamyl hexanoate	Sharp, mixed fruit odor with apple, pineapple taste	172.515	Apple, banana, pineapple, rum
isoamyl isobutyrate	Mixed fruit odor, sweet, fruity, apricot, peach taste	172.515	Pineapple, peach, apricot
isoamyl iso valerate	Choking apple odor and apple taste	172.515	Apple, banana, berry flavors
isoamyl kaurate	Fatty, oily odor with fatty taste	172.515	Coconut, fatty flavors
isoamyl 2-methylbutyrate		172.515	
isoamyl nonanoate	Nutty, fruity, apricot odor and taste	172.515	Nut, apricot, cognac, rum
isoamyl octanoate	Fruity fatty odor and taste	172.515	Banana, fruit flavors
isoamyl phenylacetate	Resinous, fruity odor with sweet taste	172.515	Honey, cocoa, chocolate, butter
isoamyl propionate	Very sweet, mixed fruit odor, sweet mixed fruit taste	172.515	Apricot, pineapple, banana, rum
isoamyl pyruvate	Estery, burnt sugar odor with rum-like taste	172.515	Rum, maple, arak, mixed fruit
isoamyl salicylate	Sweet, floral, green odor, perfumy taste	172.515	Traces in flavors, root beer
isoborneol	Camphor-like odor and taste	172.515	Traces in fruit flavors, spice blends, nut
isoborynyl acetate	Resinous, pine, camphor odor and taste	172.515	Traces in fruit flavors, berry
isobornyl formate	Earthy, camphoraceous odor and taste	172.515	Traces in fruit flavors, berry
iso bornyl isovalerate	Minty, pine odor with fruity "piney" taste	172.515	Tropical fruit flavors, berry
iso boryl propionate	Piney, camphoraceous odor and taste	172.515	Traces in fruit flavors, berry, black currant
isobutyl acetate	Solventy nail polish-like with banana taste	172.515	Banana, raspberry, strawberry, butter
isobutyl acetoacetate	Sweet, brandy-like odor with sweet, fruit taste	172.515	Berry, fruit flavors
isobutyl alcohol	Sweet, choking odor and sweet whisky taste	172.515	Rum, banana, fruit flavors
isobutyl angelate (iso buty *cis*-2-methyl-2-butenoate)	A pleasant, sweet odor with a taste similar to iso Jasmone	172.515	Essential oil imitations, chamomile
isobutyl anthranilate	A fruity, grape odor and taste	172.515	Grape, fruit flavors, tangerine
isobutyl benzoate	A medicinal, chocolate-like odor with an anise, chocolate taste	172.515	Raspberry, strawberry, fruit flavors, cherry
isobutyl-2-butenoate	A fruity odor with an excellent "jammy" taste	3432	Raspberry, blueberry, grape
isobutyl butyrate	Choking, minty, earthy, cheese odor with fruity, peach, apricot taste	172.515	Peach, apricot, berry
isobutyl cinnamate	A pleasant, fruity odor with a fatty, waxy taste	172.515	Berry flavors, currant, peach

isobutyl formate	Estery, fruity odor with rum taste	172.515	Rum, fruit flavor, whisky
isobutyl 2-furanpropionate	Fruity, rum, brandy odor sharp, fruity taste	172.515	Pineapple, fruit flavors, berry
isobutyl heptanoate	Green, fruity odor with sweet, fruity taste	172.515	Fruit flavors, wine, brandy, mixed fruit
isobutyl hexanoate	Fruity, pineapple odor and taste	172.515	Pineapple, apple, mixed fruit
isobutyl iso butyrate	Fruity, berry-like odor and taste	172.515	Berry, fruit flavors, rum
2-isobutyl-3-methoxy-pyrazine	A powerful, bell pepper odor and taste	3131	Bell pepper, potato
2-isobutyl-3-methyl-pyrazine	—	3133	—
alpha-isobutyl phenethyl alcohol	A weak, chocolate odor with a bitter chocolate taste	172.515	Chocolate flavors, caramel, spice
isobutyl phenylacetate	Floral, rose-lily odor with sweet honey taste	172.515	Honey, nut, caramel, chocolate
isobutyl propionate	Estery rum-like odor and taste	172.515	Rum flavors, lift for fruit flavors, strawberry
isobutyl salicylate	Harsh, sweet, floral odor and taste	172.515	Traces in fruit flavors, root beer
2-isobutyl thiazole	A weedy, tomato-leaf odor, upon great dilution, tomato taste	3134	Tomato, fruit flavors
isobutyraldehyde	Extremely sharp odor with fruity taste	172.515	Fruit flavors, banana, caramel
isobutyric acid	Sharp, butter-fat-like odor, cheesy taste	172.515	Butter, cheese flavors, apple, caramel
isoeugenol (1-hydroxy-2-methoxy 4-propenyl benzene)	Sweet, spicy, clove-like odor and taste	172.515	Spice flavors, banana, raspberry, strawberry, clove
isoeugenyl acetate (acetyl iso Eugenol)	Spicy, clove-like odor with a sweet spicy taste	172.515	Spice flavors, raspberry, strawberry
isoeugenyl benzyl ether	A spicy, banana-like odor with a clove, banana taste	172.515	Spice flavor, banana
isoeugenyl ethyl ether	A spicy, clove odor and taste	172.515	Spice flavors, banana, vanilla
isoeugenyl formate	Woody, perfumy odor with clove-like taste	172.515	Spice flavors
isoeugenyl methyl ether	Sweet, spicy, clove-like odor and taste	172.515	Spice blends, berry flavors, vanilla
isoeugenyl phenylacetate	A weak, clove odor at low levels, clove-like, greater than 10 ppm, honey taste	172.515	Spice flavors, vanilla, honey, traces in fruit flavors
isojasmone	Green, floral odor and taste	172.515	Traces in fruit flavors, strawberry
D, L-isoleucine (2-amino β-methyl valeric acid)	Odorless, white powder with slightly bitter taste	3295	—
D,L-isomenthone	Weedy, peppermint odor and taste	3460	Mint flavors, toothpaste, mouthwash
alpha, isomethylionone	Sweet, floral odor with a perfumy, soapy taste	172.515	Raspberry, cachou flavors
isopentylamine	An irritating, ammonia-like odor and taste	3219	Chocolate, seafood
isopropenylpyrazine	—	3296	—
iso propyl acetate	Similar to ethyl acetate, not as sharp	172.515	Rum, solvent for fruit flavors
p-isopropylacetophenone	Strong, woody, ionone-like odor and taste	172.515	Honey, fruit flavors
isopropyl alcohol	Rubbing alcohol odor and bitter taste	172.515	Fruit flavors, solvent for flavors, banana

Table 4 (continued)
SYNTHETIC FLAVOR INGREDIENTS

Name	Organoleptic characteristics	Regulation	Flavors useful in
isopropyl benzoate	Fatty medicinal, caraway odor; at low dilution, a fruity taste	172.515	Cherry, fruit flavors, berry
p-isopropylbenzyl alcohol (cumin alcohol)	Spicy, cumin odor at 20 ppm, definite cinnamon with bite	172.515	"Hot" spices, cinnamon, ginger, caraway
isopropyl butyrate	Pungent, fruity odor with sweet, strawberry taste	172.515	Strawberry, apple, pineapple, mixed fruit
isopropyl cinnamate	Sweet resinous odor, fruity, jam-like taste	172.515	Strawberry, raspberry, honey
isopropyl formate	A pleasant nondescript odor and taste	172.515	Lift for fruit flavors, plum, melon
isopropyl hexanoate	Fruity capirate odor, fruity citrus taste	172.515	Fruit flavors, pineapple
isopropyl isobutyrate	Strong pineapple odor and taste	172.515	Pineapple
isopropyl isovalerate	Sweet, estery, apple-pineapple-like odor and taste	172.515	Apple, pineapple, nut
2-isopropyl-5-methyl-2-hexenal	A woody, chocolate odor with a floral, bitter chocolate taste	3406	Chocolate, cocoa
2-isopropylphenol	Spicy, phenolic odor with smokey taste	3461	Meat, soup seasoning, smoke flavors
p-isopropylphenylacetaldehyde	Green, woody odor with fruity citrus-like taste	172.515	Citrus blends, fruit flavors
isopropyl phenylacetate	Floral, rosy, honey, chocolate-like odor with floral, chocolate taste	172.515	Chocolate, honey, butter
3-(*p*-isopropylphenyl)-propionaldehyde (cuminyl acetaldehyde)	Powerful, sweet, green, floral odor with a fruity taste at low levels	172.515	Fruit blends
isopropyl propionate	Estery, fruity, rum-like odor and taste	172.515	Rum, mixed fruit
isopropyl tiglate	—	3229	—
isopulegol (*p*-menth-8-en-3-ol)	Minty odor with bitter minty taste	172.515	Mint flavors, traces in berry flavors
isopulegone (*p*-menth-8-en-3-one)	Strong, minty, woody odor with minty, bitter taste	172.515	Mint flavors, traces in berry, mouthwash, toothpaste, artificial essential oils
isopulegyl acetate	A minty, jammy odor with a pear, jammy, minty taste	172.515	Berry flavors, fruit flavors
isoquinoline	Sweet, almond, anise odor with sweet, floral taste	172.515	Vanilla flavors
isovaleric acid	Strong, cheesy odor and taste	172.515	Cheese, butter, traces in fruit flavors
3-Keto-4-butanethio (4-mercapto-2-butanone)	—	3357	—
Lactic acids (2-hydroxy propionic acid)	Mild, creamy odor with pleasant, sour taste	182.1061	Dairy flavors, fruit flavors, butterscotch
Lauric acid (dodecanoic acid)	Very low odor with waxy, soapy taste	2614	Traces in citrus, butter

Lauric atdehyde (aldehyde C-12 lauric)	Sweet, floral odor with waxy, citrus taste	172.515	Citrus flavors, mixed fruit
Lauryl acetate (acetate C-12)	Waxy, fruity odor with citrus taste	172.515	Citrus flavors
Lauryl alcohol (alcohol C-12)	Oily, waxy odor with citrus, soapy taste	172.515	Lemon, orange, coconut, pineapple
L-Leucine (2 amino iso caproic acid)	Odorless and tasteless while solid	3297	—
Levulinic acid (3-acetyl propionic acid)	A maple, malt, cereal odor with a maltol, maple taste	172.515	Maple, butterscotch, nut, berry
d-Limonene (*d-p*-mentha-1,8-diene)	A fresh, citrus odor and taste	182.60	Citrus flavor, artificial essential oils
d-l-Limonene (dipentene)	A citrus odor and sweet, citrus taste	182.6	Essential oil, imitation citrus
Linalool (3,7 Dimethyl-1,6 octadien 3-ol)	Floral, spicy, wood odor with spicy, citrus taste	182.6	Spice flavors, citrus pineapple, peach, chocolate
Linalool oxide (2,3-epoxy-2,6 dimethyl-7 octen-6-ol)	Similar to linalool, with a minty taste	172.515	Spice flavors, fruit
Linalyl acetate	Fruity, floral, citrus, pear odor with citrus taste	182.60	Citrus, apricot, peach, pear, gooseberry
Linalyl anthranilate	Green, fruity, grape odor and taste	172.515	Grape, citrus, pineapple, lychee
Linalyl benzoate	Fruity, citrus odor and taste	172.515	Berry flavors, citrus, mixed fruit
Linalyl butyrate	Fruity, pear-citrus, odor, sweet, honey taste	172.515	Citrus, pear, peach, loganberry
Linalyl cinnamate	Sweet floral resinous odor with mild fruity flavor	172.515	Grape, berry flavors, loganberry
Linalyl formate	Citrus odor with spicy, apricot, peach taste	172.515	Apricot, peach, pineapple, apple
Linalyl hexanoate	Fruity odor with pear, pineapple taste	172.515	Pineapple, pear
Linalyl isobutyrate	Fruity, floral odor with sweet, fruity taste	172.515	Black currant fruit flavors, berry, cherry
Linalyl isovalerate	A fruity, apple-like odor and taste	172.515	Apple, peach, plum, loganberry
Linalyl octanoate	Unripe fruit odor and a peach, apricot taste	172.515	Peach, apricot, fruits citrus
Linalyl phenylacetate	Floral, sweet odor with a taste of honey	3501	Honey
Linalyl propionate	Fruity, floral odor with fruity, pineapple, pear taste	172.515	Pineapple, pear, black currant
L-Malic acid (hydroxysuccinic acid)	Colorless white solid with a sour taste	182.60	All fruit beverages
Maltol (3-hydroxy-2-methyl-411-pepane-4-one)	A sweet cotton candy odor and sweet jammy taste	172.515	Strawberry, all fruits
Maltyl isobutyrate	A sweet, berry-like odor and taste	3462	Strawberry
p-Mentha-1,8, dien-7-ol (perilla alcohol)	A spicy, oily aroma with peanut taste	172.515	Citrus, mint flavors
Menthadienol (*p*-menthan-1,8(10) dien-9-ol)	—	172.515	—
Menthadienyl acetate (*p*-menthan-1,8(10) dien-9 yl acetate)	—	172.515	—
p-Menthan-2-one (carvomenthone)	Carvone, rye bread, spearmint odor with caraway taste	3176	Rye bread
p-Mentha-8 thiol-3-one	Odor and taste of buchas	3177	Black currant, grape

Table 4 (continued)
SYNTHETIC FLAVOR INGREDIENTS

Name	Organoleptic characteristics	Regulation	Flavors useful in
p-Menth-1-ene-9-al	Spoiled minty citrus, earthy odor with sweet licorice taste	3176	Mint flavors,licorice flavors
p-Menth-1-en-3ol	Strong, minty, camphor odor and taste	3179	Mint flavor, toothpaste, mouthwash flavors
p-Menth-3-en-1-ol	—	172.515	—
1-*p*-Menthen-9-yl acetate	—	172.515	—
Menthol (hexahydrothymol) (1-methyl-4-isopropyl-cyclohexan-3-ol)	A clean, minty odor with cool, minty taste	172.515	Mint flavors, toothpaste, mouthwash, tobacco flavors
Menthone (1-methyl-4-isopropyl-cyclohexan-3-one)	A dirty peppermint odor and taste	172.515	Toothpaste flavors, mint flavors
Menthyl acetate	Earthy, weedy, fruity, berry-like odor and taste	172.515	Mint flavors, fruit flavors, berry
Menthyl isovaterate	A minty, fruity odor with a fruity, soapy taste	172.515	Mint flavors, fruit flavors
¼ + Mercapto-3-butanol	—	3502	—
3-Mercapto-2-butanone	—	3298	—
2-Mercaptomethyl pyrazine	Roasted odor with roasted meat-like taste	3299	Meat flavors, nut
3-Mercapto-2-pentanone	—	3300	—
2,3 or 10 Mercaptopinane	—	3503	—
Mercaptopropionic acid	A fatty, cheesy, sulfide odor with a fatty, bacon taste	3180	Meat, bacon flavors
p-Mentha-1,8 dien-7-ol (perillyl alcohol)	Spicy, woody odor with very sweet taste	2664	Mint flavors, fruit blends
D,L-Methionine (2 amino -methylmercapto butyric acid)	Sulfury odor with a sweet, then bitter taste	3301	—
D-neo-menthol 2-iso-butyl-5-methyl cyclohexanol)	Woody, menthol, minty odor, a lesser cooling effect	2666	Toothpaste, mouthwash, mint flavors
p-Methoxybenzaldehyde (anisic aldehyde)	A sweet, anise, vanilla, odor and taste	172.515	Vanilla, spice favors, berry
o-Methoxy benzaldehyde	A very sweet, spicy, anisic aldehyde-like odor and taste	172.515	Vanilla, spice blends, berry
o-Methoxycinnamaldehyde	A spicy, cassia odor and spicy, sweet taste	172.515	Spice flavors, cassia
2-Methoxy-3(5 and 6) isopropylpyrazine	A bell pepper odor and taste at low levels	3358	Bell pepper
Methoxyalpha methyl cinnamaldehyde	Mild, spicy, fruity odor with weak, spicy taste	3182	Spice flavors
2-Methoxy-4-methyl phenol (creosol) (4-methyl guaiacol)	Sweet phenolic odor and taste	172.515	Rum, nut, clove, vanilla

2 Methoxy-3-(1-methylpropyl)-pyrazine	A green, vegetable, bell-pepper-like odor and taste	3433	Bell pepper
2,5 or 6 methoxymethyl pyrazine	A roasted almond, hazelnut-like odor with roasted peanut taste	3183	Hazelnut, almond, peanut
(4-(*p*-Methoxyphenyl)-2-butanone (anisyl acetone)	A sweet anise-like odor and taste	172.515	Vanilla, berry flavors, licorice
1-(4-Methoxyphenyl)-4-methyl-1-pentene-3-one	A weak, rye-bread-like odor with distinct cheese taste	172.515	Cheese flavors, butter maple, vanilla, cream
1-(*p*-Methoxyphenyl)-1-penten-3-one (ethone)	Weak, creamy vanilla odor with tongue coating, cream, buttery tase	172.515	Butter, butterscotch, vanilla
1-(*p*-Methoxyphenyl)-2-pronanone (canisic ketone)	Medicinal anise odor with sweet anise taste	172.515	Vanilla, nut, cherry
Methoxy pyrazine	A none characteristic solvent note; none characteristic taste	3302	Roasted flavors
2-Methoxy-4-vinylphenol (*p*-vinylguaiacol)	Spicy, clove-like odor with very sweet taste	172.515	Vanilla, cocoa coffee
Methyl acetate	Very volatile, fruity odor with fleeting, fruity taste	172.515	Rum, brandy, whisky, lift for fruit flavors
4-Methylacetophenone (*p*-acetyl toluene)	Fruity, floral, medicinal odor; sweet, berry taste	172.515	Nut, cherry, vanilla, strawberry
1-Methyl-2 acetyl pyrrole	—	3184	—
2-Methylallyl butyrate	Powerful, fruity odor with sharp, pineapple, apple taste	2678	Pineapple, apple
Methyl anisate	Sweet, floral anise odor; spicy, green, anise taste	172.515	Licorice, root beer, melon, vanilla
o-Methyl anisole (*o*-methoxy toluene)	Sharp, fatty, walnut odor with sweet, nut-like taste	172.515	Nut flavors
p-Methyl anisole (*p*-methoxy toluene)	Sharp, fruity sweet odor with nutty taste	172.515	Blackwalnut, maple, berry
Methyl anthranilate	Fruity, vinous, grape odor and taste	182.60	Grape, citrus, loganberry
Methylated silica		3185	Free flowing agent
Methyl benzoate (oil of niobe)	A fruity, nutty, medicinal odor with a cherry taste	172.515	Cherry
alpha methylbenzyl acetate (styrollyl acetate)	A pungent, fruity, green odor with tropical fruit taste	172.515	Tropical fruit, raspberry
alpha methylbenzyl alcohol (styrollyl alcohol)	Floral, earthy, green odor; fruity taste	172.515	Tropical fruit, strawberry
alpha-Methylbenzyl butyrate (styrollyl butyrate)	A fruity, green odor with fruity taste	172.515	Apple, apricot, fruit flavors
Methyl benzyl disulfide	—	3504	—

Table 4 (continued)
SYNTHETIC FLAVOR INGREDIENTS

Name	Organoleptic characteristics	Regulation	Flavors useful in
alpha methylbenzyl formate (styrollyl formate)	A floral, green odor with a floral, citrus taste	172.515	Citrus (grapefruit)
alpha methylbenzyliso Butyrate (styrollyl isobutyrate)	Green, gassy, fruity odor tropical fruit taste	172.515	Guava, mango, pear, grapefruit
alpha Methylbenzyl propionate (styrollyl propionate)	Sweet, floral, green odor and tropical fruit taste	172.515	Tropical fruit, berry flavors
4-Methylbiphenyl	—	3186	—
2-Methyl-1-butanethiol	Mercaptan-like odor and taste	3303	
3-Methyl-2-butanethiol	Mercaptan-like odor and taste	3304	
2-Methyl-2-butenal		3407	
2-Methylbutyl isovalerate		172.515	
3-Methylbutyl 2-methyl-butanoate		3505	
2-Methylbutyl 3-methyl-butanoate		3506	
2-Methylbutyl 2-methyl-butyrate		3359	
3-Methylbutyl-2 methyl-propanoate		3507	
Methyl *p*-tertiary butylphenyl acetate	An odor suggestive of chocolate, vanilla, with a creamy milk taste	172.515	Chocolate, strawberry
2-Methyl butyraldehyde (methylethyl acetaldehyd)	A pungent aroma with a chocolate taste	172.515	Chocolate, apple
3-Methyl butyraldehyde (isovaleraldehyde)	Pungent, green chocolate-like aroma and taste	172.515	Cocoa, coffee, chocolate
Methyl butyrate	A choking cheese odor with a fatty cheese taste	172.515	Milk, cheese, apple
2-Methyl-butyric acid	An excellent cheese odor with mild cheese taste	172.515	Cheese, dairy flavors
Methyl cellulose	—	182.1480	Stabilizer
alpha-methylcinnaldehyde	Sweet, fruity, spicy, cinnamon bark odor and taste	172.515	Spice flavors, modifier for cinnamic aldehyde
p-Methyl cinnamic aldehyde	Sweet, fruity, vanilla, heliotupine odor and taste	172.515	Modifier for vanilla and spice flavors
Methyl cinnamate	Fruity, resinous odor with cherry, berry taste	172.515	Cherry, strawberry, grape
6-Methyl coumarin	A sweet, coconut odor and taste	2699	Coconut flavors, vanilla, caramel
3-Methylcrotonic acid	Sour, spicy, earthy odor spicy, sour taste	3187	Spice blends
2-Methyl-1,3-cyclohexadiene	—	172.515	—
1-Methyl-2,3-cyclohexadione	—	3305	—
3-Methyl-2-cyclohexen-1-one		3360	

3-Methyl-1-cyclopenta-decanone (muscone)	Musky odor and taste	3434	Traces in fruit flavors, pear
Methylcyclopentenolone (cyclotene)	A maple, lovage aroma and taste	172.515	Maple, smoke, butterscotch, apricot
1-Methyl-1 cyclopentene-3-one	—	3435	—
5H-5-Methyl-6,7 dihydro cyclopenta(B)pyrazine	—	3306	—
Methyl dihydrojasmonate	Fruity, jasmine-like odor; woody fruity taste	3408	Raspberry, strawberry, fruit flavors
Methyl 3,7-dimethyl-6-octeneoate	Fruity, apple, brandy odor and taste	3361	Apple, fruit flavors
Methyl disulfide (dimethyl disulfide)	A powerful, onion, cabbage, canned corn odor and taste	172.515	Meat, butter, potato, cocoa, garlic, onion
4-(3,4-Methylenedioxy-phenyl) 2-butanone	A floral, woody, cherry-pie odor with sweet taste	2701	Cherry, vanilla
Methyl ester of rosin (partially hydrogenated methyl dihydroabietate)		172.515	Physical fixative for flavors
2 Methyl-3-foranthiol	Odor and taste of roasted meat	3188	Meat flavor, nut
5-Methyl furfural	A furfural, cereal, spicy odor with a cereal taste	2702	Maple, coffee, cereal, bread
Methyl furfuryl disulfide	—	3362	—
2-Methyl-3,5 or 6 furfuryl thiopyrazine	A fried garlic odor and taste	3189	Spice blends, coffee, cocoa
Methyl 2-furoate	Earthy fruity odor with fruity, nutty taste	2703	Nut flavors, tobacco flavors, meat
2 Methyl-3 furyl acrolein	A spicy, cinnamon odor and taste	2704	Spice blends, nut, maple, walnut
3-(5-Methyl-2-furyl)-butanal	—	3307	—
6-Methyl-3,5 heptadien-2-one	Fatty, spice note with fruity, nut taste	3363	Spice flavors, nut, coconut
Methyl heptanoate	Fruity, green odor with green, sour taste	172.515	Apple, peach, apricot, berry, pineapple
2 Methyl heptanoic acid	Fatty, sour odor with fruity, sour taste	172.515	Fruit blends, coconut
6-Methyl-5-hepten-2-one	Powerful, fatty, green, citrus odor and fruity taste	172.515	Citrus, banana, pear, berry
5-Methyl-2,3 hexandione (acetyl isovaleryl)	Fruity, buttery odor with fruity, buttery taste	3190	Butter, cheese
Methyl hexanoate	A fruity, pineapple odor and taste	172.515	Pineapple, apricot
2-Methylhexanoic acid	Fatty, unpleasant, sour odor with a sour, nutty taste	3191	Apple, apricot, berry
Methyl 2-hexenoate	A fruity, apple, pineapple odor with a green, banana taste	172.515	Apple, pineapple, banana, melon
Methyl 3-hexenoate	A fruity, estery, pineapple odor and taste	3364	Pineapple, tropical fruit
5-Methyl 5-hexen-2-one	Pungent, green, grass odor and taste	3365	Citrus
5-Methyl-3-hexen-2-one	—	3409	—
Methyl p-hydroxy benzoate (methyl parabin)		172.515	Preservative

Table 4 (continued)
SYNTHETIC FLAVOR INGREDIENTS

Name	Organoleptic characteristics	Regulation	Flavors useful in
Methyl 3-hydroxyhexanoate	Estery, fruity odor with fruity taste	3508	Citrus, juice flavors, pineapple
alpha-methyl-beta-hydroxypropyl alpha-Methyl-beta mercaptopropyl sulfide	—	3509	—
Methyl alpha-ionone	Sweet, woody, floral odor with soapy, berry taste	172.515	Raspberry, strawberry, cachou flavors
Methyl beta-ionone	Sweet, woody, floral odor with berry taste	172.515	Raspberry, strawberry, blackberry, cachou
Methyl delta-ionone	Sweet, woody, floral odor with berry taste	172.515	Raspberry, strawberry, blackberry, cachou
alpha iso-methyl ionone	Floral, woody odor with berry taste	2714	Raspberry, strawberry, blackberry, cachou
Methyl isobutylate	A choking fruity, rum odor and taste	172.515	Apricot, fruit flavors
2-Methyl-3-(*p*-isopropyl phenyl) propionaldehyde (cyclamen aldehyde)	A floral, melon odor and a cantaloupe taste	172.515	Melon, tropical fruit
Methyl iso Valerate (Methyl 3-methylbutyrate)	A strong, fruity apple odor and taste	172.515	Apple, fruit flavors
Methyl jasmonate	Floral odor like jasmine; fruity, perfumy taste	3410	Berry flavors
Methyl laurate (methyl dodecanoate)	Oily, winey, fatty odor with fatty, "soupy" taste	172.515	Fatty flavors, coconut, nut
Methyl linoleate	—	3411	—
Methyl mercaptan (methanthiol)	Unpleasant cabbage odor and taste	172.515	Traces in coffee, onion, garlic flavors
Methyl *o*-methoxybenzoate (methyl-*o*-anisate)	A sweet, anisyl-like odor with berry taste	172.515	To add sweetness to a flavor berry
1-Methyl-3 methoxy-4-isopropylbenzene	Spicy, earthy odor with athymol-like taste	3436	Spice flavors, mouthwash flavors
2-Methyl-5 methoxy thiazole	A roasted, meat, peanut odor with a roasted peanut taste	3192	Meat, peanut, chocolate
Methyl n-methyl-anthranilate (dimethyl anthranilate)	An orange flower odor with a grape taste	172.515	Grape, grapefruit, citrus
Methyl 2 methylbutyrate	A fruity, choking odor with an apple, rum taste	172.515	Apple, rum, lift for fruit flavors
Methyl 4-(methylthiol)-butyrate	A fruity, pineapple odor and taste	3412	Pineapple, fruit flavors
5-Methyl-5-(Methylthiol)-furan	—	3366	—

Methyl-3-methylthio-propionate (pineapple mercaptan)	A sulfury, garlic, meat odor with a pineapple taste at low levels	2720	Pineapple
2-Methyl 3,5 or 6 (Methylthiol) pyrazine	A roasted almond, hazelnut odor and taste	172.515	Hazelnut, almond
Methyl 4 methylvalerate (methyl-4-methylpentanoate)	Fruity odor with pineapple taste	172.515	Pineapple
Methyl myristate (methyl tetradecanoate)	A fatty coconut, cognac odor with coconut taste	172.515	Solvent for flavors, coconut, honey
1-Methyl naphthalene	—	3193	—
Methyl beta-naphthyl ketone	Fruity, neroli odor with grape-like taste	172.515	Grape, strawberry, citrus, neroli
Methyl nonanoate	A fatty, fruity, odor with a nut, coconut taste	172.515	Coconut, brandy, nut, pineapple
Methyl-2-nonenoate	A green, leafy, fruity odor with a green, fatty melon taste	172.515	Melon, coconut, berry
Methyl 2 nonynoate (methyl octyne carbonate)	Pleasant, green, leaf odor with green, vegetable taste	172.515	Banana (green), cucumber, peach
2-Methyloctanal (methylhexylacetaldehyde)	Floral, fatty aldehyde odor and taste	172.515	Citrus flavors
Methyl octanoate	A fruity, citrus odor with fruity taste	172.515	Fruit flavors, pineapple, berry
Methyl *cis*-4-octenoate	A pleasant, fruity green odor with a fatty, green taste	3367	Banana, melon, tropical fruits
Methyl 2-octynoate (methyl heptine carbonate)	Powerful green, leaf odor and taste	172.515	Cucumber, banana, berry
2-Methyl pentanal	An odor reminiscent of ham and cheese with a woody, oily, ham taste	3413	Ham
4-Methyl-2,3-pentandione (acetyl isobutyryl)	Fruity, buttery odor; creamy, buttery taste	172.515	Butter, fruits, cheese
3-Methyl pentanoic acid	A fruity, cheesy, pineapple, strawberry odor with apple, strawberry taste	3437	Apple, strawberry, fruit flavors
4-Methyl pentanoic acid	Powerful sour odor and taste	3463	Cheese, fruit flavors
4 Methyl-2-pentanone (methyl isobutyl ketone)	Estery, fruity odor; sweet, fruity taste	172.515	Fruit flavors, rum cheese
2-Methyl-2-pentenal	A furfural-like odor with a cereal, nutty astringent taste	3194	Nut, cereal flavors
4-Methyl-2-pentenal	Pungent, cherry odor with almond, apricot taste	3510	Almond, apricot, fruit flavors
2-Methyl-2-pentenoic acid	A cheese odor with a fatty, cherry taste	3195	Cherry, cheese
2-Methyl-3-pentenoic acid	—	3464	—
2-Methyl-4-pentenoic acid	—	3511	—
4-Methyl-3-penten-2-one (mesityl oxide)	A peppermint, spearmint odor with an oily taste	3368	Peppermint, spearmint
3-Methyl-2-(2 pentenyl)-2-cyclopenten-1-one	Floral, fruity odor with jasmine-like taste	3196	Traces in fruit, mint flavors

Table 4 (continued)
SYNTHETIC FLAVOR INGREDIENTS

Name	Organoleptic characteristics	Regulation	Flavors useful in
beta-methylphenethyl alcohol	Floral medicinal odor with a melon taste	172.515	Melon, honey, nut
alpha methylphenethyl butyrate	Spicy, floral odor with spicy, floral, fruity taste	3197	Peach, tropical fruit
Methyl phenethyl ether	A pungent top-note with a blue-cheese mushroom taste	3198	Blue cheese, mushroom
Methyl phenylacetate	Honey odor and taste	172.515	Honey, tobacco, chocolate
3-Methyl-4-phenyl-3-butene-2-one	Fruity burnt sugar odor with fruity taste	172.515	Cherry, vanilla, nut, fruit flavors
2-Methyl-4-phenyl-2-butylacetate (dimethyl phenethyl carbinyl acetate)	Fruity green-leaf odor and taste	172.515	Fruit flavors, tea
2 Methyl-4-phenyl-2-butyl isobutyrate (Methyl phenethyl carbinyl isobutyrate)	Sweet, fruity, leaf-like odor with sweet fruity taste	172.515	Melon, plum
2-Methyl-4-phenyl-butyraldehyde	Sweet, earthy, floral odor fruity taste	2737	Raspberry, nut, fruit flavors
3-Methyl-2-phenyl-butyraldehyde	Green, fruity odor with fruity taste	172.515	Traces in fruit flavors
Methyl 4 phenylbutyrate	Floral, honey, fruity odor with sweet, fruity taste	172.515	Honey, strawberry, fruit flavors
5-Methyl-2-phenyl-2-hexenal	A floral, phenylacetaldehyde odor with chocolate taste	3199	Chocolate, cocoa flavors
4-Methyl-1-phenyl-2-pentanone (benzyl isobutyl ketone)	Woody, fruity, burnt-sugar odor; sweet, fruity, spicy taste	172.515	Raspberry, strawberry, fruit flavors
4-methyl-2-phenyl-2-pentenal	Floral, chocolate, cocoa odor with green, chocolate taste	3200	Vegetable flavors, chocolate
Methyl 3-phenlpropionate	Fruity, floral, very sweet odor with sweet, honey taste	172.515	Honey, fruit flavors, pineapple, peach
Methyl propionate	Sweet, fruity, rum-like odor with sweet, rum taste	172.515	Rum, black currant, fruit blends
5-Methyl-5-propyl-2 cyclohexen-1-one (celery ketone)	A characteristic odor of rye bread with a "brothy" celery taste	172.515	Broth flavors, celery, spice blends
Methyl propyl disulfide	Cooked garlic odor and sweet onion, garlic taste	3201	Onion, garlic imitation, spice flavors
2 Methylpropyl-3-methylbutyrate	Fruity estery, apple odor and taste	3369	Apple, fruit blends
2-(2-Methylpropyl) pyridine	—	3370	—
3-(2-Methylpropyl) pyridine	—	3371	—
2-(1-Methylpropyl) thiazole	—	3372	—
Methyl propyl trisulfide (propyl methyl trisulfide)	Green, garlic odor and taste	3308	Garlic, onion, imitation spice flavors, cocoa

2-Methyl pyrazine	A penetrating solventy peanut odor, with a chocolate taste at 30 ppm	3309	Chocolate, peanut, popcorn
Methyl-2-pyrrole ketone (acetyl pyrrole)	Fish-like odor and taste	3202	Seafood flavors
6-Methyl quinoline	Phenolic, fish, animal odor with a fishy taste	2744	Fish seafood, nut flavors
5-Methyl quinoxaline	—	3203	—
Methyl salicylate	A characteristic wintergreen aroma and taste	2745	Root beer, chewing gum flavors, fruit flavors, toothpaste
Methyl sulfide (dimethyl sulfide)	A cabbage odor and at extremely low levels, a fruity taste	172.515	Butter, cheese, coffee, cocoa, onion, fruit flavors
2-Methyl tetrahydrofuran-3-one	—	3373	—
2-Methyltetrahydro-thiophene-3-one	—	3512	—
4-Methyl-5-thiazole-ethanol	A bone marrow meat odor with excellent, boiled, meat taste	3204	Meat, chocolate, nut
5-Methyl-5-thiazole ethanol acetate	A meaty, ham odor and taste	3205	Ham, meat flavors
2-Methyl thioacetaldehyde	—	3206	—
3-(Methylthio) butanal	A tomato, cheese odor with cooked, tomato taste	3374	Tomato flavors
4-(Methylthio) butanal	—	3414	—
1-(Methylthio)-2-butanone	—	3207	—
4-(Methylthio)-2-butanone	A tomato, cheese, fish odor with oily, fish-like taste	3375	Seafood flavors
Methyl thiobutyrate	—	3310	—
Methyl thiofuroate		3311	
3-Methylthio-1-hexanol	A styrollyl acetate-like odor; with green, raspberry, rhubarb taste	3438	Fruit flavors
4-(Methylthio)-4-Methyl-2-pentanone	A solventy, fried garlic odor and taste	3376	Spice and meat flavors
Methyl(thio)methylpyrazine	A hazelnut, almond odor with a hazelnut taste	3208	Hazelnut, nut, chocolate, egg
5-(Methylthio)-2-(Methylthio) methyl penten-2-al-1	—	3483	—
5-Methyl-2-thiophene carboxaldehyde	Almond odor with sweet almond, cherry taste	3209	Cherry, almond
o-(Methylthio) phenol	Strong, irritating, sulfury odor with smokey, caramel taste at great deletion	3210	Coffee, smoke flavors
3-(Methylthio) propanol	A garlic aroma and taste	3415	Meat, broth, cheese flavors
3-Methylthiopropionaldehyde (Methional)	A garlic, onion, as a foetida odor with oily potato taste	172.515	Potato, cooked meat, onion, bread
3-Methylthiopropyliso-thiocyanate	Powerful, mustard, radish odor and taste	3312	Mustard, seasoning flavors
2-Methyl-3-tolyl-propionaldehyde	A strong, melon, honeydew odor with honeydew taste	172.515	Melon, canteloupe, honeydew
2-Methylundecanal (aldehyde C-12 MNA)	A powerful, fatty, green, citrus odor with fatty, citrus taste	172.515	Citrus flavors

Table 4 (continued)
SYNTHETIC FLAVOR INGREDIENTS

Name	Organoleptic characteristics	Regulation	Flavors useful in
Methyl 9-undecenoate	A fatty, waxy odor with a fatty burnt, mouldy taste	172.515	Cognac, rum, brandy flavors
Methyl 2-undecynoate	Earthy, green, floral odor and taste	172.515	Traces in flavors
Methyl valerate	Eastery, apple odor and taste	172.515	Apple, pineapple, chewing gum flavors
2-Methylvaleric acid	A pungent, sour odor and sweet, creamy taste	172.515	Cheese, dairy flavors
2-Methyl vinylpyrazine	—	3211	—
4 Methyl-5-vinylthiazole	A solventy, peanut odor with a nut, chocolate taste	3313	Nut, chocolate, cocoa
Monosodium glutamate	Flavor enhancer	2756	Meat, soup, flavors
Musk ambrette (2,6 dinitro-3-methoxy-1 methyl 4 *tert*-butylbenzene)	Very sweet, musky odor with bitter taste	2758	Traces in fruit flavors, cachou, peach
Myrcene (7 methyl-3 methylene-1,6 octadiene)	A resinous, terpeney odor with sweet, citrusy taste	172.515	Citrus imitations, fruit blends
Myristic aldehyde (tetradecanal)	A fatty, citrus odor with citrus taste at low levels	172.515	Citrus flavors
Myristic acid (tetradecanoic acid)	Very low, waxy odor and taste	2764	Butter, butterscotch, caramel
Myrtenol	Woody, camphor odor with minty taste	3439	Essential oil imitations
2-Naphthalenthiol (2 mercaptonaphthalene)		3314	
beta-Naphthyl anthranilate	A harsh, neroli-like odor and grapy taste	2767	Grape flavors
beta-Naphthyl ethyl ether (nerolin)	A floral odor and taste	2768	Traces in fruit flavors, grape, honey
delta-neomenthol	Woody, menthol-like odor with a minty taste	172.515	Mint flavors
Nerol (*cis*-3,7-dimethyl-2,6-octadien-1-ol)	Sweet, floral odor and taste	172.515	Fruit flavors, raspberry, strawberry, citrus
Nerolidol (3,7,11-trimethyl-1,6 dodecatrien-3-ol)	Green, fruity odor with mild apple taste	172.515	Apple, fruit blends, citrus
Neryl acetate	Fruity, floral odor with berry, citrus taste	172.515	Raspberry, citrus
Neryl butyrate	Geraniol with sourness; carrot, earthy taste	172.515	Vegetable flavors, cocoa
Neryl formate	Carrot odor with vegetable, earthy taste	172.515	Vegetable flavors, fruit flavors
Neryl isobutyrate	Sweet, floral odor with citrus taste	172.515	Citrus, strawberry
Neryl isovalerate	Fruity, peach, citrus odor with peach taste	172.515	Peach, citrus flavors
Neryl propionate	Sweet, fruity odor; sweet preserved fruit taste	172.515	Strawberry, raspberry, citrus, plum
Nitrous oxide		182.1545	Propellant and aerating agent

2,4 Nonadienal	A fatty, green, rancid, oily odor with a fatty, chicken-soup taste	3212	Meat, poultry flavor
Nona-2-*trans*,6-*cis*-dienal	A fatty, spicy, cucumber odor with excellent, cucumber taste	3377	Traces in cucumber, watermelon
2,6 Nonadienal diethyl acetal	A pleasant, cucumber aroma with cucumber, snap-bean taste	3378	Salad dressing flavors, cucumber, watermelon
2,6 Nonadien-1-ol (violet leaf alcohol)	A green, cucumber odor and taste	2780	Traces in fruit flavors, violet, berry
gamma-Nonalactone (aldehyde C-18 so-called)	A fatty coconut odor and taste	172.515	Coconut flavors, milk, cream
Nonanal (aldehyde C-9)	Powerful, fatty, floral odor with citrus taste	172.515	Orange, lemon, lime
1,3 Nonanediol acetate	A floral, jasmine, mushroom aroma, a green watermelon taste	172.515	Melon flavors
1,9-Nonanedithiol		3513	—
Nonanoic acid	A fatty coconut aroma with excellent coconut taste	172.515	Coconut, berry
2-Nonanol	Fruity, green, floral odor and taste	3315	Coconut, spice
2-Nonanone (methyl heptyl ketone)	Fatty, fruity, floral odor and taste	172.515	Cheese favors, apple
3-Nonanone (ethyl hexyl ketone)	Fatty, fruity, green, floral odor with musty taste	3440	Cheese flavors, apple, artificial essential oils
3-Nonanon-1-yl acetate	Fruity, estery odor with fruity, spicy taste	172.515	Fruit flavors; spice blends
Nonanoyl-4-hydroxy-3-methoxybenzylamide (pelargonyl vanillyamide)	An aroma suggestive of black pepper — at 1 ppm it has a considerable bite — (synthetic capsicum)	2787	Spice flavors for bite
2-Nonenal	Chicken, roasted pork aroma with cucumber taste at 1 ppm	3213	Meat, poultry flavors at low levels
trans-2-nonen-1-ol		3379	
cis-6-nonen-1-ol	Fatty, oily, fruity melon-like odor and taste	3465	Melon
Nonyl acetate (acetate C-9)	Fruity, green-leaf aroma with fatty, sweet taste	172.515	Citrus flavors
Nonyl alcohol (alcohol C-9)	A fatty, citrus aroma and with great dilution citrus taste	172.515	Orange, lemon, lime, pineapple
Nonyl isovalerate	Fatty, aliphatic alcohol, fried foods aroma with nutty taste	172.515	Nut, fatty flavors
Nonyl octanoate	Fatty, earthy, citrus odor with citrus taste	172.515	Citrus, mushroom
Nootkatone	Sweet, citrus, grapefruit odor and taste	172.515	Citrus flavors, grapefruit, fixative, blender
Ocimene (2,6 dimethyl-1,5,7-octatriene)	Sweet, terpeney odor and taste	172.515	Essential oil imitations, citrus flavors
2-*trans*-6-*trans* octadienal	Fatty, green vegetable odor with fatty green taste	3466	Citrus, poultry flavors

Table 4 (continued)
SYNTHETIC FLAVOR INGREDIENTS

Name	Organoleptic characteristics	Regulation	Flavors useful in
gamma-octalactone	Pleasant, peach, coumarin, coconut, rye bread odor and taste	172.515	Coconut, peach, apricot
delta octalactone octanal	Sweet, fatty coumarin, coconut odor with creamy taste	3214	Coconut, peach, apricot, dairy flavors
Octanol (aldehyde C-8)	A fruity, fatty aroma with citrus taste	172.515	Citrus flavors, apricot
Octanal diethyl acetal	A fatty, green pepper aroma with green, citrus taste	172.515	Citrus flavors, melon
1,8 Octanedithiol	—	3514	—
Octanoic acid (caprylic acid)	A sweat-like odor with cheese, butter taste	172.515 182.3025	Butter, dairy flavors, rum, pineapple
1-Octanol (alcohol C-8)	An orange-like aroma with fatty orange taste	172.515	Citrus, coconut, peach
2-Octanol methyl hexyl carbinol		172.515	Nut flavors
3-Octanol	Mushroom aroma with cheesy taste	172.515	Cheese, mushroom flavors
2-Octanone (methyl hexyl ketone)	Fatty, green, cheese aroma with cheese taste	172.515	Cheese flavors
3-Octanone (ethyl amyl ketone)	Sweet, fruity, earthy, cheese aroma, with cheese, mushroom taste	172.515	Cheese, mushroom, coffee
3-Octanon-1-ol	Spicy, fruity odor and taste	172.515	Spice flavors, traces in fruit flavors
2-Octenal	Penetrating fatty, meaty green aroma with cucumber, chicken taste, depending on concentration	3215	Chicken, cucumber
1-Octen-3-ol (amyl vinyl carbinol)	An earthy, mushroom aroma and taste	172.515	Mushroom, earthy flavors
cis-3-Octen-1-ol	—	3467	
3-Octen-2-one	A mushroom, coconut aroma with coconut taste	3416	Coconut, nut, mushroom
trans-2-octen-1 yl Acetate	Green, earthy, mushroom odor and fruity, mushroom taste	3516	Mushroom
1-Octen-3 yl acetate	Fatty, green, earthy mushroom aroma less than 5 ppm, mushroom taste	172.515	Mushroom
trans-2-octen-1-yl butanoate		3517	
Octyl acetate (acetate C-8)	A weedy, citrus aroma with citrus taste	172.515	Citrus flavors, peach, raspberry
3-Octyl acetate	A flowery, citrus aroma and taste	172.515	Citrus flavors
Octyl butyrate	Oily, fried aroma with citrusy taste	172.515	Citrus flavors, melon, pineapple
Octyl formate	Fruity, green, floral odor and taste	172.515	Citrus, mixed fruit
Octyl 2-furoate	Low, fruity, caramel-like odor with fatty taste	3518	Fruit flavors

Octyl heptanoate	Pleasant, fatty, coconut with fruity, waxy taste	172.515	Fruit flavors, coconut
Octyl isobutyrate	Fruity, geen, earthy vegetable odor, green, floral taste	172.515	Citrus flavors, mixed fruit, grape, melon
Octyl isovalerate	Fatty, waxy, coconut aroma with creamy, coconut taste	172.515	Coconut flavors
Octyl octanoate	A fatty, coconut aroma with oily taste	172.515	Coconut, poultry, nut flavors, pineapple
Octyl phenylacetate	Floral citrusy odor with honey-citrus taste	172.515	Honey, citrus blends, berry
Octyl propionate	Sweet, fatty, fruity odor with fatty, fruity taste	172.515	Citrus blends, fruit flavors, melon
Oleic acid (9-octadecenoic acid)	Mild, fatty odor and taste	2815	Butter, cheese, banana, meat
3-Oxobutanal dimethyl acetal (3 ketobutyraldehyde-dimethyl acetal)	Sweet, spearmint aroma and taste	3381	Spearmint, fried potato flavor
Palmitic acid (hexadecanoic acid)	Odorless and tasteless	2832	Mouthfeel in certain flavors; butter, cheese
omega-pentadecalactone (angelicalactone)	A powerful, musk aroma and taste	172.515	Traces in flavors
2,4 pentadienal		3217	
2,3 pentandione (acetyl propionyl)	Fatty, cooked, butter aroma, creamy, milky, waxy taste	172.515	Butterscotch, caramel traces in fruits
2-Pentanol	Volatile, fruity, winey, fusel aroma; low taste	3316	Fruit flavors for lift, chocolate
2-Pentanone (methyl propyl ketone)	Pungent, fruity odor with fruity, banana taste	172.515	Banana, pineapple fruit blends
2-Pentenal	A sharp penetrating furfural-like aroma with cherry taste	3218	Cherry, spice, apple
4-Pentenoic acid (allyl acetic acid)	Woody, cheesy aroma with a fruity, cheesy taste, depending on concentration	172.515	Cheese, butter, fruit
1-Penten-3-ol	A pungent mustard aroma with tingling aftertaste	172.515	Fruit flavors for lift, spice flavors for bite
1-Penten-3-one (ethyl vinyl ketone)	A pungent mustard aroma with a pungent-bite in taste	3382	Fruit flavors for lift, spice flavors for bite
3-Penten-2-one	A pungent mustard, horseradish odor and taste	3417	Traces in spice flavors
2-Pentyl furan	An earthy, rooty aroma with a carrot, spoiled-terpene taste	3317	Wherever "earthy" notes are applicable
Pentyl-2-furyl ketone	A fruity-berry aroma; very low taste	3418	Fruit flavors, raspberry
2-Pentyl pyridine	—	3383	—
Perillaldehyde (*p*-mentha-1,8 dien-7-al)	A spicy, cinnamic aldehyde-like aroma, with oily, peanut taste	172.515	Spice flavors, peanut
Perillyl acetate	A fruity, woody, raspberry aroma with an ionone raspberry taste	172.515	Raspberry
Perillyl alcohol	A spicy, flowery aroma with oily, peanut taste	2264	Spice, peanut

Table 4 (continued)
SYNTHETIC FLAVOR INGREDIENTS

Name	Organoleptic characteristics	Regulation	Flavors useful in
Alpha-phellandrene (2-methyl-5-isopropyl-1,3-cyclohexadiene)	Spicy, black pepper odor with spicy, minty taste	172.515	Essential oil imitations, citrus, spice
Phenethyl acetate (2 phenylethyl acetate)	Floral, fruity, honey odor with floral, honey taste	172.515	Fruits, honey, peach, vanilla
Phenethyl alcohol (β-phenylethyl alcohol)	Green, floral, earthly, aroma with floral, dough-like taste	172.515	Honey, bread flavors, berry, peach
Phenethylamine	Amine, fishy aroma with a distinct fishy taste	3220	Seafood flavors, clam
Phenethyl anthranilate (2 phenylethyl anthranilate)	A floral fruity aroma with honey, grape taste	172.515	Honey, grape flavor, character fixative for grape
Phenethyl benzoate (2 phenylethyl benzoate)	Resinous, mild honey odor with fruity, berry taste	172.515	Honey, berry flavors
Phenethyl butyrate (2 phenylethyl butyrate)	Fruity, buttery, floral odor with fruity-honey taste	172.515	Strawberry, raspberry, peach, honey
Phenethyl cinnamate (2 phenylethyl cinnamate)	Fruity, resinous odor with fruity, sweet taste	172.515	Peach, apricot, cherry, strawberry
Phenethyl formate (2 phenylethyl formate)	Leafy, green, floral note; leafy, fruity taste	172.515	Strawberry, apple, greengage
Phenethyl-2-furoate (2 phenylethyl-2-furoate)	An earthy, mushroom odor and taste	2865	Mushroom, vegetable flavors
Phenethyl hexanoate (2 phenylethyl hexanoate)	Floral, rose, dough-like aroma with soapy floral taste	3221	Honey, pineapple, fruit flavors
Phenethyl isobutyrate (2 phenylethyl isobutyrate)	A floral, fruity odor with fruity taste	172.515	Pineapple, strawberry, greengage fruit mixtures
Phenethyl isovalerate (2 phenylethyl isovalerate)	Fruity, resinous odor; sweet green, fruity taste	172.515	Apple, pineapple, pear, peach, fruit mixtures
Phenethyl-2-methylbutyrate (2 phenylethyl-2-methylbutyrate)	A fatty rose aroma with rosy flour or dough-like taste	172.515	Honey
Phenethyl octanoate (2 phenylethyl octanoate)	A fruity, brandy-like odor and taste	3222	Cheese, mushroom
Phenethyl phenylacetate (2 phenylethyl phenylacetate)	Very sweet, floral, honey odor; sweet, fruity taste	172.515	Honey, cherry
Phenethyl propionate (2-phenylethyl propionate)	Fruity, spicy odor with fruity, berry-like taste	172.515	Raspberry, strawberry, fruit blends, honey

Phenethyl salicylate (2 phenylethyl salicylate)	Faint, resinous clove-like odor with honey, apricot taste at low levels	172.515	Peach, apricot, honey
Phenethyl senecioate (phenethyl 3-methylcrotonate)	A pleasant, winey, dough aroma with a wine, raisin, rum taste	172.515	Sherry flavors, bread
Phenethyl tiglate (2 phenylethyl tiglate)	Sweet, floral, leafy green odor with nutty taste	172.515	Nut flavors
Phenol	A phenolic, medicinal odor and taste	3233	Smoke flavors
Phenoxyacetic acid	A weak, vinegar odor with honey taste	172.515	Honey, fruit flavors
2-Phenoxyethyl isobutyrate	Fruity, flowery odor with sweet, honey-like taste, similar to phenylacetates	172.515	Honey, peach, tutti frutti
Phenylacetaldehyde	Pungent, green, floral odor with almond-cherry taste	172.515	Cherry, apricot, peach
Phenylacetaldhyde 2,3-butylene glycol acetal	Floral, earthy odor with fruity, earthy taste	172.515	Cherry, apricot, peach
Phenylacetaldehyde di-isobutyl acetal	Floral, green odor with fruity taste	3384	Cherry, apricot, peach
Phenylacetaldehyde dimethyl acetal	Pungent, green, earthy odor with green, spicy taste	172.515	Cherry, apricot, peach, plum
Phenylacetaldehyde glyceryl acetal	Sweet, floral odor with green, fruity taste	172.515	Fruit flavors
Phenylacetic acid	An animal, honey-like aroma and taste	172.515	Honey, beer flavors
4-Phenyl-2-butanol (phenylethyl methyl carbinol)	Fruity, melon, spice aroma with spicy melon taste	172.515	Melon flavors
2-Phenyl-2-butenal	—	3224	—
4-Phenyl-3-buten-2-ol	—	172.515	—
4-Phenyl-3-buten-2-one (benzilidene acetone)	Sweet, creamy coumarin odor and taste	172.515	Strawberry, cherry, peach, chocolate, nut, vanilla
4-Phenyl-2-butylacetate	Spicy, citronella-like aroma with spicy taste	172.515	Spice flavors, fruits, peach
2-Phenyl-3-carbethoxy furan	—	3468	—
Phenyl disulfide (diphenyl disulfide)	—	3225	—
Phenyl-3-methyl-3-pentanol	Pleasant, floral odor with fruity, green taste	172.515	Fruit flavors
3-Phenyl-4-pentenal	A somewhat floral, chocolate odor and taste	3318	Cocoa, chocolate

Table 4 (continued)
SYNTHETIC FLAVOR INGREDIENTS

Name	Organoleptic characteristics	Regulation	Flavors useful in
2 Phenyl-4-pentenal	Weedy, green odor with tomato taste	3519	Tomato, vegetable blends
1-Phenyl-1,2 propandione	—	3226	—
1-Phenyl-1-propanol	Floral resinous odor with honey taste	172.515	Honey, fruit blends
3-Phenyl-1-propanol	Floral, sweet resinous odor; sweet preserved fruit taste	172.515	Peach, apricot, strawberry, nut flavors
2-Phenylpropionaldehyde (alpha-methyl phenylacetaldehyde)	A green, earthy, leaf aroma with melon taste	172.515	Melon, apricot, peach, cherry, berry
3-Phenylpropionaldehyde (benzylacetaldehyde)	Green, earthy odor; sweet spicy taste	172.515	Cherry, almond, spice
2-Phenylpropionaldehyde (dimethyl acetal)	Spicy, green fruity aroma; earthy taste at low levels	172.515	Mushroom, spice, nut flavors
3-Phenylpropionic acid (hydrocinnamic acid)	A very weak aroma with a fruity, cinnamate type flavor with acid	172.515	Fixative for flavors, cheese
3-Phenylpropyl acetate	Pleasant honey aroma and taste	172.515	Honey, peach, apricot, gooseberry
2-Phenylpropyl butyrate	A fruity aroma with a spicy, black pepper, vegetable taste	172.515	Fruit, spice flavors, plum
3-Phenylpropyl cinnamate	Fruity, resinous odor with sweet taste	172.515	Chocolate, cocoa, caramel, coconut
3-Phenylpropyl formate	A fatty, citrus aroma and taste	172.515	Citrus flavors, currant
3-Phenylpropyl hexanoate	Sweet, green, fruity odor with fruity, pineapple taste	172.515	Pineapple, peach, tropical fruit
2-Phenylpropyl isobutyrate	Fruity, apricot, peach aroma with spicy, sweet, cherry taste	172.515	Peach, apricot, cherry
3-Phenylpropyl isobutyrate	Spicy, fruity melon odor with fruity, spicy taste	172.515	Honey, melon flavors, apple, pear, pineapple
3 Phenylpropyl isovalerate	Fruity, melon aroma with honey, melon, blueberry taste	172.515	Honey, melon, berry flavors, apple
3 Phenylpropyl propionate	Perfumy odor with sweet, green, fruity taste	172.515	Apricot, vanilla
2-(3-phenylpropyl)-tetrahydrofuran	Sweet, burnt-sugar odor with sweet, fruity taste	172.515	Maple, caramel, honey
Phosphoric acid	Acidulant	182.1073	Cola beverages
alpha-pinene (2,6,6 trimethylbicyclo-(3.1.1)-2-heptene)	Piney, turpentine-like odor and taste	172.515	Spice flavors, nutmeg imitations, citrus imitations
beta-pinene (nopinene)	Piney, turpentine-like odor and taste	172.515	Spice flavors (nutmeg), citrus imitations
Pine tar 0.1		172.515	

Pinocarveol	Piney, anise-like odor and taste	172.515	Spice flavors
Piperidine (hexahydropyrzidine)	Animal odor with spicy taste	172.515	Spice, fish flavors
Piperine	Odorless with burning taste	172.515	Black pepper imitations, spice blends
Piperitenone (1-methyl-4-isopropylidene 1-cyclohexen-2-one)	Minty odor and taste	172.515	Mouthwash, toothpaste flavor
Piperitenone oxide	—	172.515	—
d-Piperitone (1-methyl-4-isopropyl-1-cyclohexen-3-one)	Minty odor and taste	172.515	Spice blends, toothpaste flavors
Piperonal (heliotropine)	A sweet, vanilla, cherry pie-like aroma and taste	182.60	Vanilla, cherry flavors
Piperonyl acetate (heliotropyl acetate)	A mild heliotropine aroma and taste	172.515	Vanilla, raspberry, stawberry (jam)
Piperonyl isobutyrate (heliotropyl isobutyrate)	Sweet spicy odor, jammy taste	172.515	Cherry, raspberry, strawberry flavors
Polylimonene	A fixative for flavors	172.515	—
Polysorbate 20	Emulsifier - solubilizer	172.515	—
Polysorbate 60	Emulsifier - solubilizer	172.515	—
Polysorbate 80	Emulsifier - solubilizer	172.515	—
Potassium acetate	Buffer	172.515	—
Potassium sorbate	Preservative	182.3640	—
L-Proline	Amino acid	182.5650	—
1,2-Propanedithiol		3520	
Propanethiol (*n*-propylmercaptan)	Powerful sulfur-like odor with onion, cabbage taste at low levels	172.515	Durian flavors, meat, pet food flavors
Propenyl guaethol (1-ethoxy-2-hydroxy-4-propenylbenzene) (vanatrope)	A sweet, phenolic, anise-like odor with sweet taste	172.515	Vanilla imitation, chocolate
Propenyl propylsulfide	Cooked onion odor and taste	3227	Onion imitations, spice blends
Propionaldehyde	Choking odor with green, cocoa, coffee taste	172.515	Apple, cherry, chocolate, onion
Propionic acid	Pungent sour odor and taste	182.3081	Cheese butter flavors, apple
Propiophenone	Acetophenone cherry, watermelon aroma and taste below 5 ppm	3469	Traces in cherry, honey
Propyl acetate	Pleasant, solventy aroma; estery taste gets pungent at 30 ppm	172.515	Solvent for fruit flavors, lift pear, currant
Propyl alcohol	Sickening, sweet odor; rubbing alcohol odor and taste	172.515	Fruit flavors
p-Propyl anisole (dihydro anethol)	A pleasant fennel, anise odor and taste	172.515	Root beer, mouthwashes
Propyl benzoate	A fruity, somewhat berry-like aroma with licorice, methol taste	172.515	Fruit flavors, cherry
Propyl butyrate	Fruity, choking aroma with tutti frutti taste	172.515	Fruit blends, pineapple apricot

Table 4 (continued)
SYNTHETIC FLAVOR INGREDIENTS

Name	Organoleptic characteristics	Regulation	Flavors useful in
Propyl cinnamate	Typical cinnamate aroma with fruity, cognac taste	172.515	Fruit flavors, berry
Propyl disulfide (dipropyl disulfide)	An onion, garlic, scallion odor and taste	172.515	Salad dressing flavors, onion, garlic imitations
Propylene glycol	Solvent	182.1666	—
Propylene glycol alginale	Stabilizer	2941	—
Propylene glycol dibenzoate	Weighing agent for citrus oils	3419	—
Propylene glycol stearate	Emulsifier	3942	—
Propyl formate	A pungent, rum odor with a weak rum taste at 10 ppm	172.515	Rum, fruit flavors for lift, apple
Propyl 2-furanacrylate	Fruity, woody, pineapple odor and taste at 20 ppm	172.515	Pineapple, coffee
Propyl 2-furoate	Fruity, woody, mild, pineapple odor and taste	2946	Pineapple, fruit, jamminess, chocolate
Propyl gallate	Antioxidant, preservative	2947	—
Propyl heptanoate	A green, fruity odor and taste	172.515	Fruit flavors, coffee, cognac, rum
Propyl hexanoate	A fruity, pineapple aroma and fruity taste	172.515	Fruit flavors, pineapple, loganberry
Propyl *p*-hydroxybenzoate	Preservative	172.515	—
3-Propylidenephthalide	An excellent, celery-herb-like aroma with brothy taste	172.515	Spice flavors, soup, and gravy flavors
Propyl isobutyrate	A fruity, pineapple aroma and taste	172.515	Fruit flavors, pineapple, banana
Propyl isovalerate	Fruity, choking aroma with apricot, dried-fruit taste	172.515	Fruit flavors, apricot, banana, apple, strawberry
alpha-Propylphenethyl alcohol (1-phenyl-2-pentanol)	Sweet, green odor with fruity, green taste	172.515	Fruit flavors
o-Propylphenol	Spicy medicinal odor and taste	3522	Smoke flavors
Propyl phenylacetate	A honey aroma and taste	172.515	Honey, butter, caramel
Propyl propionate	A powerful, rum aroma with rum taste	172.515	Rum, fruit flavors for lift, banana, cherry, melon
Propyl thioacetate	—	3385	—
Pulegone (*n*-menth-4(8)-en-3-one)	Weedy, minty odor and taste	172.515	Mint flavors artificial essential oils, peppermint
Pyrazinyl ethanethiol	—	3230	—
Pyrazinyl methyl sulfide	—	3231	—

Pyridine	A powerful, fish-like aroma with amine taste	172.515	Seafood flavors, smoke flavors, chocolate
2-Pyridinemethanethiol	A pyridine aroma at high levels, a prickling taste on tongue	3232	Smoke flavors
Pyroligneous acid	A pungent smoke odor and taste	172.515	Smoke flavors, meat, fish, butterscotch
Pyrrole	Sweet sickening odor; sweet taste	3386	Tobacco flavors
Pyrrolidine (tetrahydropyrrole)	Powerful, ammoniacal odor and taste	3523	Tobacco flavors
Pyruvaldehyde (2 ketopropionaldehyde)	Irritating odor with caramel-like taste	172.515	Coffee, maple, caramel, honey
Pyruvic acid (2-ketopropionic acid)	A roasted HVP-like aroma with maple, chestnut flavor	172.515	Maple, coffee
Quinine bisulfate	White crystals, odorless with bitter taste	2975	Bitters, tonic
Quinine hydrochloride	White, odorless crystals with bitter taste	2976	Bitters, tonic
Quinine sulfate	White, odorless crystals with bitter taste	2977	Bitters, tonic
Quinoline (benzo (6) pyridine)	Powerful nauseating odor with pyridine-like taste	3470	Fish, seafood flavors vanilla
Rhodinol (3,7 dimethyl-7-octen-1-ol)	A floral, rose-like odor and sweet taste	172.515	Strawberry, grapefruit flavors for sweetness
Rhodinyl acetate	Floral, rose odor with sweet fruity, berry taste	172.515	Raspberry, strawberry, coconut, cachou
Rhodinyl butyrate	Sweet, fruity, floral odor with fruity-apple taste	172.515	Apple, pear, strawberry, bilberry
Rhodinyl formate	Leafy, green floral odor with fruity taste	172.515	Apricot, strawberry, raspberry, peach, cherry
Rhodinyl isobutyrate	Fruity sweet odor with very sweet taste	172.515	Pineapple, apple, pear, peach, raspberry
Rhodinyl isovalerate	Fruity, floral odor; sweet woody taste	172.515	Berry flavors, cherry, tobacco flavors
Rhodinyl phenylacetate	Floral honey odor and taste	172.515	Honey, peach
Rhodinyl propionate	Sweet fruity odor and taste	172.515	Raspberry, strawberry, plum, honey
Rum ether (ethyl oxyhydrate)	A characteristic sweet rum aroma and taste	172.515	Rum, whisky, butter, eggnog
Saccharine, sodium salt	Sweetening agent	2997	Dietetic beverages
Salicylaldehyde (*o*-hydroxy benzaldehyde)	Pungent, phenolic odor with spicy, nut taste	172.515	Nut, cassia
alpha-,beta, santalol	Sweet, woody aroma with nutty taste	172.515	Nut, raspberry, traces in fruit flavors
Santalyl acetate	Sweet, sandalwood odor and taste	172.515	Pear, apricot, pineapple, traces in fruit flavors
Santalyl phenylacetate	Sweet, heavy, honey odor with woody, honey taste	172.515	Caramel, honey, traces in fruit
Skatole (3-methyl indole)	An animal-note odor with over-ripe fruit taste	172.515	Cheese, grape, berry, nut, egg
Sodium acetate	Buffer	182.1721 3024	—
Sodium benzoate	Preservative	3025	Up to 0.1%

Table 4 (continued)
SYNTHETIC FLAVOR INGREDIENTS

Name	Organoleptic characteristics	Regulation	Flavors useful in
Sodium citrate	Buffer, sequestrant	182.1751 3026	—
Sodium hexametaphosphate	Sesquestrant	3027	—
Sorbitan monostearate	Emulsifier solubilizer	172.515	—
Sorbic acid	Preservative	182.3089	—
d-Sorbitol	Sweetening agent	3029	—
Stearic acid (octadecanoic acid)	Odorless with waxy taste	3035	Butter, coconut, softener in chewing gum
Styrene (phenylethylene)	Resinous, floral odor and taste	172.515	Essential oil imitations
Succinic acid	Odorless with sour, acid taste	182.1091	Fruit, wine flavors
Sucrose octaacetate	Odorless crystals with extremely bitter taste	172.515	Bitters, spice, gingerale
Sulfur dioxide	Preservative, bleaching agent	182.3862 3039	—
Tannic acid	Odorless with dry astringent taste	172.515	
Tartaric acid	Acidulant	182.1099 3044	Fruit beverages, grape
alpha-terpinene (1-Methyl-4-isopropyl-1,3-cyclohexadiene)	Citrus, lemon odor and taste	172.515	Artificial essential oils lemon, mint
gamma-terpinene (1-methyl-4-isopropyl-1,4-cyclohexadiene)	Citrus, lemon odor and taste	172.515	Artificial essential oils
Alpha-terpineol (*p*-menth-1-en-8-ol)	Sweet, floral odor with lime taste	172.515	Lime, lemon, spice flavors, peach
beta terpineol (8,9)-*p*-menthen-1-ol)	Woody, earthy odor and taste	6156	Traces in lemon, lime, gingerale, spices
Terpinolene (*p*-Menth-1,4(8)-diene)	A pine-like aroma with a terpene taste	172.515	Artificial essential oils, citrus, fruit
Terpinyl acetate	Mild flowery citrus peel odor; spicy, citrus-like taste	172.515	Citrus, spice, cherry, raspberry, peach
Terpinyl anthranilate	Sweet, neroli odor and sweet, fruity taste	172.515	Citrus, grape
Terpinyl butyrate	Flowery, sour aroma with "meaty" pineapple, apple taste	172.515	Pineapple, apple, plum
Terpinyl cinnamate	Very sweet, resinous odor; spicy, fruity taste	172.515	Grape, spice flavors
Terpinyl formate	A mild floral aroma with oily, mushroom taste	172.515	Fruit flavors
Terpinyl isobutyrate	Flowery aroma with fatty, black pepper taste	172.515	Fruit flavors, spice
Terpinyl isovalerate	Piney citrus odor with sweet, fruity taste	172.515	Apple, fruits, tobacco
Terpinyl propionate	A weak odor with taste ranging from cognac-coconut	172.515	Fruit flavors, coconut, cognac

4,5,67-tetrahydro-3,6-dimethyl-benzofuran (menthofuran)	A weedy peppermint aroma and taste	3235	Peppermint, mint flavors
Tetrahydrofurfuryl acetate	No characteristic aroma; persistant bite in back of throat	172.515	Solvent for flavors, honey, maple, chocolate
Tetrahydro furfuryl alcohol	Pleasant, fruity, solventy odor and taste residual; nut taste	172.515	Coffee, peanut, pecan
Tetrahydro furfuryl butyrate	Mild, fruit aroma very low flavor value	172.515	A good solvent for fruit flavors
Tetrahydro furfuryl cinnamate	Sweet, resinous, spicy odor and taste	3320	Solvent for spice, flavors
Tetrahydrofurfuryl propionate	Fruity, solventy odor; hint of peanut in taste	172.515	Honey, maple, chocolate; solvent for flavors
Tetrahydropseudo ionone (6,10-dimethyl-9-undecene-2-one)	Woody, resinous odor; sweet fruity taste	172.515	Berry flavors, peach, apple
Tetrahydrolinalool (3,7-dimethyl octan-3-ol)	A coriander, spice aroma and taste	172.515	Spice flavors, berry flavors
Tetrahydro-4-methyl-2-(2-methyl propen-1-yl)-pyran (rose oxide)	Green, geranium odor with rose taste	3236	Artificial geranium and rose oils
5,6,7,8-tetrahydro-quinoxaline	—	3321	—
Tetramethyl ethylcyclo-hexenone	Spicy, nut, burnt-sugar odor with fruity, caramel taste	172.515	Butterscotch, caramel, rum
1,5,5,9-tetramethyl-13-oxatricyclo (8.3.0.0) (4,9) tridecane	—	3471	—
2,3,5,6 Tetramethyl pyrazine	A mild, pyrazine, chocolate aroma, milk chocolate taste at 20 ppm	3237	Chocolate
Thiamine hydrochloride (vitamin B_1)	A white, crystalline powder with vitamin odor, salty meaty, taste	182.5875 3322	—
2-Thienyl disulfide	Distinct rubbery odor with unpleasant earthy taste	3323	Traces in meat flavors
2-Thienyl mercaptan	Burnt-rubber, roasted coffee odor with burnt or roasted taste	172.515	Coffee roasted flavors
2,2′-(Thiodimethylene)-difuran	A coffee, meat aroma with a mushroom taste	3238	Mushroom, coffee
Thiogeraniol	A mercaptan-like aroma; upon great dilution, grapefruit	3472	Citrus flavors, grapefruit
4-Thujanol (jabinene hydrate)	Spicy, woody odor with pungent, spicy taste	3239	Artificial essential oils, black pepper, black currant
Thymol (5-methyl-2-isopropyl phenol)	A medicinal, phenolic odor and taste	172.515	Mouthwash flavors, peppermint, spice, citrus
Tolualdehyde glyceryl acetal	Very low odor with a bitter, almond taste	172.515	Cherry, almond, vanilla
Tolualdehydes	Sweet, benzaldehyd-like odor and taste	3068	Cherry, almond, peach, spice

Table 4 (continued)
SYNTHETIC FLAVOR INGREDIENTS

Name	Organoleptic characteristics	Regulation	Flavors useful in
o-Toluenethiol (2-methyl thiophenol)	Unpleasant, phenolic aroma which when greatly diluted, tastes like meat	3240	Meat, coffee
p-Tolylacetaldehyde	A watermelon, cherry-like odor and taste	172.515	Watermelon, cherry, nut, honey
o-Tolyl acetate (acetyl *o*-cresol)	Floral, fruity, medicinal odor with fruity taste	172.515	Cherry, honey, caramel, butter
p-Tolyl acetate (acetyl p-cresol)	Floral, fruity, green odor; sweet, fruity taste	172.515	Banana, cherry, nut, fruit blends
4-(*p*-Tolyl)-2-butanone	Sweet, floral, fruit odor sweet, fruity taste	172.515	Raspberry, strawberry
p-Tolyl isobutyrate	A "sweaty" horse, zoo aroma and taste	172.515	Traces in fruit flavors
p-Tolyl laurate	Very low odor with fatty, soapy, phenolic taste	172.515	Nut flavors
p-Tolyl 3-methylbutyrate	—	3387	—
p-Tolyl phenylacetate	A horse-stable odor and oily taste suggestive of nut	172.515	Nut, honey, caramel, butter
2-(*p*-Tolyl)-propionaldehyde	Strong, sweet, woody, green odor and taste	172.515	Traces in spice flavors
Triacetin (glyceryl triacetate)	A fruity, acetic acid aroma and taste	182.1901 2007	Solvent for flavors
Tributyl acetylcitrate	A weak, winey, fruity odor, very low taste	172.515	Solvent for flavors
Tricalcium phosphate	Odorless and tasteless powder	3081	Free flowing agent buffer
2-Tridecanone (methyl undecyl ketone)	Fatty, waxy nutty odor with coconut taste	3388	Coconut, nut
2-Tridecenal	A fatty, citrus aroma and taste	172.515	Citrus flavors
Triethyl citrate	A pleasant, fruity odor with a bitter taste	182.1911 3083	Citrus, berry, cherry solvent for flavors
Trimethyl amine	A powerful ammonia, fish odor and taste of "old" fish	3241	Fish, seafood flavors
p-alpha, alpha-trimethyl benzylalcohol (*p*-cyonen-8-ol)	A nut-like aroma with a distinct walnut taste	3242	Essential oil imitations nut, berry, flavors
4-(2,6,6-trimethyl cyclohexa-1,3-dienyl)butene-2-en-4-one (damascenone)	A fruity, woody, cinnamon-bark aroma with woody, fruity taste	3420	Fruit flavors, jamminess
4-(2,6,6-trimethylcyclo-hexa-1-enyl) but-2en-4-one (β-damascone)	Fruity, soapy, ionone aroma and taste	3243	Fruit flavors, jamminess
2,6,6-trimethyl cyclohexa-1,3 dienyl methanal (safranal)	A powerful, saffron odor and taste	3389	Rum flavors, traces in fruit flavors, spice
2,2,6-Trimethylcyclo hexanone	Sweet, honey-like, tobacco odor with sweet taste	3473	Tobacco, honey flavors

2,6,6-trimethyl-1-cyclo hexen-1-acetaldehyde	—	3474	—
2,6,6-Trimethylcyclo hexa-2-ene-1,4-dione	—	3421	—
4-(2,6,6-Trimethylcyclo hex-1-enyl) but-2-en-4 one	—	3243	—
3,5,5-trimethyl hexanal	A spicy, linalool, coriander top note with oily, coriander taste	3524	Spice flavors, chocolate
3,5,5-trimethyl-1-hexanol	A green, minty, melon aroma with tobacco like taste	3324	Tobacco flavors
1,3,3-trimethyl-2-norbornanyl acetate (fenchyl acetate)	A sweet, pine-like odor with camphor-like taste	3390	Spice, mint flavors
2,2,4-trimethyl-1,3-oxacyclopentane	—	3441	—
2,4,5-trimethyl, delta-3-oxazoline	—	3525	—
2,6,10-trimethyl-2,6,10-pentadecatrien-14-one (farmesyl acetone)	Sweet, floral, creamy odor and taste	3442	Artificial essential oils
2,3,5 trimethyl pyrazine	A molasses, rum aroma with a chocolate, cocoa taste	3244	Chocolate, cocoa, roast nuts
2,4,5-trimethylthiazole	A roasted, meaty aroma with corresponding taste	3325	Roasted flavors, meat, nut, cocoa
Trithioacetone	A sulfur, meaty aroma with a woody-minty taste	3475	Spearmint, peppermint imitations
2,4 Undecadienal	Fatty aldehyde, chicken odor with a chicken, cucumber taste depending on concentration	3422	Poultry flavors, cucumber
2,3-Undecadione (acetyl nonyryl) (acetyl pelargonyl)	Fatty, waxy, citrus aroma with fatty taste	172.515	Citrus, berry, butter
gamma-undecalactone (aldehyde C-14) so-called	Fatty, fruity, peach, apricot odor and taste	172.515	Apricot, peach, milk, coconut
Undecanal (aldehyde C-11)	Waxy, fatty, floral odor with citrusy taste	3092	Citrus, lime, lemon, orange
Undecanaic acid	Very little odor with a fatty taste	3245	Coconut, rum, brandy
2-Undecanol	Fruity, oily odor with mild citrus taste	3246	Coconut, rum, brandy
2-Undecanone (methyl nonyl ketone)	A fatty, fruity citrus aroma and taste	172.515	Citrus, cheese, coconut nut, rue
9-Undecenal	Sweet, citrus odor and taste	172.515	Citrus flavors, nut
10-Undecenal	Fatty, citrus odor and taste	172.515	Citrus flavors, fruit
2-Undecenal	A strong citrus odor and taste	172.515	Citrus flavors
10-Undecenoic acid	A waxy, citrus aroma with a fatty, waxy, coconut taste	3247	Coconut, rum, brandy
Undecen-l-ol	Sweet, waxy, floral odor with citrusy taste	172.515	Citrus, mint flavors
10-Undecen-l-yl acetate	Fatty, floral, honey odor fruity, citrus taste	172.515	Citrus

Table 4 (continued)
SYNTHETIC FLAVOR INGREDIENTS

Name	Organoleptic characteristics	Regulation	Flavors useful in
Undecyl alcohol	Fruity, floral, citrus odor fruity, citrus taste	172.515	Citrus
Valencene	Fatty, aldehyde-like, extracted orange aroma citrusy taste (chemical intermediate for production of nootketone)	3443	Fixative, blender for artificial essential oils
Valeraldehyde	Pleasant, chocolate aroma and taste	172.515	Chocolate, coffee, nut flavors
Valeric acid	Powerful, pungent, acid odor; fruity taste	172.515	Butter, cheese, butterscotch, strawberry, rum
gamma-valerolactone	A coumarin, coconut odor, fruity, coconut taste	3103	Peach, coconut, vanilla
d,l-Valine (2-aminoisovaleric acid)	Odorless white powder with sweet taste	182.5925 3444	Amino acid
Vanillin (4-hydroxy-3-methoxy benzaldehyde)	Characteristic vanilla flavor	182.60	Vanilla, chocolate, root beer, butter
Vanillin acetate (acetyl vanilla)	Mild, vanilla odor, much weaker than vanillin	172.515	Vanilla flavors, spice
Veratraldehyde (3,4 dimethoxy benzaldehyde)	A woody, earthy, vanilla aroma, with heliotropine taste	172.515	Vanilla, nut
Verbenol (2-pinen-4-ol)	—	172.515	—
o-Vinylanisole	Sassafras, root beer aroma with sweet, anise taste	3248	Vanilla, root beer
2,6-Xylenol (2,6 dimethoxy phenol)	A powerful, smokey odor; smoke taste at low levels	3249	Smoke flavors
Zingerone 4-(4 hydroxy-3-methoxyphenyl)-2-butanone)	A sweet, spicy aroma and taste	172.515	Spice, root beer, raspberry
Vitamin U (*d-l*-(3 amino-3-carboxy-propyl)dimethyl sulfonium chloride)	Weedy aroma with vegetable taste	3445	Vegetable, tomato

(coffee), *trans*-2-*cis*-6-nonadienal (cucumber), 1-octene-3-ol (mushroom), methyl-2-pyridyl ketone (popcorn), and diacetyl (butter). These items are characteristic of the named flavors, but by themselves do not make a good or complete flavor. The use of too much of these will give an unbalanced flavor and ruin the overall effect. The solvent, of course, is important in making the flavor applicable to the specific food substrate. (In an actual preparation, the solvent system is selected first since the choice of aromatics will depend significantly on the solvent. For example, if high heat stability is desired and a vegetable oil is selected as solvent, one cannot at some later stage use propylene glycol as the diluent for more potent aromatics, which will then be added as a solution to the flavor.)

The problem that exists today, if it can be called a problem, is the rapidity with which new aromatic chemicals have appeared for flavor use. After the invention of gas chromatography in 1952, the identification, synthesis, and use of new aromatics increased sharply. In 1960, there were somewhere between 50 to 100 major aromatics; currently there are in excess of 1500. In 1960, the pyrazines and thiazoles were unknown as flavor ingredients and not used as such in flavors. However, flavorists in the early 1960s were using oil galbanum, neroli, and petitgrain, and unaware that these ingredients contained pyrazines. (I recall eluting oil neroli from a chromatography column and finding it had a roasted-pepper-like note). As flavor chemists isolated and identified more flavor-relavent materials from natural products, more derivatives appeared on the market. For example, 2,5-dimethyl pyrazine was identified as a flavor-relavent compound in chocolate. Within a short period thereafter, ethyl dimethyl pyrazine, methyl methoxy pyrazine, and methyl thiopyrazine appeared on the market. Some of these were nature identical, others were not. A problem currently facing the flavorist is having access to these materials. Some are covered by composition-of-matter or use patents. Others are made exclusively for captive use by certain flavor houses. However, through special groups such as the Flavor Chemists Society and the Chemical Sources Association, Inc., this issue is being resolved.

Many of these materials have characteristic threshold values in the ppm range, and yet they exhibit a pronounced effect. The characteristic threshold is the concentraton at which the flavor character of the aromatic is evident. Usually, this is at the ppm level. However, attempting to decide which of these is suitable for use in a composition by tasting and smelling the chemicals directly is no longer a simple matter. One cannot give a nasal appraisal to a bottle of the aromatic and conclude that it will fit well into a particular flavor composition. For example, the odor of 2,5-dimethyl thiophene in concentrated form is perhaps the best substitute for the odor of nitrobenzene (shoe polish). Yet, it is useful in many flavors at low levels. Consequently, the flavorist must spend considerably more time in studying the materials. Who could have predicted a few years ago that oil galbanum had a pyrazine in it? More importantly, who could say that it might hold the key for potato flavors? Methional is another example. It was identified in potato, later in meat and cheese, and its full range of uses is still not fully explored.

The chemistry of some of these new aromatics is also an issue that must be considered when using them in finished flavor formulations. Many are characterized as having functional substituent groups. Some of these are highly reactive and display catalyzing or inhibiting effects on other flavor components. Consequently, the flavor chemist must be very selective in their use because artificial flavors can get better and closer to natural targets only through an understanding of how they interact with one another and the food substrate. They must be judiciously employed. The influx into the flavor field of highly competent and skillful researchers has made these new aromatics available, but more importantly, they are lending insight into what makes many natural flavors unique.

FLAVOR FORMS

Flavors are presented to the market place in three primary forms, i.e., liquids, powders, and pastes. Within the three primary forms, there are several subgroups. The subgroups can be called flavor forms. Examples of the various forms include:

- Liquids—water soluble, oil soluble, and emulsions.
- Powders—spray dried and extended (plated).
- Pastes—fat based, starch, carbohydrate based, and protein based.

The liquid flavors are, by far, the most numerous. In terms of flavor fidelity, flavors in liquid form are perhaps superior to those in powdered and paste forms. The liquid form is an important asset in the majority of applications since they more readily disperse into a flavor base. They also disperse more uniformly throughout the food system, thus eliminating the tendency to form hot spots (localized concentrations or pockets of flavor).

Liquid Flavor Forms

Extracts

Flavor extracts are solutions of flavors in ethyl alcohol or some other food-grade solvent. The bases for these flavors are essential oils, fruit extracts, botanical extracts, and imitations of the same.

Alcoholic Extracts

This is probably the oldest form (method) of flavoring. It probably began by macerating or steeping of spices in wines. Since the first alcohol came from the distillation of wine, alcohol (so it would seem) is the obvious solvent.

A typical extract might contain:

Nutmeg extract

Oil nutmeg	2.0 fl oz
Ethyl alcohol	95.0 fl oz

Water quantity sufficient to one gallon.

The advantages of this type of flavor are that it can easily be measured and it is dispersible in water. However, extracts have several drawbacks: they are usually very weak and lack strength, the cost and availability of alcohol can be a problem, and they tend to evaporate before and during heat processing.

Nonalcoholic Extracts

Propylene glycol and glycerine can be used as solvents for extracts since they are available as food grade, are relatively inexpensive, and show low volatility. However, they do not have the wide range of solubility displayed by alcohol. Orange, lemon, and anise oils are insoluble in propylene glycol. One important fact concerning extracts should be kept in mind: dissolving a liquid flavor, spice, or other essential oil in alcohol does not necessarily make it co-soluble in water. When alcoholic extracts are added to water, the flavor oils will usually come out of solution. They may disperse, but eventually the oil/water phases will separate, especially if the oil concentration in the extract is high. Citrus oils can be modified to remove the water insoluble terpenes, which tends to render them more water soluble. Spice oils and flavors are more difficult to modify in order to improve their solubility characteristics.

Oil Soluble Extracts

These are flavors incorporated into vegetable oil or an oil-soluble solvent, such as benzyl alcohol, as the carrier or vehicle. Again, the solvents are rather inexpensive. They display low volatility, and for use in bakery flavors show solubility in shortening, which is a very important property. Cake batter is an oil-in-water (o/w) emulsion, i.e., the fat is dispersed as tiny droplets in an aerated aqueous phase. Therefore, the flavor that can be dissolved or dispersed in the fat is better dispersed throughout the cake batter. Since most of the flavor losses incurred in baking is due to volatilization, it would seem that the flavor, if oil soluble, will be better absorbed, distributed, and protected from volatilization.

Emulsions

An emulsion is a two-phase system consisting of two immiscible liquids, one being dispersed in the other as fine droplets (globules). The names given to the liquid that is dispersed are internal, dispersed, or discontinuous phase. The liquid in which the droplets are dispersed is called the continuous or external phase. The phases in a stable emulsion are held together with an emulsifying agent such as gum arabic (acacia). The emulsifying agent serves two functions: (1) assists in the development of small particles and (2) maintaining them. Milk is an emulsion. It is an example of an oil-in-water emlusion (o/w) in which the fat is dispersed in the aqueous phase. Margarine is an example of a water-in-oil emulsion (w/o), that is, water dispersed as small droplets throughout the oil phase. The largest use of flavor emulsions is for the beverage industry, both carbonated and still. One important fact that should be kept in mind is that an emulsion is dilutable with its continuous phase. What this means is that an oil/water emulsion is dilutable with water.

Another type of flavor emulsion produced by the flavor industry is the so-called baker's emulsion. This product is difficult to categorize. It is indeed an emulsion, but has the appearance and flow characteristics of a paste. It is considerably more viscous than a beverage emulsion. Baking emulsions are quite popular and easy to prepare. The flavor is added to propylene glycol, a gum (usually gum tragacanth) added to form a slurry, and finally diluted with water. The blend is then mechanically homogenized. A viscous liquid is obtained initially and this is practically transformed into a solid upon standing. Good emulsions are not generally obtained with gum tragacanth since the dispersed particles are fairly large (10 to 15 μm). However, excellent suspensions are obtained with gum tragacanth. In baker's emulsions, the viscosity of the emulsion prevents the separation of the phases. The reason for this becomes obvious when one applies Stokes Law:

$$V = \frac{D^2 g({}^{d}d\text{-}dc)}{18\,\mu}$$

where: V = the velocity (cm/sec) of separation of the disperse particles, ^{d}d = specific gravity of the dispersed phase (oil or flavor), dc = specific gravity of the continuous phase (water), D = diameter of dispersed particles (cm), g = acceleration due to gravity (981 cm/sec^2), and u = viscosity of emulsion (g/cm-sec).

As the viscosity term (expressed in the demonimator) increases, the emulsion becomes more viscous and the rate of separation is reduced. The following is a typical example of a baker's emulsion:

Oil lemon	16.0 fl oz
Gum tragacanth	3.0 Av oz
Gum acacia	1.0 Av oz
Propylene glycol	16.0 fl oz
Water g.s. to 1 U.S. gallon	

The propylene glycol is used as a preservative while the gum acacia is used to regulate the viscosity of the gum tragacanth. The emulsion can be kneaded into the dough without the danger of it running off. The advantages of such emulsions are that they are readily produced and can be easily dispersed in a product. A disadvantage of such emulsions is that they are fairly concentrated and must be used with caution. Also, in making foam-type cakes, such as sponge or angel cake, care must be used when adding the emulsion since it may cause the foam to collapse or fall back when beaten with the egg or egg whites.

Powdered Flavor Forms

Extended (Plated) Flavors

These are simple dispersions of a flavor on a dry carrier. The flavor principle may be a liquid (essential oil, oleoresin, or aromatic chemical) or solids (vanillin or heliotropine). The dry carrier can be salt (for chips), dextrose (for cake mixes), flour (in bakery goods), or rusk (in meats). Extended flavor compositions are simple to prepare and usually inexpensive. Typically, the carrier is placed in a suitable blender and the liquid sprayed into the agitated carrier. When extending solids, the ingredients are simply dry blended with one another. The following is an example of an extended, powdered, artificial vanilla composition:

Vanillin	5.0
Ethyl vanillin	1.5
Heliotropine	0.2
Starch	10.0
Dextrose	83.3
Total	100.0

Dextrose in the above composition functions as the carrier and grinding medium for the heliotropine crystals.

The main function of the carrier is to distribute the flavor throughout a product. As mentioned previously, this helps to eliminate uneven flavor dispersion (hot spots). The chief value of products such as vanilla powders is that they offer an economical way to distribute strongly flavored aromatics. Vanillin, by itself, is difficult to disperse uniformly in food products and there is a high risk of overflavoring. Extended flavors need not be confined to vanillin and its limitations. Any flavor can be made into an extended flavor; cherry, orange, grape, etc. Some requirements that need to be considered include which carrier is appropriate, the nature of the flavor, stability, particle size, flow characteristics (anticaking agent), etc. There are certain extended flavors that function particularly well, e.g., vanilla sugars, cinnamon powders, and dry solubles. Another class of flavors that works well as extended flavors are those high in natural botanical extracts, such as foenugreek and molasses. These do not spray dry very well and the only way to get full flavor from them is by extension onto suitable carriers.

Extended flavors are a way of presenting liquid flavors in dry form. It is also a way to disperse liquid flavors into a dry product without wetting it. It is a way to disperse dry powders in a dry product to achieve uniformity of flavor. An advantage of extended (plated) flavors is that they usually contain sufficient aroma. They are economical and easy to produce and give good distribution of flavor in a product. However, they do have two drawbacks: evaporation of flavor and low flavor concentration levels. Because the flavor is distributed over an extensive surface area, several changes can take place. Important deleterious changes include evaporation or flavor loss and oxidation (rancidity) of the flavor. The concentration of the active flavor constituents is by necessity relatively low, because if it is too high the finished product will be moist and cause clumping and other mixing problems.

Spray Dried Flavors

Spray drying is a process by which the solids are recovered from a solution or slurry by spraying the liquid into a stream of drying gas (usually air) under conditions which permit recovery of a dry granular product. In a typical procedure, the flavor oil is emulsified with gum acacia, dextrin, modified starch, or other gums. It is important that the flavor oil is dispersed in the liquid as small droplets. The emulsion is then fed through an atomizer which sprays it out as minute droplets into a current of hot air. Because the droplets are so small, they dry almost instantly. Evaporation is usually so rapid that the temperature of the droplet falls below that of the air current. The low temperature and rapid drying are important because most flavors are heat labile and easily oxidized. The most important feature of spray drying is that the emulsifying colloids (gum acacia, dextrin, or modified starch) on drying form a thin film around the flavor, protecting it from oxidation and evaporation. The largest use for spray dried flavors is in cake mixes, beverage powders, and gelatin desserts.

The advantages of spray drying are that it produces small, usually granular, free-flowing powders which are easy to weigh and disperse. The flavors are locked in somewhat, that is, protected from exposure to air. Provided they are stored properly and kept free from moisture, they will retain their flavor value for extended periods. The use of spray-dried flavors eliminates the expense of overflavoring food products so often needed when using extended flavors. It is not necessary to make up for losses of flavor due to the loss of volatile components that occurs during storage or preparation of mixes. Fresh packages of extended flavors are often too high in flavor value while older packages are weak and unbalanced. There is better flavor fixation with the use of spray drying and these problems are rarely encountered. Spray drying greatly reduces the amount of flavor used in a composition.

As for the disadvantages of spray drying, there are few (at least for the flavor user) other than cost. The fact that they can only be produced with flavor concentrations in the 15 to 30% w/w range might be an additional disadvantage. However, from the point of view of the flavor manufacturer, there are also some disadvantages to spray drying. The equipment is not inexpensive and the additional processing (handling) and the R & D investment in time and knowledge necessary to make flavors applicable for spray drying constitute added costs.

Before concluding this section on powdered flavors, perhaps the comparison of extended and spray-dried flavors might be useful. They include:

- Stability—Stability of flavor character, stability against flavor loss and stability against oxidative changes are greatly increased in spray dried flavors. For example, spray dried forms of benzaldehyde are stable for years while benzaldehyde extended on a carrier surface can oxidize in seconds to benzoic acid.
- Concentration—A higher concentration of flavor principle is possible with spray dried compositions. Salt can hold about 2% (by weight) flavor without becoming wet. Dextrose, because of its greater surface area in proportion to its mass, can hold up to 5% without becoming wet. The use of anticaking agents helps, but essentially it is difficult to obtain high concentrations of flavor on extended formulations. The effective flavor concentration for spray dried compositions is about 20% by weight as compared to about 6% by weight for extended formulations.

Chemical Reactions—This can readily occur on extended flavors. For example, diacetyl (2,3-butandione) is a yellowish-green liquid. When extended on dextrose, it will turn colorless in a few minutes.

Paste Flavor Forms

Paste forms of flavor have just recently appeared from various flavor houses. In the past, they were offered principally by those producing protein hydrolysates (autolyzed yeasts, hydrolyzed vegetable protein pastes, etc.). The flavors are generally distributed in low-melting fats or hydrogenated vegetable oils; animal fats such as lard and beef tallow have also been employed in some formulations. Usually, they are blends of fat, salt, hydrolyzed vegetable protein, artificial and natural flavorings with some water soluble substance such as sucrose and dextrose. The fat is melted and the other ingredients are added while the fat is still fluid. It is allowed to cool with or without agitation; sometimes the product is further homogenized. Some consider paste flavors difficult to handle, cumbersome, and expensive to make except in large quantities. They are usually weak in flavor value and require usage in food at relatively high levels. At times, the quantities needed are so large that they become a basic ingredient in the final product as opposed to just a flavor additive. Often, problems of autoxidation of the fat are encountered and they are easily contaminated by other odors during storage via adsorption onto the lipid. Meat flavors currently appearing on the market probably constitute the largest flavor type produced in paste form. Paste flavors are blends of fat, starch, gums, and other carbohydrates, nucleotides, monosodium glutamate, hydrolyzed vegetable proteins, and other flavorng solids. Again, they are usually weak in flavor value and because of this are sold inexpensively. Manufacturers of soups, gravy bases, and bouillon cubes are the major users of this type of flavor.

CREATING FLAVOR COMPOSITIONS

We have seen that a flavor composition may be composed of both natural and synthetic materials chosen from an array of over 1500 items. The permutations and combinations are enormous and one can justifiably ask how does one go about the business of using this information to create flavors? One approach is to first analytically segregate flavor into its component parts. Analytically, a flavor composition is composed of two essential parts: the flavor relevant portion and the diluent.

The flavor portion consists of three distinct flavor items: a flavor character item, a flavor contributory item, and a flavor differential item.

The combined functions of the flavor portion of a composition are to:

- Simulate a flavor, whether it be fruit, meat, spice, and whether it stem from natural or artificial sources.
- Maintain flavor (character fixation).
- Enhance product acceptability.

The diluent portion of the flavor composition may take several forms, including:

- Water solubility; i.e., the addition of a diluent or carrier that renders the flavor portion water soluble. These include such diluents as ethanol, propylene glycol, and polysorbate 80.
- Oil solubility; i.e., diluents intended to render the flavor portion compatible with nonaqueous systems. These include such diluents as fats, vegetable oils,and other nonaqueous solvents, such as benzyl alcohol.
- Bulk fillers (extenders), such as sugar, salt, and vegetable gums.

The functions of the diluent portion include:

- Applicability—it must make the flavor applicable to the medium in which it is to be used (e.g., water solubility and water or oil dispersibility). It must disperse the flavor throughout the product.
- Cost regulation—the solvents, because they are used in fairly large quantities, should be inexpensive and can be used to regulate costs of the flavor portion.
- Strength regulation—the diluent portion, depending on the amount used, regulates to a large degree the strength of the flavor.
- Fixation—the solvent adds a degree of physical fixation.
- Color carrier—the solvent can be used to carry other igredients other than flavor into the food; e.g., color, preservative, etc.

At this juncture, it is important to define with greater accuracy the three factors comprising the flavor portion of the composition:

Flavor Character Item

This is an ingredient whose aroma and/or taste is clearly reminiscent of the named flavor. It may be an essential oil, botanical extract, organic chemical, and/or combinations of these that help create the desired flavor. The following are some examples of flavor character items:

Flavor character item	Character contributed
Oil Tagette	Apple
Oil Davana	Raspberry, strawberry
Beta-Ionone	Raspberry
Gamma-Undecalactone (aldehyde C-14)	Peach, apricot
Ethyl Methyl Phenyl Glycidate (aldehyde C-16)	Strawberry
3-Methyl Butylacetate	Banana
Fufuryl Thiopropionate Furfuryl Mercaptan	Coffee
2-Methyl-3-(-p-isopropyl phenyl)-propronaldehyde	Melon
Gamma-Nonalactone	Coconut
Nona-2-*trans*-6-*cis* dienal	Cucumber

Figure 6 illustrates some selected character impact items of known chemical composition.

Flavor Contributing Item

These are ingredients that help to create the named flavor. The item is not necessarily (and by itself) reminiscent of the named flavor, but when used in conjunction with flavor character items, tends to bring it closer to the named flavor. It can be an essential oil, botanical extractive, aromatic chemical, and/or combinations of these and be natural or synthetic. Some examples include:

- Acetaldehyde; in orange flavors, acetaldehyde helps create naturalness, fruitiness, and juiciness.
- Ethyl butyrate; in strawberry, it imparts naturalness. It provides the lift in grape flavors, together with the enhancement of naturalness.
- Adehyde C-16; in apple flavors, it imparts fruitiness.
- Esters; the combination of various esters imparts all fruits with naturalness.

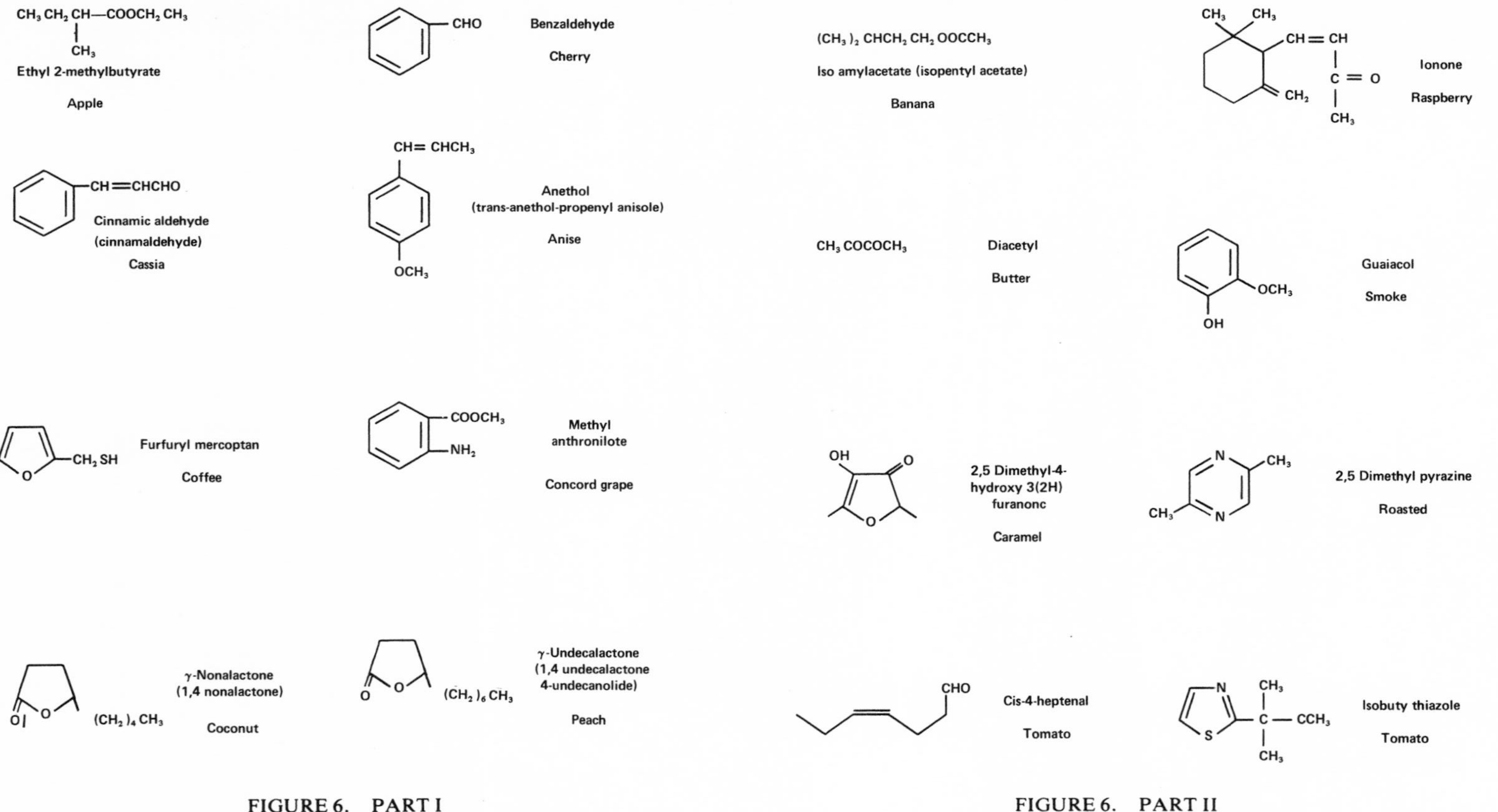

FIGURE 6. PART I

FIGURE 6. PART II

FIGURE 6. Selected character impact items.

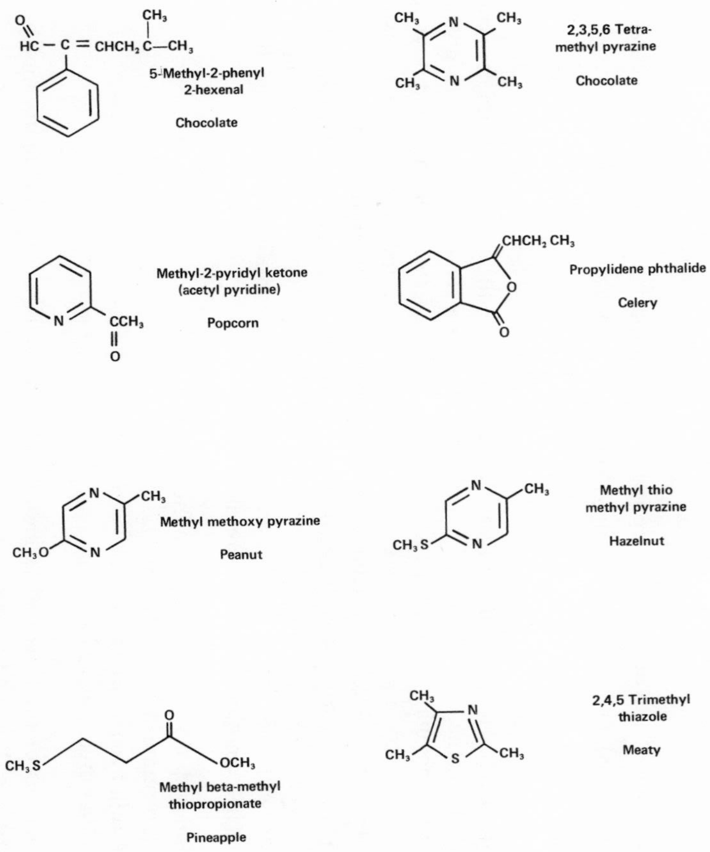

$HC(=O)-C(C_6H_5)=CHCH_2CH(CH_3)-CH_3$
5-Methyl-2-phenyl 2-hexenal
Chocolate

2,3,5,6 Tetra-methyl pyrazine
Chocolate

Methyl-2-pyridyl ketone (acetyl pyridine)
Popcorn

$CHCH_2CH_3$
Propylidene phthalide
Celery

CH_3O, CH_3
Methyl methoxy pyrazine
Peanut

CH_3S, CH_3
Methyl thio methyl pyrazine
Hazelnut

$CH_3S-CH_2CH_2-C(=O)-OCH_3$
Methyl beta-methyl thiopropionate
Pineapple

2,4,5 Trimethyl thiazole
Meaty

FIGURE 6. PART III

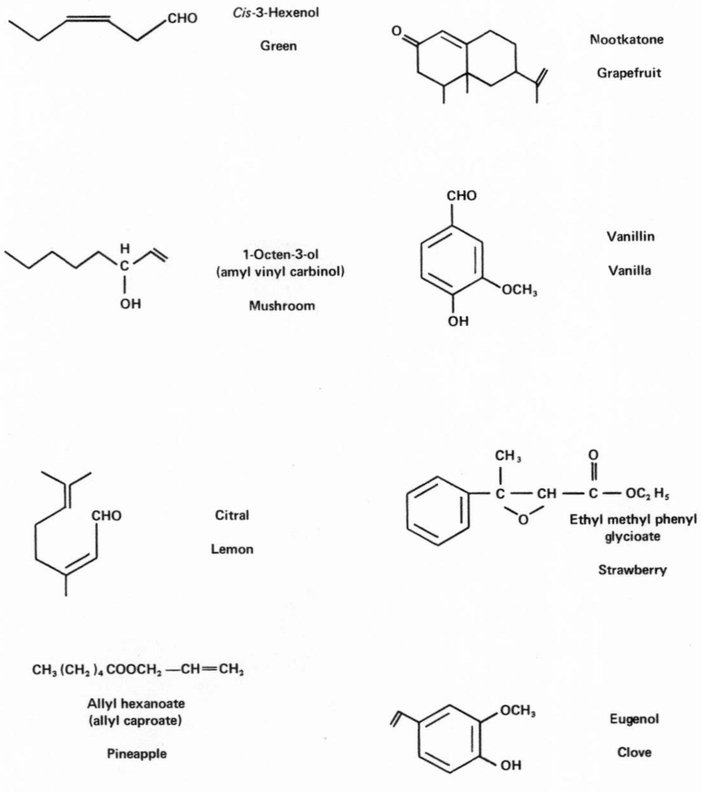

Cis-3-Hexenol
Green

Nootkatone
Grapefruit

1-Octen-3-ol (amyl vinyl carbinol)
Mushroom

Vanillin
Vanilla

Citral
Lemon

$C_6H_5-C(CH_3)(-O-)CH-C(=O)-OC_2H_5$
Ethyl methyl phenyl glycioate
Strawberry

$CH_3(CH_2)_4COOCH_2-CH=CH_2$
Allyl hexanoate (allyl caproate)
Pineapple

Eugenol
Clove

FIGURE 6. PART IV

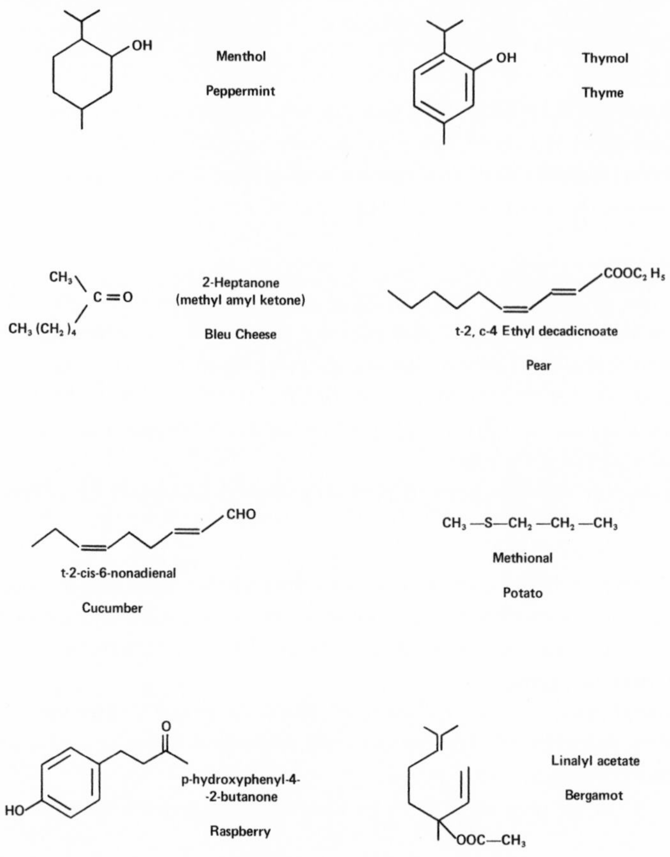
OH
Menthol
Peppermint
OH
Thymol
Thyme
CH_3
C=O
$CH_3(CH_2)_4$
2-Heptanone
(methyl amyl ketone)
Bleu Cheese
$COOC_2H_5$
t-2, c-4 Ethyl decadicnoate
Pear
CHO
t-2-cis-6-nonadienal
Cucumber
CH_3—S—CH_2—CH_2—CH_3
Methional
Potato
O
HO
p-hydroxyphenyl-4-
-2-butanone
Raspberry
OOC—CH_3
Linalyl acetate
Bergamot

FIGURE 6. PART V

The functions of flavor-contributing items are to impart a naturalness and fidelity of flavor (trueness), enhance acceptability, and to aid in establishment of the character items.

Flavor Differential Item

These are ingredients or combinations thereof that have little, if any, character reminiscent of the named flavor. These items are added to a flavor compound to give it individuality, imagination, and difference. These are the items a flavorist employs to create special effects, e.g., sweetness, lift, fixation, nuance, undertone, and aftertaste. They are the items which often characterize or distinguish the products of one flavor house from another. Also, it adds considerable difficulty to anyone trying to duplicate the flavor; used in extremely small quantities, they render duplication almost impossible. It is often difficult to determine and rationalize the significance of flavor differential items in a composition. For example:

- Vanillin and oxanone in a peppermint flavor
- Allyl capronate in vanilla flavor
- Oil Jasmin in a strawberry flavor
- Oil Olibanum in pineapple flavor

The functions of the flavor differential items are to impart individuality, imagination, and personality to a flavor composition.

Fixation

Character Fixation

This is the use of relatively high-boiling ingredients at high concentrations, i.e., above their threshold values. Examples include: vanillin, maltol, oxanone, and heliotropine. They are usually used in combination. Since they are used above their thresholds, they maintain their strength and character when diluted.

Physical Fixation

This is the use of relatively high-boiling materials to decrease the vapor pressure, thereby raising the boiling point and making the flavor constituents less heat labile. They are usually employed when flavors must be used in certain processing applications that require temperatures above the boiling point of water. Materials employed in physical fixations are usually very high-boiling solvents and may include such items as vegetable oil and butyl stearate.

It is often possible for one item to perform several functions in a flavor. For example:

- Ethyl oleate in a butter flavor is a flavor contributory item and a diluent.
- Benzyl alcohol in a nut flavor is a flavor contributory item and a diluent.
- Certain vegetable oils in coconut flavors are flavor character items, flavor contributory, and a diluent.
- Ethyl oenanthate is a flavor character item for grape (*Vitis vinifera*); a flavor contributory for grape (*Vitis lambrusca*), rum, and brandy; and a flavor differential item in coconut.
- Vanillin is a flavor character item in vanilla flavors and a flavor differential item in grape.
- Cinnamic aldehyde is a flavor character item in cassia or cinnamon and a flavor differential in grape.
- gamma-Hexalactone is a flavor character item in coconut and flavor contributory in peppermint and peach.

What is the significance of looking at the creation of a flavor in the above manner? It gives one a starting point for analysis of a flavor and the format for flavor development. It serves as a basis for discrimination, i.e., the examination of the component parts of a flavor. The flavorist can localize his thoughts and ideas on the different portions of a flavor composition with the ability to revise and change — to create new flavors. It gives the creative flavorist an unlimited number of future flavors. Above all, it gives us a starting point to learn about flavors. By tasting and categorizing the flavor materials into the different categories (character, contributory, and differential), one can begin to learn about flavors and how to create flavor compositions without access to a preset formula.

This categorizing of the parts of a flavor is important because it is the only way to retain flavor experiences. By giving the materials a label (name), we can recall them better. With memory and retention, we tend to recall items in the same category together, and since we encode our experiences as verbal labels, it follows that anything to which we give a name helps us to retain and recall our experiences — in this case our flavor experiences.

All this presupposes that there has been, or there will be, exposure to the flavor materials. It also suggests that the flavorist must taste and smell these materials and that he has personally placed them into his own notions of various categories. Since it is the individual flavorist who will make the flavors, he must know and believe that a particular substance will make a contribution. Because someone says a flavor material tastes or smells like a certain flavor does not mean that it does so for everyone. It is indispensible for the flavorist to see, taste, and smell the flavor materials, and mentally categorize each. It is important to form as many and as diverse associations as one can and, lastly, one must try to utilize them to make flavors. It is not a skill mastered overnight — but the reward is well worth the effort.

THE FUTURE OF FLAVORS

While forcasting future trends within an industry whose growth is linked to advancing technology is always a less than precise task, certain aspects can be generally envisioned. With little doubt, any statement must include the interaction of technology with government regulatory trends which stress human safety.

From a purely technological viewpoint, certain trends appear obvious. For example, highly sophisticated analytical techniques have unlocked many of the complexities of naturally occurring flavor materials. At this juncture, many important essential oils and extractions can be reconstituted; some partially, others nearly so. Such reconstituted products offer distinct advantages over natural products, including a potential reduction in cost, freedom from the whims of agriculture, and the elimination of ingredients which are irrelevant to flavor or in some instances appear to hamper the functionality of the natural product. On this latter issue, for example, gamma-terpinene in lemon oil appears to serve no useful function, but it does cause spoilage of the product. A synthesized lemon oil reconstituted without this product renders the oil significantly more stable. Other examples of this sort are well known to the flavorist.

Some future trends are clear. These include:

- The use of fermentation processes to produce useful flavors. Closely related processes include the use of isolated enzymes to drive complex reactions, e.g., aldehyde production in fats with lipooxidases, carbohydrate modification using glucose isomerase, etc.
- The use of by-products from food manufacturing processes, e.g., whey, yeast, and hydrolyzed meat proteins. Flavor research shows that nonvolatile portions of flavor products contribute significantly to overall sensory characteristics.
- The increased use of new flavors. This is especially true of certain tropical fruits not available a few years ago, e.g., passion fruit, guava, jackfruit, kiwi, mango, durian, etc. These will provide interesting diversions from the usual fruital aromas.
- The use of partial flavors or precursors, i.e., incomplete flavor, but which develop full characteristics during cooking or processing operations.
- Reaction flavors, i.e., Maillard products. This is often the ultimate duplication of nature's way of achieving certain flavors, e.g., cooked beef, pork, chicken, lamb, mushroom, chocolate, coffee, etc.

Sooner or later one must come to the conclusion that flavors are a very necessary part of our food supply. Along with this realization, one must also be aware that both natural and artificial flavors differ in name only: both are composed of chemicals, both have advantages and disadvantages in their use — but both are necessary. Natural flavors have the current advantage in that they present less labeling problems when they are used in foods. The real advantage to the use of natural flavors is that they contain substances (precursors) that appear to continually evolve flavor while cooking. With artificial flavors, we find that they generally are less expensive, have greater storage stability, can have much higher strength, and are usually more thermally stabile. Artificial flavors are generally not subject to crop failures or economic manipulations to regulate their prices. They are easier to control, that is, their quality and composition. As mentioned previously, an advantage of artificial flavors is that they can simulate the sensory characteristics of natural products without the concomitant use of flavor irrelevant matrials.

In the last few years, we have seen what could be called breakthroughs in the area of flavor creation and flavor application. Foods and flavors that a short while ago were nonexistent are now commercially available. When one notes the scientific and technological advances to date, there is a tendency to assume that they will continue — but this is not a certainty. As flavor research unravels nature's secrets, it makes the consuming public more and more aware of the complexity of a flavor, but at the same time it gives them cause for concern. Words like pyridines, pyrazines, thiazoles, furanones, and mercaptans appear in print — and to the uninitiated they don't "taste" too good. If one is to utilize these newly discovered materials to make new and novel flavors, one must not only have access to them, but must have regulatory permission to use them. However, flavorists are continually faced with increasing restrictions in the use of these new materials. Undoubtedly, today is not the most opportune time to seek regulatory approval for new materials. The concern for human safety is real and increasing. Should this trend continue, it could significantly hamper future flavor development. Even when it is possible to get new materials approved, it is an extremely costly and time-consuming venture. This aspect is expected to get worse rather than better. At a time when flavors are being asked to render top performance under extremely difficult processing conditions (e.g., extrusion, retorting, and dehydrations), the use of certain new flavor materials would be extremely helpful, but their use is being restricted. Ultimately, it would seem that the development of future flavors will be closely allied to government regulations.

A SELECTED GLOSSARY OF USEFUL TERMS FOR THE FLAVORIST

Absolute — A concrete that is extracted with alcohol to give an alcohol soluble concentrate.

Acetal — The product obtained from the reaction (addition) of 2 mol of alcohol to an aldehyde in the presence of acid. A dialkyl ether, e.g. $CH_3CHO + 2CH_3CH_2OH \rightarrow CH_3CH(OCH_2H_3)_2 + H_2O$

Alcohol — Ethyl alcohol (ethanol); usually used as a 95% by volume solution.

Antioxidant — An additive to retard oxidation; usually a steorically hindered phenol (see autoxidation).

Aroma — The odor, fragrance of a flavor.

Aroma chemical — Any chemical that has aroma or flavor properties. Not to be confused with the chemists' definition of an aromatic compound which contains the benzene ring structure.

Artificial flavor — Any substance the function of which is to impart flavor, which is derived from a spice, fruit, or fruit juice, vegetable or vegetable juice, edible yeast, herb, bark, bud, root, leaf, or similar plant material, meat, fish, or poultry, eggs, dairy products, or fermentation products thereof. Legally, the addition of any flavoring material to food not native to that substance.

A.S.T.A. — American Spice Trade Association, New York, N.Y.

A.S.T.M. — American Society for Testing Materials, Philadelphia, Pa.

Autoxidation	A series of spontaneous and degrative reactions that take place in essential oils when they are exposed to air, light, heat, or metallic ions.
Baumé (Bé)	A scale used to express the sugar concentration of a syrup. A scale to determine the specific gravity of a liquid (syrup).
Blender	A material that when added to a substance appears to bring various flavor characteristics together. A blender may or may not introduce a flavor of its own, e.g., vanilla can act as a blender.
Botanical	Any plant material such as herbs, roots, seeds, leaves, flowers, etc., which can be used for flavoring as such or a flavor that can be isolated from it.
Bottlers' extract	A flavor for the bottling industry; usually carbonated beverages. It usually appears in ½ to 2 fluid oz/gal strength, i.e., ½ to 2 fluid oz are added to a gallon of sugar syrup to make a bottlers' syrup.
Bottlers' syrup	A dilutable syrup containing sugar, flavor, acidulants, and color, which is diluted with carbonated water to make a finished beverage.
Brix	A scale used to express sugar concentrations; usually refers to percent soluble sugar solids by weight.
Browning reaction	(see Maillard Reaction)
Buffer	A salt of an organic acid — added to maintain pH.
Bulking	Mixing of one or more lots of the same flavorful materials to produce a uniform product. The entire crop of an essential oil may be bulked to assure uniformity.
Carrier	The diluent or solvent for a flavor, it may be a liquid or solid, e.g., gum acacia is the carrier for spray-dried flavors and alcohol, propylene glycol are the carriers for many flavors.
CFR—21	(21 CFR) Code of Federal Regulations Title 21, containing all regulations pertaining to food and drugs.
Citrus flavors	Flavors made from the oils and juices of the citrus fuits, e.g., orange, lemon, lime, tangerine, grapefruit, mandarin, and bergamot. Also, synthetic ingredients used to simulate these flavors.
Clouding agent	A flavoring adjunct used to create a translucent or opaque appearance in citrus drinks.
Cold pressing	A process for expressing citrus essential oils by pressure without the use of heat. A process for pressing or squeezing out the oil from the rind of a fruit (see also Expression.)
Comminution	The process of grinding or breaking into small fragments.
Compounds	A flavoring mixture composed of two or more substances. These substances may be natural or synthetic, they may be combinations of aromatic chemicals, essential oils, oleoresins, extracts, or other blends. These should not be confused with the chemical definition: two elements chemically combined.
Concentrated fruit juices	A fruit juice in which the water has been partially removed by some form of evaporation, e.g., distillation. It is usually concentrated to the strength of 5 to 6 times the single-fold juice. The essences are usually added back to the concentrate before use.

Concrete — A semisolid mixture containing the essential oil and fatty, waxy materials obtained after extracting the plant tissue, especially flowers, with various solvents.

Concentrated or folded citrus oils — Essential oils, such as lemon oil, in which part of the terpene fraction has been removed either by distillation or solvent extraction. The process produces essential oils of greater strength and character, with greater alcohol solubility. They also usually show better stability to oxidation.

Decoction — A solution made by boiling the material to be extracted with a solvent; usually followed by filtration.

Distillate — A clear, flavorful liquid produced from fruits, herbs, roots, etc. by distillation. Also the condensed product separated by distillation.

Distillation — The separation of a more volatile part of a substance from those less volatile by vaporizing and subsequent condensation. Two types are generally used; steam and fractional distillation.

Distilled oil — Oil separated from a botanical material by distillation methods.

Dry solubles — Natural spice oils and/or oleoresins extended on a soluble, dry, edible carrier.

Emulsion — A system containing two immiscible liquids in which one is dispersed as very small droplets or globules throughout the other.

Encapsulation — A process by which a particle is coated with a partially impermeable layer to retard evaporation and/or chemical reaction. Basically, the material to be protected is wrapped in a wall of impervious material which serves to lock-in or entrap the volatile compounds.

Enhancer — An ingredient that is added to supplement, enhance, or modify the original taste and/or aroma of a food without imparting a characteristic taste or aroma of its own, e.g., maltol and ethyl maltol.

EOA — Essential Oil Association of the USA Inc., New York, N.Y.

Essence — A concentrated fragrance or flavorant. In some countries, essence is used to designate volatile oils, but in the U.S. this term is commonly applied to alcoholic solutions of volatile oils.

Essential oils — A volatile, odoriferous substance obtained from plant material through steam distillation. The essential oil normally has the characteristic flavor of the plant from which it was derived. An essential oil is differentiated from a fixed oil, which is not steam distillable.

Expression — A process using pressure to obtain an essential oil, usually out of the rind of citrus fruits without the use of heat.

Extract — A solution otained by passing alcohol or an alcohol-water mixture through a substance. An example would be vanilla extract. Household extracts found on grocery shelves such as almond, lemon, etc. are usually essential oils dissolved in an alcohol-water mixture.

Extended flavor — Dispersion of a flavor on a dry carrier. They may be liquids (flavors, spice oils, or oleoresins), solids (vanillin or heliotropine), and can be natural or artificial. The dry carrier is usally an anhydrous material, such as dextrose. These flavors are also referred to as plated flavors (see also Dry Solubles).

FCC — Food Chemical Codex. A compendium of food ingredients and additives. It lists specifications and prescribes minimum require-

	ments of purity for an appropriate grade of food chemical. It was prepared by the Food Protection Committee, National Academy of Science, National Research Council.
FDA	Food and Drug Administration. A regulatory body and branch of the Department of Health, Education and Welfare of the U.S. Federal Government.
FEMA	Flavor Extract Manufacturers Association, Washington, D.C.
Fixative	A material of low volatility (such as vegetable oil) that retards the evaporation of volatile substances when both of these materials are in common solution.
Flavor chemist (flavorist)	An artistic, scientific worker in the food and flavor field who creates finished flavors from basic aroma chemicals and essential oils.
Fold	A term used to designate the strength of a material, e.g., a twofold oil has twice the flavor value of a single-fold oil.
Food additive	All substances the intended use of which results or may reasonably be expected to result, directly or indirectly, either in their becoming a component of food or otherwise affecting the characteristics of food. (Title 21 Part 112.1 (e).)
Food Additive Amendment of 1958	This is an amendment to the 1938 Act. It provides for protection of public health by requiring proof of safety before a substance may be used in foods. It permits the use in foods of substances which are safe at the levels of intended use. It forbids the issuance of any regulation permitting the use of any substance in any amount whatever which "is found to induce cancer when ingested by man or animal, or if it is found after tests which are appropriate for the evaluation of safety of food additives to induce cancer in man or animal." (This is the so-called Delaney Clause.)
Food Drug and Cosmetic Act of 1906	This act made illegal the adulteration of foods and drugs that are sold interstate. It did not,however, regulate foods and drugs that were sold intrastate. This was left to the indivdual states.
Food Drug and Cosmetic Act of 1938	This act contained many of the provisions of the 1906 act. It still is the basic legislation currently enforced. Some of its important provisions are: • It defined a food. • It prohibited adulteration of foods. • It made labeling truthful and mandatory. • It defined Standards of Identity for certain food products.
Freeze-drying	A concentration process by which water is removed by freezing and subsequent sublimation under vacuum.
GMP	Good Manufacturing Practice—the quantity of a substance added to food does not exceed the amount reasonably required to accomplish its intended physical, nutritional, or other technical effect in the food, and... The quantity of a substance that becomes a component of a food as a result of its use in manufacturing processes or packaging of food, and which is not intended to accomplish any physical or other technical effect in the food itself, shall be reduced to the extent reasonably possible. (Title 21, Part 182.1 (b) (1) (2).)
GRAS list	Generally Recognized as Safe—for flavors. Used under conditions

	of intended use in accordance with Good Manufacturing Practice as defined by the Food Drug and Cosmetic Act section 182.1.
Herb	Plants of which the leaves or stem and leaves are used for food or medicine, or for their scent or flavor.
Household extract	A flavoring extract designed for home use, e.g., for cakes and cookies, etc.
Infusion	An aqueous solution obtained by steeping a substance in water below its boiling point and then straining.
I.O.F.I.	International Organization of the Flavor Industry, Geneva, Switzerland.
182.1	Section of the U.S. Food Additive regulations put forth by the FDA listing GRAS substances. It contains very few artificial flavoring substances specifically named, but it does contain an extensive list of spices, seasonings, essential oils, oleoresins, and natural extractives.
172.510	Section of CFR (Code of Federal Regulations) put forth by the FDA listing natural flavoring additives for use in foods, under GMP.
172.515	Section of CFR—put forth by FDA listing permitted synthetic flavoring additives for use in foods, under GMP.
Isolate	A chemical or fraction obtained from a natural substance, e.g., citral can be isolated from oil of lemongrass, menthol from mentha arvensis, and eugenol from clove. It is a pure substance that is separated from a natural product or an essential oil.
Key (specialty)	Usually similar to a compound, only not generally finished. A key, although not complete, carries the major part of the flavor load so that only a few substances are needed to complete the flavoring.
Maceration	To steep or soak in a solvent for the purpose of extraction.
Maillard reaction	Flavor production by nonenzymatic browning in food. Proceeds mainly from reactions of reducing sugars with amines, amino acids, peptides, and proteins.
Masking agent	An ingredient capable of covering or at least making more acceptable an unpleasant odor or taste in a food or pharmaceutical.
Menstruum	The medium in which a substance is dissolved—a solvent.
Mercaptan	A hydrosulfide or compound containing the radical—SH. Prefix: Mercapto, Suffix: thiol, e.g., Benzyl mercaptan (Toluene thiol) $C_6H_5CH_2SH$; Methanethiol (Methyl mercaptan) CH_3SH; 2-Mercapto thiosphene (Thiopenethiol) C_4H_3S-SH.
Middle note or main note	The main characteristic of a flavor.
Modifier	An ingredient which influences, but does not change materially, the flavor and odor characteristics of a flavor.
N.F.	National Formulary. A compendium of drugs and some flavor materials, providing specifications regarding quality, purity, and strength under the auspices of the American Pharmaceutical Association.
Nature identical flavoring substances	Substances chemically isolated from natural materials or obtained synthetically that are chemically identical to substances present in natural products intended for human consumption.
Natural flavors	The essential oil, oleoresin, essence or extractive, protein hydrolysate, distillate, or any product of roasting, heating, or enzymolysis

which contains the flavoring constituents derived from a spice, fruit or fruit juice, vegetable or vegetale juice, edible yeast, herb, bark, bud, root, leaf, or similar plant material, meat, seafood, poultry, eggs, dairy products, or fermentation products thereof whose significant function in food is flavoring rather than nutrition. (21 CFR 1.12 (a), (3).)

Note — A distinct flavor or odor characteristic. For example, many strawberry or raspberry flavors have a seedy note or a green note.

Oleoresin — An extractive of plant material consisting of the essential oil, if any, and the other principles of the plant consisting not only of the flavor, but also the fatty oils, soluble resins, gums, and other principles. The essential oil of ginger, for example, does not impart any pungency to a product; oleoresin ginger does.

Oxygenated compounds — These are aromatic substances containing carbon, hydrogen, and oxygen such as alcohols, aldehydes, ketones, acids, esters, and other substances. These are the flavorful components of essential oils.

Panel test — Sensory evaluations of foods or flavors to determine acceptability, difference, and/or similarity and sometimes identify.

Percolation — A process of extracting the soluble constituents of a substance by a slow stream of solvent; also a filtering through the material.

Plated flavors — (See Extended Flavors, Dry Solubles.)

Redistilled (rectified) oils — These are oils that have undergone a second distillation (rectification) in addition to the distillation that separated the oil from the plant, to remove certain undesirable fractions. Rectified peppermint oil is an example: the sulfide fraction is removed to make the oil cleaner in top-note.

Resinoid — An exuadate of a plant material with flavor and odor value. An extraction of gums and balsams, sometimes containing a diluent to allow ease of handling, if the product is viscous.

Schiff base — Condensation product of aromatic amines and aliphatic aldehydes.
$PhNH_2 + CH_3CHO \rightarrow PhN = CHCH_3 + H_2O$

Scoville units — An organoleptic measurement used to describe the intensity of capsicum by threshold dilution. See EOA specifications on Oleoresin Capsicum.

Seasoning — Natural or artificial compounds that enhance, modify, or supplement the taste impact of the food, without being recognizable themselves.

Sesquiterpene — A hydrocarbon found in the higher boiling fractions of many essential oils. General formula: $(C_5H_8)_3$.

Sesquiterpeneless oil — An essential oil, such as lemon oil, in which all the terpene and sesquiterpene fractions have been removed.

Shelf-life — The stability of a food or flavor under specific conditions of storage.

Solubilizer — An ingredient capable of dispersing an otherwise insoluble flavor in an aqueous phase.

Solubilized flavor — A flavor material not normally soluble in a particular medium made soluble with the use of solubilizers, e.g., Polysorbate 80.

Solvent — A liquid material such as ethyl alcohol, glycerine, or propylene glycol which is capable of dissolving, distributing, or otherwise putting into solution with itself various flavoring materials. (Also called diluent.)

Society of Flavor Chemists Inc. The professional organization of Flavor Chemists—with headquarters in New York City.

Spice Any aromatic vegetable substance in the whole, broken, or ground form; whose significant function in food is seasoning rather than nutritional; that is true to name, and from which no portion of any volatile oil or other flavoring principle has been removed, e.g., allspice, clove, etc. (21 CFR 1.12 (a) (2).)

Spray drying A process by which solids are recovered from a liquid solution or slurry by spraying into a stream of drying gas or air—and the dry product recovered.

Standard of identity A regulation under the Food Drug and Cosmetic Act which specifies the composition and labeling of many foods in the U.S., e.g., 21 CFR 51 — Canned Vegetables.

Stokes Law The rate of phase separation in an emulsion. The settling velocity of spherical particles:

$$V = \frac{D^2 g(d - d_1)}{18\,n}$$

V = Rate of emulsion phase separation in cm/sec.
D = Diameter of spherical particle in cm.
d = Density of emulsified particle in g/cm^3.
d_1 = Density of whole emulsion in g/cm^3.
g = Acceleration due to gravity (981 cm/sec/sec).
n = Viscosity of the whole emulsion in g/cm-sec.

Strecker degradation The deamination, decarboxylation of alpha-amino acids to produce aldehydes or ketones of one carbon less than the original amino acid.

$$\underset{\text{alanine}}{CH_3CH(NH_2)COOH} \longrightarrow \underset{\text{acetaldehyde + carbon dioxide + ammonia}}{CH_3CHO + CO_2 + NH_3}$$

Synthetic Not naturally produced. Menthol synthetically produced is chemically identical to menthol obtained from peppermint oil, but it cannot be called natural. It is synthetically produced, therefore, it is artificial. However, it can be called nature identical.

Taste Can have two meanings depending on the part of speech it is used as, e.g., taste as a verb: To perceive by the sense of taste. Taste as a noun: usually refers to the four basic taste qualities that are perceived by the sense of taste, namely, bitter, salt, sour, and sweet (sometimes one might include metallic).

Terpene A hydrocarbon component found in essential oils. Terpenes have little flavor value. They are insoluble in water and very prone to autoxidation. For example, orange oil contains over 90% limonene, a common terpene. General formula: ($C_{10}H_{16}$).

Terpeneless oil An essential oil such as orange or lemon oil, in which all of the terpene fraction has been removed. These oils are stronger in character and strength and are more water soluble than the oils they are derived from. They are also more stable to oxidation.

Tincture A dilute water-alcohol solution of an essential oil, aromatic sub-

	stance, or vegetable material. A solution of a plant principle in alcohol or alcohol-water menstruum.
Top note	The first note normally perceived when a flavor is smelled. Usually a topnote is relatively volatile and suggests identity.
Throw	The extent of dilution of a bottlers' syrup with carbonated water. It is usually expressed as the volume of syrup plus the volume of carbonated water, e.g., one plus five.
True fruit extract	A flavor derived entirely from the fruit bearing the characterizing name.
True fruit flavor WONF	A flavor whose flavor strength and taste is derived from concentrated juice of the characterizing name, e.g., cherry, grape, etc. and the remaining flavor from other natural flavoring materials.
U.S.D.A.	U.S. Department of Agriculture. They are responsible for the approval of ingredients and labeling of meat and poultry products.
USP	United States Pharmacopeia — a compendium which sets forth standards of purity, quality, and strength for drugs and some flavoring materials.
Volatile oils	(See Essential Oils.)

APPENDIX 1

Subpart C—Solvents, Lubricants, Release Agents and Related Substances

§ 173.210 Acetone.

A tolerance of 30 parts per million is established for acetone in spice oleoresins when present therein as a residue from the extraction of spice.

§ 173.220 1,3-Butylene glycol.

1,3-Butylene glycol (1,3-butanediol) may be safely used in food in accordance with the following prescribed conditions:

(a) The substance meets the following specifications:

(1) 1,3-Butylene glycol content: Not less than 99 percent.

(2) Specific gravity at 20/20° C: 1.004 to 1.006.

(3) Distillation range: 200°–215° C.

(b) It is used in the minimum amount required to perform its intended effect.

(c) It is used as a solvent for natural and synthetic flavoring substances except where standards of identity issued under section 401 of the act preclude such use.

§ 173.230 Ethylene dichloride.

A tolerance of 30 parts per million is established for ethylene dichloride in spice oleoresins when present therein as a residue from the extraction of spice; *Provided, however,* That if residues of other chlorinated solvents are also present the total of all residues of such solvents shall not exceed 30 parts per million.

§ 173.240 Isopropyl alcohol.

Isopropyl alcohol may be present in the following foods under the conditions specified:

(a) In spice oleoresins as a residue from the extraction of spice, at a level not to exceed 50 parts per million.

(b) In lemon oil as a residue in production of the oil, at a level not to exceed 6 parts per million.

(c) In hops extract as a residue from the extraction of hops at a level not to exceed 2.0 percent by weight: *Provided,* That,

(1) The hops extract is added to the wort before or during cooking in the manufacture of beer.

(2) The label of the hops extract specifies the presence of the isopropyl alcohol and provides for the use of the hops extract only as prescribed by paragraph (c)(1) of this section.

§ 173.250 Methyl alcohol residues.

Methyl alcohol may be present in the following foods under the conditions specified:

(a) In spice oleoresins as a residue from the extraction of spice, at a level not to exceed 50 parts per million.

(b) In hops extract as a residue from the extraction of hops, at a level not to exceed 2.2 percent by weight; *Provided,* That:

(1) The hops extract is added to the wort before or during cooking in the manufacture of beer.

(2) The label of the hops extract specifies the presence of methyl alcohol and provides for the use of the hops extract only as prescribed by paragraph (b)(1) of this section.

§ 173.255 Methylene chloride.

Methylene chloride may be present in food under the following conditions:

(a) In spice oleoresins as a residue from the extraction of spice, at a level not to exceed 30 parts per million; *Provided,* That, if residues of other chlorinated solvents are also present, the total of all residues of such solvents shall not exceed 30 parts per million.

(b) In hops extract as a residue from the extraction of hops, at a level not to exceed 2.2 percent, *Provided,* That:

(1) The hops extract is added to the wort before or during cooking in the manufacture of beer.

(2) The label of the hops extract identifies the presence of the methylene chloride and provides for the use of the hops extract only as prescribed by paragraph (b)(1) of this section.

(c) In coffee as a residue from its use as a solvent in the extraction of caffeine from green coffee beans, at a level not to exceed 10 parts per million (0.001 percent) in decaffeinated roasted coffee and in decaffeinated soluble coffee extract (instant coffee).

§ 173.270 Hexane.

Hexane may be present in the following foods under the conditions specified:

(a) In spice oleoresins as a residue from the extraction of spice, at a level not to exceed 25 parts per million.

(b) In hops extract as a residue from the extraction of hops, at a level not to exceed 2.2 percent by weight; *Provided,* That:

(1) The hops extract is added to the wort before or during cooking in the manufacture of beer.

(2) The label of the hops extract specifies the presence of the hexane and provides for the use of the hops extract only as prescribed by paragraph (b)(1) of this section.

§ 173.290 Trichloroethylene.

Tolerances are established for residues of trichloroethylene resulting from its use as a solvent in the manufacture of foods as follows:

Decaffeinated ground coffee.	25 parts per million.

APPENDIX 1 (continued)

Decaffeinated soluble (instant) coffee extract.	10 parts per million.
Spice oleoresins	30 parts per million (provided that if residues of other chlorinated solvents are also present, the total of all residues of such solvents in spice oleoresins shall not exceed 30 parts per million).

APPENDIX 2

Subpart B—Specific Food Labeling Requirements

§ 101.22 Foods; labeling of spices, flavorings, colorings and chemical preservatives.

(a)(1) The term "artificial flavor" or "artificial flavoring" means any substance, the function of which is to impart flavor, which is not derived from a spice, fruit or fruit juice, vegetable or vegetable juice, edible yeast, herb, bark, bud, root, leaf or similar plant material, meat, fish, poultry, eggs, dairy products, or fermentation products thereof. Artificial flavor includes the substances listed in §§ 172.515(b) and 182.60 of this chapter except where these are derived from natural sources.

(2) The term "spice" means any aromatic vegetable substance in the whole, broken, or ground form, except for those substances which have been traditionally regarded as foods, such as onions, garlic and celery; whose significant function in food is seasoning rather than nutritional; that is true to name; and from which no portion of any volatile oil or other flavoring principle has been removed. Spices include the spices listed in § 182.10 of this chapter, such as the following:

Allspice	Marjoram
Anise	Mustard flour
Basil	Nutmeg
Bav leaves	Oregano
Caraway seed	Paprika
Cardamon	Parsley
Celery seed	Pepper, black
Chervil	Pepper, white
Cinnamon	Pepper, red
Cloves	Rosemary
Coriander	Saffron
Cumin seed	Sage
Dill seed	Savory
Fennel seed	Star aniseed
Fenugreek	Tarragon
Ginger	Thyme
Horseradish	Turmeric
Mace	

Paprika, turmeric, and saffron or other spices which are also colors, shall be declared as "spice and coloring" unless declared by their common or usual name.

(3) The term "natural flavor" or "natural flavoring" means the essential oil, oleoresin, essence or extractive, protein hydrolysate, distillate, or any product of roasting, heating or enzymolysis, which contains the flavoring constituents derived from a spice, fruit or fruit juice, vegetable or vegetable juice, edible yeast, herb, bark, bud, root, leaf or similar plant material, meat, seafood, poultry, eggs, dairy products, or fermentation products thereof, whose significant function in food is flavoring rather than nutritional. Natural flavors include the natural essence or extractives obtained from plants listed in §§ 182.10, 182.20, 182.30, 182.40, and 182.50 of this chapter, and the substances listed in § 172.510 of this chapter.

(4) The term "artificial color" or "artificial coloring" means any "color additive" as defined in § 8.1(f) of this chapter.

(5) The term "chemical preservative" means any chemical that, when added to food, tends to prevent or retard deterioration thereof, but does not include common salt, sugars, vinegars, spices, or oils extracted from spices, substances added to food by direct exposure thereof to wood smoke, or chemicals applied for their insecticidal or herbicidal properties.

(b) A food which is subject to the requirements of section 403(k) of the act shall bear labeling, even though such food is not in package form.

(c) A statement of artificial flavoring, artificial coloring, or chemical preservative shall be placed on the food, or on its container or wrapper, or on any two or all of these, as may be necessary to render such statement likely to be read by the ordinary individual under customary conditions of purchase and use of such food.

(d) A food shall be exempt from compliance with the requirements of section 403(k) of the act if it is not in package form and the units thereof are so small that a statement of artificial flavoring, artificial coloring, or chemical preservative, as the case may be, cannot be placed on such units with such conspicuousness as to render it likely to be read by the ordinary individual under customary conditions of purchase and use.

(e) A food shall be exempt while held for sale from the requirements of section 403(k) of the act (requiring label statement of any artificial flavoring, artificial coloring, or chemical preservatives) if said food, having been received in bulk containers at a retail establishment, is displayed to the purchaser with either (1) the labeling of the bulk container plainly in view or (2) a counter card, sign, or other appropriate device bearing prominently and conspicuously the information required to be stated on the label pursuant to section 403(k).

(f) A fruit or vegetable shall be exempt from compliance with the requirements of section 403(k) of the act with respect to a chemical preservative applied to the fruit or vegetable as a pesticide chemical prior to harvest.

(g) A flavor shall be labeled in the following way when shipped to a food manufacturer or processor (but not a consumer) for use in the manufacture of a fabricated food, unless it is a flavor for which a standard of identity has been

APPENDIX 2 (continued)

promulgated, in which case it shall be labeled as provided in the standard:

(1) If the flavor consists of one ingredient, it shall be declared by its common or usual name.

(2) If the flavor consists of two or more ingredients, the label either may declare each ingredient by its common or usual name or may state "All flavor ingredients contained in this product are approved for use in a regulation of the Food and Drug Administration." Any flavor ingredient not contained in one of these regulations, and any nonflavor ingredient, shall be separately listed on the label.

(3) In cases where the flavor contains a solely natural flavor(s), the flavor shall be so labeled, e.g., "strawberry flavor", "banana flavor", or "natural strawberry flavor". In cases where the flavor contains both a natural flavor and an artificial flavor, the flavor shall be so labeled, e.g., "natural and artificial strawberry flavor". In cases where the flavor contains a solely artificial flavor(s), the flavor shall be so labeled, e.g., "artificial strawberry flavor".

(h) The label of a food to which flavor is added shall declare the flavor in the statement of ingredients in the following way:

(1) Spice, natural flavor, and artificial flavor may be declared as "spice", "natural flavor", or "artificial flavor", or any combination thereof, as the case may be.

(2) An incidental additive in a food, originating in a spice or flavor used in the manufacture of the food, need not be declared in the statement of ingredients if it meets the requirements of § 101.100(a)(3).

(3) Substances obtained by cutting, grinding, drying, pulping, or similar processing of tissues derived from fruit, vegetable, meat, fish, or poultry, e.g., powdered or granulated onions, garlic powder, and celery powder, are commonly understood by consumers to be food rather than flavor and shall be declared by their common or usual name.

(4) Any salt (sodium chloride) used as an ingredient in food shall be declared by its common or usual name "salt."

(5) Any monosodium glutamate used as an ingredient in food shall be declared by its common or usual name "monosodium glutamate."

(6) Any pyroligneous acid or other artificial smoke flavors used as an ingredient in a food may be declared as artificial flavor or artificial smoke flavor. No representation may be made, either directly or implied, that a food flavored with pyroligneous acid or other artificial smoke flavor has been smoked or has a true smoked flavor, or that a seasoning sauce or similar product containing pyroligneous acid or other artificial smoke flavor and used to season or flavor other foods will result in a smoked product or one having a true smoked flavor.

(i) If the label, labeling, or advertising of a food makes any direct or indirect representations with respect to the primary recognizable flavor(s), by word, vignette, e.g., depiction of a fruit, or other means, or if for any other reason the manufacturer or distributor of a food wishes to designate the type of flavor in the food other than through the statement of ingredients, such flavor shall be considered the characterizing flavor and shall be declared in the following way:

(1) If the food contains no artificial flavor which simulates, resembles or reinforces the characterizing flavor, the name of the food on the principal display panel or panels of the label shall be accompanied by the common or usual name of the characterizing flavor, e.g., "vanilla", in letters not less than one-half the height of the letters used in the name of the food, except that:

(i) If the food is one that is commonly expected to contain a characterizing food ingredient, e.g., strawberries in "strawberry shortcake", and the food contains natural flavor derived from such ingredient and an amount of characterizing ingredient insufficient to independently characterize the food, or the food contains no such ingredient, the name of the characterizing flavor may be immediately preceded by the word "natural" and shall be immediately followed by the word "flavored" in letters not less than one-half the height of the letters in the name of the characterizing flavor, e.g., "natural strawberry flavored shortcake," or "strawberry flavored shortcake".

(ii) If none of the natural flavor used in the food is derived from the product whose flavor is simulated, the food in which the flavor is used shall be labeled either with the flavor of the product from which the flavor is derived or as "artificially flavored."

(iii) If the food contains both a characterizing flavor from the product whose flavor is simulated and other natural flavor which simulates, resembles or reinforces the characterizing flavor, the food shall be labeled in accordance with the introductory text and paragraph (i)(1)(i) of this section and the name of the food shall be immediately followed by the words "with other natural flavor" in letters not less than one-half the height of the letters used in the name of the characterizing flavor.

(2) If the food contains any artificial flavor which simulates, resembles or re-

APPENDIX 2 (continued)

inforces the characterizing flavor, the name of the food on the principal display panel or panels of the label shall be accompanied by the common or usual name(s) of the characterizing flavor, in letters not less than one-half the height of the letters used in the name of the food and the name of the characterizing flavor shall be accompanied by the word(s) "artificial" or "artificially flavored", in letters not less than one-half the height of the letters in the name of the characterizing flavor, e.g., "artificial vanilla", "artificially flavored strawberry", or "grape artificially flavored".

(3) Wherever the name of the characterizing flavor appears on the label (other than in the statement of ingredients) so conspicuously as to be easily seen under customary conditions of purchase, the words prescribed by this paragraph shall immediately and conspicuously precede or follow such name, without any intervening written, printed, or graphic matter, except:

(i) Where the characterizing flavor and a trademark or brand are presented together, other written, printed, or graphic matter that is a part of or is associated with the trademark or brand may intervene if the required words are in such relationship with the trademark or brand as to be clearly related to the characterizing flavor; and

(ii) If the finished product contains more than one flavor subject to the requirements of this paragraph, the statements required by this paragraph need appear only once in each statement of characterizing flavors present in such food, e.g., "artificially flavored vanilla and strawberry".

(iii) If the finished product contains three or more distinguishable characterizing flavors, or a blend of flavors with no primary recognizable flavor, the flavor may be declared by an appropriately descriptive generic term in lieu of naming each flavor, e.g., "artificially flavored fruit punch".

(4) A flavor supplier shall certify, in writing, that any flavor he supplies which is designated as containing no artificial flavor does not, to the best of his knowledge and belief, contain any artificial flavor, and that he has added no artificial flavor to it. The requirement for such certification may be satisfied by a guarantee under section 303(c)(2) of the act which contains such a specific statement. A flavor used shall be required to make such a written certification only where he adds to or combines another flavor with a flavor which has been certified by a flavor supplier as containing no artificial flavor, but otherwise such user may rely upon the supplier's certification and need make no separate certification. All such certifications shall be retained by the certifying party throughout the period in which the flavor is supplied and for a minimum of three years thereafter, and shall be subject to the following conditions:

(i) The certifying party shall make such certifications available upon request at all reasonable hours to any duly authorized office or employee of the Food and Drug Administration or any other employee acting on behalf of the Secretary of Health, Education, and Welfare. Such certifications are regarded by the Food and Drug Administration as reports to the government and as guarantees or other undertakings within the meaning of section 301(h) of the act and subject the certifying party to the penalties for making any false report to the government under 18 U.S.C. 1001 and any false guarantee or undertaking under section 303(a) of the act. The defenses provided under section 303(c)(2) of the act shall be applicable to the certifications provided for in this section.

(ii) Wherever possible, the Food and Drug Administration shall verify the accuracy of a reasonable number of certifications made pursuant to this section, constituting a representative sample of such certifications, and shall not request all such certifications.

(iii) Where no person authorized to provide such information is reasonably available at the time of inspection, the certifying party shall arrange to have such person and the relevant materials and records ready for verification as soon as practicable: *Provided*, That, whenever the Food and Drug Administration has reason to believe that the supplier or user may utilize this period to alter inventories or records, such additional time shall not be permitted. Where such additional time is provided, the Food and Drug Administration may require the certifying party to certify that relevant inventories have not been materially disturbed and relevant records have not been altered or concealed during such period.

(iv) The certifying party shall provide, to an officer or representative duly designated by the Secretary, such qualitative statement of the composition of the flavor or product covered by the certification as may be reasonably expected to enable the Secretary's representatives to determine which relevant raw and finished materials and flavor ingredient records are reasonably necessary to verify the certifications. The examination conducted by the Secretary's representative shall be limited to inspection and review of inventories and ingredient records for those certifications which are to be verified.

APPENDIX 2 (continued)

(v) Review of flavor ingredient records shall be limited to the qualitative formula and shall not include the quantitative formula. The person verifying the certifications may make only such notes as are necessary to enable him to verify such certification. Only such notes or such flavor ingredient records as are necessary to verify such certification or to show a potential or actual violation may be removed or transmitted from the certifying party's place of business: *Provided,* That, where such removal or transmittal is necessary for such purposes the relevant records and notes shall be retained as separate documents in Food and Drug Administration files, shall not be copied in other reports, and shall not be disclosed publicly other than in a judicial proceeding brought pursuant to the act or 18 U.S.C. 1001.

(j) A food to which a chemical preservative(s) is added shall, except when exempt pursuant to § 101.100 bear a label declaration stating both the common or usual name of the ingredient(s) and a separate description of its function, e.g., "preservative", "to retard spoilage", "a mold inhibitor", "to help protect flavor" or "to promote color retention".

(Secs. 402, 403, 409 701(a), 702, 703, 704, 52 Stat. 1046, 1047, 1048–1049 as amended, 1055, 1056–1057 as amended; 21 U.S.C. 342, 343, 348, 371(a), 372, 373, 374.)

APPENDIX 3

Subpart F—Flavoring Agents and Related Substances

§ 172.510 Natural flavoring substances and natural substances used in conjunction with flavors.

Natural flavoring substances and natural adjuvants may be safely used in food in accordance with the following conditions.

(a) They are used in the minimum quantity required to produce their intended physical or technical effect, and in accordance with all the principles of good manufacturing practice.

(b) In the appropriate forms (plant parts, fluid and solid extracts, concretes, absolutes, oils, gums, balsams, resins, oleoresins, waxes, and distillates) they consist of one or more of the following, used alone or in combination with flavoring substances and adjuvants generally recognized as safe in food, previously sanctioned for such use, or regulated in any section of this part.

Common name	Scientific name	Limitations
Aloe	*Aloe perryi* Baker, *A. barbadensis* Mill., *A. ferox* Mill., and hybrids of this sp. with *A. africana* Mill. and *A. spicata* Baker.	
Althea root and flowers	*Althea officinalis* L.	
Amyris (West Indian sandalwood).	*Amyris balsamifera* L.	
Angola weed	*Roccella fuciformis* Ach	In alcoholic beverages only.
Arnica flowers	*Arnica montana* L., *A. fulgens* Pursh, *A. sororia* Greene, or *A. cordifolia* Hooker.	Do.
Artemisia (wormwood)	*Artemisia* spp	Finished food thujone free.[1]
Artichoke leaves	*Cynara scolymus* L	In alcoholic beverages only.
Beeswax, white (Cire d'abeille).	*Apis mellifera* L.	
Benzoin resin	*Styrax benzoin* Dryander, *S. paralleloneurus* Perkins, *S. tonkinensis* (Pierre) Craib ex Hartwich, or other spp. of the Section *Anthostyrax* of the genus *Styrax*.	
Blackberry bark	*Rubus*, Section *Eubatus*.	
Boldus (boldo) leaves	*Peumus boldus* Mol	Do.
Boronia flowers	*Boronia megastigma* Nees.	
Bryonia root	*Bryonia alba* L., or *B. diocia* Jacq	Do.
Buchu leaves	*Barosma betulina* Bartl. et Wendl., *B. crenulata* (L.) Hook. or *B. serratifolia* Willd.	
Buckbean leaves	*Menyanthes trifoliata* L	Do.
Cajeput	*Melaleuca leucadendron* L. and other *Melaleuca* spp.	
Calumba root	*Jateorhiza palmata* (Lam.) Miers	Do.
Camphor tree	*Cinnamomum camphora* (L.) Nees et Eberm	Safrole free.
Cascara sagrada	*Rhamnus purshiana* DC	
Cassie flowers	*Acacia farnesiana* (L.) Willd.	
Castor oil	*Ricinus communis* L	
Catechu, black	*Acacia catechu* Willd.	
Cedar, white (aborvitae), leaves and twigs.	*Thuja occidentalis* L	Finished food thujone free.[1]
Centuary	*Centaurium umbellatum* Gilib	In alcoholic beverages only.
Cherry pits	*Prunus avium* L. or *P. cerasus* L	Not to exceed 25 p.p.m. prussic acid.
Cherry-laurel leaves	*Prunus laurocerasus* L	Do.
Chestnut leaves	*Castanea dentata* (Marsh.) Borkh	
Chirata	*Swertia chirata* Buch.-Ham	In alcoholic beverages only.
Cinchona, red, bark	*Cinchona succirubra* Pav. or its hybrids	In beverages only; not more than 83 p.p.m. total cinchona alkaloids in finished beverage.
Cinchona, yellow, bark	*Cinchona ledgeriana* Moens, *C. calisaya* Wedd., or hybrids of these with other spp. of *Cinchona*.	Do.
Copaiba	South American spp. of *Copaifera* L	
Cork, oak	*Quercus suber* L., or *Q. occidentalis* F. Gay	In alcoholic beverages only.
Costmary	*Chrysanthemum balsamita* L	Do.
Costus root	*Saussurea lappa* Clarke.	
Cubeb	*Piper cubeba* L. f.	
Currant, black, buds and leaves.	*Ribes nigrum* L.	
Damiana leaves	*Turnera diffusa* Willd	
Davana	*Artemisia pallens* Wall.	

APPENDIX 3 (continued)

Common name	Scientific name	Limitations
Dill, Indian	*Anethum sowa* Roxb. (*Peucedanum graveolens* Benth et Hook., *Anethum graveolens* L.).	
Dittany (fraxinella) roots	*Dictamnus albus* L.	In alcoholic beverages only.
Dittany of Crete	*Origanum dictamnus* L.	
Dragon's blood (dracorubin)	*Daemonorops* spp.	
Elder tree leaves	*Sambucus nigra* L.	In alcoholic beverages only; not to exceed 25 p.p.m. prussic acid in the flavor.
Elecampane rhizome and roots	*Inula helenium* L.	In alcoholic beverages only.
Elemi	*Canarium commune* L. or *C. luzonicum* Miq.	
Erigeron	*Erigeron canadensis* L.	
Eucalyptus globulus leaves	*Eucalyptus globulus* Labill.	
Fir ("pine") needles and twigs	*Abies sibirica* Ledeb., *A. alba* Mill., *A. sachalinensis* Masters or *A. mayriana* Miyabe et Kudo.	
Fir, balsam, needles and twigs	*Abies balsamea* (L.) Mill.	
Galanga, greater	*Alpinia galanga* Willd.	Do.
Galbanum	*Ferula galbaniflua* Boiss. et Buhse and other *Ferula* spp.	
Gambir (catechu, pale)	*Uncaria gambir* Roxb.	
Genet flowers	*Spartium junceum* L.	
Gentian rhizome and roots	*Gentiana lutea* L.	
Gentian, stemless	*Gentiana acaulis* L.	Do.
Germander, chamaedrys	*Teucrium chamaedrys* L.	Do.
Germander, golden	*Teucrium polium* L.	In alcoholic beverages only.
Guaiac	*Guaiacum officinale* L., *G. santum* L., *Bulnesia sarmienti* Lor.	
Guarana	*Paullinia cupana* HBK.	
Haw, black, bark	*Viburnum prunifolium* L.	
Hemlock needles and twigs	*Tsuga canadensis* (L.) Carr. or *T. heterophylla* (Raf.) Sarg.	
Hyacinth flowers	*Hyacinthus orientalis* L.	
Iceland moss	*Cetraria islandica* Ach.	Do.
Imperatoria	*Peucedanum ostruthium* (L.) Koch (*Imperatoria ostruthium* L.).	
Iva	*Achillea moschata* Jacq.	Do.
Labdanum	*Cistus* spp.	
Lemon-verbena	*Lippia citriodora* HBK.	Do.
Linaloe wood	*Bursera delpechiana* Poiss. and other *Bursera* spp.	
Linden leaves	*Tillia* spp.	Do.
Lovage	*Levisticum officinale* Koch.	
Lungmoss (lungwort)	*Sticta pulmonacea* Ach.	
Maidenhair fern	*Adiantum capillus-veneris* L.	Do.
Maple, mountain	*Acer spicatum* Lam.	
Mimosa (black wattle) flowers	*Acacia decurrens* Willd. var. *dealbata*.	
Mullein flowers	*Verbascum phlomoides* L. or *V. thapsiforme* Schrad.	Do.
Myrrh	*Commiphora molmol* Engl., *C. abyssinica* (Berg) Engl., or other *Commiphora* spp.	
Myrtle leaves	*Myrtus communis* L.	In alcoholic beverages only.
Oak, English, wood	*Quercus robur* L.	Do.
Oak, white, chips	*Quercus alba* L.	
Oak moss	*Evernia prunastri* (L.) Ach., *E. furfuracea* (L.) Mann, and other lichens.	Finished food thujone.
Olibanum	*Boswellia carteri* Birdw. and other *Boswellia* spp.	
Opopanax (bisabolmyrrh)	*Opopanax chironium* Koch (true opopanax) of *Commiphora erythraea* Engl. var. *Llabrescens*	
Orris root	*Iris germanica* L. (including its variety *florentina* Dykes) and *I. pallida* Lam.	
Pansy	*Viola tricolor* L.	In alcoholic beverages only.
Passion flower	*Passiflora incarnata* L.	
Patchouly	*Pogostemon cablin* Benth. and *P. heyneanus* Benth.	
Peach leaves	*Prunus persica* (L.) Batsch	In alcoholic beverages only; not to exceed 25 p.p.m. prussic acid in the flavor.
Pennyroyal, American	*Hedeoma pulegioides* (L.) Pers.	
Pennyroyal, European	*Mentha pulegium* L.	
Pine, dwarf, needles and twigs	*Pinus mugo* Turra var. *pumilio* (Haenke) Zenari.	
Pine, Scotch, needles and twigs	*Pinus sylvestris* L.	
Pine, white, bark	*Pinus strobus* L.	In alcoholic beverages only.
Pine, white oil	*Pinus palustris* Mill., and other *Pinus* spp.	
Poplar buds	*Populus balsamifera* L. (*P. tacamahacca* Mill.), *P. candicans* Ait., or *P. nigra* L.	Do.
Quassia	*Picrasma excelsa* (Sw.) Planch, or *Quassia amara* L.	
Quebracho bark	*Aspidosperma quebracho-blanco* Schlecht, or (*Quebrachia lorentzii* (Griseb).	*Schinopsis lorentzii* (Griseb.) Engl.
Quillaia (soapbark)	*Quillaja saponaria* Mol.	
Red saunders (red sandal wood).	*Pterocarpus santalinus* L.	In alcoholic beverages only.
Rhatany root	*Krameria triandra* Ruiz et Pav. or *K. argentea* Mart.	
Rhubarb, garden root	*Rheum rhaponticum* L.	Do.
Rhubarb root	*Rheum officinale* Baill., *R. palmatum* L., or other spp. (excepting *R. rhaponticum* L.) or hybrids of *Rheum* grown in China.	
Roselle	*Hibiscus sabdariffa* L.	Do.
Rosin (colophony)	*Pinus palustris* Mill., and other *Pinus* spp.	Do.

APPENDIX 3 (continued)

Common name	Scientific name	Limitations
St. Johnswort leaves, flowers, and caulis.	*Hypericum perforatum* L.	Hypericin-free alcohol distillate form only; in alcoholic beverages only.
Sandalwood, white (yellow, or East Indian).	*Santalum album* L.	
Sandarac	*Tetraclinis articulata* (Vahl.), Mast	In alcoholic beverages only.
Sarsaparilla	*Smilax aristolochiaefolia* Mill., (Mexican sarsaparilla), *S. regelii* Killip et Morton (Honduras sarsaparilla), *S. febrifuga* Kunth (Ecuadorean sarsaparilla), or undetermined *Smilax* spp. (Ecuadorean or Central American sarsaparilla).	
Sassafras leaves	*Sassafras albidum* (Nutt.) Nees	Safrole free.
Senna, Alexandria	*Cassia acutifolia* Delile	
Serpentaria (Virginia snakeroot).	*Aristolochia serpentaria* L.	In alcoholic beverages only.
Simaruba bark	*Simaruba amara* Aubl	Do.
Snakeroot, Canadian (wild ginger).	*Asarum canadense* L.	
Spruce needles and twigs	*Picea glauca* (Moench) Voss or *P. mariana* (Mill.) BSP.	
Storax (styrax)	*Liquidambar orientalis* Mill. or *L. styraciflua* L.	
Tagetes (marigold)	*Tagetes patula* L., *T. erecta* L., or *T. minuta* L. (*T. glandulifera* Schrank).	As oil only.
Tansy	*Tanacetum vulgare* L.	In alcoholic beverages only; finished alcoholic beverage thujone free.[1]
Thistle, blessed (holy thistle)	*Onicus benedictus* L.	In alcoholic beverges only.
Thymus capitatus (Spanish "origanum").	*Thymus capitatus* Hoffmg. et Link.	
Tolu	*Myroxylon balsamum* (L.) Harms.	
Turpentine	*Pinus palustris* Mill. and other *Pinus* spp. which yield terpene oils exclusively.	
Valerian rhizome and roots	*Valeriana officinalis* L.	
Veronica	*Veronica officinalis* L.	Do.
Vervain, European	*Verbena officinalis* L.	Do.
Vetiver	*Vetiveria zizanioides* Stapf	Do.
Violet, Swiss	*Viola calcarata* L.	
Walnut husks (hulls), leaves, and green nuts.	*Juglans nigra* L. or *J. regia* L.	
Woodruff, sweet	*Asperula odorata* L.	Do.
Yarrow	*Achillea millefolium* L.	In beverages only; finished beverage thujone free.[1]
Yerba santa	*Eriodictyon californicum* (Hook, et Arn.) Torr.	
Yucca, Joshua-tree	*Yucca brevifolia* Engelm	
Yucca, Mohave	*Yucca schidigera* Roezl ex Ortgies (*Y. mohavensis* Sarg.).	

[1] As determined by using the method (or, in other than alcoholic beverages, a suitable adaptation thereof) in sec. 9.091, "Official Methods of Analysis of the Association of Official Agricultural Chemists," 10th Edition (1965).

APPENDIX 4

§ 169.175 Vanilla extract.

(a) Vanilla extract is the solution in aqueous ethyl alcohol of the sapid and odorous principles extractable from vanilla beans. In vanilla extract the content of ethyl alcohol is not less than 35 percent by volume and the content of vanilla constituent, as defined in § 169.3 (c), is not less than one unit per gallon. The vanilla constituent may be extracted directly from vanilla beans or it may be added in the form of concentrated vanilla extract or concentrated vanilla flavoring or vanilla flavoring concentrated to the semisolid form called vanilla oleoresin. Vanilla extract may contain one or more of the following optional ingredients:

(1) Glycerin.

(2) Propylene glycol.

(3) Sugar (including invert sugar).

(4) Dextrose.

(5) Corn sirup (including dried corn sirup).

(b) (1) The specified name of the food is "Vanilla extract" or "Extract of vanilla".

(2) When the vanilla extract is made in whole or in part by dilution of vanilla oleoresin, concentrated vanilla extract, or concentrated vanilla flavoring, the label shall bear the statement "Made from __________" or "Made in part from __________", the blank being filled in with the name or names "vanilla oleoresin", "concentrated vanilla extract", or "concentrated vanilla flavoring", as appropriate. If the article contains two or more units of vanilla constituent, the name of the food shall include the designation "__________ fold", the blank being filled in with the whole number (disregarding fractions) expressing the number of units of vanilla constituent per gallon of the article.

(3) Wherever the name of the food appears on the label so conspicuously as to be easily seen under customary conditions of purchase, the labeling required by paragraph (b) (2) of this section shall immediately and conspicuously precede or follow such name, without intervening written, printed, or graphic matter.

§ 169.176 Concentrated vanilla extract.

(a) Concentrated vanilla extract conforms to the definition and standard of identity and is subject to any requirement for label statement of optional ingredients prescribed for vanilla extract by § 169.175, except that it is concentrated to remove part of the solvent, and each gallon contains two or more units of vanilla constituent as defined in § 169.3(c). The content of ethyl alcohol is not less than 35 percent by volume.

(b) The specified name of the food is "Concentrated vanilla extract __________ fold" or "__________ fold concentrated vanilla extract", the blank being filled in with the whole number (disregarding fractions) expressing the number of units of vanilla constituent per gallon of the article. (For example, "Concentrated vanilla extract 2-fold".)

§ 169.177 Vanilla flavoring.

(a) Vanilla flavoring conforms to the definition and standard of identity and is subject to any requirement for label statement of optional ingredients prescribed for vanilla extract by § 169.175, except that its content of ethyl alcohol is less than 35 percent by volume.

(b) The specified name of the food is "Vanilla flavoring".

§ 169.178 Concentrated vanilla flavoring.

(a) Concentrated vanilla flavoring conforms to the definition and standard of identity and is subject to any requirement for label statement of optional ingredients prescribed for vanilla flavoring by § 169.177, except that it is concentrated to remove part of the solvent, and each gallon contains two or more units of vanilla constituent as defined in § 169.3(c).

(b) The specified name of the food is "Concentrated vanilla flavoring __________ fold" or "__________ fold concentrated vanilla flavoring", the blank being filled in with the whole number (disregarding fractions) expressing the number of units of vanilla constituent per gallon of the article. (For example, "Concentrated vanilla flavoring 3-fold".)

§ 169.179 Vanilla powder.

(a) Vanilla powder is a mixture of ground vanilla beans or vanilla oleoresin or both, with one or more of the following optional blending ingredients:

(1) Sugar.

(2) Dextrose.

(3) Lactose.

(4) Food starch (including food starch-modified as prescribed in § 172.892 of this chapter).

(5) Dried corn sirup.

(6) Gum acacia.

Vanilla powder may contain one or any mixture of two or more of the anticaking ingredients specified in paragraph (b) of this section, but the total weight of any such ingredient or mixture is not more than 2 percent of the weight of the finished vanilla powder. Vanilla powder contains in each 8 pounds not less than one unit of vanilla constituent, as defined in § 169.3(c).

APPENDIX 4 (continued)

(b) The anticaking ingredients referred to in paragraph (a) of this section are:

(1) Aluminum calcium silicate.

(2) Calcium silicate.

(3) Calcium stearate.

(4) Magnesium silicate.

(5) Tricalcium phosphate.

(c) (1) The specified name of the food is "Vanilla powder ---------- fold" or "---------- fold vanilla powder", except that if sugar is the optional blending ingredient used, the word "sugar" may replace the word "powder". The blank in the name is filled in with the whole number (disregarding fractions) expressing the number of units of vanilla constituent per 8 pounds of the article. However, if the strength of the article is less than 2-fold, the term "---------- fold" is omitted from the name.

(2) The label of vanilla powder shall bear the common names of any of the optional ingredients specified in paragraphs (a) and (b) of this section that are used, except that where the alternative name "Vanilla sugar" is used for designating the food it is not required that sugar be named as an optional ingredient.

(3) Wherever the name of the food appears on the label so conspicuously as to be easily seen under customary conditions of purchase, the labeling required by paragraph (c)(2) of this section shall immediately and conspicuously precede or follow such name, without intervening written, printed, or graphic matter.

§ 169.180 Vanilla-vanillin extract.

(a) Vanilla-vanillin extract conforms to the definition and standard of identity and is subject to any requirement for label statement of optional ingredients prescribed for vanilla extract by § 169.-175, except that for each unit of vanilla constituent, as defined in § 169.3(c), contained therein, the article also contains not more than 1 ounce of added vanillin.

(b) The specified name of the food is "Vanilla-vanillin extract ---------- fold" or "---------- fold vanilla-vanillin extract", followed immediately by the statement "contains vanillin, an artificial flavor (or flavoring)". The blank in the name is filled in with the whole number (disregarding fractions) expressing the sum of the number of units of vanilla constituent plus the number of ounces of added vanillin per gallon of the article. However, if the strength of the article is less than 2-fold, the term "---------- fold" is omitted from the name.

§ 169.181 Vanilla-vanillin flavoring.

(a) Vanilla-vanillin flavoring conforms to the definition and standard of identity and is subject to any requirement for label statement of optional ingredients prescribed for vanilla-vanillin extract by § 169.180, except that its content of ethyl alcohol is less than 35 percent by volume.

(b) The specified name of the food is "Vanilla-vanillin flavoring ---------- fold" or "---------- fold vanilla-vanillin flavoring", followed immediately by the statement "contains vanillin, an artificial flavor (or flavoring)". The blank in the name is filled in with the whole number (disregarding fractions) expressing the sum of the number of units of vanilla constituent plus the number of ounces of added vanillin per gallon of the article. However, if the strength of the article is less than 2-fold, the term "---------- fold" is omitted from the name.

§ 169.182 Vanilla-vanillin powder.

(a) Vanilla-vanillin powder conforms to the definition and standard of identity and is subject to any requirement for label statement of optional ingredients prescribed for vanilla powder by § 169.179, except that for each unit of vanilla constituent as defined in § 169.3 (c) contained therein, the article also contains not more than 1 ounce of added vanillin.

(b) The specified name of the food is "Vanilla-vanillin powder ------------ fold" or "---------- fold vanilla-vanillin powder", followed immediately by the statement "contains vanillin, an artificial flavor (or flavoring)". If sugar is the optional blending ingredient used, the word "sugar" may replace the word "powder" in the name. The blank in the name is filled in with the whole number (disregarding fractions) expressing the sum of the number of units of vanilla constituent plus the number of ounces of added vanillin per 8 pounds of the article. However, if the strength of the article is less than 2-fold the term "---------- fold" is omitted from the name.

REFERENCES

1. **Arctander, S.,** *Perfume and Flavor Chemicals,* (Aroma Chemicals), Monclair, N. J., 1969.
2. **Arctander, S.,** *Perfume and Flavor Materials of Natural Origin,* Elizabeth, N. J., 1960.
3. **Bedoukian, P. Z.,** *Perfumery Synthetics and Isolates,* D Van Nostrand, New York, 1951.
4. **Bedoukian, P. Z.,** *Perfumery and Flavoring Synthetics,* Elsevier, New York, 1967.
5. *Code of Federal Regulations,* 21 Food and Drugs Parts 10—99 Rev; April 1, 1976, U.S. Government Printing Office, Washington, D.C.
6. **Durrans, T. H.,** *Chemistry and Industry,* 1937, 1129.
7. Food Additives and Contaminants Committee, Report on the Review of Flavourings in Food, Her Majesty's Stationery Office, London, 1976.
8. *Food Chemical Codex,* 1st ed., Pub. No. 1406, National Academy of Science, N.R.C., Washington, D.C., 1966.
9. **Furia, T. E. and Bellanca, N.,** *Fenaroli's Handbook of Flavor Ingredients,* 2nd ed., Vol. 1 and 2, CRC Press, Cleveland, Ohio, 1975.
10. *Givaudan Index,* 2nd ed., Givaudan-Delawanna, New York, 1961.
11. **Guenther, E.,** *The Essential Oils,* Vol. 1—6, D Van Hostrand, New York, 1948.
12. **Jacobs, M. B.,** *Synthetic Food Adjuncts,* D Van Nostrand, New York, 1947.
13. **Langenau, E.,** *Oleoresins for the Flavor Chemists,* American Perfumer and Aromatics, 1959, 37.
14. **Littlejohn, W. P.,** *Flavours,* 1940, 8.
15. *Lists of Volatile Compounds in Food,* Central Institute for Nutrition and Food Research, TNC, Zeist, Netherlands.
16. **Macleod, D.,** *A Book of Herbs,* Gerald Duckworth, Ltd., London, 1968.
17. *Merck Index,* 9th ed., Merck, Rahway, N. J.
18. **Merory, J.,** *Food Flavoring,* AVI Publishing, Westport, Conn., 1960.
19. Natural Flavouring Substances, Their Sources, and Added Artificial Flavouring Substances; Council of Europe 1974 Maisonneuve S.A. France.
20. *Oleoresin Handbook,* Fritzsche, Dodge and Olcott, Inc., New York, 1974.
21. *Perfumer's Handbook,* Fritzsche Bros., New York, 1944.
22. **Saldarini, A. V.,** *Flavor Industry,* 8/9, 1974, 247.
23. *United States Dispensitory,* Osol, Farrar, and Pratt, Vol. 1, 2,Lippincott, New York, 1960.

SYNTHETIC FOOD COLORS

James E. Noonan and Harry Meggos

INTRODUCTION

In the years since 1970, food colors, especially the synthetic organic colors, have received tremendous publicity — nearly all of it bad. Food colors have been a prime target of consumer activists. The activists seize on poorly conducted studies and cry that food colors are for cosmetic reasons only and deny any benefit in assessing the risk to benefit ratio. Food processors openly advertise their products to be natural and free of colors; hyperkinesis and learning difficulties are attributed to colors (without substantive proof).

The news media report on the activists' allegations in a sensational fashion, not bothering to take the time required to do a balanced reporting. It seems that Congress and the FDA interpret this as being of concern to the consumers; and as in the FD&C Red #2 incident, the FDA acted out of concern for the public concern. In many quarters, the question is being raised as to whether colors are needed at all in foods. Many believe that the attack on colors will be followed by an attack on flavors in an attempt to reach the ultimate targets: confections and convenience foods.

There are no hard data supporting the need and utility of colors in food. Elimination of food colors would be the end of many traditional foods, carbonated beverages, gelatin desserts, hard candy, and many others. Some feel the American consumer may soon tire of being told what they can and cannot eat. Much of the upheaval in evaluation of the need and safety of colors and other food additives must be attributed to the science of toxicology and safety testing. Colors have been the guinea pigs for new modifications in testing before a history of standard response can be developed. An example of this is the addition of in utero expose as a part of a chronic feeding study. It is likely that there will always be colors in food; however, it is difficult to know what kind they will be — certified colors, uncertified colors, natural colors, or non-absorbable colors.

LEGISLATIVE AND REGULATORY HISTORY OF COLOR ADDITIVES

Color additives for food represent a unique and special category of food additives. They have historically been so considered in legislation and regulation. The current legislation governing the regulation and use of color additives in the U.S. is the Food, Drug and Cosmetic Act of 1938, as amended by the Color Additives Amendments of 1960. To fully understand the present color additive situation and to judge the future, it is necessary to have a feeling for the history of color additives. In tracing the history of color additives in the U.S., it is extremely important to establish the difference between legislation and regulation. Congress creates the law; government agencies generate the regulations. The law can only be amended by an act of Congress, while regulations can theoretically be changed by petitioning, or in the case of a proposed regulation, by filing comments. In addition, proposed regulations are often issued allowing a time period for comments by interested parties. During the comment period, requests can also be made for a public hearing. The last resort following refusal of a hearing, or an unfavorable decision in a regulatory hearing, is litigation in the U. S. District Court.

The Food and Drug Act of 1906

In 1900, some 80 dyes were in use in the U.S. for coloring food. There were no regulations regarding the nature and purity of these dyes and the same batch of color used to dye cloth might find its way into candy. The first comprehensive legislation was the Food and Drug Act of 1906, which listed seven dyes which were permitted for use in foods. These seven were chosen after a thorough study of the colors in current use and only those colors of known composition, which had been examined physiologically, with no unfavorable results, were listed.[1]

The seven original permitted colors were:

- Orange I (later FD&C Orange #1)
- Erythrosine (later FD&C Red #3)
- Ponceau 3R (later FD&C Red #1)
- Amaranth (later FD&C Red #2)
- Indigotine (later FD&C Blue #2)
- Naphthol Yellow (later FD&C Yellow #1)
- Light Green (later FD&C Green #2)

A system was set up for certification of batches of these colors to acknowledge conformance with chemical specifications. Color certification was under the Department of Agriculture and was not mandatory. However, color manufacturers soon found it advantageous to obtain certification. The original list of colors left much to be desired to fulfill all the needs of the food industry and in the years prior to 1938, additional colors were added — but only after physiological testing. The chronology of these added colors is as follows:

- 1916 — Tartrazine (later FD&C Yellow #5)
- 1918 — Yellow AB (later FD&C Yellow #2) and Yellow OP (later FD&C Yellow #3)
- 1922 — Guinea Green (later FD&C Green #1)
- 1927 — Fast Green (later FD&C Green #3)
- 1929 — Ponceau SX (later FD&C Red #4), Sunset Yellow (later FD&C Yellow #6), and Brilliant Blue (later FD&C Blue #1)

The Food, Drug and Cosmetic Act of 1938

This Act superceded the Act of 1906 and broadened the scope of certified colors, creating three categories: FD&C Colors, D&C Colors, and External D&C Colors. Under the new Act, the common names of the dyes were not employed, but color prefixes and numbers were used. Also, certification was mandatory and was placed under the jurisdiction of the Food and Drug Administration (FDA). Before being listed under the 1938 Act, the 15 food colors were again scrutinized for their toxicological effect. At the hearings for this legislation, Dr. H. O. Calvery, Chief of the Division of Pharmacology of the FDA, reported that the safety conclusions were based on a 3-year study of the dyes. He also stated there were no known causes of harm from the use of these colors, some of which had been in use as certified colors for 32 years. Dr. Calvery also testified that the term "harmless and suitable for use in foods" must be judged in the light of the amounts used. Under the 1938 Act, the burden of safety for color additives rested with the Food and Drug Administration.

Following the passage of the Food, Drug and Cosmetic Act of 1938, the situation was peaceful until the early 1950s when three unfavorable incidents occurred as the result of overuse of color in candy and on popcorn. These overuses resulted in a number of cases of diarrhea in children. The colors involved were FD&C Orange #1 and FD&C Red #32. These incidents, coupled with chronic toxicity animal feeding studies by FDA, led to the delisting of FD&C Orange #2, FD&C Red #32.

The suggestion that quantity limitations be established for colors was met with objections that the law did not empower the Secretary of Health, Education and Welfare to set quantity limitations and that harmless meant zero toxicity, that is, harmless in any amounts used. This issue was litigated in the courts with a final Supreme Court ruling[3] that the FDA, under the 1938 law, did not have the authority to set quantity limitations. Following soon thereafter were delistings of FD&C Yellow #1, FD&C Yellow #2, and FD&C Yellows #3 and 4, the remaining oil-soluble colors.

The Supreme Court decision had rendered the old law obsolete and unworkable and colors were being delisted even though the FDA admitted they were not endangering the public health. It was evident that remedial legislation was required if food colors were to survive. Records from the hearings held prior to the passage of new legislation showed dramatically that both Congress and the FDA were convinced that synthetic colors in food were necessary and desirable. With the support and advice of the food color manufacturers and the FDA, Congress passed the Color Additives Amendments to the 1938 Act[4] and these become law July 12, 1960. The new Amendments were chiefly a relief measure designed to correct the inflexibility of the old law. The Amendments consist of two parts (or Titles). Title I is the so-called permanent part; Title II, the temporary part.

Title I, the more important section, set up uniform rules for all permitted colors, both certified and uncertified. The term "color additive" is defined as "any dye, pigments or other substance capable of coloring a food, drug or cosmetic or any part of the human body". The term "coal tar color" is eliminated from the law. This is fortunate as the term has, for years, been an undeserved stigma. The law provides for the listing of color additives which must be certified and color additives exempt from certification. The term "natural colors" formerly and improperly applied to the uncertified color additives which are of synthetic as well as natural origin, was also eliminated. The Secretary of Health, Education, and Welfare was given the authority to decide whether a color additive should be classified as certified or uncertified. The law allows the Secretary to list color additives for specific uses, and also to set conditions and tolerances (limitations) on the use of color additives. A color additive can, if required, be given a zero tolerance, meaning that it cannot be used.

Title I also included a cancer clause similar to the Delaney Clause, which states that a color additive cannot be listed for any use whatsoever if it is found to induce cancer when ingested by man or animal. This law is softened somewhat by permitting, in cases involving the cancer clause, the appointment of an advisory committee to serve as a fact-finding body. The advisory committee reports its findings to the Secretary, who makes his decision. He is not bound to follow the recommendation of the advisory committee.

Title II, the so-called temporary part, was designed to permit the use of current color additives pending the completion of scientific investigations needed to determine the safety of these materials for permanent listing. The provisional listing of the color additives had an original closing date of 2½ years after passage of the law. However, the Secretary of Health, Education, and Welfare was given the power to grant extensions of the closing dates, which he has done many times. In retrospect, the Color Additives Amendments have been salutary, however, serious problems developed in later years because of delays in moving some of the colors from the provisional listing to permanent listing.

Under the new law, the onus of establishing the safety of the provisionally listed colors rested on the color manufacturers and other interested persons; however, in 1957, prior to the passage of the Color Additives Amendments, the FDA had initiated feeding studies on all of the currently used FD&C colors. These were all long-term, chronic studies in rats and dogs, and in some instances, mice. The dog studies were normally for a 2-year duration, however, with 4 of the certified colors, the studies in dogs were extended for 7 years, supposedly to investigate the feasibility in general of such studies. Most of these feeding tests were run at feeding levels of 5, 2, 0.5% of the total diet. Because these in-house FDA studies were underway, there was no need in 1960 for the color industry to initiate studies, however, considerable analytical work was done by color industry chemists to tighten and hone the chemical specifications in preparation for submission of petitions for permanent listings.

Thus, in 1960, the colors in current use were provisionally listed pending completion of the FDA studies, to be followed by petitioning by industry for permanent listing. Unfortunately, FD&C Red #1, a fairly important scarlet red color, did not long enjoy its provisional listing and in November 1960, 5 months after listing under the Color Additives Amendments, its provisional listing was terminated because of liver damage in rats, mice, and dogs at the lowest level of feeding — 0.5% of the diet. This was the fifth food color which had been delisted since 1953. While the food industry was concerned about these delistings, these five colors did not have the chemical structures that prevailing toxicological theory hypothecated for safety. In addition, there were still sufficient colors and the remaining colors did conform with these theories of safety. Concern mounted, however, in 1965 when the provisional listing of FD&C Red #4 was terminated. This action was based on rather poorly conducted studies initiated in 1957, the results of which were not considered important in 1961, but suddenly became critical when action was taken against FD&C Red #4 by France.

Following completion of the FDA safety studies in 1964, the certified food color manufacturers jointly, through the Certified Color Industry Committee (CCIC), initiated their filing of petitions for permanent listing. In February of 1965, the petition for permanent listing of FD&C Yellow #5 was submitted and filed by the FDA 1 month later in March. Filing means that the FDA publishes a notice in the *Federal Register* that they will evaluate a certain petition. They are given a 90-day period from that date, which can be extended another 90 days if necessary, to complete the evaluation of the petition.

In subsequent years through mid-1968, the petitions for permanent listing of the other certified food colors were submitted and filed by the FDA. Because of the lack of economic importance, FD&C Green #1 (Guinea Green) and FD&C Green #2 (Light Green), industry did not petition for extension of these colors and they were automatically deleted from the list of color additives in 1966. Also, in 1966 Orange B, a new color not in use prior to 1960, was permanently listed for coloring sausage casings at maximum rate of 150 ppm. The permanent listing of Orange B for a specific use with maximum quantity limitations illustrates the new flexibility under the Color Additives Amendments of 1960.

Following the termination of the original 2½-year time period for provisional listing, the FDA extended the closing date and at regular intervals extended the date into the future, generally, for 1 year at a time. Lack of action on the petitions for permanent listing was due in part to a dispute (and subsequent court case)[5] between the FDA and the Toilet Goods Association (TGA) over the interpretation of the Color Regulations regarding premarketing clearance for cosmetics. However, in July of 1969, FD&C Yellow #5, FD&C Red #3, and FD&C Blue #1 were permanently listed for use in foods and ingested drugs, with limitations only for good manufacturing practice. Pending settlement of the cosmetic issue, provisional listing for these colors was maintained for topical drug and cosmetic uses, as well as for the lakes of these colors. The alumina hydrate lakes of the FD&C colors, which are insoluble pigment forms of the dyes, had been approved for use in 1959 prior to the passage of the Color Additives Amendments.

It is difficult to determine why the remaining colors (except FD&C Violet #1) could not have been permanently listed at this time, and the FDA was, in later years, criticized for not taking prompt action on the petitions. Some questions regarding the safety of FD&C Violet #1, which had been added to the approved list in 1950, justifiably prevented positive action on the permanent listing of this color. Because of the concern about the safety of FD&C Violet #1, the Commissioner of the Food and Drug Administration referred the issue to an advisory committee in June 1971. The judgment of the advisory committee was that FD&C Violet #1 was not carcinogenic, but that additional studies should be conducted, and recommended that this color be kept on the provisional list pending completion of additional studies. In 1970, the Allied Chemical Co. submitted a petition for permanent listing of a new food color, Allura Red. This color was an orange-red, similar in shade to FD&C Red #4 and designed as a replacement for Red #4.

Safety Problems — 1970

At the beginning of the year 1970, the FD&C color situation seemed quite rosy and it appeared that during the year the remaining colors, with the exception of FD&C Violet #1, would achieve permanent listing. However in late 1970, abstracts of work by two different groups of Russian scientists on FD&C Red #2 appeared in the literature. One paper reported on studies which suggested carcinogenic response from this color,[6] while the second study reported embryotoxic effects from FD&C Red #2.[7] In February 1971, FDA reported they were initiating embryotoxicity studies on FD&C Red #2 and stated that industry would be required to run teratology and reproduction studies on all ingested certified colors — both FD&C and D&C. They completely discounted the Russian carcinogenicity study based on the enormity of favorable carcinogenicity data on this color. Investigation of embryotoxicity generally involves the use of two types of studies — teratology and multigeneration reproduction. Teratology observes the effects of a compound on the development of the fetuses in the uterus. The procedure consists of mating the male and female, dosing the female during a certain period of gestation, and finally, prior to birth, examining the uterus for fetal resorptions and the pups for deformities. A teratology study requires approximately 3 months for completion and evaluation. A reproduction study investigates the number and viability of the pups during three generations of animals. Such a study requires approximately 2 years.

In response to the FDA request, the color manufacturers and interested trade associations responded promptly and created an interindustry committee to initiate and monitor the teratology and reproduction studies on the ingested certified colors. This was a mammoth undertaking as the total number of colors was 25, and the state of the art of teratology and reproduction testing, while not in its childhood, was not a mature scientific procedure, having been of somewhat minor concern until the thalidomide incident.

In mid-1971, the FDA reported that their teratology studies on FD&C Red #2,[8] following the protocol of the Russian study, had shown statistically significant embryotoxic effects. Thus, while all of the ingested colors were under the question of embryotoxicity, the focus of regulatory activity was directed toward FD&C Red #2. The events in the FD&C Red #2 embryotoxicity issue are depicted in the following chronology:

Date	Event
September 11, 1971	Notice in the Federal Register stating the FDA proposed to allocate the use of FD&C Red #2 to lower levels and initiating a FD&C Red #2 survey. The order also stated the official requirements that teratology and reproduction studies be conducted on all of the ingested certified colors and stated required completion dates for these studies
November 1971	FDA requested that a National Academy of Sciences/National Research Council (NAS/NRC) appoint a committee to review the safety of FD&C Red #2. NAS/NRC declined
December 9, 1971	NAS/NRC reverses its stand and agrees to review the safety of FD&C Red #2 and the methodology of teratology and reproduction testing
February 11, 1972	NAS subcommittee met to review the data on the safety of FD&C Red #2. Part of these data was the industry-sponsored teratology study on FD&C Red #2 in rats and rabbits which contradicted the Russian study and the FDA study, in that the data showed no embryotoxic effects
June 13, 1972	Release of NAS subcommittee report. This body reviewed all previous studies on this color, including the Russian study, the FDA in-house teratology study, and the industry sponsored teratology work. The NAS committee report stated: "None (of these tests) warrants the conclusion that Red #2 in normal usage constitutes a hazard to human health"
July 4, 1972	Publication in the *Federal Register* of a proposed order which would limit FD&C Red #2 to 30 ppm in food. It was believed this action was based on new data from FDAs in-house reproduction study which had not been available at the time of the NAS committee review. FDA claims unfavorable results from an embryotoxicity study which was an off-shoot of their reproduction study work
August 1972	Industry requested access to these undisclosed test results under the Freedom of Information Act. After study of the data, industry toxicologists claimed the results were not statistically significant
November 27, 1972	The NAS subcommittee submitted an addendum to their original report which defended their original conclusions. The report stated: "In the case of Red #2, the approach adopted by FDA in their effort to elicit teratogenic effects was not a standard, recognized, accepted procedure and its validity is open to question"
December 1972	Publication of another Russian study[9] alleging gonadotoxic and embryotoxic effects of Amaranth (common name of the dye, which may or may not be of the purity specifications of FD&C Red #2)
January 1973	FDA reports the embryotoxicity data, which were the basis of the July 4, 1972 Order, were not statistically significant; however, FDA still intends to limit the use of the color
May 16, 1973	Meeting of a newly appointed ad hoc advisory committee on FD&C Red #2. This group met to investigate the divergent results obtained in the many teratology studies on FD&C Red #2. One of the problems of evaluation of the studies was that different protocols of testing had been employed. In some of the studies, the color had been administered to the test animals by gavage (oral intubation), in others in the feed. In addition, the times of administration had varied from 0 to 19 days to 6 to 19 days of gestation. After deliberation, this committee recommended triplicated studies in rats which would explore the effects of method and time of administration of FD&C Red #2. Studies were to be conducted by FDA in-house, by the National Center for Toxicological Research and in an industrial laboratory under sponsorship of the color manufacturers
June 6, 1974	The ad hoc advisory committee met to review the data from the triplicated studies. These data appeared to show no embryotoxicity effects from FD&C Red #2. Presented to the committee also, were the results from a similar study conducted by the Health Protection Branch of Canada. The Canadian study was free of unfavorable effects
January 1975	Release of the advisory committee report clearing FD&C Red #2 from any implications of embryotoxicity. The report proposed the establishment of a limit of 2.0 mg/kg/person/day or up to 150 mg/person/day. This would be a very generous limit as the estimated consumption of FD&C Red #2 per capita, per day, is 17.7 mg. The committee report also proposed that FD&C Red #2 be limited to a maximum of 50 ppm in beverages

During the years 1971 to 1975, the teratology and reproduction studies on the other ingested certified colors had been completed with no unfavorable results. During this period, extensions for the provisionally listed colors had been made from time to time.

It required 4 years of great activity and massive testing to resolve the embryotoxicity question, and FD&C Red #2 narrowly escaped being banned on basis of the Russian study and the early FDA study, which later was found to be in error. During this 4-year period, FD&C Red #2 had been under attack by the consumer activists, the press, and a number of Congressional investigating committees; and in mid-1975, the General Accounting Office (GAO) issued a report very critical of the handling of color regulations by the FDA with particular emphasis on the many extensions of the provisional listings of the colors. The GAO urged immediate action to ban or list the provisionally listed colors which, at this time, were FD&C Red #2, FD&C Yellow #6, FD&C Blue #2, and FD&C Green #3.

About the time the embryotoxicity question was resolved, it was learned that in spite of the earlier denunciations of the Russian carcinogenicity study, the FDA had initiated a similar study in rats. This study was completed in mid-1975; however, the study was flawed in that some of the dosed animals had been mixed and many of the animals which had died during the study had decomposed prior to being taken for tissue section, thus rendering many of the slides unreliable.

Apparently to relieve the pressures, the Commissioner of the Food and Drug Administration appointed a standing toxicology advisory committee, whose first assignment was to review the safety of FD&C Red #2. This committee was a multidisciplined group of scientists; only a small number were cancer experts. The committee reviewed all of the past toxicological data on FD&C Red #2 and met to deliberate the safety issue in mid-November 1975. This group denegrated most of the studies on FD&C Red #2 performed in the 1950s and 1960s, claiming that they were deficient by current standards in that insufficient numbers of animals were used, or insufficient histopathology was done. The committee confirmed that the color was not embryotoxic, but would not go on record that it had been proved to be safe. However, they refused to declare that FD&C Red #2 was a carcinogen. Thus, no solid recommendation to the FDA came from this meeting, however, prior to reconvening, a pathologist member was to review slides from a number of the studies and the biostatistician member was to reanalyze the data from the flawed FDA rat study.

In late December 1975, 5 weeks or so following the meeting of the toxicology advisory committee, the biostatistician submitted a report to the Commissioner in which he claimed that by isolating only the terminally sacrificed females in the highest dosage group, he found a statistically significant occurrence of malignant tumors. The Commissioner did not reconvene the toxicology advisory committee, but called together a working group of scientists within the Department of Health, Education, and Welfare to assess the issue. No definitive recommendation came from the working group, but it is obvious from their report that the new statistical analysis was considered only preliminary, and they felt more studies should be made on this before a conclusion could be drawn. Under pressure, the Commissioner stated in mid-January 1976 that he would terminate the provisional listing of FD&C Red #2. Two color manufacturers and another interested party brought action in the Federal District Court, claiming arbitrary and capricious action on the part of the Commissioner. A temporary restraining order was obtained, however, the Court later refused to issue a permanent injunction and ruled in favor of the Commissioner. This action was later sustained in the Court of Appeals.[11]

Thus, on February 10, 1976, the provisional listing of FD&C Red #2 was terminated and this color was banned from use. Soon thereafter, on April 9, 1976, the Food and

Drug Administration denied the petition for the permanent listing of FD&C Red #2. This action allowed for the requesting of a hearing on the issue of the safety of FD&C Red #2 and the Certified Color Manufacturers' Association (CCMA) filed and was granted a hearing. The hearing, before an administrative law judge, was held April 6, 1977. The CCMA had the task of establishing the safety of FD&C Red #2; however, the FDA was not required to state their criteria for safety. The hearing lasted 4 days and was continued on June 13 and concluded that same day. Final briefs were submitted September 19, 1977.

On March 30, 1978, the administrative law judge published his decision that the CCMA had not established that FD&C Red #2 was safe. The CCMA will file exceptions to that initial decision and the Commissioner will then render his decision. It is unlikely the Commissioner will reverse the initial decision and the next step would be an appeal in the Federal District Court by the CCMA.

In 1972 and 1973, when there was an eminent danger that FD&C Red #2 would be banned or severely limited in use because of the embryotoxicity question, many color users switched to the use of FD&C Red #40. FD&C Red #40 is quite different in shade and was never designed to be a replacement for FD&C Red #2; however, many users switched to the less desirable color in their products.

The toxicological studies, run in the late 1960s, which were the basis for the permanent listing of FD&C Red #40 in the U.S., were judged to be inadequate by the Food and Agriculture Organization/World Health Organization (FAO/WHO) of the United Nations and by the regulatory agencies of a number of foreign countries. In order to obtain listing in these other countries, the chronic feeding study in rats was repeated with an updated protocol; and a chronic mouse study was also initiated. Both studies started in mid-1975.

In early 1976, almost simultaneous with the termination of the provisional listing of FD&C Red #2, early results of the FD&C Red #40 mouse study were reported to the FDA showing an early onset of lymphomas in the dosed animals. This caused real concern and the FDA requested the sacrifice of a large number of the mice to determine the extent of the problem. In order to investigate this further, a second mouse study was started utilizing a larger number of animals in the dosed groups and in the controls. Careful monitoring of both studies was required with monthly reports to the FDA. In early December 1976, the FDA formed a working group of scientists from the Bureau of Foods, National Center for Toxicological Research, and the National Cancer Institute to review the studies being conducted on FD&C Red #40. The first meeting of this group was held December 16, 1976 and an interim report from the group was released January 19, 1977. The group concluded that the results from the first mouse study were inconclusive and were not likely to yield definitive results; they recommended that no decision be made on the safety of FD&C Red #40 until the conclusion of the second mouse study in mid-1978.

Following termination of the first mouse study and the rat study, the working group was reconvened on December 19, 1977 to assess the safety of FD&C Red #40. There were no unfavorable findings on the rat study; and in the mouse study, the final results showed no significant difference in lymphomas in the dosed animals. The ongoing second study did not reveal an early onset of lymphomas.

The report of the working group issued in mid-April 1978 concluded that the tests provided no evidence that FD&C #40 is carcinogenic either through tumor incidence or tumor acceleration. The second mouse study will be completed in late summer 1978. In mid-1976, the Food and Drug Administration was faced with a dilemma regarding the provisionally listed colors and to a lesser degree, with the permanently listed colors. At that time, the breakdown of the colors, with respect to their listings, was as follows:

Provisional listing	Permanent listing
FD&C Red #4 (for coloring maraschino cherries)	FD&C Yellow #5
FD&C Yellow #6	FD&C Red #3
FD&C Blue #2	FD&C Blue #1
FD&C Green #3	FD&C Red #40 (and lake)
Lakes of FD&C Colors (except FD&C Red #40)	Orange B (for coloring sausage casings)
	Citrus Red #2 (for coloring skins of oranges)

The provisional listing of FD&C Red #2 had been terminated and the Toxicology Advisory Committee had earlier attacked the relevancy of the safety studies on Red #2 conducted in the late 1950s and early 1960s. The studies supporting the provisionally listed colors, as well as the colors permanently listed, were of the same vintage and subject to the same questions. The General Accounting Office (GAO) in their report in the fall of 1975 had severely criticized the FDA for continuation of the provisionally listed colors for 15 years, and had in effect, issued an ultimatum to permanently list the colors or to terminate their provisional status.

On September 23, 1976, just prior to the closing date for the provisionally listed colors, the FDA published an order in the *Federal Register*:[12]

- Terminating the provisional listing of FD&C Red #4 for coloring maraschino cherries and short-term ingested drugs
- Extending the closing date for the provisionally listed colors until December 31, 1976, after which the Commissioner proposed to postpone the closing date until December 31, 1980, during which time chronic toxicity studies would have to be run by the petitioners and other interested parties
- Suggesting that similar chronic studies be conducted on the permanently listed FD&C colors

The order stated that the Commissioner was taking the action on FD&C Red #4 because questions regarding the safety of this color had not been resolved and the need for an additional toxicity study had been known to the industry sponsors for several years.

The CCMA had earlier stated their willingness to conduct additional studies on the provisionally listed colors as well as studies on the permanently listed colors in response to the September 23 publication.

On Janury 7, 1977, the Food and Drug Administration extended the provisional listings from December 31, 1976 to January 31, 1977, stating this extension was necessary to provide additional time to complete scientific and legal review of the comment received since September 23, 1976. On February 4, 1977, the final order on the postponement of closing dates was published, changing the closing date to January 31, 1981. Immediately thereafter, two activist groups, the Health Research group and the Public Citizen group, filed a complaint in the U.S. District Court for declaratory injunctive relief seeking an order directing the Department of Health, Education, and Welfare to withdraw the regulation extending the listing for the provisionally listed colors. Subsequently, the Cosmetic, Toiletries and Fragrance Association, (CTFA) and the CCMA requested and were granted the right to intervene in this litigation on the part of the Food and Drug Administration. The Federal District Court decision in October 1977 supported the action of FDA.

As a result of the September 23, 1976 proposal, the CCMA initiated work on preparation of testing protocols and selection of test material preparatory to starting chronic feeding studies in rats and mice. The rat studies were initiated on six colors in July 1977 and mouse studies started in mid-1978. Studies had been planned on FD&C Red #2; Orange B, and Carmoisine, a candidate for addition to the list; however, these were delayed because of questions involving the presence of beta-naphthylamine in trace amounts in these colors.

Hyperkinesis and Food Colors

In 1973, a great deal of concern and publicity surfaced as a result of an hypothesis suggested by Dr. Ben F. Feingold, Chief Emeritus of the Department of Allergy at the Kaiser Permanente Medical Center. Dr. Feingold claimed that hyperactivity and learning difficulties are caused by salicylate and salicylate-like natural compounds in food additives, especially artificial colors and flavors, and that a diet free of such substances dramatically reduces hyperactivity. Dr. Feingold's hypothesis was based on subjective clinical observations often made by parents and teachers without the benefit of controls to insure objectivity.

After presentation of his hypothesis at a number of medical symposia, Dr. Feingold attracted the attention of the public and the press. He testified before several Congressional committees and one of his addresses was reprinted in the Congressional Record. Further attention was focused on this issue as a result of Dr. Feingold's book, *Why Your Child is Hyperkinetic* (Random House, 1975). As a consequence of his theory, Dr. Feingold has suggested to the FDA the use on food packages of a symbol to indicate the absence of synthetic flavors and colors, and he has urged that color additives be identified in food products by name.

In response to the questions raised and the concern created, several advisory groups and panels of scientists have evaluated the evidence cited in support of Dr. Feingold's hypothesis. In general, all agreed that there were no controlled studies that have demonstrated that hyperkinesis is related to ingestion of food additives and that hyperactive children have not been shown to improve significantly on a diet free of salicylates and food additives. All agreed that controlled studies should be conducted. A number of controlled studies have been run, and at least one is in the planning stages. Results from two of the studies which have been completed do not support the Feingold thesis.

FD&C Yellow #5 — Allergenicity and Food Labeling

Since 1960, there have been a number of publications reporting allergenic reactions from ingestion of FD&C Yellow #5. These reports propose an association between aspirin intolerance and FD&C Yellow #5 intolerance. Based on these reports, the Food and Drug Administration published a proposed order, February 4, 1977, in the *Federal Register*, requiring the declaration of FD&C Yellow #5 in the ingredient statement for packages of human food. This labeling requirement does not apply to use in animal feeds and pet foods. The effective date for this portion of the order is 1 year after the date of publication of the final order in the *Federal Register*. This proposed order also prohibited the use of FD&C Yellow #5 in certain types of prescription and over the counter (OTC) drugs as well as requiring a warning statement regarding FD&C Yellow #5.

It was felt that the listing of FD&C Yellow #5 in the ingredient statement for human food products would have minimum impact on use of FD&C Yellow #5. However, following publication of this proposed order, there was a flurry of action by food processors to seek replacement of FD&C Yellow #5. There being no other FD&C colors for this shade, many investigated the uncertified color additives — annatto, tumeric, etc. for use.

In a letter dated January 3, 1978, the Commissioner of the Food and Drug Administration, Donald Kennedy, urged food processors to voluntarily identify all food colors by name in the ingredient statement of the product. According to the Commissioner, the FDA lacks authority to require such labeling. Currently, all color additives, certified and uncertified, are designated on the label as artificial color.

Safety Testing Requirements for Certified Color Additives and Uncertified Color Additives

One of the features of the Color Additives Amendments of 1960 was the equal treatment of the synthetic colors and the so-called natural colors. This intent is obvious from the records of the Committee Hearings prior to the enactment of the legislation. Part of that record states:[13] "There is a need for making applicable to all color uses and all types of color — whether they be coal tar colors or others — the same pretesting requirements and, where necessary for the protection of color users and consumers, the same requirements for certification of colors to assure their purity and identity with those listed as safe. At present there are no provisions for the certification of non-coal tar color additives as such, other than food additives."

In another part of this record, there is the following statement by Secretary Arthur S. Fleming: "From the point of view of determining safety of use, there is no sound scientific basis for distinguishing between a color additive extracted from a plant, animal or mineral source and one which is synthesized with a chemical structure which will bring it under the term 'coal tar color'. The Bill would therefore establish common ground rules for all such colors."

This legislative intent has not followed through in respect to the treatment of the certified color additives and uncertified color additives; and the questions of safety in use have focused primarily on the certified color additives. There is minimal data on the composition and safety testing for many of the uncertified color additives; and in fact, a number of these colors were listed strictly on the initiative of the Secretary of Health, Education, and Welfare. There is growing concern on the part of the FDA regarding this lack of safety testing data and it seems quite certain that in their cyclically review of generally recognized as safe (GRAS) items, that the uncertified color additives will be questioned. As a result of the banning of FD&C Red #2 and the questions raised on FD&C Red #40, color users have, in some instances, shifted to the less functional and more expensive uncertified red food colors. There is some question as to whether this is a movement to a safer position for the future.

THE INTERNATIONAL FOOD COLOR SITUATION

Hopes of having universally accepted food colors, uniform protocols for safety testing, and consistent interpretation of toxicological data exist only as an idealistic dream. In spite of the high cost of conducting safety studies, there is senseless duplication of studies, either because of lack of agreement on protocol, time pressures brought about by political or activist pressures within a country, or simple nationalistic postures by regulatory bodies. Many countries have active participation in the evaluative work the Joint FAO/WHO Expert Committee, but choose to accept or reject the recommendations of this group.

The only universality in regard to food colors on a world-wide basis is the inordinate focus on their safety. In this respect, they probably receive a degree of attention far in excess of their likelihood of hazard. The status of approved synthetic food colors in any country is in such a constant state of flux that maintenance of a list for the various countries is well nigh impossible. To add to the confusion, a number of countries may approve a color, but restrict its use to certain foods. Table 1 lists the approved colors in various countries as of January 1, 1977. The chemical structures of these colors are shown in Table 2.

The United Kingdom and the EEC

Table 1 also designates those colors which were on the original EEC list. The original countries in the EEC were Germany, France, Italy, Netherlands, Belgium, and Luxembourg. The later members are the United Kingdom, Ireland, and Denmark. Entry of these new states caused problems of synchronizing the U.K. list with the EEC list; at the same time it spurred actions to review and upgrade the original EEC list of 18 synthetic food colors.

For this purpose the EEC Commission created an advisory group — the Scientific Committee for Food (SCF). Based on recommendations from the SCF, the EEC Commission issues directives to the member states, who must incorporate the content of the directive into their national regulation or legislation. The member state must list in their food laws those colors approved by the EEC Commission, however, the individual country may list those foods in which the colors may or may not be used. Although the practice is unusual, a country may restrict the use of a color to only one food, in effect negating the directive. For example, France chose to limit the use of Amaranth (FD&C Red #2) to caviar only, a use for which it had previously not been employed. However, it is the eventual objective of the EEC Commission to designate all the foods in which particular colors can be used.

As part of the review of the original EEC list, the SCF, in 1975, set ADIs (acceptable daily intake) for the colors and recommended the removal and addition of a number of colors. The classifications were as follows:

1. Colors for which an ADI could be established and which are therefore toxicologically acceptable for use in food within these limits:

Color	EEC No.	ADI mg/kg
Erythrosine	E 127	0—2.5
Indigotine	E 132	0—5
Red 2 G	—	0—0.1
Sunset Yellow FCF	E 110	0—2.5
Tartrazine	E 102	0—7.5

2. Colors for which a temporary ADI could be established and which are toxicologically acceptable for use in food within these limits until December 31, 1978:

Color	EEC No.	Temporary ADI mg/kg
Amaranth	E 123	0—0.75
Carmoisine	E 122	0—2
Brilliant Black PN	E 151	0—0.75
Brilliant Blue FCF	—	0—2.5
Brown FK	—	0—0.05
Food Green S	E 142	0—5
Patent Blue V	E 131	0—2.5
Ponceau 4R	E 124	0—0.15
Quinoline Yellow	E 104	0—0.5

Table 1

Color	C.I. 1971 No.	E.E.C. No.	E.P.T.A. countries				E.E.C. countries												
			Austria	Norway	Sweden	Switzerland	Belgium[a]	Denmark[a]	France[a]	Germany(W)[a]	Italy[a]	Netherlands[a]	U.K.[a]	Australia[a]	Canada	Finland	Japan	S. Africa[a]	U.S.
Reds																			
Allura Red	16035							b											b
Ponceau 4R	16255	E124	b	b,d	b,c	b	b	b	b	b	b	b	b	b		b,d	b,f	b	
Carmoisine	14720	E122	b			b	b	b	b	b	b	b	b	b				b	
Amaranth	16185	E123	b	b,d,g	b,f,g	b	b	b	b	b	b	b	b	b	b	b,c,d	b,e,f	b	
Erythrosine BS	45430	E127	b,h	b,d	b	b,h	b	b	b	b	b	b	b	b	b	b,d	b,e,f	b	
Red 2G	18050												b					b	
Red 6B	18055		b																
Red FB	14780		b											b					
Fast Red E	16045		b			b								b					
Ponceau 6R	16290	E126	b																
Scarlet GN	14815	E125	b			b								b					
Ponceau SX	14700		b												b,j				
Acid Fuchsine	17200		b																
Eosine	45380																		
Rose Bengale	45440													b			b		
Oranges and Yellows																			
Orange G	16230												b					b	
Orange RN	15970																	b	
Orange GGN	15980	E111	b											b					
Oil Yellow GG	11920																	b	
Tartrazine	19140	E102	b	b,d,g	b,c,g	b	b	b	b	b	b	b	b	b	b	b,d	b,e	b	
Yellow 2G	18965												b	b					
Sunset Yellow FCF	15985	E110	b	b,d	b,c	b	b	b	b	b	b	b	b	b	b	b,d	b,e	b	
Acid Yellow	13015	E105	b			b								b					
Quinoline Yellow	47005	E104	b	b,d	b	b	b	b	b	b	b	b	b			b,d			
Chrysoine S	14270	E103	b																

Table 1 (continued)

Color	C.I. 1971 No.	E.E.C. No.	E.P.T.A. countries: Austria	Norway	Sweden	Switzerland	E.E.C. countries: Belgium[a]	Denmark[a]	France[a]	Germany(W)[a]	Italy[a]	Netherlands[a]	U.K.[a]	Australia[a]	Canada	Finland	Japan	S. Africa[a]	U.S.
Greens																			
Green S (Brilliant Green BS)	44090	E142					b	b	b	b	b	b	b	b				b	
Guinea Green B	42085																		
Fast Green FCF	42053				b			b							b	b	b,c	b	
Blues																			
Indigo Carmine	73015	E132	b	b,d	b	b	b	b	b	b	b	b	b	b	b	b,d	b,e	b	
Indanthrene Blue	69800	E130	b	b												b			
Patent Blue V	42051	E131	b	b	b	b	b	b	b	b	b	b	b			b			
Brilliant Blue FCF	42090							b,l					b	b	b		b,e		
Violets																			
Violet BNP	42580													b					
Violet 6B	42640																		
Browns																			
Brown FK	—												b	b,k					
Chocolate Brown PB	—												b	b					
Chocolate Brown KT	20285												b	b					
Blacks																			
Black BN	28440	E151	b			b	b	b	b	b	b	b	b	b					
Black 7984	27755	E152			b,c,i														

[a] Al and Ca lakes (salts) of all permitted colors also authorized.
[b] Means a color is permitted.
[c] Colors proposed for removal.
[d] Al and Ca lakes (salts) of this color also permitted.
[e] Al lakes (salts) of this color also permitted.
[f] Manufacturers have imposed a voluntary ban on use of this color.
[g] Declare by name.
[h] Whole fruit only.
[i] Only in fish roe (proposed for removal).
[j] For maraschino cherries only, maximum permitted level 150 ppm.
[k] Not permitted in New South Wales.
[l] Permitted until 12/31/77 only.

Table 2

Reds

Name	CI No.	EEC No.	Structure
Allura Red (FD&C Red #40)	16035	—	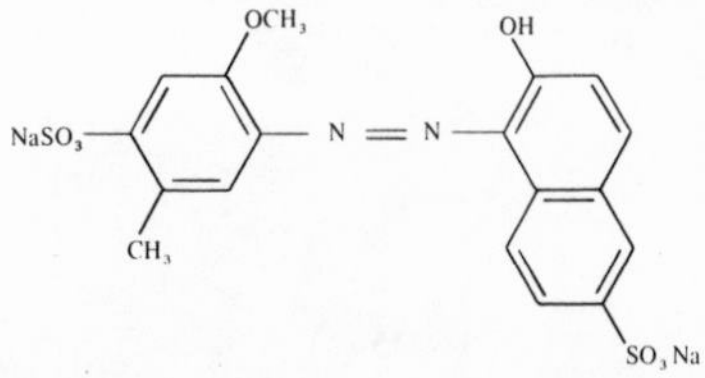
Ponceau 4R	16255	E124	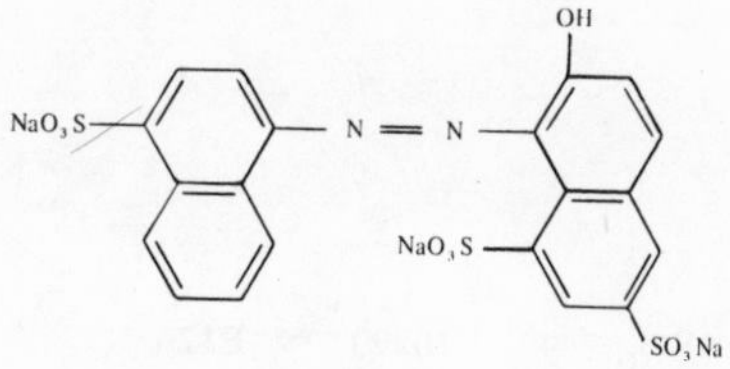
Carmoisine	14720	E122	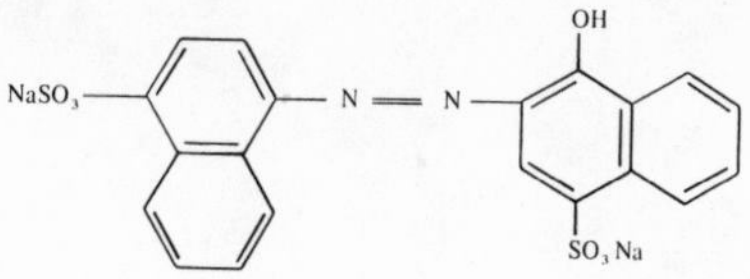
Amaranth	16185	E123	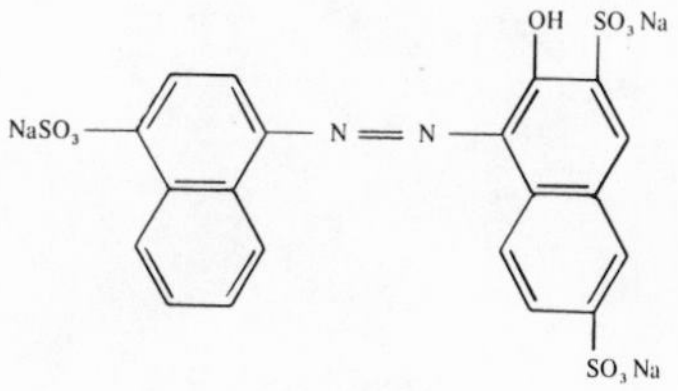
Erythrosine BS (FD&C Red #3)	45430	E127	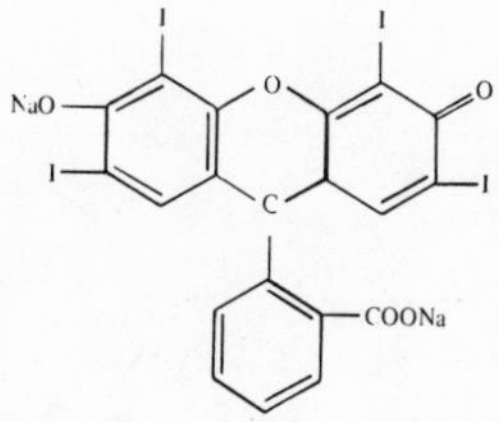
Red G2	18050	—	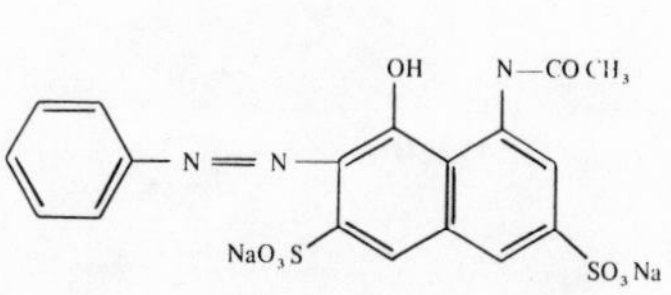
Red 6B	14780	—	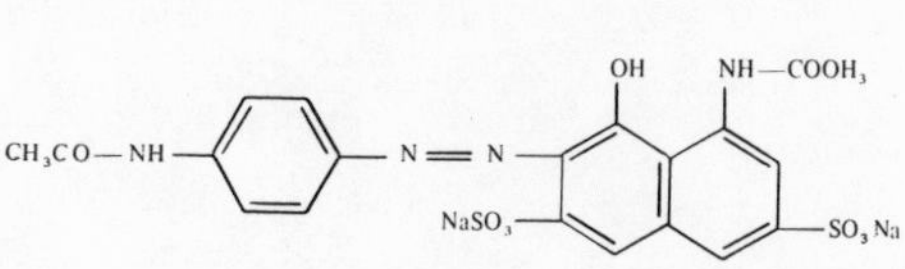

Table 2 (continued)

Name	CI No.	EEC No.	Structure
Red FB	14780	—	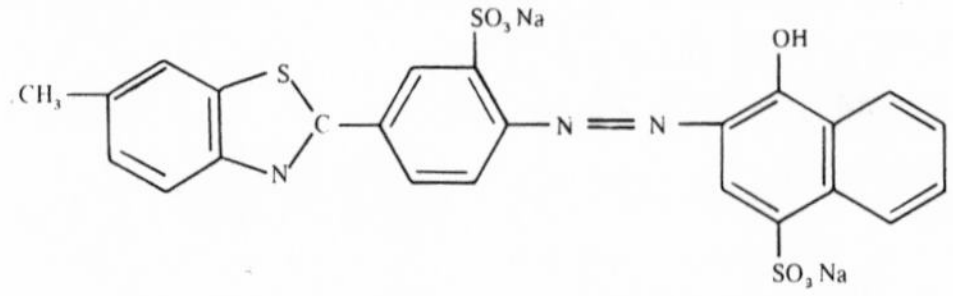
Fast Red E	16045	—	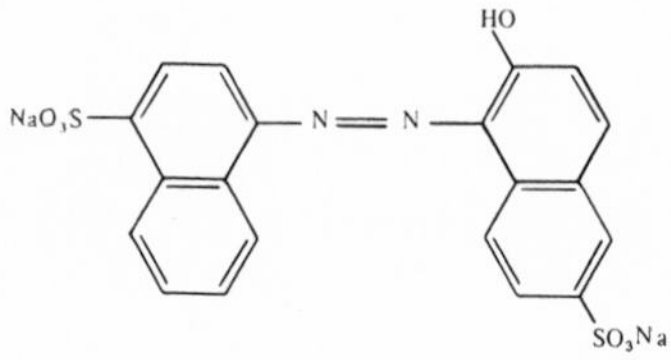
Ponceau 6R	16290	E126	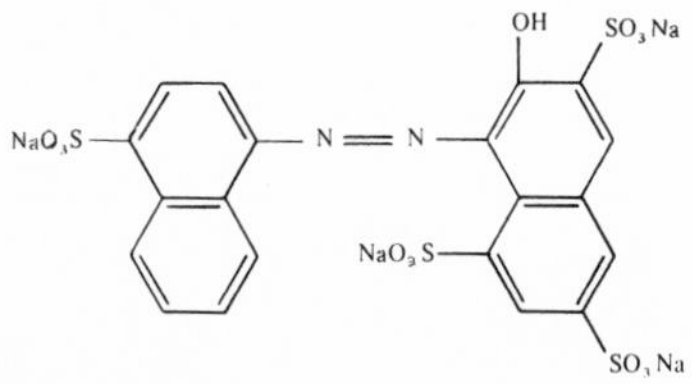
Scarlet GN	14815	E125	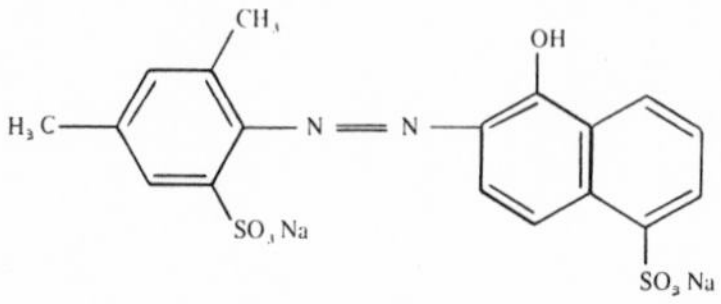
Ponceau SX	14700	—	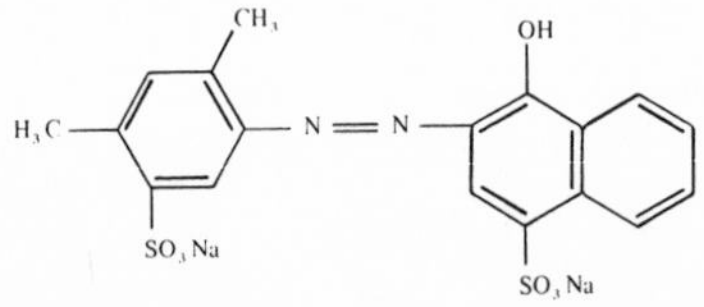
Acid Fuchsine	17200	—	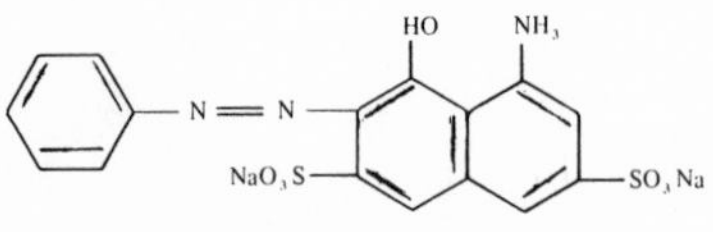
Eosine	45380	—	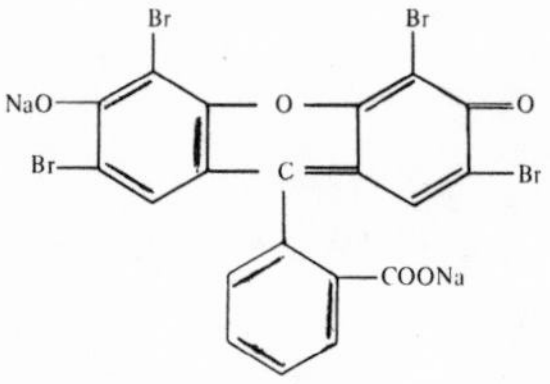

Table 2 (continued)

Name	CI No.	EEC No.	Structure
Rose Bengale	45440	—	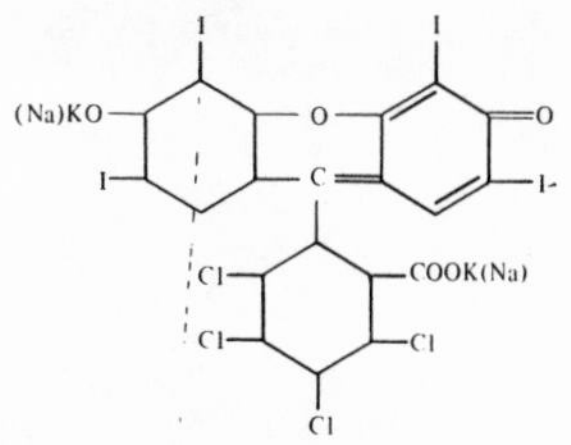
			Oranges and yellows
Orange G	16230	—	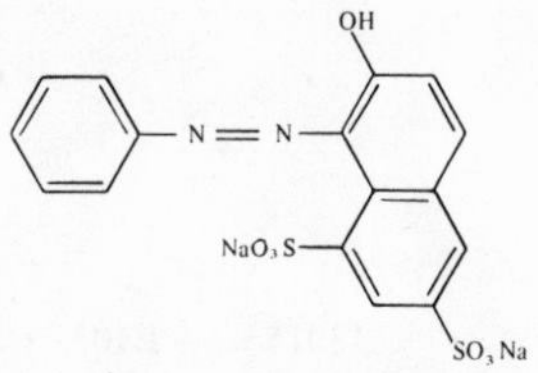
Orange RN	15970	—	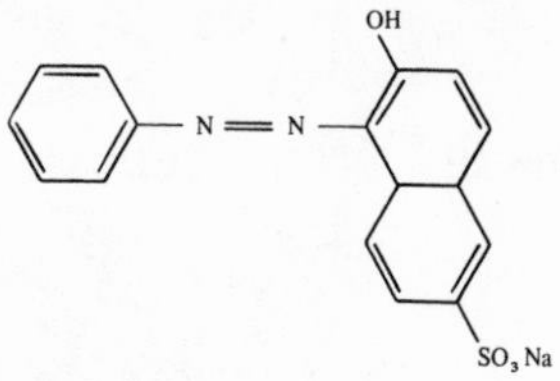
			Oranges and blends
Orange GGN	15980	E111	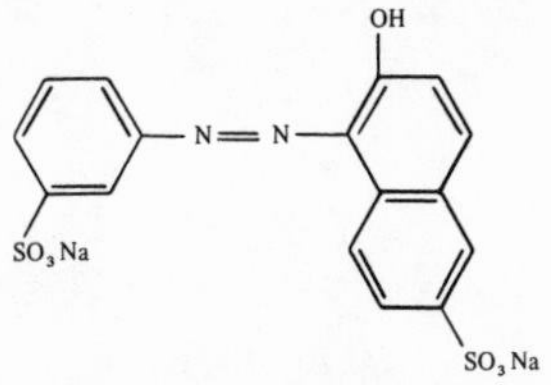
Oil Yellow GG	11920	—	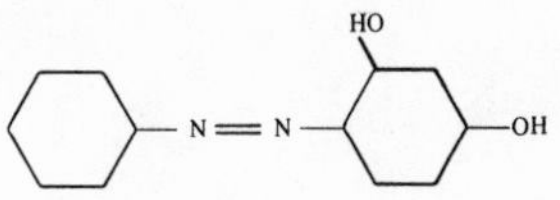
Tartrazine (FD&C Yellow #5)	19140	E102	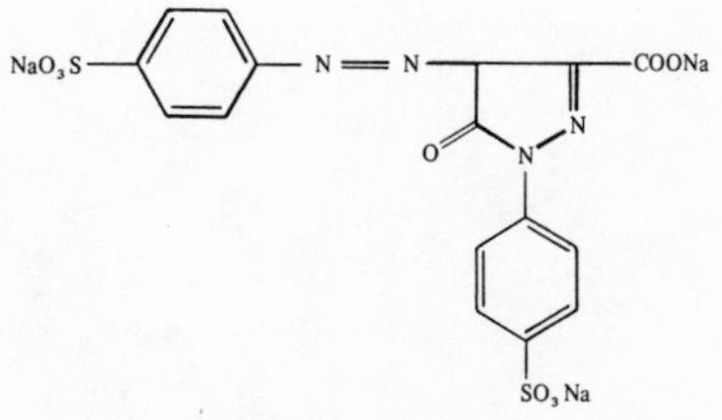

Table 2 (continued)

Name	CI No.	EEC No.	Structure
Yellow 2G	18965	—	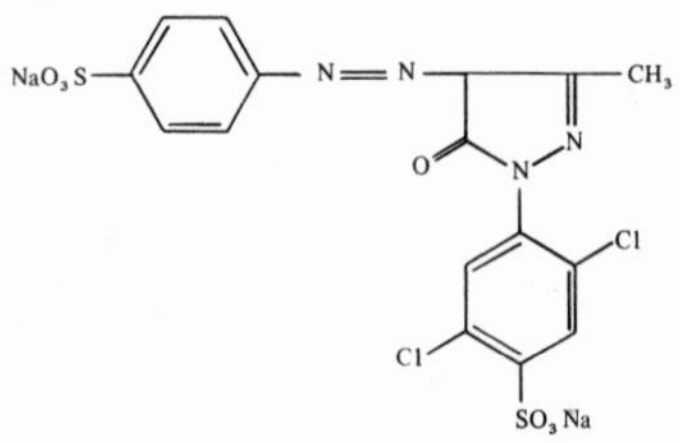
Sunset Yellow FCF (FD&C Yellow #6)	15985	E110	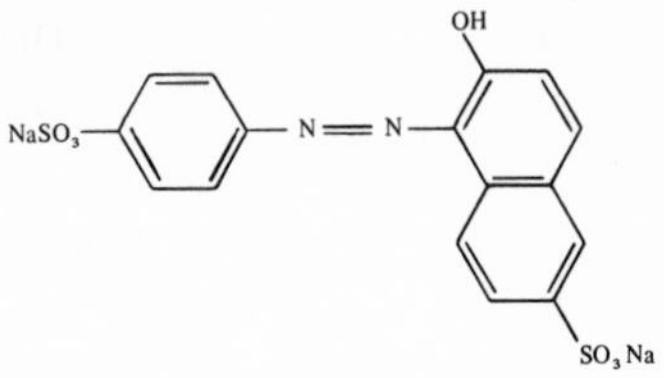
Acid Yellow	13015	E105	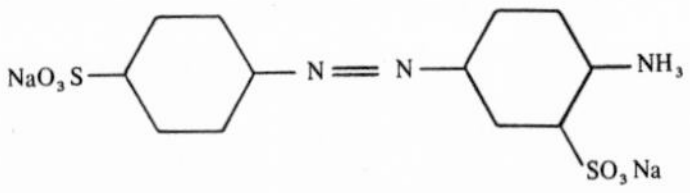
Quinoline Yellow	47005	E104	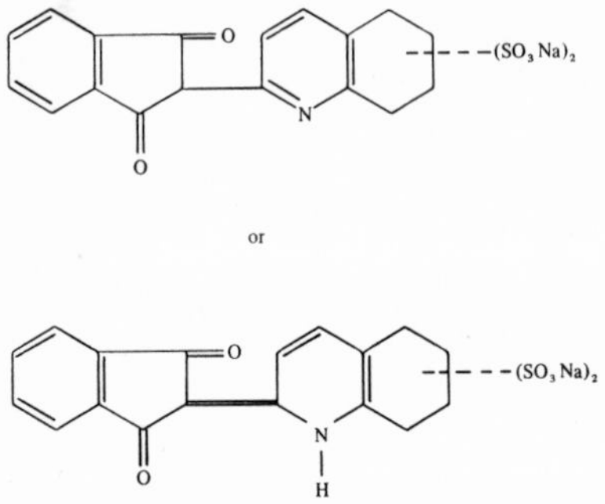 or
Chrysoine S	14270	E103	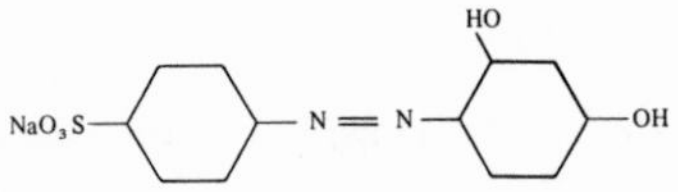
Greens			
Green S	44090	E142	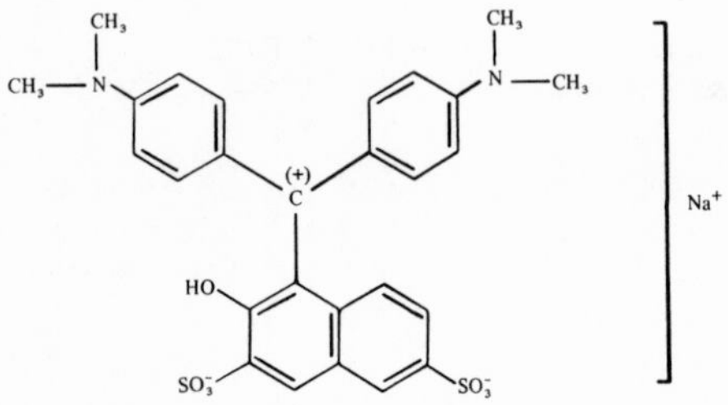

Table 2 (continued)

Name	CI No.	EEC No.	Structure
Guinea Green B	42085	—	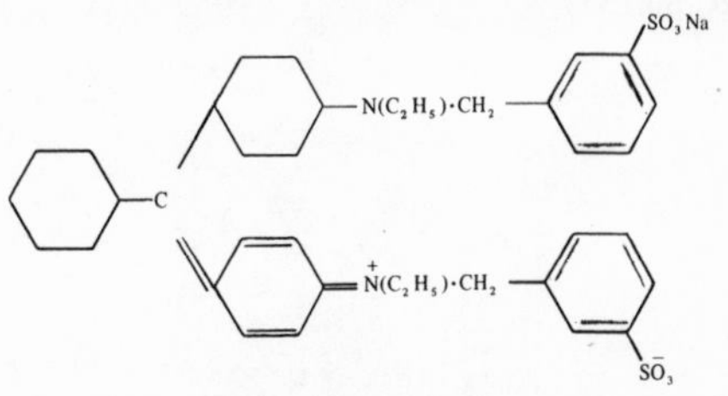
Fast Green FCF (FD&C Green #3)	42053	—	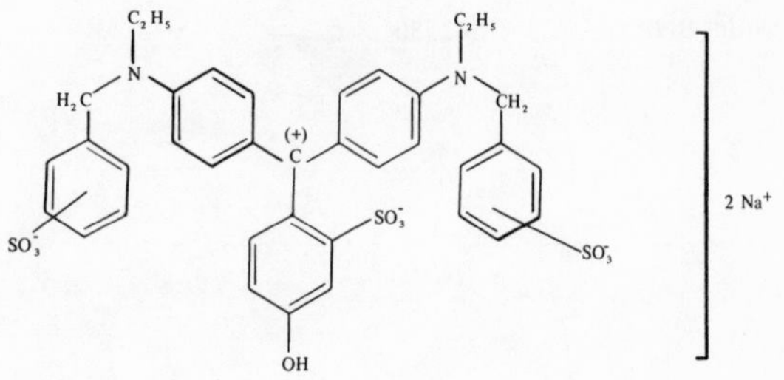

Blues

Name	CI No.	EEC No.	Structure
Indigo Carmine (FD&C Blue #2)	73015	E132	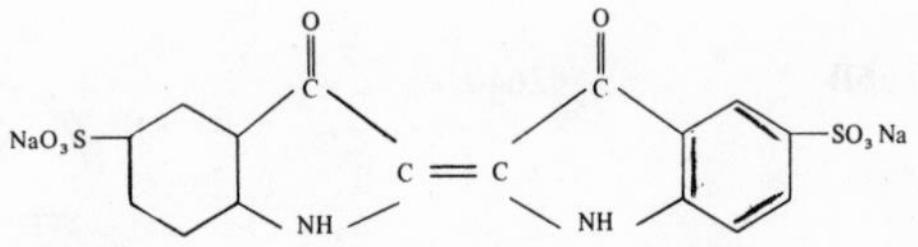
Indanthrene Blue	69800	E130	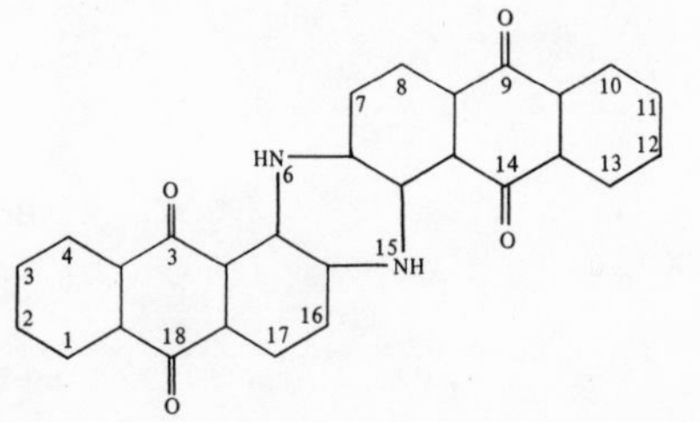
Patent Blue V	42051	E131	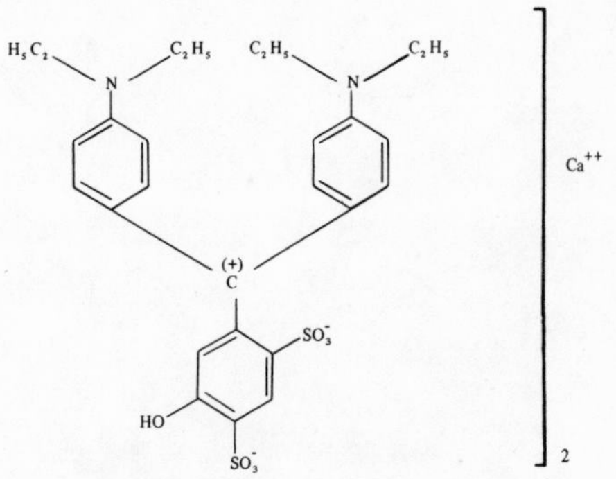

Table 2 (continued)

Name	CI No.	EEC No.	Structure
Brilliant Blue FCF (FD&C Blue #1)	42090	—	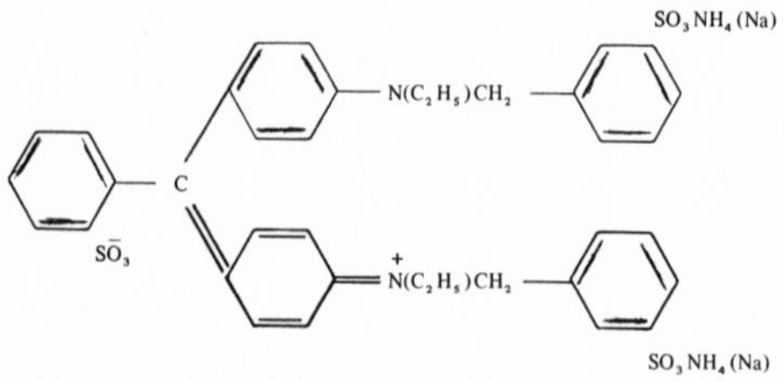
Violets			
Violet BNP	42580	—	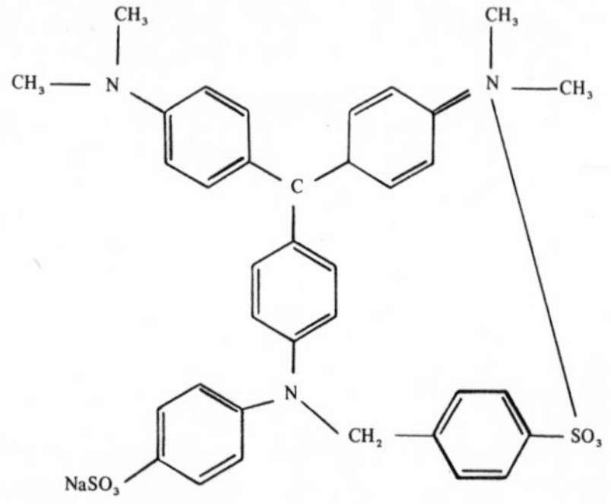
Violet 6B	42640	—	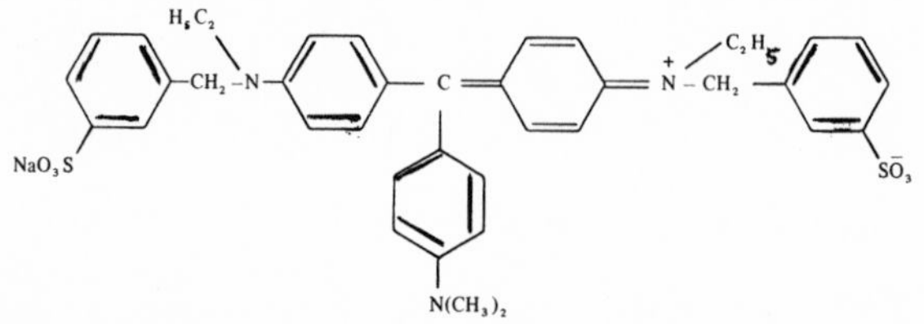
Browns			
Brown FK	—	—	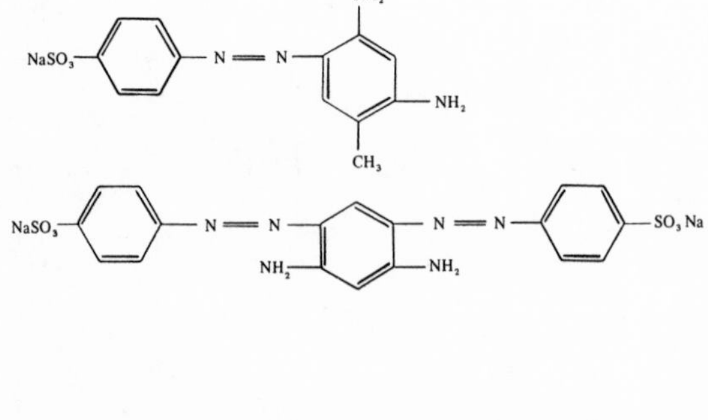

Table 2 (continued)

Name	CI No.	EEC No.	Structure
Chocolate Brown FB	—	—	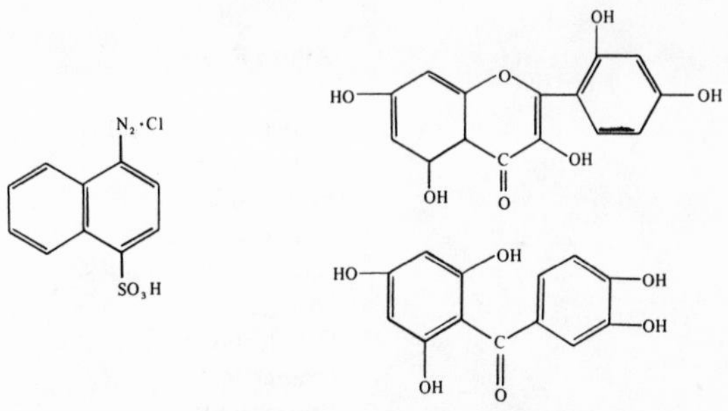

Blacks

Name	CI No.	EEC No.	Structure
Black BN	28440	E151	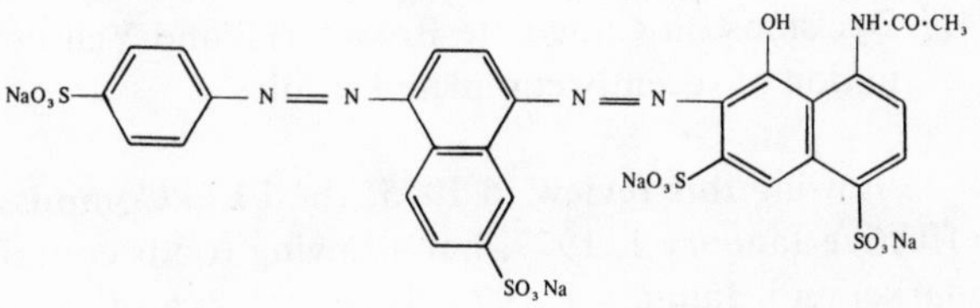
Black 7984	27755	E152	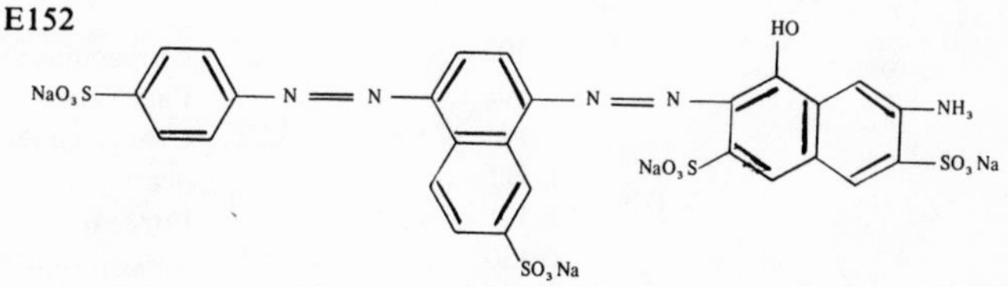

3. Colors for which an ADI could not be established and which are not toxicologically acceptable for use in food:

Color	EEC No.
Allura Red	—
Black 7984	E 152
Chocolate Brown FB	—
Chrysoine S	E 103
Fast Red E	—
Fast Yellow AB	—
Indanthrene Blue RS	E 130
Orange G	—
Orange GGN	E 111
Orange RN	—
Ponceau 6R	E 126
Scarlet GN	E 125
Violet 6B	—

4. Decisions on Chocolate Brown HT and Yellow 2G were postponed pending evaluation of recently completed studies

Following this review in 1975, the EEC Commission deleted seven synthetic colors effective January 1, 1977, but allowing foods containing these colors to remain on the market until January 1, 1978. These colors are:

E 103	Chrysonine
E 105	Fast Yellow AB
E 111	Orange GGN
E 125	Scarlet GN
E 126	Ponceau 6R
E 130	Indanthrene Blue
E 152	Black 7984

The EEC Commission has also ordered that a variety of studies be run on certain colors before codifying the final EEC list. These studies are to be completed by December 31, 1978. This is an unrealistic date; however, it is believed that it would be extended if studies were in progress. The studies required by color are shown in Table 3. This table also lists the ADI established by Joint Expert Committee on Food Additives (JECFA), if different from that set by SCF.

In the U.K., an umbrella organization of trade associations, color manufacturers, and color-using industries is pursuing the initiation and funding of the studies on most of these colors. The Color Subcommittee of the CIA (Chemical Industries Association) comprising the food color manufacturers are funding and conducting studies on:

- Amaranth
- Carmoisine
- Ponceau 4R
- Brilliant Blue FCF
- Chocolate Brown HT
- Green S

Studies on other colors are being conducted by interested users.

It will be a number of years until a comprehensive final list is dictated. At that time, all the listed colors can be used in all the EEC countries. However, nationalistic desires

Table 3
EEC COLORS — SFC DIRECTIVES ON TESTING

Color	ADI[a]	Temporary ADI	JECFA ADI	Studies required: Short term	Long term	Reproduction	Embryotoxicity	Teratology	Metabolic
Quinoline Yellow		0—0.5	0—0.5		x				x
Yellow 2G		0—0.01	0—0.01		x				x
Tartrazine	0—7.5		0—7.5						
Sunset Yellow	0—2.5		0—5						
Ponceau 4R		0—0.15	0—0.125		x	x			x
Erythrosine	0—2.5		0—2.5						
Red 2G	0—0.1		0—0.1						
Azorubine (Carmoisine)		0—2	0—2		x				x
Amaranth		0—0.75	0—0.75		x	x			
Indigo Carmine	0.5		0.5						
Brilliant Blue FCF		0—2.5	0—12.5						x
Patent Blue V		0—5	Withdrawn	x	x			x	x
Green S		0—5	Withdrawn		x	x	x	x	x
Brilliant Black BN		0—0.75	0—2.5			x	x	x	x
Brown FK		0—0.05	0—0.05		x	x		x	
Chocolate Brown HT		0—2.5	0—2.5			x			x

[a] ADI in mg/kg of body weight.

will be satisfied in those countries that can dictate the foods in which the colors can be used and the rate of use. Thus, individual countries can, in actuality, virtually ban a listed color.

The Canadian list closely parallels the U.S. approved list except that Allura Red (FD&C Red #40) has as yet not been approved, and Amaranth (FD&C Red #2) and Ponceaux SX (FD&C Red #4) are still on the Canadian list. In fact, the Health Protection Branch of Canada issued a news release at the time the U.S. banned Amaranth severely criticizing the FDA for their action. Canadian regulations state limits in foods for the approved colors. These are as follows:

- 300 ppm singly or in combination for:
  Amaranth (FD&C Red #2)
  Sunset Yellow (FD&C Yellow #6)
  Erythrosine (FD&C Red #3)
  Tartrazine (FD&C Yellow #5)
  Indigotine (FD&C Blue #2)
- 100 ppm singly or in combination for:
  Fast Green FCF (FD&C Green #1)
  Brilliant Blue FCF (FD&C Blue #1)
- 150 ppm in fruit peel, glace fruits, and maraschino cherries

TECHNICAL ASPECTS OF THE CERTIFIED COLOR

The presently certified color additive list consists of two major classifications: dyes and lakes. Some of the dyes have been in use since 1900. In the U.S., the lakes have been approved for use in foods allowed only since 1959. The dyes manifest their coloring power by being dissolved, while the lakes are insoluble pigments and color by dispersion. Table 4 shows the physical differences between dyes and lakes.

Later in this chapter we will discuss the properties and applications of FD&C dyes and compare them to the FD&C lakes.

FD&C DYES

The FD&C dyes are all water-soluble compounds and comprise four classes, shown in Table 5. Specifications and procedures for certification are published under the regulations promulgated by the FDA.

The primary or straight colors, which include the lakes, require certification. At one time, blends required certification, but regulations published in December 1964 eliminated certification measures.

The FD&C dyes have to meet strict standards as specified by the FDA. The specifications for Yellow #5 shown below are typical of those for certified color additives:

FD&C Yellow #5 — specifications

- Trisodium salt of 3-carboxy-5-hydroxy-1-*p*-sulfophenyl-4-*p* sulfophenylazo-pyrazole
- Sum of chlorides and sulfates (as sodium salts) and volatile matter (at 135°C), not more than 13.0%
- Water-insoluble matter, not more than 0.2%
- Phenylhydrazine-*p*-sulfonic acid, not more than 0.1%
- Other uncombined intermediates, not more than 0.2%

Table 4
PHYSICAL DIFFERENCES BETWEEN DYES AND LAKES

Properties	Lakes	Dyes
Solubility	Insoluble in most solvents	Soluble in water, propylene glycol, and glycerine
Method of coloring	By dispersion	By being dissolved
Pure dye content	Generally 10—40%	Primary colors 90—93%
Rate of use	0.1—0.3%	0.01—0.03%
Particle size	Average 5 μm	12—200 mesh
Stability		
Light	Better	Good
Heat	Better	Good
Coloring strength	Not proportional to pure dye content	Directly proportional to pure dye content
Shade	Varies with pure dye content	Constant

Table 5
CLASSIFICATION OF FD&C DYES

Azo dyes	Triphenylmethane dyes
FD&C Yellow #5	FD&C Blue #1
FD&C Yellow #6	FD&C Green #3
FD&C Red #4	
FD&C Red #40	
Fluorescein Type	Sulfonated Indigo
FD&C Red #3	FD&C Blue #2

- Subsidiary dyes, not more than 1.0%
- Lead, not more than 10 ppm
- Arsenic, not more than 3 ppm

CERTIFIED COLOR FORMS

Dyes are available in many forms — powder, granular, plating colors, wet-dry (blends), diluted (cut blends), liquid (aqueous), liquid (nonaqueous), and paste. The best form for any specific use will be dictated by the nature of the product to be colored, the process conditions, and volume of color used. Table 6 lists the range of pure dye percentages and the advantages and disadvantages of the various forms of color.

PROPERTIES OF FD&C DYES

Solubility

The FD&C dyes are water-soluble and are insoluble in nearly all organic solvents. Water solubility of most colors is quite high and in most application methods, solubility is usually no problem. FD&C Blue #2 (Indigotine) is the exception to this, and often FD&C Red #3 does not readily dissolve in dairy-based products.

For systems where anhydrous conditions are a consideration, glycerine and propylene glycol are used as solvents. In general, the colors are more soluble in glycerine than in propylene glycol. Most are only very slightly soluble in ethyl alcohol, but use is often made of the reasonable solubility in alcohol of FD&C Red #3, FD&C Blue

Table 6
ADVANTAGES AND DISADVANTAGES OF VARIOUS FORMS OF CERTIFIED COLOR ADDITIVES

Form	Pure dye (%)	Advantage	Disadvantage
Powder (primary color)	88—93	Ease of dissolving Suitable for dry mixes	Dusty
Granular (primary color)	88—93	Dustless, free flowing	Slower dissolving Not suitable for dry mixes
Wet dry blends	90	No flashing in dry blends when wetted	—
Aqueous liquid colors	1—6	Ready to use, ease of handling, accurate measurement	More costly than dry color
Cut blends	22—85	Permits larger weighings with more accuracy for small amounts of added color	More costly
Nonaqueous liquid colors	1—8	May be used in fatty material	More costly
Paste	4—10	May be used in products in which water is limited	Costly
Plating colors	88—93	Gives good depth to dry mixes	Not available in all primary colors

Table 7
SOLUBILITIES OF FD&C COLORS IN VARIOUS SOLVENTS

	Solubility in oz per gal of solvent			
FD&C colors	Water 70°F	Glycerine	Propylene glycol	Alcohol 95%
Red 3	16	30¾	29	2½
Red 4	9½	5¼	1½	Tint
Red 40	28	4	2	Tint
Yellow 5	17½	28	12	Tint
Yellow 6	23	14½	2½	Tint
Blue 1	25	37½	53	2
Blue 2	1½	½	Trace	Tint
Green 3	23	14½	14½	½

#1, and FD&C Green #3. Solubilities in glycerine, propylene glycol, and alcohol are shown in Table 7.

It would be most helpful to have on the approved list of FD&C Colors several oil-soluble colors, particularly a yellow color. However, the search for these has been futile and does not seem to portend much hope. The sulfonic acid groups confer water solubility to the molecule and likewise reduces oil solubility. Unsulfonated dyes would be oil-soluble, but lack of sulfonation seems to go along with high toxicity. More light will be shed on this phenomena as more metabolic studies are conducted.

Good coloring technique recommends that the dyes be solubilized before addition to the colored product. However, it is often possible, where water is added in the process, to add the dry color to the batch and depend upon the added moisture and heat to dissolve the color in processing.

Stability

In general, the certified food colors can be said to be stable for most uses in food.

In the dry state, no degradation has been noted (other than loss in dye strength by moisture absorption) in samples stored for 15 years.

With the exception of FD&C Blue #2 and FD&C Red #3, the light stability of the dyes in food products is good. Even FD&C Blue #2, which fades badly in solution, shows good stability in a number of products — candy and baked goods. Likewise, FD&C #3, which has limited light stability in coatings, shows excellent stability in retorted products.

The two areas in which the majority of the certified food colors show instability are in combination with reducing agents and retorted protein materials. The azo and triphenylmethane dyes are easily reduced to colorless compounds. Contact with metals such as zinc, tin, aluminum, and copper was formerly a large factor in color fading; however, with current progress in food processing, contact with these metals is minimal and the chief source of color fading is contact with reducing agents, such as ascorbic acid.

Color fading of certified colors in carbonated and still beverages is caused by ascorbic acid incorporated as a flavor anti-oxidant and as a source of vitamin C. The use of ethylenediaminetetraäcetic acid (ETDA) in beverages has proved somewhat successful in inhibiting the effects of ascorbic acid on dyes, but not in sufficient degree to eliminate it completely. The reductive action of ascorbic acid seems to be catalyzed by light and canned colored products containing certified color can possess reasonable shelf life.

In retorted protein foods, most of the certified food colors lack stability. However, FD&C Red #3 and to a lesser extend FD&C Blue #1 are stable. Canned dog foods can be colored using these colors; however, there is only limited use for this application because of the suitability of the very low cost iron oxides permitted in pet foods.

USE OF COLOR ADDITIVES

Food colors are unique in their applications as they are used in only minute or micro quantities in finished food products. Only such ingredients as flavors and vitamin additives approach the low levels of use of food colors. For example, most colored foods contain approximately 0.005 to 0.03% certified color by weight.

The Certified Color Industry Committee surveyed from their sales records the use of color by the various segments of the color-consuming industries.[17] Table 5 shows the concentrations, both in range and maximum of color used, in the various categories of processed foods. Table 8 lists the various food products that often exceed the maxima shown in Table 9. Each group of food products has its own coloring techniques and methods of color applications. These will be discussed in detail.

BEVERAGES

Beverages constitute one of the largest users of certified colors. Some of the difficulties involved with the use of food colors in beverages, are can corrosion and fading.

Fading due to ascorbic acid was discussed previously. Ascorbic acid is often a desired ingredient in beverages either for its vitamin activity or as a flavor protector, in which case it protects by scavenging the oxygen in the headspace air. In instances where ascorbic acid is used, uncertified color additives should be used. In fact, beverage use was the impetus for development of the β-carotene beadlet, which is a water-dispersible form of this oil-soluble color. In instances where it is desirable to use a flavor antioxidant and certified color, use of glucose oxidase-catalase as an oxygen scavenger is recommended.

Table 8
FOOD PRODUCTS REPORTED TO CONTAIN COLOR IN EXCESS OF THE MAXIMA LISTED IN TABLE 5[a]

Category	Product	Color concentration (maximum ppm)[a]
Candy and confections	Tableted candy	500
	Circus peanuts	600
Dessert powders	Chocolate pudding	1000
Bakery goods	Cake decorations	1000
	Sugar wafers	800
	Chocolate cookies and cake	800
	Ice cream cones (other than yellow)	900
Ice cream and sherbets	Very dark chocolate, including ice cream bar coatings	600
Miscellaneous	Spray-dried cheese	700—2000
	Spices	1000
	Fruit rings	600
Snack foods	Cheese-flavored items	750

[a] These specific products will not always exceed maximum values as listed in Table 10, but under certain conditions, the higher concentrations are necessary.

From Certified Color Industry Committee, *Food Technol.*, 22(8), 14, 1968. With permission.

Table 9
PROCESSED FOODS IN WHICH CERTIFIED FD&C COLORS ARE USED

Category	Color concentration (ppm) Range	Average
Candy and confections	10—400	100
Beverages (liquid and powdered)	5—200	75
Dessert powders	5—600	140
Cereals	200—500	350
Pet foods	100—400	200
Bakery goods	10—500	50
Ice cream and sherbets	10—200	30
Sausage (surface)	40—250	125
Snack foods	25—500	200

From Certified Color Industry Committee, *Food Technol.*, 22(8), 14, 1968. With permission.

Although present in canned carbonated beverages at only 50 to 150 ppm, the azo dyes (FD&C Red #40, FD&C Yellow #5, and FD&C Yellow #6) appear to have a causative effect on can corrosion; the rate of corrosion is directly proportional to the azo dye concentration.[18,19] However, inclusion of azo dyes at rates of 50 ppm or less will result in canned products with sufficient shelf life (9 to 12 months). For many products, 50 ppm is sufficient for good tinctorial values, but inclusion of caramel color is often desirable to raise the color depth. The nonazo certified dyes do not produce can corrosion. The uncertified color additives seem to have no effect on rate of corrosion.

Canned carbonated drinks and canned still drinks seem to represent two different systems. In canned carbonated beverages, the chief problem is can corrosion, and color fading is minimal. In still drinks, corrosion is usually of no significance, but color fading is a problem. The presence of juice in a drink seems to lessen the color fading. in still canned drinks. Much is known about canned beverages, but a great deal has been gleaned empirically and often defies scientific explanation. Glass-bottled beverages present few problems except those brought on by product abuse. Fading of colors by over-exposure to sunlight is a problem. This often produces a green-colored grape soda as a result of preferential fading of the FD&C Red #40. Such fading, however, can be looked upon as an advantage, as it serves as an indication of product abuse as the flavor certainly has reached a point of degradation before the color faded. Most fruit-type flavored beverages contain certified color additives, while colas and root beers are colored with special grades of caramel.

With the banning of Red #2 in February of 1976, many dark red and purple shades for beverages are difficult to obtain. The yellow-orange content of FD&C Red #40 renders such colors a dirty or brownish cast. Table 10 lists color formulas and amounts for various flavored soft drinks.

BAKERY PRODUCTS

Colors are employed in dough products, cookies, sandwich fillings, icings, and coatings. Because of the high moisture content of doughs and batters, little problem of color addition exists. However, obtaining the depth of color can often be a problem. This is particularly true in dark chocolate pieces, and use of certified color alone results in excessive color use. In such goods, combinations of certified colors with uncertified color additives have proved successful. Caramel is added with various combinations of certified colors. Caramel color has also found extensive use in coloring rye and pumpernickel bread.

Ice cream cones at one time were colored with water-soluble annatto; however, in recent years, there has been a switch to the use of certified colors which are less expensive and offer greater variety of shades. Cone-color blends generally contain FD&C Yellow #5 and FD&C Yellow #6.

Sandwich fillings and icings represent systems which require special coloring techniques and these techniques depend upon the composition. Formulations containing little or no moisture, sugar, and fat (shortening) necessitate dye solutions in propylene glycol or glycerine or FD&C lakes. Icings, particularly in the small bakery, are colored with paste colors which are compositions containing glycerine, propylene glycol, and sugar.

DAIRY PRODUCTS

Nearly all ice creams and sherbets contain artificial color. Chocolate ice cream is often the exception. Annatto is used to a considerable extent in vanilla ice cream, but as in the coloring of cones, certified colors (egg shades) are increasingly popular. Because of the small amounts of color used and because of convenience, liquid colors are often used. Ice cream presents very few problems of color stability except in instances where excessively high bacterial counts are present. Again, the fading of the color is an indicator of trouble.

Cheese is a product in which the certified colors are not sufficiently stable and an-

Table 10
COLOR AND CONCENTRATION IN CARBONATED BEVERAGES

Flavor	Color	Parts	Concentration (ppm)
Orange	FD&C Yellow #6	100	75
	FD&C Yellow #6	96	50
	FD&C Red #40	4	
Cherry	FD&C Red #40	99.5	
	FD&C Blue #1	0.5	100
Grape	FD&C Red #40	80	75
	FD&C Blue #1	20	
Strawberry	FD&C Red #40	100	60
Lime	FD&C Yellow #5	95	20
	FD&C Blue #1	5	
Lemon	FD&C Yellow #5	100	20
Cola	Caramel color	100	400
Root beer	Caramel color	100	400

natto and β-carotene are the desired colorants. Likewise, margarine and butter are colored chiefly with β-carotene and oil-soluble annatto. At one time it was felt the FD&C lakes would perform satisfactorily in margarine, but differences exist in the appearance of melted margarine containing lakes. The affinity of the lakes for protein causes eggs fried in margarine colored with lakes to have a yellow appearance.

CANDY

It is difficult to imagine many candies, particularly hard candy, without color. Hard candy represents a system low in moisture (1 to 2%), and aqueous color solutions are added with reluctance. Paste colors and other nonaqueous plastic materials are used. The temperature at which hard candy is added to the slab is somewhat destructive to color. Color should be added at the lowest temperature that will permit adequate distribution. Continuous hard candy operations utilize concentrated aqueous color solutions.

Compound or summer coatings cannot be colored with aqueous color solutions. Such use results in color specking or changes in physical properties of the coating. Nonaqueous color solutions using propylene glycol or glycerine have found use for this application, but lake dispersions are being used in increasing amounts. Pink coatings are much in demand, and FD&C Red #3 (dye) is fugitive to light. FD&C Red #3 lake produces attractive coatings with increased light stability over the dye.

PET FOODS

It should be remembered that by definition in the Food, Drug, and Cosmetic Act, a "food is anything consumed by man or animal". Consequently, pet foods, if colored, must contain approved color additives.

Pet foods consist of three chief types; dry extruded, wet-pack or canned and semi-moist. The original pet food marketed was the wet-pack or canned food. Iron oxide, because of its low price, has been traditionally used in canned pet food as the colorant. In addition, the less-expensive certified color did not withstand the retorting operation in the preparation of such products. In late 1966, the provisional listing of iron oxide as a color additive for pet foods was terminated and work at that time showed that certain certified color combinations based on FD&C Red #3 showed good stability in processing. However in July 1968, iron oxide was permanently listed for use in cat and dog foods at 0.25% by weight maximum. The certified colors are used extensively in dry extruded pet foods — yellow, red, and brown being the most popular colors. Rates of 0.01 to 0.03% are generally used, and the powdered color may be added to the dry mix before extruding. A more acceptable method is to dissolve the dye in some of the water used to condition the feed before extrusion. An interesting use is made of color in the gravy-type pet foods. This consists of extruded pet food to the surface of which is added a dry powder consisting of a gum, dry color (brown), and an inert material. When water is added to the food, the gum imparts viscosity to the liquid. This together with the dissolved color resembles a natural gravy. Because of the differential migration and dissolving of the individual primary colors, a wet-dry or monoblend brown is recommended.

The semimoist products are generally colored with little difficulty. Red dye blends are generally used, but lakes also find application. Since the banning of Red #2, FD&C Red #40 has been used to color semimoist products.

TABLETTED AND PAN-COATED PRODUCTS

Tabletting is an operation which has its roots in the pharmaceutical industry. Of late, there has been a great emphasis on tabletted products in the confectionery industry. Tablets are of two main types: compressed uncoated tablets and coated tablets, which are also formed by compression and afterwards treated with a thin colored coating. It should be noted that there are important economic factors which apply to pharmaceutical tablets as opposed to confectionery tablets. Pharmaceutical tablets are quite expensive and are sold by numerical units. As a result, the pharmaceutical items generally demand a higher degree of elegance.

Compressed tablets are prepared by exerting a high pressure upon a material of desired particle size and distribution in a tabletting press. To obtain the desired particle size and distribution, the powdered material is first granulated. A granulation can be prepared in a number of ways, the most important being slugging and addition of a granulating liquid while mixing. In slugging, the powdered material is run through a tablet press, producing a rather crude tablet. This tablet is then ground and screened to produce the granulation. A wet-type granulation is made by adding an aqueous solution of a binder (gum acacia, sugar, etc.) to the powdered material in a mixer. After agglomeration takes place, the material is dried, ground, and screened. The traditional method of coloring granulations was to dissolve a dye in the granulating solution. However, wicking or chromatographic migration often occurs on drying and greater amounts of color appear on the surface of the dry material or with dye blends, the dyes stratify into multicolored bands. Such granulations produce compressed tablets having a mottled appearance. This can be minimized, or in some instances eliminated, by use of the lakes. The lakes are also useful in preparing granulations by slugging. Here a dry mix is maintained throughout the processing and no drying is required. Also, in wet granulations, the lakes, because of their insolubility, do not migrate and stratify during drying to the same extent as dyes.

The coating of tablets has long been steeped in artistry, but is slowly yielding to scientific investigation. The most popular coating material is sucrose, but there are many film coatings which make use of various film forming gums and resins. Likewise, the coating systems may be either transparent or opaque. In the pan-coating operation, a revolving pan is charged with tablets. Numerous uncolored subcoatings are applied by adding to the pan, sufficient amounts of syrup, sufficient to uniformly wet and cover the surface area of the tablets. Air is blown over the wetted tablets to speed up the evaporation of solvent. When a smooth subcoat has been obtained, color coats are added. When dyes are used, as many as 30 or 40 coats are applied. Coatings colored with dyes are transparent and the depth of color depends on the concentration of the dye and the number of coats. With transparent coatings, tablet surface imperfections will show up as mottling. When opaque coatings are used, i.e., dyes or lakes with titanium dioxide, the shade is independent of the number of coats (after five or six), and the color depth depends upon the concentration of dye or lake and the concentration of titanium dioxide. Mottling effects are minimized by use of opaque coatings.

Though they are pigments, the FD&C lakes are very transparent and used alone produce coatings similar to dyes. Film coatings are generally based on an organic solvent containing a gum or resin plus plasticizers. There are many different film coating systems, most of which are nonaqueous and demand the use of the FD&C lakes. Several reviews on coatings are available describing manufacturing details.[20,21]

DRY MIX PRODUCTS

Colors for dry mixes should impart maximum color to the dry product and if a blend, dissolve without flashing, i.e., without showing the individual component colors. To obtain maximum color in a dry mix, dissolved color can be added in solution. This operation, however, necessitates some sort of moisture removing operation and often dry mixes cannot tolerate moisture additions. A number of primary colors are available in plating grade form. They show superior coloring power when distributed throughout a dry mix.

The difficulty of color flashing can be eliminated by the use of wet-dry blends (monoblends). These are prepared by dissolving the colors to form a solution and drying the solution. In this fashion, a blend of FD&C Red #40 and FD&C Blue #1 will dissolve from a dry mix product as a purple rather than showing streakings of red and blue. Such blends are popular for use in gelatin desserts and puddings. There is no reason to use a plating grade color or a monoblend if the color is applied to the food product by means of a color solution.

Cake, doughnut, and pancake mixes were formerly colored with oil-soluble colors (FD&C Yellows #3 and #4). Good depth of color was obtained in the dry mix because the dyes were soluble in the shortening. When these oil-soluble colors were delisted, the switch to water-soluble colors was disappointing as they did not impart sufficient color to the dry product even though the color in the batter was satisfactory. Lakes help to solve the problem, as they adequately color both the dry product and the batter. In some instances, combinations of dyes and lakes have been employed for these products. Table 11 lists a number of uses for the FD&C dyes and compares them with FD&C lakes.

AVOIDING COLOR TROUBLES

It is easy to obtain good results with very little trouble when using certified food coors, but a knowledge of conditions which cause poor coloring can help avoid many pitfalls. Most color troubles can be classified as precipitation, dull effects, spotting, or fading.

Precipitation

When color precipitates from a color solution or from liquid products, it is an indication that the solubility limit of the color has been overstepped. In a liquid product

Table 11
SUITABILITY OF DYES AND LAKES

Product	Water soluble colors	Lakes
Hard candy	Suitable	Not suitable
Striped candies (mints, etc.)	Suitable	Reduces color bleed
Compound coatings	Require dissolving in glycerine or propylene glycol	Easily incorporated
Sandwich cookie fillings	Suitable	Easily incorporated
Icings, marshmallow coatings	Suitable — Red #3 not light fast	For pink using Red #3 — the lake is more light stable
Sugar coatings — gum tablets	Suitable	Suitable
Inks, plastics, food can linings, films	Not suitable	Suitable
Cake mixes	Suitable — do not contribute much color to the dry mix	Suitable
Carbonated beverages	Suitable	Not suitable
Compressed tablets	Suitable — requires granulation operation	Very good
Solid fats and waxes	Not suitable	Suitable
Dry mix products — soft drink powders, gelatin desserts	Suitable	Used in combination with dyes
Chewing gum	Suitable	Suitable — leave colored cud

it may mean there is not enough solvent (such as water) present to keep the color in solution.

Precipitation may also be due to chemical action between the color and something else in the solution. For example, calcium ions (from hard water, gum arabic, etc.) cause formation of insoluble calcium salts of many certified dyes. Acids cause precipitation of FD&C Red #3. Low temperatures may also cause colors to precipitate, particularly from a concentrated solution. If there is a throwout due to chilling, the remedy is to move the container into a warm area, shaking occasionally until the solution is clear. Color solutions of ordinary 4-oz/gal strength are much less troublesome than concentrated solutions when a drop in temperature occurs.

Dull Effects

When certified food colors produce dull, unattractive effects instead of bright, pleasing shades, the trouble most frequently is that too much color has been used.

Exposure of colors to high temperatures also tends to dull shades as well as actually destroy or fade colors. As a rule in cooked products, color should be added at the latest possible stage in the process, and cooking should not be continued after the color is in the batch.

Spotting

Specking and spotting sometimes occur in the coloring of confectionery and bakery products. This is usually due to sediment in the liquid color or may be caused by failure to completely dissolve the color when making the solution. The remedy is to make certain the color is completely dissolved and the solution should be filtered to remove all sediment.

Specking and spotting may also occur when water solutions have been used to color products in which the fat content is too high. Dispersion is impossible because water is incompatible with fats. The remedy may be use of another color solvent, such as glycerine or propylene glycol. In extreme cases, a blending agent such as lecithin may be necessary.

Fading

When colors fade in a product, the cause can usually be traced to one of the following:

- Strong light (sunlight; ultraviolet)
- Metals (tin, zinc, aluminum, iron, etc.)
- Microorganisms (molds and bacteria)
- Excessive heating
- Chemical oxidizing and reducing agents
- Strong acids and alkalies

Action of Light

Light causes most colors to fade in some degree depending on relative light fastness and the nature and intensity of the light. Ultraviolet light, abundant in sunlight, is especially destructive.

Colored products should be protected as well as possible from sunlight. The ability of colors to withstand ordinary strong light, such as might be encountered in display windows, varies considerably with the nature of the colors. FD&C Green #3 and FD&C Blue #1 are reasonably light fast, while FD&C Blue #2 and FD&C Red #3 are not. The nature of the product may also decrease light fastness of colors, especially if there is some ingredient in the product which has a tendency to react chemically with the color (e.g., traces of metals). This tendency is greatly increased by exposure to light.

In selecting colors for products subject to considerable light exposure, it is safest to eliminate colors known to have poor light fastness, such as the two mentioned above. Moreover, any selected colors should be thoroughly tested in the product for light fastness under normal conditions of storage and exposure.

Action of Metals

Some metals are a prime cause of fading in certified colors. Contact of color solutions and colored products with such active metals as aluminum, zinc, tin, and iron should be avoided. This applies equally to dissolving, handling, and storage vessels, pipe lines, fittings, and product containers and closures. Technically, these metals act as reducing agents. Galvanized iron equipment must also be avoided, since this is zinc-coated iron.

Copper should not be used for dissolving, handling, or storage of color solutions because of the possibility of copper contamination. No trouble is experienced, however, when color is added to products that are practically neutral while being processed in copper kettles (as in candy manufacture), so long as there is no prolonged exposure to the metal. The action of copper, different from that of tin and zinc, has a darkening effect on FD&C Yellow #5. The reaction may be observed when products colored with this dye undergo progressive darkening on prolonged contact with copper.

Action of Microorganisms

Molds and certain kinds of bacteria (reducing bacteria) cause partial or complete fading of certified colors. This is not altogether unfortunate, for such microorganisms render most products unfit for use, and fading of the color often serves as a warning that they are growing and multiplying in the product. Where molds are not present in sufficient masses to be apparent to the eye, their presence can be ascertained by examination under a microscope.

Reducing bacteria may be introduced into a product by way of the processing water. Water that contains such bacteria can be presumed to be contaminated and should not be used.

All vessels and accessories for dissolving, handling, and storage of color solutions should be thoroughly cleaned before use. It should also be remembered that water solutions of color which are to be stored more than 2 days before use must contain a preservative to prevent growth of molds.

Excessive Heating

Certified colors differ in their ability to withstand heat, but temperatures prevailing in most cooking processes are not ideal for colors as a general rule. The safest procedure where color must be mixed with hot materials is to add the color at the latest possible moment in the process, preferably after the batch has cooled considerably below cooking temperature.

Chemical Oxidizing and Reducing Agents

Oxidizing agents such as ozone, chlorine, and hypochlorites will discharge the color of most dyes. Hypochlorite solutions like those used for bleaching and disinfecting will quickly decolorize water solutions of FD&C Red #40, FD&C Yellow #5, and FD&C Yellow #6. FD&C Green #3, FD&C Blue #1, and FD&C Red #3 have a limited resistance to this oxidizing agent so long as the solution is not acid, but these colors are immediately decolorized on acidifying, even with weak acids, such as acetic. Colors, color solutions, and colored products should always be kept away from bleaching agents. Certified colors are not easily faded by moderate chemical reducing agents such as sulfur dioxide, except in the presence of metals.

Strong Acids and Alkalies

The problem of coloring strongly acid materials is not often encountered, but occasionally color is desired in certain decidedly alkaline products. Fastness of certified colors to acids and alkalies varies depending on other ingredients in a product. In an acid or alkaline medium, the activity of other fading agencies, notably metals, may be greatly increased. Quite apart from fading or color destruction, it should be remembered that shades of many dyes vary considerably depending upon whether they are in acid or alkaline media.

Use of modern process equipment and proper water treatment has eliminated many of the problems associated with the use of color additives. These difficulties have been minimized to a great extent by increased knowledge of the colors. Table12 summarizes the more common coloring problems with dyes and the causes.

FD&C LAKES

The FD&C lakes, which were admitted to the approved list of certified color additives in 1959, comprise an important class of color additives. Their use is growing at a very rapid rate. The Color Regulations define FD&C lakes as "Extension on a substratum of alumina, of a salt prepared from one of the water soluble straight colors by combining such color with the basic radical aluminum or calcium." The alumina hydrate or aluminum hydroxide substratum is insoluble so what is produced is an insoluble form of the dye — a pigment. Dyes color by being dissolved in a solvent; pigments color by dispersion. Previous to the admission of lakes, insoluble forms of the dyes were made by adsorbing them on insoluble materials such as startch, cellulose, flour, etc. These can be highly colored materials, but are of low coloring power and generally perform poorly as pigments.

The pigmentary properies of the shade of alumina hydrate lakes depend a great deal upon the preparation of the alumina hydrate and the processing variables (tempera-

Table 12
PROBLEMS WITH DYES

Problem	Cause
Precipitation from color solution or colored liquid food	Exceeded solubility limit
	Insufficient solvent
	Chemical reaction
	Low temperatures, especially for concentrated color solution
Dulling effects instead of bright pleasing shades	Excessive color
	Exposure to high temperatures
Specking and spotting during coloring of bakery and confectionery products	Color not completely dissolved while making a solution
	Employed liquid color containing sediment
	Attempted dispersion in an aqueous color solution in products containing excessive fat
Fading due to light	Colored products not protected from sunlight
Fading due to metals	Color solutions or colored products were in contact with certain metals (zinc, tin, aluminum, etc.) during dissolving, handling, or storing
Fading due to microorganisms	Color-preparing facilities not thoroughly cleaned to avoid contaminating reducing organisms
Fading due to excessive heat	Processing temperature too high
Fading due to oxidizing and reducing agents	Contacted color with oxidizers such as ozone or hypochlorites or reducers, such as SO_2 and ascorbic acid
Fading due to strong acids or alkalis	Presence of such strong chemicals during the coloring of certain foods
Fading due to retorting with protein material	Color is unstable under these conditions
Poor shelf life with colored canned carbonated beverages	Used in excessive amount of certified azo-type dye

tures, pH, etc.) must be carefully controlled. Even the agitation, the concentration, and rate of addition of the reactants bear on the physical properties of the lake. Lake-making is steeped in artistry, but depends upon careful process control to insure uniformity.

The physical properties of water soluble colors (dyes) are of small significance in most products. There is, for example, a choice between granular or powdered forms of dye. Dissolved, both forms produce equivalent solutions. Dyes from different manufacturers are interchangeable for most uses. This is not always true with lakes, even though they may be in the same pure dye content. Shades of difference can exist from manufacturer to manufacturer.

The tinctorial strength of a water-soluble color is proportional to the pure dye content. Thus, two units of a 45%-pure dye color are equivalent in color strength and shade to one unit of a 90%-pure dye color. This is not true with the lakes. One unit of a 24%-pure dye lake is not equivalent in strength of shade to two units of a 12%-pure dye-color. Yellow #5 Lake, for example, at 5%-pure dye is lemon-yellow; at 15%-pure dye is egg-yellow; and at 25%-pure dye is orange-yellow.

Having aluminum hydroxide as substratum, the lakes are insoluble in nearly all solvents. When suspended in water there is a slight bleed of dye into the water, but for most food applications, this is of no consequence. The lakes are stable in the pH range of 3.5 to 9.5. Outside this range, the substratum breaks down, releasing the dye.

The color regulations classify the FD&C lakes as straight colors, subject to certification. Also, they must be made from dyes which have been previously certified. In most cases, the color regulations specify a minimum of 85%-pure dye for primary dyes. However, most lots will be in the range of 90 to 93% pure dye. The FD&C Lakes do not have a specified minimum dye content and range from 10 to 40% in pure dye.

By their very nature, lakes are suitable for coloring products which cannot tolerate water and products where use of water is undesirable.

The FD&C dyes are water soluble and are insoluble in organic solvents, fats, and oils. Often, the coloring of oils can be done conveniently with the lakes. However, in spite of their great transparency as pigments, a clear, colored material, free of opacity, is generally not possible.

APPLICATIONS FOR THE FD&C LAKES

When the FD&C lakes were added to the list of approved colors in 1959, many of today's color applications could not have been anticipated. New varieties of lakes have been developed to meet the constant demand for new product applications. Because lakes are absorption compounds and relatively inert, they are extremely adaptable. New uses are being found almost daily. Through good dispersion, an almost infinite variety of products can be colored.

Bakery products such as icing, sandwich cookie fillings, cake, and doughnut mixes can be colored easily with FD&C lakes. Normally, good mixing, such as that obtained with a Hobart Planetary mixer, will disperse the lakes sufficiently. It is better, though not always necessary, to predisperse the lakes in an hydrogenated vegetable oil. (See following section for additional information.) Hard-fat coatings (or summer coatings) and hard-candy wafers which were difficult to color with the water soluble dyes can have bright, vivid, light-resistant shades using the lakes.

Chewing gum is another product where lakes are a true asset. The color remains in the cud and retains its bright, attractive appearance without staining the mouth. Dry mix products such as beverage and dessert powders and compressed tablets show excellent colors using lakes. Often, in reconstituted products like dry drink mixes, a lake/dye combination works the best. The lake colors the dry mix and the dye colors the finished hydrated product. In addition, the lakes, normally insoluble, release the water soluble dye due to the low pH of the reconstituted product. They then contribute additional color to the finished drink without clouding the solution. In gelatin dessert powders, lakes can not only be used to color the dry mix, but also the finished hydrated product. In this application, the opacity normally associated with lakes is not a concern due to the fact that the refractive index of the lake is nearly identical to that of the gelatin, resulting in a transparent, uncloudy appearance.

In canned pet foods, lakes have proven useful because of their inertness to high temperatures and protein interaction (both causes of fading in FD&C water soluble dyes). Also, where moisture contents are low as in semimoist pet foods, the lake pigments perform well.

Traditionally, sugar-coated tablets have been colored with syrup solutions containing varying amounts of the water soluble dyes. Using this method, it was not uncommon for as many as 50 to 60 coats to be applied to obtain the desired shade. Using the lakes, with or without Titanium Dioxide, attractive shades can be achieved with as few as 8 to 12 coats.

Pigment Dispersion

Dispersion of pigments is an operation not generally well known in the food industry. Dispersion is merely the distribution of the pigment throughout the material to be colored. It is generally considered that pigment dispersion is —

- Breaking up of the agglomerates

- Wetting and coating each particle with the vehicle in cases where the material to be colored is a liquid (vehicle)

To accomplish this, energy is required and high shear forces must be applied. Dispersion in such fields as inks and paints is referred to as grinding, though it is doubtful if much particle fracture is involved. In these industries, very fine grinds in the vehicles are required and ball mills, pebble mills, and colloid mills find use. For most food and pharmaceutical applications, high shear mixers such as the Kady Mill, the Eppenbach Homomixer, Premier Dispersator, Daysolver and Cowles Dissolver are adequate for lake dispersion in liquids.

Many lake applications demand dispersion in dry materials. In crystalline materials, simple mixing sets up sufficient attritive forces to obtain dispersion. In noncrystalline powders, such as starch or flour, it is often necessary to prepare a concentrated premix of 5 to 10 parts powder to one part of lake, followed by grinding through a hammer mill. This concentrated premix is then readily incorporated into the final mix.

Recent investigation into the relationship of the lake particle size and tinctorial strength have led to the development of fine-ground lakes. Fine-ground lakes have an average particle size of 1 μm as compared with about 5 μm for the conventionally ground product. In some instances, considerable increase in tinctorial strength is obtained by fine grinding.

COLOR SPECIFICATION AND QUALITY CONTROL

It should be remembered that color manufacturers carefully assay their product before submitting it for certification. Refusal of certification means loss of the certification fee of 15 cents/lb. Likewise, the Color Certification Laboratory of the FDA assays the material before issuing a certification certificate. Thus, it is usually unnecessary for the color user to repeat the assays dealing with the certification specifications (lead, sodium chloride, etc.). It is prudent, however, to determine that the proper primary colors have been supplied or that the blend is of the proper shade; i.e., proper composition. Visual comparison is quite helpful in checking the shade of colors and spectrophotometric analysis indicates color strength. These comparisons should be made against standard samples in similar dilute solutions. Solution concentrations suitable for both visual and spectrophotometric analysis (in 1 cm cells) are as follows for the various dyes:

FD&C Yellow #5	10 mg/ℓ
FD&C Yellow #6	10 mg/ℓ
FD&C Blue #1	4 mg/ℓ
FD&C Green #3	4 mg/ℓ
FD&C Red #3	7 mg/ℓ
FD&C Blue #2	10 mg/ℓ
FD&C Red #40	10 mg/ℓ

These solutions should be buffered to avoid any shade differences due to pH.

Quality control checking of lakes is more complex. Dispersions of test and standard samples in various materials such as shellac can be prepared and drawdowns compared or test colorings of small laboratory amounts of the finished product can be made. For such comparison and for quality control of finished products, many color differences (tri-stimulus comparators) are available. These have proved only mildly successful, but in the future they should become more helpful.

Knowledge of color additive technology in the food industry has grown significantly in recent years and fewer application problems are being referred to the color manufacturers. This is a salutary occurrence; however, this same period has been a time of industry confusion regarding the regulatory status of color additives. Hopefully, this too will be overcome shortly as the result of more convincing scientific data.

NONABSORBABLE POLYMERIC FOOD COLORS

In 1972, a new company, Dynapol (Palo Alto, California) was formed to pursue a concept for developing safer food additives. The concept was that additives of sufficiently large molecular size (polymers) would not pass through the intestinal wall and would therefore not pass into the blood stream. Consequently, the large molecule would not contact the usual target organs — liver, kidneys, etc.[15] However, in order for such additives to be nonabsorbable, they would have to maintain their molecular integrity and be resistant to the enzymatic processes in the gastrointestinal tract. They should not be subject to metabolic breakdown into fragments of low molecular weight.

Additionally, they should maintain their molecular integrity during the stresses of food processing and service history including light and heat exposure. Dynapol chose as their target food additives — colors, antioxidants, and sweeteners.

Acceptance Criteria

In approaching the formidable task of developing the nonabsorbable colors, the following criteria were set:

- Low (less than 1%) absorption through the intestinal tract
- Good stability toward heat, light, and food processing environments
- Suitable tinctorial strength
- Good functional and application properties
- Compatibility with food systems
- Blendability so as to produce attractive shades throughout the spectrum
- Cost in use below that of uncertified color additives (natural colors)

Initial plans were to develop dyes similar in shade to the current certified colors. It was felt that four such colors would be desirable — a magenta red similar to FD&C Red #2, a yellow similar to FD&C Yellow #5, a blue similar to FD&C Blue #1, and an orange similar to FD&C Yellow #6.

Early Work

Early work aimed at leashing the FD&C azo dyes to stable polymer backbones was successful and produced water-soluble polymeric dyes with excellent physical and chemical properties and spectral values similar to their monomeric counterparts. However, when these polymer-leashed azo chromophores were exposed to the gut microflora, it was found that the azo linkage was reductively cleaved, producing an absorbable low molecular weight moiety. This work, however, showed that the bond linking the chromophore to the polymer backbone was extremely stable.[15]

Search for Stable Chromophores

It was evident from this work that chromophores more stable than the azo type currently used in food dyes were necessary. Consequently, a broad range of dye classes were screened for stable candidates with initial emphasis on red chromophores having shade characteristics similar to FD&C Red #2. Success was achieved with the development in late 1976 of two red polymeric dyes, Poly R^{TM}-478 and Poly R^{TM}-481, using anthrapyridone chromophores. Not long thereafter, a yellow polymeric dye with spectral characteristic similar to FD&C Yellow #5, Poly Y^{TM}-607, was developed utilizing a nitroaniline chromophore.

Poly R^{TM}-478, the first nonabsorbable red dye developed by Dynapol, was a magenta red considerably more bluish than FD&C Red #2. Poly R^{TM}-481, a result of careful molecular engineering, was less bluish than Poly R^{TM}-478, but slightly bluer than FD&C Red #2. Poly Y^{TM}-607 is similar in shade to FD&C Yellow #5, being slightly less greenish. The physical, chemical, biological, and applications in food systems was reviewed by Furia in 1977.[22]

Nonabsorbability Established

The nonabsorbability of these colors was established by feeding ^{14}C-radiolabeled polymeric dyes to rodents. Urine, feces, blood, and expired carbon dioxide were collected and the level of radioactivity measured to determine the fate of the polymeric dye. Table 13 compares absorption vs. ^{14}C-radiolabeled FD&C dyes fed to rats. Subsequent studies with radiolabeled polymeric dyes fed to other species produced absorption levels of the same magnitude.

Stability Confirmed

The molecular integrity through the gastrointestinal tract was confirmed by the absorption studies. The next hurdle was that the colors in food systems did not degrade prior to ingestion. The Polydyes were tested in a number of model systems. Table 14 illustrates the thermal stability in solutions at various pHs and as a dry powder.

Particular emphasis was placed on carbonated beverage and still-drink systems containing high levels of ascorbic acid, which is highly destructive toward the FD&C colors. Table 15 shows the comparative light stability following exposure to 2200 Langleys of sunlight (about 5 bright summer days). It is worth noting that all the products containing the Polydyes were found after exposure to stress to be negative in the Ames mutagenicity test.

Blendability

An important criterion for qualification as a suitable polymer color is the ability to produce bright, attractive shades in combination with other Polydyes. This property was investigated by developing matching shades for a large number of standard blends of FD&C colors; some of the target shades contained FD&C Red #2. Table 16 lists the combination and concentration of Poly R^{TM}-481 and Poly Y^{TM}-607 needed to match these standard blends in clear solutions. Since no blue Polydye is yet available, FD&C Blue #1 was employed with the Polydyes in blends requiring a blue dye.

Tinctorial Strength

As might be expected, because of the lower percentage chromophore, the Polydyes have lower tinctorial strengths. When used in blends, the strength of the Polydyes vary from 1:1 to 1:3 relative to the FD&C dyes.

Termination of Work on Poly R^{TM}-478

Poly R^{TM}-478 was the first polymeric dye developed by Dynapol which met acceptance criteria as a nonabsorbable color. The target shade for a red Polydye was, of course, FD&C Red #2. And while Poly R^{TM}-478 was a bright and attractive color, it was considerably bluer in shade than FD&C Red #2.

Table 13
COMPARATIVE ABSORPTION OF ^{14}C-RADIOLABELED POLYDYES AND ^{14}C-RADIOLABELED FD&C DYES

Dye	Absorbed[a] (%)
Poly R^{TM}-478	0.33
Poly R^{TM}-481	0.22
Poly Y^{TM}-607	0.17
FD&C Yellow #6[b]	47.4
FD&C Red #2[b]	10.7
FD&C Yellow #59[b]	23.8

[a] Of oral dose

[b] Intact dye and metabolites.

Table 14
THERMAL STABILITY OF POLYMERIC DYES

Dye system	Temperature (°C)	Time	Weight % chromophore loss[a]	
			Poly Y^{TM}-607	Poly R^{TM}-481
pH 2.5 or 7.0[b]	22	1 year	0.3	
pH 2.5 or 7.0[b]	45	1 month	0.3	
pH 2.5 or 7.0[b]	59	30 min	None	None
Neat powder	200	20 min	<0.2	0.3

[a] This level of chromophore loss has no effect on observe color properties.

[b] 0.05 *M* phosphate buffer.

From Furia, T. E., *Food Technol.*, 31(5), 34, 1977. With permission.

However, with the banning of FD&C Red #2 and the questioning of the safety of FD&C Red #40, the need for a stable red food color became intense. While confident that other polymeric red dyes closer in shade to FD&C Red #2 could be developed, it was decided to move into the limited production phase and to initiate safety studies on Poly ^{TM}-478. These actions would allow the shake down of production operations as well as obtaining valuable data confirming the safety of this type of color. Consequently, short-term studies were completed, suitable color was prepared under manufacturing conditions, and the chronic studies initiated.

Development of Poly R^{TM}-481 followed soon thereafter. This color employed the same polymer backbone as Poly R^{TM}-478 and had a structurally similar chromophore. It was less bluish than Poly R^{TM}-478 and only slightly bluer than FD&C Red #2. It was quickly moved into the production/testing phase.

After a number of months of evaluating production and economics issues, functional and blending properties and the safety aspects of Poly R^{TM}-481, it was decided that Poly R^{TM}-478 was coloristically redundant and was eliminated from the program in favor of Poly R^{TM}-481. However, much valuable information had been obtained by proceeding with Poly R^{TM}-478.

Table 15
COMPARATIVE SUNLIGHT STABILITY OF POLYMERIC DYES AND CERTIFIED AZO DYES

	Fade at λ max after 2200 Langleys exposure (%)		
Dye solutions	Phosphate pH 2.5	Carbonate beverage pH 3.2	Citrate and ascorbate[a] pH 3.0
Poly Y™-607	0	2	30
FD&C Yellow 5	26	41	75
Poly R™-481	15	1	10
FD&C Red 40	22	28	89
Amaranth[b]	20	36	100

[a] 350 ppm ascorbic acid.
[b] Formally FD&C Red #2.

From Bellanca, N. and Leonard, W., Jr., *Current Aspects of Food Colorants,* Furia, T. E., Ed., CRC Press, Cleveland, 1977, 49. With permission.

Table 16
TYPICAL BLENDS FOR CLEAR AND TRANSLUCENT BEVERAGE SHADES

		Concentration of component dyes in finished beverage (ppm)			
Target shade	Type	Poly R™-481	Poly Y™-607	FD&C Blue 1	Total
Cherry	I	180			180
	II	81	125		206
Strawberry	I	155	5		160
	II	131	26		157
	III	54	147		201
Raspberry	I	200			200
	II	119	14		133
Orange	I	20	40		60
	II	9	109		118
	III	12	102		114
	IV	9	89		98
Lemon			31		31
Lime			23	1.5	24.5
Emerald			78	19	97
Grape	I	80	11	4	95
	II	188	7	5	200
	III	35	16	3	54
Cola	I	49	39	4	92

Progress on Safety Studies

In spite of the presumption of safety due to their nonabsorbabilty, the Polydyes will be subjected to the full-range of safety studies required for conventional additives plus a number of additional adsorption studies. Protocols established for these safety studies were submitted to FDA prior to initiation of testing. Table 17 summarizes the results of safety tests completed to date on Poly R™-481 and Poly Y™-607.

Table 17
SUMMARY OF SAFETY EVALUATIONS TO DATE FOR POLYMERIC DYES

Tests	Poly Y™-607	Poly R™-481
Acute oral LD_{50}		
Mice	>10 g/kg	>10 g/kg
Rats	>10 g/kg	>10 g/kg
Dogs	>10 g/kg	>10 g/kg
2-week maximum tolerated dietary dose, rats	>5%	>5%
13-week dietary study, rats	>5%	
Skin and eye irritation,[a] rabbits	Nonirritant	Nonirritant
Ames mutagenicity, *Salmonella*/microsome	Nonmutagenic	Nonmutagenic
Intestinal absorption, % dosed radioactivity[b]		
Rat	0.18	0.22
Mouse	0.23	0.33
Guinea pig	0.24	0.06
Rabbit	0.51	0.22

[a] 16 CFR, Part 1500.41.
[b] ^{14}C-radiolabeled polymeric dyes.

The limiting time factor on submission of color additive petitions is the completion date of the most lengthy of the tests, the chronic studies, i.e., lifetime studies in rats and mice. The expected completion date for these studies is during the first quarter of 1981.

Lakes of Polymeric Dyes

The Polydyes form lakes with the alumina hydrate substratum. The lakes produced show considerably less bleed in aqueous systems which will be an advantage over the FD&C lakes where migration is a problem. Alumina lakes with unusually high loadings are possible with the Polydyes and since they are strongly bound, the manufacture of lakes from solutions of dye blends is possible.

Application Properties of the Nonabsorbable Colors

Table 18 lists the solubilities of Poly Y^{TM}-607 and Poly R^{TM}-481 in a variety of solvents. Concentrated aqueous solutions can become quite viscous; however, 3% solutions are easily handled.

The polymeric dyes have been shown to be applicable to a large number of food products. Experience thus far indicates that the handling characteristics are sufficiently similar to those of the water-soluble food colors currently utilized by the food industry to permit users to adopt them without major changes in traditional unit operations. However, since these colors do not migrate across membranes, foods which are traditionally colored by diffusion processes must be handled by alternate means. A wide range of food products — carbonated beverages, hard candy, ice cream, gelatin desserts — have been successfully colored. The Polydyes are compatible with most food additives; however, care must be exercised in respect to limits to the concentration of some cations — calcium ion in excess of 1200 ppm may cause haziness. Two properties of the Polydyes which will be helpful in a number of food applications are the excellent heat and light stability and lack of can corrosion for canned carbonated beverages.

Table 18
SOLUBILITY OF POLYDYES

Solvents	Solubility, g/100 mℓ	
	Poly Y^{TM}-607	Poly R^{TM}-481
Water	>20	>10
Glycerine	>20	>10
Propylene glycol	0.03	>10
25% Ethanol	>10	>10
50% Ethanol	>10	>10
Vegetable oil	Nil	Nil

[a] By gel permeation chromatography; peak MW relative to polystyrene sulfonate.
[b] To a concentration of 3%.

Early indications are that the Polydyes will possess a high level of safety, will effectively color a broad variety of food systems, and confer improved stability in use and during storage. It is hoped that soon after 1981 a new color system of nonabsorbable dyes will be approved and available to the U.S. food industry.

REFERENCES

1. **Hesse, B. C.,** Coal-tar colors used in food products, Bureau of Chemistry, U.S. Department of Agriculture Bull. 147, 1912.
2. Public Law No. 717, 75th U.S. Congress, 1938.
3. U.S. Supreme Court, 358 U.S. 153, Dec. 15, 1958.
4. Public Law 86-618, 86th U.S. Congress, 1960.
5. *TGA* vs. *Finch,* 63 Civ. 3349, Southern District, New York Nov. 26, 1969.
6. **Andrianova, M. M.,** Carcinogenic properties of red food dyes Amaranth, Ponceau SX and Ponceau 4R, *Vopr. Pitan.,* 29(5), 61, 1970.
7. **Shtenberg, A. I. and Gavrelenko, E. V.,** Influence of the food dye Amaranth upon the reproductive function and development of progeny in tests on albino rats, *Volpr. Pitanya,* 29(2), 66, 1970.
8. **Collins, T. F. X. and McLaughlin, J.,** Teratology studies on food colorings Part I Embryotoxicity of Amaranth (FD&C Red No. 2) in rats, *Food Cosmet. Toxicol.,* 10, 619, 1972.
9. **Shetenberg, A. I. and Gavrelenko, E. V.,** Gonadotoxic and embryotoxic action of the food dye — Amaranth, *Vopr. Pitanya,* 5, 28, 1972.
10. **Khera, K. S., Przybylski, W., and McKinley, W. P.,** Implantation and embryonic survival in rats treated with Amaranth during gestation, *Food Cosmet. Toxicol.,* 12, 507, 1974.
11. *CCMA* vs. *Mathews,* U. S. Court of Appeals, District of Columbia No. 76-1120.
12. *Fed. Regist.,* 41(186), 41852, 1976.
13. Report of the Committee on Interstate and Foreign Commerce of the House of Representatives on H. R. 7624, House Report No. 1761.
14. Report of the E.E.C. Scientific Committee for Food 2635/VI/75-E, 1975.
15. **Brown, J. P.,** Reduction of polymeric azo dyes by cell suspensions of enteric bacteria, Abstr. Annu. Meeting, Am. Soc. Microbiology, page 23, Abstract No. 172, 1976.
16. **Bellanca, N. and Leonard, W., Jr.,** Nonabsorbable polymeric dyes for food, in *Current Aspects of Food Colorants,* Furia, T. E., Ed., CRC Press, Cleveland, 1977, 49.
17. Certified Color Industry Committee, Guidelines for good manufacturing practice: use of certified FD&C colors in food, *Food Technol.,* 22(8), 14, 1968.

18. **Dean, G. E.,** A shelf life factor in canned soft drinks, *Am. Soft Drink J.,* February, pages 36 and 62, 1966.
19. **Martin, L. E.,** Antioxidants and sequestering agents to increase product-container shelf life of carbonated beverages, *Am. Soft Drink J.,* March, 1965.
20. **Clarkson, R.,** Tablet Coating, Drug & Cosmetic Industry, New York, New York.
21. **Little, A. and Mitchell, K. A.,** *Tablet Making,* 2nd ed., Northern Publishing, Liverpool, 1963.
22. **Furia, T. E.,** Nonabsorbable polymeric food colors, *Food Technol.,* 31(5), 34, 1977.

INDEX

A

Abies
alba, 334
balsamea, 334
mayriana, 334
sachalinesis, 334
sibirica, 334
Absolute, definition, 319
Acacia
calechu, 333
decurrens, 334
farnesiana, 333
Acer spicatum, 334
Acesulfames, 219—221
Acetal, 256, 319
Acetaldehyde, 256, 313
Acetaldehyde benzyl methoxyethyl acetal, 256
Acetaldehyde butyl phenethyl acetal, 256
Acetaldehyde diethyl acetal, 256
Acetaldehyde dimethyl acetal, 267
Acetaldehyde phenethyl prupyl acetal, 256
Acetal R, 256
Acetanisol, 256
Acetate C-8, 294
Acetate C-9, 293
Acetate C-10, 265
Acetate C-12, 283
Acetic acid, 256
Acetic ether, 269
Acetoin, 256
Acetone, 256, 327
Acetophenone, 256
Acetyl butyryl, 276
Acetyl *o*-cresol, 304
Acetyl *p*-cresol, 304
3-Acetyl-2,5 dimethylfuran, 256
2-Acetyl-3,5 (and 6) dimethyl-pyrazine, 256
2-Acetyl-3-ethyl-1, 4 diazine, 256
2-Acetyl-3-ethylpyrazine, 256
Acetyl eugenol, 273
Acetyl furan, 274
Acetyl isobutyryl, 289
Acetyl isoeugenol, 273, 281
Acetyl isovaleryl, 287
Acetyl methyl carbinol, 256
Acetyl nonyryl, 305
Acetyl pelargonyl, 305
3-Acetyl propionic acid, 283
Acetyl propionyl, 295
2-Acetyl pyrazine, 256
Acetyl pyridine, 315
2-Acetyl pyridine, 256
3-Acetyl pyridine, 256
Acetyl pyrrole, 291
2-Acetyl thiazole, 256
p-Acetyl toluene, 285
Acetyl valeryl, 275
Acetyl vanilla, 306
Achillea
millefolium, 335
moschata, 334
Acid fuchsine, 351, 354
Acid fungal protease, 75
Acid Yellow, 351, 356
Acidulants, dry beverage base, 148
Acidulants, salad dressing, 155
Aconitic acid, 256
Active oxygen method (AOM), 18
Adehyde C-16, 313
Adiantum capillus-teneris, 334
Adipic acid, 150, 153, 256
Adjuzyme®, 73
Aerobacter aerogenes, 73, 84
Agar, in frozen desserts, 160
beta-Alanine, 256
Albizzia lophanta, 113
Alcohol, definition, 319
Alcohol C-6, 277
Alcohol C-7, 276
Alcohol C-8, 294
Alcohol C-9, 293
Alcohol C-10, 265
Alcohol C-12, 283
Aldehyde C-6, 276
Aldehyde C-7, 275
Aldehyde C-8, 294
Aldehyde C-9, 292
Aldehyde C-10, 265
Aldehyde C-11, 305
Aldehyde C-12 lauric, 283
Aldehyde C-12 MNA, 291
Aldehyde C-14, 305, 313
Aldehyde C-16, 271, 313
Aldehyde C-18, 292
Aldehydic oils, 244
Algins, in frozen desserts, 160
Alkaline protease, 75
Allium
cepa, 232
sativum, 232
Allspice, 232
Allura red, 351, 353
Allyl acetic acid, 295
p-Allylanisole, 268
Allyl anthranilate, 256
Allyl butyrate, 257
Allyl caproate, 257, 315, 316
Allyl caprylate, 257
Allyl cinnamate, 257
Allyl cyclohexaneacetate, 257
Allyl cyclohexanebutyrate, 257
Allyl cyclohexanehexanoate, 257
Allyl cyclohexanepropionate, 257
Allyl 3-cyclohexylpropionate, 257
Allyl cyclohexanevalerate, 257
Allyl disulfide, 257, 265
Allyl 2-ethyl butyrate, 257

Allyl 2-furoate, 257
4-Allyl guaiacol, 273
Allyl heptanoate, 257
Allyl 2,4-hexadienoate, 257
Allyl hexanoate, 257, 315
Allyl alpha-ionone, 257
Allyl iso-thiocyanate, 257
Allyl iso-valerate, 257
Allyl mercaptan, 257
4-Allyl-2-methoxyphenol, 273
4-Allyl-2-metxy-phenyl formate, 273
Allyl methyl disulfide, 257
Allyl methyl trisulfide, 257
Allyl nonanoate, 257
Allyl octanoate, 257
Allyl phenoxy acetate, 257
Allyl phenylacetate, 257
Allyl propionate, 257
Allyl sorbate, 257
Allyl sulfhydrate, 257
Allyl sulfide, 257, 265
Allyl thiopropionate, 257
Allyl tiglate, 258
Allyl T-2-methyl-2-butenoate, 258
Allyl trisulfide, 265
Allyl 10-undecenoate, 258
Allyl undecylenate, 258
4 Allyl veratrole, 273
Aloe, 333, 251
  africana, 333
  barbadensis, 333
  ferox, 333
  perryi, 333
  spicata, 333
Alpinia galanga, 334
Althea officinalis, 333
Althea root, 333
4-Ally-2-methoxy phenyl, 273
Amaranth, 340, 350, 351, 353, 361
Ambrettolide, 276
American® rennet, 73
Amino acetic acid, 275
Aminoacylase, 79
d-l-(3 Amino-3-carboxy-propyl) dimethyl sulfonium chloride, 306
2 Aminoglutaric acid, 275
2-Amino iso caproic acid, 283
2-Aminoisovaleric acid, 306
2-Amino-3-mercaptopropionic acid, 265
2-Amino β-methyl valeric acid, 281
Aminopeptidase, see Exopeptidase
2-Amino propionic acid, 256
Ammonium iso valerate, 258
Ammonium sulfide, 258
2-Amyl-5, 258
Amyl acetate, 258
Amyl alcohol, 258
α-Amylase
  baking industry, 85—86
  brewing, 86—88
  classification, 81—85
  in cereal products, 68
  in dairy products, 66
  marketed in U.S., 73—77
β-Amylase
  in cereal, 68
  industrial uses, 81—88
  marketed in U.S., 77
Amylase lipase, 75, 76
Amylases
  classification, 60
  commercial uses, 82
    baking, 85—86
    brewing, 86—88
    dextrose production, 82—84
    minor uses, 88
    syrup production, 81—82, 84
  fungal, 81, 82
  in vegetables and fruit, 69
  marketed in U.S., 73, 74, 75
  thermo stability, 86
Amyl butyrate, 258
Amyl caproate, 258
Amyl caprylate, 258
alpha-Amyl cinnamaldehyde, 258
alpha-Amyl cinnamaldehyde dimethyl acetal, 258
alpha Amyl cinnamyl acetate, 258
alpha-Amyl cinnamyl alcohol, 258
alpha-Amyl cinnamyl formate, 258
alpha-Amyl cinnamyl isovalerale, 258
Amyl formate, 258
Amyl-2-furoate, 258
Amyl heptanoate, 258
Amyl hexanoate, 258
Amyl octanoate, 258
Amyloglucosidase, see Glucoamylase
Amyloglucosidase Novo 150AMG 150L, 75
Amyl vinyl carbinol, 294, 315
Amyris balsamifeva, 333
Amyris, 333
Anethol, 238, 314
Anethole, 258
Anethum
  graveolens, 232, 334
  sowa, 334
Angelicalactone, 295
Angola weed, 333
Animal extracts, 251—253
Anise, 232, 233, 314
Anisic aldehyde, 284
Anisole, 258
Anisyl acetate, 258
Anisyl acetone, 285
Anisyl alcohol, 258
Anisyl butyrate, 259
Anisyl formate, 259
Anisyl phenylacetate, 259
Anisyl propionate, 259
Antioxidants
  addition limits, 51
  commercial
    animal fat, 17—18
    essential oils, 21
    fish products, 21

food products with high fat content, 18—20
low fat foods, 20—21
meat products, 21
vegetable oils, 26—27
comparative activity, 41
definition, 319
food approved, 25, 30, 32, 35—36
from natural sources, 36—39
lipid oxidation, 21—24
measurement of oxidative stability
AOM, 48
free fatty acids, 48—49
objective methods, 49—50
oven storage test, 49
oxygen bomb test, 49
quantitative measurements, 46—47
separation techniques, 44—46
shelf storage test, 49
2-thiobarbituric Acid Test (TBA), 48
mechanism of action, 24—25
methods of determining quantities in food, 44—47
new, 40—42
oxidation, 16—17
regulations, 50—51
selection, 17—21
synergism with chelates, 42—44
technology, 13
AP
antioxidant activity, 41
Apis mellifera, 333
Apium graveolens, 232
Apple, 313, 314
Aristolochia serpentaria, 335
Arnica
cordifolia, 333
fulgens, 333
montana, 333
sororia, 333
pallens, 333
species, 333
Arnica flowers, 333
Aroma, definition, 319
Artemisia, 333
Arthrobacter, 97
Artichoke leaves, 333
Artificial flavor, definition, 319
Artificial mustard oil, 257
Asarum canadense, 335
Ascorbic acid, 26—27, 259
Ascorbic acid oxidase, 69
Ascorbyl palmitate, 26—27
Aspartame
discovery, 193—194
potency relative to sucrose, 194, 197
production, 195
properties, 194—196
stability, 197
structural considerations, 199—200, 201
Aspartyl-phenylalanine diketopiperazine, 198
Aspergillus
flavusoryzae, 76
niger, 73—77, 91, 95, 107
oryzae, 73—77, 99, 101
species, 73, 74
Asperula odorata, 335
Asperzyme®, 73
A.S.T.A., definition, 319
A.S.T.M., definition, 319
Atherosclerosis, 16
Autooxidation
definition, 320
of lipids, 21—23

B

Bacillus
coagulans, 91
licheniformis, 75, 83
subtilis, 73—77, 83
Bactamyl®, 74
Bacterial amylase F, 73
Bacterial amylase Novo®, 75
Bacterial proteases, 103
Baked goods
low calories
compensating ingredients, 161—162
prototype formulas, 162
sugar replacement, 161
Baking
amylases, 85—86
lipoxidase, 107, 108
nonnutritive sweetners, 144
proteases, 101—102
vanilla, 253
BAN 120L, 75
BAN 360S, 75
Banana, 313, 314
Barosma
betulina, 333
crenulata, 333
serratifolia, 333
Basil, 231
Baumé, definition, 320
Bay, 231
Bé, see Baumé
Beeswax, white, 333
Benzaldehyde, 238, 259, 314
Benzaldehyde dimethyl acetal, 259
Benzaldehyde glyceryl acetal, 259
Benzaldehyde propyleneglycol acetal, 259
Benzenethiol, 259
Benzilidene acetone, 297
Benzodihydropyrone, 266
2-Benzofurancarboxaldehyde, 259
Benzoic acid, 259
Benzoin, 259
Benzoin resin, 333
Benzophenane, 259
Benzo (b) pyridine, 301
2,3-Benzopyrrole, 279
Benzothiazole, 259
Benzylacetaldehyde, 298

Benzyl acetate, 259
Benzyl acetoacetate, 259
Benzyl alcohol, 259
Benzyl benzoate, 259
Benzyl buthylether, 259
Benzyl butyrate, 259
Benzyl cinnamate, 259
Benzyl dimethyl carbinyl acetate, 267
Benzyl dimethyl carbinyl butyrate, 267
Benzyl dimethyl carbinyl formate, 267
Benzyl-2,3-dimethyl crotanate, 259
Benzyl dipropyl ketone, 260
Benzyl disulfide, 259
Benzyl ethyl ether, 259
Benzyl formate, 260
3-Benzyl-4-heptanone, 260
Benzyl-*o*-hydroxybenzoate, 260
Benzyl isobutyl ketone, 290
Benzyl iso butyrate, 260
Benzyl iso valerate, 260
Benzyl mercaptan, 260
Benzyl methyltiglate, 259
Benzyl phenylacetate, 260
Benzyl propionate, 260
Benzyl salicylate, 260
Benzylthiol, 260
Benzyl T-2-methyl-2-butenoate, 260
Benzyl tiglate, 260
Bergamot, 316
Beverage bases, dry, low-calorie, 148
Beverages
  aspartame, 197
  carbonated, low calorie
    carbonation, 146—147
    flavor selection, 145—146
    sugar replacement, 145
    typical formulations, 146
  FD&C dyes, 365—367, 368
  FD&C lakes, 375
  fruit juices, 249—251
  terpeneless oils, 242
  typical dye blends, 380
BHA
  addition limits, 51
  analytical methods, 44—47
  antioxidant activity, 41
  antioxidant for vegetable oil, 26—27
  antioxidants, 16—17
  breakfast cereals, 33
  chemical structure, 16
  mechanism of action, 24
  physical properties, 28
  safety, 13
  solubility, 28
  synthetic flavor, 261
  vegetable oils, 39
BHT
  addition limits, 51
  analytical methods, 44—47
  antioxidant activity, 41
  antioxidant for vegetable oils, 26—27
  antioxidants, 16—17
  breakfast cereals, 33
  chemical structure, 16
  food approved, 25, 26
  lemon essence oil, 43
  physical properties, 28—29
  restriction, 13
  solubility, 29
  synthetic flavor, 261
  vegetable oils, 39
Biphenyl, 260
Birch tar oil, 260
Bisabolene, 260
Bisabolmyrrh, 334
Bis-(2,5-dimethyl-3-furyl) disulfide, 260
Bis (2-furfuryl) disulfide, 260
Bis (2-furfuryl) sulfide, 260
Bis (2 methyl-3-furyl)-disulfide, 260
Bis (2-methyl-3-furyl) tetrasulfide, 260
Black 7984, 352, 359
Blackberry bark, 333
Black BN, 352, 359
Black pepper, 231, 232, 247
Black wattle flowers, 334
Blanching
  process of, 65, 70, 71
Blender, definition, 320
Bleu cheese, 316
Boldus leaves, 333
Borneol, 260
Bornyl alcohol, 260
Bornyl acetate, 260
Bornyl formate, 260
Bornyl iso valerate, 260
Bornyl valerate, 261
Boronia flowers, 333
Boronia megastigma, 333
Boswelliacarteri,
  species, 334
Botanical, definition, 320
Botanical extracts, 251—253
Bottler's extract, definition, 320
Bottlers' syrup, definition, 320
β-Bourbonene, 261
Brabender® amylograph, 68
Brassica nigra, 232
Breakfast cereals
  aspartame, 197
Brewing
  amylases, 85—88
  proteases, 102—103
Brewnzyme®, 75
Brilliant Black BN, 350, 361
Brilliant Blue, 340
Brilliant Blue FCF, 350, 352, 358, 361
Brilliant Green BS, 352, 356, 361
Brix, definition, 320
Bromelain, 71, 73, 74, 103
Brominated vegetable oil, 261
Brown FK, 354, 352, 358, 361
Browning reaction
  antioxidants, 37—38
  definition, 320

Bryonia
alba, 333
diocia, 333
Bryonia root, 333
Buchu leaves, 333
Buckbean leaves, 333
Bud® Chips, 73
Buffer, definition, 320
Bulking, definition, 320
Bulnesia sarmienti, 334
Bursera
delpechiana, 334
species, 334
2,3-Butandione, 265
2,3-Butandithiol, 261
1-Butanethiol, 261
1-Butanol, 261
2-Butanone, 261
Butan-3-one-zyl butanoate, 261
Butter, 314
Butter acids, 261
Butter esters, 261
Butter starter distillate, 261
2-Butyl-5, 262
Butyl acetate, 261
Butyl acetoacetate, 261
Butyl alcohol, 261
Butylamine, 261
Butyl anthranilate, 261
Butylated hydroxyanisole, see BHA
Butylated hydroxytoluene, see BHT
2-Butyl-2-butenal, 261
Butyl butyrate, 261
Butyl butyrllactate, 261
Butyl caproate, 261
alpha-Butylcinnamaldehyde, 261
Butyl cinnamate, 261
2-sec-Butylcyclohexanone, 261
Butyl 2-decenoate, 261
Butyl decylenate, 261
1,3 Butylene glycol, 261, 327
sec-Butyl ethyl ether, 261
Butyl ethyl malonate, 261
Butyl formate, 261
Butyl heptanoate, 261
Butyl hexanoate, 261
mono-*tert*-Butylhydroquinone, see TBHQ
Butyl *p*-hydroxybenzoate, 262
3-Butylidene pathalide, 262
Butylisobutyrate, 262
Butyl iso valerate, 262
Butyl lactate, 262
Butyl laurate, 262
Butyl levulinate, 262
N-Butyl mercaptan, 261
n-Butyl-2-methylbutyrate, 262
Butyl octadecanoate, 262
Butyl parasept, 262
Butyl phenylacetate, 262
3-n-Butylphthalide, 262
Butyl propionate, 262
Butyl sebacate, 266
Butyl stearate, 262
Butyl sulfide, 262
Butyl 10-undecenoate, 262
Butyl valerate, 262
Butyraldehyde, 262
Butyric acid, 262
Butyroin, 279

C

Cadinene, 262
Caffeine, 262
Cajeput, 333
Calcium acetate, 262
Calcium cyclamate, 127, 129, 130
Calcium saccharin
properties, 132, 135, 136, 137
structure, 128
Calumba root, 333
Camphene, 262
d-Camphor, 262
Camphor tree, 333
Canarium
commune, 334
luzonicum, 334
Candy
FD&C dyes, 368
FD&C lakes, 375
Canisic ketone, 285
Capalase®, 73
Capric acid, 265
Caproic acid, 276
Capryllic acid, 294
Capsicum
annum, 38, 232
fratescens, 232
Capsicum, 232
Capsicum NF, 246
Caramel, 314
Carbohydrases
cellulases and hemicellulases, 94—96
dextranase, 96
disaccharide splitting enzymes, 88—91
fruits and vegetables, 69
glucose isomerase, 96—98
lysozyme, 96
marketed in U.S., 74, 75, 76
naringinase, 93
pectic enzymes, 91—93
starch-splitting
α-amylase, 80
β-amylase, 80—81
glucoamylase, 80, 81
Carboxymethyl cellulose (CMC)
in baked goods, 161
synthetic flavors, 263
with aspartame, 197
Carboxypeptidase, see Exopeptidase
Cardamom, 232
Cardamom elettaria, 232
Carmoisine, 350, 351, 353, 361

β-Carotene
chemical structure, 15
in fats, 15
Carotenoids, 15
Carrageenan
in frozen desserts, 160
in salad dressing, 155
Carrier, definition, 320
4-Carromenthenol, 263
Caruacrol, 263
Caruacryl ethyl ether, 263
Carveol, 263
Carvomenthone, 283
d, l-Carvone, 263
D-Carvone, 238
L-Carvone, 238
cis-Carvone oxide, 263
Carvyl acetate, 263
Carvyl propionate, 263
beta-Caryophyllene, 263
Caryophyllene alcohol, 263
Caryophyllene alcohol acetate, 263
beta-Caryophyllene oxide, 263
Caryophyllus aromaticus L, 232
Cascara sagrada, 333
Cassia, 314
Cassia acutifolia, 335
Cassia flowers, 333
Castanea dentata, 333
Castoreum, 251, 253
Castor oil, 333
Catalase
classification, 60
fruit and vegetables, 69
marketed in U.S., 73, 74, 75
production, 71
uses, 111—112
Catalyst
definition, 58
Catecholase, 71
Catechu, black, 333
Catechu, pale, 334
Cedar, white, 333
Cedarwood oil alcohols, 263
Cedarwood oil terpenes, 263
Celery, 232, 246, 315
Celery ketone, 290
Cellulases
in frozen desserts, 160
uses, 74, 94—96
Cellulose, definition, 94
Cellulose gums
Cellzyme®, 77
Centaurium umbellatum, 333
Centuary, 333
Cephalin, 14
α-Cephalin, 15
Cereal products
enzymes in, 67—69
Cereflo® 200L, 76
Cerevase®, 76
Cetraria islandica, 334
Cetyl alcohol, 276
CFR-21 definition, 320
Cheesemaking, 99—100
Chelation, 23
Cherry, 314
Cherry-laurel leaves, 333
Cherry pits, 333
Chestnut leaves, 333
Chilco®, 75
Chirata, 333
Chlorogenic acid, 216
Chlorophyllase, 69
Chlorosucrose, 222
Chocolate, 315
Chocolate Brown Ht, 361
Chocolate Brown PB, 352, 359
Chocolate Brown kT, 352
Cholesterol, 16
Chrysanthemum balsamita, 333
Chrysoin S, 356
Chymo-Set®, 75
Chymotrypsin, 71
Cinchona
calisaya, 333
ledgeriana, 333
succirabra, 333
Cinchona, red, 333
Cinchona, yellow, 333
1,4-Cineole, 263
1,8-Cineol, 272
Cinnamaldehyde, 263
Cinnamaldehyde ethylene glycol acetal, 263
Cinnamic acid, 263
Cinnamic aldehyde, 238, 314
Cinnamom camphora, 333
Cinnamomum zeylanicum, 232
Cinnamon, 231, 232
Cinnamyl acetate, 263
Cinnamyl alcohol, 263
Cinnamyl anthranilate, 263
Cinnamyl benzoate, 263
Cinnamyl butyrate, 263
Cinnamyl cinnamate, 263
Cinnamyl formate, 263
Cinnamyl isobutyrate, 264
Cinnamyl iso valerate, 264
Cinnamyl phenylacetate, 264
Cinnamyl propionate, 264
Cire d'abeille, see Beeswax, white
Cistus, 334
Citral, 238, 248, 264, 315
Citral diethyl acetal, 264
Citral dimethyl acetal, 264
Citral propylene glycol acetal, 264
Citric acid, 26—27, 150, 153, 264
Citridic acid, 256
Citronella, 238
Citronellal, 264
Citronellic acid, 267
D,L-Citronellol, 264
Citronelloxyacetaldehyde, 264
Citronellyl acetate, 264

Citronellyl butyrate, 264
Citronellyl formate, 264
Citronellyl isobutyrate, 264
Citronellyl phenylacetate, 264
Citronellyl propionate, 264
Citronellyl valerate, 264
Citrus aurantium, 93
Citrus flavors, definition, 320
Citrus oils
  expression, 233
  storage, 243—244
Civet, 251, 252
Clarase®, 74
Clouding agent, definition, 320
Clove, 231, 232, 315
Cocoa, 251
Coconut, 314
Coffee, 251, 314
Cold pressing, definition, 320
Colophony, see Rosin
Color additives
  certified
    advantages and disadvantages, 364
    FD&C dyes, 362—365
    FD&C lakes, 373—376
    in processed food, 366
    solubilities, 364
  international situation
    EEC approved colors, chemical structures, 353
    EEC list of approved colors, 351—352
    EEC-SFC testing directives, 361
    United Kingdom and EEC, 350, 360—362
  legislative and regulatory history
    FD&C Yellow #5, 348—349
    Food and Drug Act 1906, 339—340
    Food, Drug and Cosmetic Act of 1938, 340—343
    hyperkinesis, 348
    safety problems - 1970, 343—348
    safety testing requirements, 349
  nonabsorbable polymeric
    absorption comparison with FD&C dyes, 379
    safety evaluations, 381
    solubility, 382
    sunlight stability, 380
    thermal stability, 379
  quality control, 376—377
  specifications, 376—377
  troubles, 370—373, 374
  uses
    beverages, 365—367, 368
    candy, 368
    dairy products, 367—368
    dry mix products, 370
    pan-coated products, 369—370
    pet foods, 368—369
    tabletted products, 369—370
Comminution, definition, 320
Commiphora
  abyssinica, 334
  erythraea Engl. var. *llabrescens*, 334
  molmol, 334
  species, 334
Compound 20, 209
Compounds, definition, 320
Concentrated citrus oils, definition, 321
Concentrated fruit juices, definition, 320
Concord grape, 314
Concrete, definition, 321
Confections
  aspartame, 197
Coniothyrium diplodiella, 92
Contaminants, 1
Convertit®, 77
Cookerzyme®, 73
Copaiba, 333
Copaifera, 333
Coriander, 232, 233
Coriandrum savitun, 232
Coridothymus capitalus, 232
Cork, oak, 333
Cornstarch, 155
Costmary, 333
Costus root, 333
Creosol, 284
o-Cresol, 264
p-Cresol, 264
Crucifera, 112
Cubeb, 232, 333
Cubeb NF, 246
Cucumber, 316
Cumin alcohol, 282
Cuminaldehyde, 264
Cuminyl acetaldehyde, 282
Currant, black, 333
Cyclamates
  baked goods, 161—162
  carbonated beverages, 145—147
  canned fruits, 148—149
  dry beverage bases, 147—148
  flavor characteristics, 141
  frozen desserts, 159—161
  gelatin desserts, 149—152
  principals of application, 141—145
  properties, 126—128, 129, 130
  puddings, 152—154
  relative sweetness, 138—140
  safety, 132, 135
  salad dressing, 154—157
Cyclamen aldehyde, 288
Cyclamic acid, 127, 129, 130
Cycloheptadeca-9-en-1-one, 264
Cyclohexalamine, 135
Cyclohexamethyl acetate, 264
Cyclohexanacetic acid, 264
Cyclohexyl acetate, 264
Cyclohexyl anthranilate, 264
Cyclohexyl butyrate, 265
Cyclohexyl cinnamate, 265
Cyclohexyl formate, 265
Cyclohexyl iso valerate, 265
Cyclohexyl propionate, 265
Cyclopentanthiol cyclopentyl mercaptan, 265

Cyclotene, 287
p-Cymene, 265
Cynara scolymus, 333
p-Cyonen-8-ol, 304
L-Cysteine, 265
Cytophaga, 84

D

Dactylium dendoides, 110
Daemonorops, 334
Dairy products
  FD&C dyes, 367—368
  FD&C lakes, 375
  enzymes in, 66
Damasceneone, 304
β-Damascone, 304
Damiana leaves, 333
Davana, 333
Dawe's catalase, 73
Dawe's glucose oxidase, 73
DEAE-Sephadex®, 79
2-*trans*, 4-*trans*-Decadienal, 265
gamma-Decalactone, 265
delta-Decalactone, 265
Decanal, 265
Decanal dimethyl acetal, 265
1-Decanal, 265
2-Decenal, 265
4-Decenal, 265
Decanoic acid, 265
3-Decen-2-one, 265
Decoction, definition, 321
Decyl acetate, 265
Decyl butyrate, 265
Decyl propionate, 265
DeeO®, 75
Dehydrases, 60
Dehydrodihydroionol, 265
Dehydrodihydroionone, 265
Dehydrogenses, 60
Desserts
  aspartame, 197
Dex-lo®, 76
Dextranase, 96, 114
Dextrinase® A, 74
Dextrins
  in baked goods, 161
Dextrose
  from starch, 82—84
Diacetyl, 265, 314
Diallyl disulfide, 257, 265
Diallyl sulfide, 257, 265
Diallyl trisulfide, 265
p-Diamine oxidase, 66
Diastase, 60, 73
Diastatic supplement, 73
Diazyme®, 75
Dibenzyl disulfide, 259
Dibenzyl ether, 266
Dibenzyl ketone, 268
Di-(butan-3-one-1-yl) sulfide, 266
4,4-Dibutyl-gamma-butyrolactone, 266
Dibutyl sebacate, 266
Dictamnus albus, 334
Dicyclohexyl disulfide, 266
Didodecyl, 26—27
Diethyl malate, 266
Diethyl malonate, 266
2,3-Diethyl-5-methyl-pyrazine, 266
2,3-Diethyl pyrazine, 266
Diethyl sebacate, 266
Diethyl succinate, 266
Diethyl tartrate, 266
2,5-Diethyl tetrahydrofuran, 266
Difurfuryl disulfide, 260
Difurfuryl ether, 266
Difurfuryl sulfide, 260
Diglycerides
  in fats, 15
Dihydro anethol, 299
Dihydrocarveol, 266
Dihydrocarvone, 266
Dihydrocarvyl acetate, 266
Dihydrochalcone
  analogues, 207
  relative sweetness, 208—210
  sensory evaluation, 214—215
  simplified, new, 207, 213
  structure-taste relationship, 211—212
  taste quality, 208—210
Dihydrocoumarin, 266
4,5-Dihydro-3(2H)thiophene, 266
5,7-Dihydro-2-methyl-thienos (3,4-D) pyrimidine, 266
Dilauryl thiodipropionate, 17, 26—27
Dill, Indian, 334
Dill seed, 232
1,2-Dimethoxy-4-allylbenzene, 273
3,4 Dimethoxy benzaldehyde, 306
m-Dimethoxy benzene, 266
p-Dimethoxy benzene, 266
1,1-Dimethoxyethane, 267
2,6-Dimethoxy phenol, 267, 306
3,4-Dimethoxy-1-vinyl-benzene, 267
Dimethyl acetal, 298
2,4-Dimethyl acetophenone, 267
2,5-Dimethyl-3-acetylfuran, 256
1,4-Dimethyl-4-acetyl-1-cyclohexane, 266
2,4-Dimethyl-5-acetyl-thiazole, 267
Dimethyl anthranilate, 288
2,4-Dimethyl benzaldehyde, 267
p-alpha-Dimethylbenzyl alcohol, 267
Dimethyl benzyl carbinol, 267, 7
alpha, alpha-Dimethylbenzyl isobutyrate, 267
3,4-Dimethyl-1,2-cyclopentadione, 267
3,5-Dimethyl-1,2-cyclopentadione, 267
2,5-Dimethyl-2,5, dihydroxy-1,4-dithione, 267
Dimethyl disulfide, 287
2,5-Dimethyl-3-furanthiol, 267
2,6-Dimethyl-4-heptanol, 267
2,6-Dimethyl-5-heptenal, 267

Dimethyl hydroquinone, 266
3,7-Dimethyl-7-hydroxoctanal, 278
2,5-Dimethyl-4-hydroxy 3(2H) furanonc, 314
Dimethyl ketone, 256
2,6-Dimethyl-10-methylene-2,6,11-dodecatrienal, 267
2,2-Dimethyl-3-methylene-norbornane, 262
3,7 Dimethyl-2,6-octadienal, 264
3,7 Dimethyl-1, 6-octadien 3-ol, 283
cis-3,7-Dimethyl-2,6-octadien-1-ol, 292
trans-3, 7, Dimethyl-2,6-octadien-1-ol, 274
3,7-Dimethylocta-2,6-dienyl-2-ethylbutanoate, 267
2,6-Dimethyloctanal, 267
3,7-Dimethyl-1,7 octanediol, 278
3,7-Dimethyl-1-octanol, 267
3,7-Dimethyl octan-3-ol, 303
2,6-Dimethyl-1,5,7-octatriene, 293
3,7-Dimethyl-6-octenal, 264
3,7-Dimethyl-6-octenoic acid, 267
3,7-Dimethyl-7-octen-1-ol, 301
7-Dimethyl-6-octen-1-ol, 264
6,10-Dimethyl-3-oxa-9-undecenal, 264
2,4-Dimethyl-2-pentenoic acid, 267
alpha-alpha-Dimethylphenethyl acetate, 267
alpha-alpha-Dimethyl-phenethylalcohol, 267
alpha, alpha-Dimethyl-phenethyl butyrate, 267
Dimethyl phenethyl carbinyl acetate, 290
alpha-alpha-Dimethyl phenethyl formate, 267
2,6-Dimethyl phenol, 268
2,5-Dimethyl pyrazine, 267, 314
2,6-Dimethylpyrazine, 268
Dimethyl resorcinol, 266
p-alpha Dimethylstyrene, 268
Dimethyl succinate, 268
Dimethyl sulfide, 291
4,5-Dimethyl thiazole, 268
2,5-Dimethyl-3-thiofuroyl-furan, 268
2,5-Dimethyl-3-thioiso-valeryl furan, 268
Dimethyl trisulfide, 268
6,10-Dimethyl-9-undecene-2-one, 303
2,4-Dimethyl-5-vinyl-thiazole, 268
2,6-Dinitro-3-methoxy-1 methyl 4 *tert*-butylbenzene, 292
Dioctadecyl thiodipropionate, 26—27
Dioscoreophyllum cumminsii, 188
Dipentene, 283
Dipeptide sweeteners, see Sweeteners, dipeptide
Diphenyl disulfide, 297
Diphenyl ketone, 259
1,3-Diphenyl-2-propanone, 268
Diproprionyl, 276
Dipropyl disulfide, 300
Dipropyl ketone, 275
Dipropyl trisulfide, 268
Disaccharide splitting enzymes, 88—90
Disodium phosphate, 268
Disodium succinate, 268
Distillate, definition, 321
Distillation
  definition, 321
  essential oils, 236—237
  terpeneless oils, 239—241
Distilled oil, definition, 321
2,8-Dithianon-4-en-4-carboxaldehyde, 268
2,2′ (Dithiodimethylene)-difuran gamma-dodecalactone, 268
Dittany of Crete, 334
Dittany roots, 334
delta-Dodecalactone, 268
Dodecanoic acid, 282
2-Dodecenal, 268
Dodecyl isobutyrate, 268
Dracorubin, 334
Dragon's blood, 334
Dry solubles, definition, 321

E

Econozyme®, 76
Ecuadorean sarsaparilla, 335
EDTA, 23
Egg drying
  glucose oxidase, 108—109
Eggs
  in salad dressing, 155
Egg yolks, salad dressings, 155
Elder tree leaves, 334
Elecampane rhizome and roots, 334
Elemi, 334
Emporase®, 73
Emulsin, 60
Emulsions, 309—310, 321
Encapsulation, definition, 321
Endopeptidase, 98
Endothia parasitica, 76, 100
Enfleurage, 237—238
Enhancer, definition, 321
Environmental Protection Agency, 4
Enzobake®, 73
Enzopharm®, 73
Enzymatic browning, 71
Enzyme 201, 77
Enzyme 4511-3, 76
Enzyme action
  influencing factors, 61—64
Enzymes
  assay methods, 64—65
  carbohydrases
    amylases, 80—88
    cellulases and hemicellulases, 94—96
    dextranase, 96
    disaccharide splitting, 88—91
    glucose isomerase, 96—98
    lysozyme, 96
    naringinase, 93
    pectic enzymes, 91—93
    starch-splitting, 80
  chemical properties, 58—60
  classification, 60
  flavor enzymes, 112—114
  heat inactivation, 66
  hydrolysis, 62

immobilized, 78—80
inactivation by blanching, 65, 70, 71
industrial
applications, 72, 78—80
classes, 80
production, 71—72
lipases, 105—107
management, 65—71
cereal products, 67—69
dairy products, 66
meat, 66—67
vegetables and fruits, 69—71
marketed in U.S., 73—77
naming, 60
oxidoreductases, 107—112
pH optima, 63
proteases, 99—105
reaction
activators, 63
enzyme substrate concentration, 61—62
influencing factors, 61
inhibitors, 63—64
temperature, 62—63
regeneration after inactivation, 66
substrate complex, 60—61
temperature optima, 62, 63
theory, 58—65
Enzyme W, 76
Enzyme WC-8, 76
Enzymolysis, 237
EOA, definition, 321
Eosine, 351, 354
EPA, 4
2,3-Epoxy-2,6 dimethyl-7 octen-6-ol, 283
Erigeron canadensis, 334
Erigeron, 334
Eriocitrin DHC, 210
Eriodictyon californicum, 335
Erythrobic acid, 268
Erythrosine, 340, 350, 361
Erythrosine BS, 351, 353
Essence, definition, 321
Essential oils
chemical composition, 240, 242—243
definition, 233, 321
distillation, 236—237
dry spice equivalents, 234
enfleurage, 237—238
extraction, 238—239
fermentation, 237
flow chart, 235
storage, 243—244
Esterase, 60, 75, 77
Esterase lipase, 77
Esters, 313
Estragole, 268
Ethanal, 256
1,2-Ethanedithiol, 269
Ethanoic acid, 256
Ethone, 285
Ethanol, 269
p-Ethoxybenzaldehyde, 269
2 Ethoxy-*p*-cymene, 263
3-Ethoxy-4-hydroxy benzaldehyde, 272
1-Ethoxy-2-hydroxy-4-propenyl-benzone, 299
o-(Ethoxymethyl)phenol, 269
Ethoxyquin, 25
2-Ethoxythiazole, 269
2-Ethyl (or methyl), 271
Ethyl acetate, 269
Ethyl acetoacetate, 269
Ethyl-2-acetyl-3-phenyl-propionate, 269
1-Ethyl-2-acetyl pyrrole, 269
Ethyl aconitate, 269
Ethyl acrylate, 269
Ethyl alcohol, 269
Ethyl amyl ketone, 294
Ethyl-*p*-anisate, 269
Ethyl anthranilate, 269
Ethyl benzoate, 269
Ethyl benzoylacetate, 269
Ethyl benzylacetoacetate, 269
alpha-Ethyl benzylbutyrate, 269
Ethyl brassylate, 269
Ethyl *trans*-2-butenoate, 269
Ethyl butrate, 313
2-Ethyl butyl acetate, 269
Ethyl butyl ketone, 275
2-Ethyl butyraldehyde, 269
Ethyl butyrate, 269
2-Ethyl butyric acid, 269
Ethyl caprate, 270
Ethyl caproate, 270
Ethyl caprylate, 271
Ethyl cellulose, 269
Ethyl cinnamate, 269
Ethyl crotonate, 269
Ethyl cyclohexane-propionate, 269
Ethyl-*trans*-2, *cis*-4-decadienoate, 270
t-2, *c*-4-Ethyl decadienoate, 316
Ethyl decanoate, 270
2-Ethyl-3, (5 or 6)-dimethyl-pyrazine, 270
3-Ethyl-2,6-dimethylpyrazine, 270
Ethyl-2, 4-dioxyhexanoate, 270
Ethyl dodecanoate, 271
Ethylenediaminetetraacetic, see EDTA
Ethylene dichloride, 327
Ethylene oxide, 270
Ethylenimine polymers, 40
Ethyl-2-ethyl-3-phenyl-propionate, 270
Ethyl formate, 270
2-Ethyl furan, 270
Ethyl-2-furan propionate, 270
4-Ethylguaiacol, 270
Ethyl heptanoate, 270
2-Ethyl-2-heptenal, 270
Ethyl hexadecanoate, 272
Ethyl-2,4-hexadienoate, 272
Ethyl hexanoate, 270
Ethyl-3-hexenoate, 270
2-Ethyl-1-hexanol, 270
Ethyl hexyl ketone, 293
Ethyl hydrocinnamate, 272
Ethyl-3-hydroxybutyrate, 270

3-Ethyl-2-hydroxy-2-cyclopenten-1-one, 270
2-Ethyl-2-hydroxy-4-methylcyclopent-2-en-1-one, 270
5-Ethyl-2-hydroxy-3-methylcyclopent-2-en-1-one, 270
5-Ethyl-3-hydroxy-4-methyl-2(5H)-furanone, 270
Ethyl-2-hydroxypropionate, 271
2 Ethyl-3-hydroxy-4H-pyran-4-one, 271
Ethyl isobutyrate, 271
N-Ethyl-2-isopropyl-5-methylcyclohexane carboxamide, 271
Ethyl iso valerate, 271
Ethyl lactate, 271
Ethyl laurate, 271
Ethyl levulinate, 271
Ethyl malate, 266
Ethyl malonate, 266
Ethyl maltol, 271
Ethyl-2-mercaptopropionate, 271
Ethyl-*p*-methoxybenzoate, 269
4-Ethyl-2-methoxyphenol, 270
Ethyl-*trans*-2-methyl-2-butenoate, 272
Ethyl-2-methylbutyrate, 271, 314
Ethyl-2-methylpentanoate, 271
Ethyl-2-methyl-3-pentenoate, 271
Ethyl-2-methyl-4-pentenoate, 271
Ethyl methyl phenyl glycidate, 271, 313, 315
2-Ethyl-5-methylpyrazine, 271
3-Ethyl-2-methylpyrazine, 271
Ethyl-3-methylthiopropionate, 271
Ethyl myristate, 271
Ethyl nitrite, 271
Ethyl nonanoate, 271
Ethyl-2-nonynoate, 271
Ethyl octadecanoate, 271
Ethyl octanoate, 271
Ethyl-*cis*-4-octenoate, 272
Ethyl octyne carbonate, 271
Ethyl oleate, 272
Ethyl oxhydrate, 301
Ethyl-3-oxobutanoate, 269
Ethyl palmitate, 272
Ethyl pelargonate, 271
p-Ethylphenol, 272
Ethyl phenylacetate, 272
Ethyl-4-phenylbutyrate, 272
Ethyl-3-phenylglycidate, 272
2-Ethyl phenylmercaptan, 272
Ethyl-3-phenylpropionate, 272
Ethyl propenoate, 269
Ethyl propionate, 272
Ethyl propyl ketone, 277
2-Ethylpyrazine, 272
3-Ethylpyridine, 272
Ethyl pyruvate, 272
Ethyl salicylate, 272
Ethyl sebacate, 266
Ethyl sorbate, 272
Ethyl stearate, 271
Ethyl succinate, 266
Ethyl tartrate, 266
Ethyl tetradecanoate, 271
Ethyl thioacetate, 272
2-Ethylthiophenol, 272
Ethyl tiglate, 272
Ethyl (*p*-tolyoxy) acetate, 272
2-Ethyl-1, 3,3-trimethyl-2-norbornanol, 272
Ethyl undecenoate, 272
Ethyl-10-undecenoate, 272
Ethyl valerate, 272
Ethyl vanillin, 272
Ethyl vinyl ketone, 295
Eucalyptol, 272
Eucalyptus globulus, 334
Eucalyptus globulus leaves, 334
Eugenol, 238, 273, 315
iso-Eugenol, 273
Eugenyl acetate, 273
iso-Eugenyl acetate, 273
Eugenyl benzoate, 273
Eugenyl formate, 273
Eugenyl methyl ether, 273
Evernia
  furfuracea, 334
  prunastri, 334
Exopeptidase, 98
Expression, definition, 321
Extended flavor, definition, 321
Extractase®, 74
Extraction, 238—239
Extracts, 308—309, 321

F

FAO, 1—2
FAO/WHO Codex Alimentarius, 1—2
FAO/WHO Experts Committee on Food Additives, 9
Farmesyl acetone, 305
Farnesol, 273
Fast Green, 340
Fast Green FCF, 352, 357
Fast Red E, 351, 354
Fats, 14—16
FCC, definition, 321
FDA, definition, 4, 322
FD&C Blue #1, 342, 362, 364, 368, 370, 376
FD&C Blue #2, 345, 362, 364, 376
FD&C dyes
  absorption comparison with polymeric dyes, 379
  classification, 362—363
  color specification, 376—377
  physical differences between dyes and lakes, 363
  problems, 374
  properties, 363—365
  quality control, 376—377
  suitability, 371
  uses, 365—370
FD&C Green #1, 342, 362
FD&C Green #2, 342
FD&C Green #3, 345, 364, 376

FD&C lakes
  applications, 375—376
  color specification, 376—377
  properties, 373—375
  quality control, 376—377
  suitability, 371
FD&C Orange#1, 340
FD&C Orange #2, 340
FD&C Red #1, 342
FD&C Red #2, 343, 348—362
FD&C Red #3, 342, 362, 363, 364, 376
FD&C Red #4, 342, 363, 364
FD&C Red #32, 340
FD&C Red #40, 343, 363, 364, 368, 370, 376
FD&C Violet #1, 343
FD&C Yellow #1, 341
FD&C Yellow #2, 341
FD&C Yellow #3, 341, 370
FD&C Yellow #4, 341, 370
FD&C Yellow #5, 342, 348—349, 362-364, 367, 368, 376
FD&C Yellow #6, 245, 362—364, 367, 368, 376
Federal Food, Drug, and Cosmetic Act, 3
Federal Insecticide, Fungicide, and Rodenticide Act, 4
FEMA, definition, 322
D-Fenchone, 273
Fenchyl acetate, 305
Fenchyl alcohol, 273
Fenugreek, 251
Fermentation, 237
Fermcolase®, 74
Fermcozyme®, 74
Fermex®, 76
Fermlipase PL, 74
Fermvertase®, 74
Ferula
  galbaniflua, 334
  species, 334
Ficin, 71, 74, 100, 103
Ficus carica, 100
Fir, balsam, 334
Fir needles and twigs, 334
Fixative, definition, 322
Flavor chemist, definition, 322
Flavor enzymes, 112—114
Flavors
  classification
    animal extracts, 251—253
    botanical extracts, 251—253
    essential oils, 233—244
    fruit juices, 249—251
    isolates, 248—249
    oleoresins, 244—248
    species and herbs, 230—233
  composition
    natural compounds, 230—253
    synthetic compounds, 253, 255, 307
  creating compositions
    character items, 313
    contributing item, 313, 316
    differential item, 316
    fixation, 317
  definition, 229
  emulsions, 309—310
  extracts, 308—309
  forms
    liquid, 308—310
    paste, 312
    powdered, 310—312
  functions, 230
  future trends, 318—319
  glossary, 319—326
  plated, 310—311
  spray dried, 311—312
  synthetic colors
    avoiding troubles, 370—373, 374
    certified, color forms, 363
    certified, technical aspects, 362—363, 364
    international situation, 349—362
    legislative and regulatory history, 339—349
    nonabsorbable polymeric, 377—382
    properties of FD&C dyes, 363—365
    quality control, 376—377
    specification, 376—377
    use, 365—370
  synthetic ingredients, 256—306
Fold, definition, 322
Food
  common ingredients, 140
  labeling requirements, 329—332
  synthetic colors
Food Additive Amendment of 1958, 6—7, 322
Food additives, 322
  definition, 1—3, 322
  functions, 8—9
  indirect, 7—8
  legal considerations, 1—11
  regulatory outlook, 10—11
  safety, 9—10
  U.S. legal status, 3—7
Food and Drug Administration, 4
Food and Agriculture Organization, 1—2
Food Drug and Cosmetic Act of 1906, 322
Food Drug and Cosmetic Act of 1938, 322
Food Green S, 350
Formic acid, 273
2-Formyl benzofuran, 259
2-Formyl-6,6-dimethyl bicyclo (3,1,1) hept-2-ene, 273
Fraxinella roots, 334
Freeze-drying, definition, 322
Fresh-N®, 76
Fritzsche®-Dodge Olcott, Inc., see Gemini® base
Fromase®, 77
Fruit
  aspartame, 197
  canned, low calorie, 148—149
  enzymes in, 69—71
  flavor enzymes, 112—114
  juices, 249—251
Fumaric acid, 150, 153, 273
Fungal alphaamylase, 73, 74

Fungal amylase VAC, 73
Fungal lipase, 74
Fungal protease, 74, 103
Fungamyl®, 75
2-Furaldehyde, 273
2-Furanacrolein, 274
2-Furanmethanethiol, 273
2-Furanmethanethiol formate, 273
Furfural, 273
Furfuryl acetate, 273
Furfuryl acetone, 274
Furfuryl alcohol, 273
Furfuryl ether, 266
2-Furfurylidene butynaldehyde, 273
Furfuryl isopropyl sulfide, 273
Furfuryl mercaptan, 273, 313, 314
Furfuryl 2-methyl-butyrate, 273
Furfuryl methyl ether, 273
Furfuryl methyl sulfide, 273
alpha-Furfuryl octanoate, 273
alpha-Furfuryl pentanoate, 274
Furfuryl propionate, 274
N-Furfurylpyrrole-1-(2-furfuryl) pyrzole, 274
Furfuryl thioacetate, 274
Furfuryl thiopropionate, 274, 313
Furyl acrolein, 274
4-(2-Furyl)-3-butene-2-one, 274
2-Furyl methyl ketone, 274
(2-Furyl)-2-propanone, 274
1-Furyl-2-propanone, 274
Fusel oil, refined, 274

G

Galactose oxidase, 110—111
α-Galactosidase, see Melibiase
Galanga, greater, 334
Galbanum, 334
1-Galloyl glycerol, 40
Gambir, 334
Garlic, 232
Gas-Chrom® Q, 47
Gelatin
  in desserts, 149—152, 160
  in salad dressing, 155
Gel systems
  nonnutritive sweetners, 144
Gemini® base, 38
Genet, 238
Genet flowers, 334
Gentian, 251
Gentiana
  acaulis, 334
  lutea, 334
Gentian rhizomes and roots, 334
Gentian, stemless, 334
Geraniol, 274
Geranyl acetate, 74, 274
Geranyl acetoacetate, 2
Geranyl acetone, 274
Geranyl benzoate, 274
Geranyl butyrate, 274
Geranyl-2-ethlbutyrate, 267
Geranyl formate, 274
Geranyl hexanoate, 274
Geranyl iso butyrate, 274
Geranyl iso valerate, 274
Geranyl phenylacetate, 274
Geranyl propionate, 274
Geranyl tiglate, 238
Germander, chamaedrys, 334
Germander, golden, 334
Ginger, 231, 232
Ginger NF, 246
β-Glucanase, 73, 74, 76
Glucanase GV, 74
Glucoamylase
  marketed in U.S., 73, 75, 76
  starch conversion, 82—84
  syrup production, 81—82
Glucose isomerase, 76, 80, 96—98
Glucose oxidase, 73, 74, 75, 107—110
Glucose oxidase catalase, 108
Glucose pentaacetate, 274
α-Glucosidase, see Maltase
β-Glucosidases, 75
Glucox®, 74
l-Glutamic acid, 275
Glycerine, 275
Glycerol, 275
Glycerol ester of wood resin, 275
Glyceryl monooleate, 275
Glyceryl monostearate, 275
Glyceryl triacetate, 275, 304
Glyceryl tribenzoate, 275
Glyceryl tributyrate, 275
Glyceryl tripropanate, 275
Glycine, 275
Glycyphyllin, 210
GMP, definition, 322
Grapefruit, 315
GRAS list, definition, 5—7, 322
Green, 315
Green S, 352, 356, 361
Guaiac, 334
Guaiacol, 314
Guaiacum
  officinale, 334
  santum, 334
Guaiacyl acetate, 275
Guaicyl phenylacetate, 275
Guaiene, 275
Guaiol acetate, 275
5′-Guanosine monophosphate (GMP), 112
Guarana, 334
Guar gum
  in frozen desserts, 160
Guassia, 251
Guiacol, 275
Guinea Green, 340
Guinea Green B, 352, 357
Gum arabic

in salad dressing, 155
with aspartame, 197
Gum guaiac
antioxidants for vegetable oils, 26—27
Gum karaya
in salad dressing, 155
Gum tragacanth
in salad dressing, 155

H

Haw, black, bark, 334
Hazelnut, 315
Hazyme®, 76
Hedeoma pulegioides, 334
Hei-Pep®, 76
Heliotropine, 299
Heliotropyl acetate, 299
Heliotropyl isobutyrate, 299
Hemicellulases, 94—96
Hemicellulase CE-100®, 75
Hemlock needles and twigs, 334
2,4-Heptadienal, 275
gamma-Heptalactone, 275
Heptanal, 275
Heptanal glyceryl acetal, 275
2,3-Heptandione, 275
Heptanoic acid, 275
2-Heptanol, 275
3-Heptanol, 275
2-Heptanone, 275, 316
3-Heptanone, 275
4-Heptanone, 275
2-Heptenal, 275
4-Heptenal, 276
cis-4-Heptenal, 314
4-Heptenal diethyl acetal, 276
2-Hepten-4-one, 276
3-Hepten-2-one, 276
trans-3-Heptenyl acetate, 276
trans-3-Heptenyl-2-methyl propionate, 276
Heptyl acetate, 276
Heptyl alcohol, 276
Heptyl butyrate, 276
Heptyl cinnamate, 276
Heptyl formate, 276
2-Heptyl furan, 276
Heptylidene acetone, 265
Heptyl iso butyrate, 276
3-Heptyl-5-methyl-2-(3H)-furanone, 276
Heptyl octanoate, 276
Herbs
definition, 323
flavors, 231—233
oleoresin equivalents, 232—233
Hesperetin DHC, 210
Hesperidin DHC, 209
1-Hexadecanol, 276
Hexadecanoic acid, 295
omega-6-Hexadecenlactone, 276
trans, trans-2,4-Hexadienal, 276
Hexadioic acid, 256
Hexahydropyrzidine, 299
Hexahydrothymol, 284
delta-Hexalactone, 276
gamma-Hexalactone, 276
Hexanal, 276
3-Hexenyl-2-methyl-butanoate, 277
3-Hexenyl 3-methyl-butanoate, 277
3-Hexenyl 2-methyl-butyrate, 277
3-Hexenyl phenyl acetate, 277
2-Hexyl-5, 278
Hexyl acetate, 277
2-Hexyl-4-acetoxy-tetrahydrofuran, 277
Hexyl alcohol, 277
N-Hexyl-2-butenoate, 277
Hexyl crotonate, 277
Hexyl butyrate, 277
alpha-Hexylcinnamaldehyde, 277
Hexyl formate, 277
Hexyl 2-furoate, 277
Hexyl hexanoate, 277
2-Hexylidene cyclo-pentanone, 278
Hexyl isobutyrate, 278
Hexyl isovalerate, 278
Hexyl-2-methylbutanoate, 278
Hexyl-3-methylbutanoate, 278
Hexyl-2 methylbutyrate, 278
Hexyl octanoate, 278
Hexyl phenylacetate, 278
Hexyl propionate, 278
Hibiscus sabdarifia, 334
Holy thistle, see Thistle, blessed, 335
Honduras sarsaparilla, 335
Horehound, 251
Household extract, definition, 323
HT-Amylase®, 74
HT-Proteolytic®, 75
Hyacinthus orientalis, 334
Hyacinth flowers, 334
Hydrangea
macrophylla, 192
opuloides, 192
Hydrocinnamic acid, 298
Hydrolases, 60
o-Hydroxy benzaldehyde, 301
4-Hydroxybutanoic acid lactone, 278
1-Hydroxy-2-butanone, 278
Hydroxycitronellal, 278
Hydroxycitronellal diethyl acetal, 278
Hydroxycitronellal dimethyl acetal, 278
Hydroxycitronellol, 278
2-Hydroxy-2-cyclohexen-1-one, 278
2 Hydroxy *p*-cymene, 263
4-Hydroxy-2,5-dimethyl-3(2H) furanone, 278
6-Hydroxy-3,7 dimethyl octanoic acid lactone, 278
4-Hydroxy-3-methoxy benzaldehyde, 306
N-(4-Hydroxy-3-methoxybenzyl)-8-methyl-6-nonenamide, 278
N-(4-Hydroxy-3 methoxybenzyl) nonanamide, 278
1-Hydroxy-2-methoxy 4-propenyl benzene, 273, 281

4-Hydroxymethyl-2,6-di-*tert* butylphenol
  antioxidant for vegetable oils, 26—27
  chemical structure, 30
  food approved, 30
2-Hydroxymethyl-6, 6 dimethyl bicyclo(3.1.1) hept-2-enyl formate, 279
3-(Hydroxymethyl)-2-octanone, 279
3-Hydroxy-2-methyl-411-pepane-4-one, 283
Hydroxynonanoic acid delta-lactone, 279
5-Hydroxy-4-octanone, 279
4-Hydroxy-3-pentenoic acid lactone, 279
2-Hydroxy-2-phenyl acetophenone, 259
4-(p-Hydroxy phenyl)-2-butanone, 279
p-Hydroxyphenyl-4-2-butanone, 316
2-Hydroxy propionic acid, 282
Hydroxysuccinic acid, 283
6-Hydroxy-,2,5,7,8-tetramethyl-chroman-2 carboxylic acid, see Trolox C
4 Hydroxytoluene, 264
o-Hydroxytoluene, 264
2-Hydroxy-3,5,5 trimethyl-2-cyclohexenone, 279
5-Hydroxyundecanoic acid lactone, 279
Hypericum perforatum, 335

I

Iceland moss, 334
Imperatoria ostruthium, 334
Indanthrene Blue, 352, 357
Indigo Carmine, 352, 357, 361
Indigotine, 340, 350
Indole, 279
Infusion, definition, 323
5′-Inosine monophosphate (IMP), 112
Inula helenium, 334
Invertase
  classification, 69
  marketed in U.S., 73, 74, 76
  uses, 88—89
I.O.F.I., definition, 323
Ion-exchange resins, 40
Ionone, 248, 314
alpha-Ionone 4-[2,6,6-trimethyl-2-cyclohexen-lyl]-3-buten-2-one, 279
beta-Ionone, 313
beta-Ionone (4-[2,6,6-trimethyl-1-cyclo-hexen-lyl]-3 buten-2 one, 279
gamma-Ionone, 279
Irgazm, 73
Irgazym, 73
Iris
  florentina, 334
  germanica, 334
  pallida, 334
alpha-Irone, 279
Iso amylacetate, 279, 314
Isoamyl acetoacetate, 279
Isoamyl alcohol, 279
Isoamylase, 84
Isoamyl benzoate, 279
Isoamyl cinnamate, 279
Iso amyl formate, 279
Isoamyl-2-furanbutyrate, 279
Isoamyl-2-furan propionate, 279
Iso amyl hexanoate, 280
Isoamyl isobutyrate, 280
Isoamyl iso valerate, 280
Isoamyl kaurate, 280
Isoamyl 2-methylbutyrate, 280
Isoamyl nonanoate, 280
Isoamyl octanoate, 280
Isoamyl phenylacetate, 280
Isoamyl propionate, 280
Isoamyl pyruvate, 280
Isoamyl salicylate, 280
Isoborneol, 280
Isoborynyl acetate, 280
Isobornyl formate, 280
Isobornyl isovalerate, 280
Iso boryl propionate, 280
Iso butyl acetate, 280
Iso butyl acetoacetate, 280
Isobutyl alcohol, 280
Isobutyl angelate, 280
Isobutyl anthranilate, 280
Isobutyl benzoate, 280
Isobutyl-2-butenoate, 280
Isobutyl butyrate, 280
Isobutyl cinnamate, 280
Isobutyl formate, 281
Isobutyl 2-furanpropionate, 281
Isobutyl heptanoate, 281
Isobutyl hexanoate, 281
Isobutyl iso butyrate, 281
2-Isobutyl-3-methoxy-pyrazine, 281
Iso butyl *cis*-2-methyl-2-buteroate, 280
2-Iso-butyl-5-methyl cyclohexanol, 284
2-Isobutyl-3-methyl-pyrazine, 281
alpha-Isobutyl phenethyl alcohol, 281
Isobutyl phenylacetate, 281
Isobutyl propionate, 281
Isobutyl salicylate, 281
2-Isobutyl thiazole, 281
Isobutyraldehyde, 281
Isobutyric acid, 281
Isobuty thiazole, 314
Isoeugenol, 281
Isoengenyl acetate, 281
Isoeugenyl benzyl ether, 281
Isoeugenyl ethyl ether, 281
Isoeugenyl formate, 281
Isoeugenyl methyl ether, 281
Isoeugenyl phenylacetate, 281
Isojasmone, 281
Isolate, 248—249, 323
D,L-Isoleucine, 281
D,L-Isomenthone, 281
Isomerases, 60
alpha-Isomethylionone, 281
Isopentyl acetate, 314
Isopentylamine, 281
Isopropenylpyrazine, 281
Isopropyl acetate, 281

p-Isopropylacetophenone, 281
Isopropyl alcohol, 281, 327
p-Isopropylbenzaldehyde, 264
Isopropyl benzoate, 282
p-Isopropylbenzyl alcohol, 282
Isopropyl butyrate, 282
Isopropyl cinnamate, 282
Isopropyl citrate, 26—27
Isopropyl formate, 282
Isopropyl hexanoate, 282
Isopropyl isobutyrate, 282
Isopropyl isovalerate, 282
2-Isopropyl-5-methyl-2-hexenal, 282
2-Isopropylphenol, 282
p-Isopropylphenylacetaldehyde, 282
Isopropyl phenylacetate, 282
3-(*p*-Isopropyl phenyl)-propionaldehyde, 282
Isopropyl propionate, 282
Isopropyl tiglate, 282
Isopulegol, 282
Isopulegone, 282
Isopulegyl acetate, 282
Isoquinoline, 282
Isovaleraldehyde, 286
Isovaleric acid, 282
Italase®, 73
Iva, 334

J

Jabinene hydrate, 303
Jamaica, 246
Jams, jellies, & preserves
  low calorie
    acid adjustment, 158
    gelling ingredients, 157—158
    nonnutritive sweetners, 158
    preservatives, 158
    processing, 157
    syneresis, 158
    typical formulations, 159
Japanese mint oil, 248
Jasmine, 238
Jateorhiza palmata, 333
Juglans
  nigra, 335
  regia, 335

K

Katal, 65
Keritinase, 114
3-Keto-4-butanethio, 282
3 Ketobutyraldehyde-dimethyl acetal, 295
6-Keto-1,4 dioxane, 258, 262, 278
2-Ketopropionaldehyde, 301
2-Ketopropionic acid, 301
Key, definition, 323
Kinase® K, 75
Kinase M, 75
Klebsiella aerogenes, see *Aerobacter aerogenes*
Klerzyme®, 77
Krameria
  argentea, 334
  triandra, 334
Kynurenine derivatives, 221—222

L

Labdanum, 334
Laccase, 71
Lactase
  classification, 60
  marketed in U.S., 74, 75, 77
  uses, 88—90
Lactase LP, 77
Lactic acids, 282
Laurel leaf, 232
Lauric acid, 282
Lauric aldehyde, 283
Laurus nobilis, 232
Lauryl acetate, 283
Lauryl alcohol, 283
Leaf alcohol, 277
Lecithin
  in fats, 14
  synergist for vegetable oils, 26—27
α-Lecithin, 15
Lemon, 315
Lemon-verbena, 334
L-Leucine, 283
Levisticum officinale, 334
Levalinic acid, 283
Liberase®, 73
Licorice, 251
Ligases, 60
Light Green, 340
d-Limonene, 283
d,l-Limonene, 283
Linaloe wood, 334
Linalool, 238, 283
Linalool oxide, 283
Linalyl acetate, 238, 283, 316
Linalyl anthranilate, 283
Linalyl benzoate, 283
Linalyl butyrate, 283
Linalyl cinnamate, 283
Linalyl formate, 283
Linalyl hexanoate, 283
Linalyl isobutyrate, 283
Linalyl isovalerate, 283
Linalyl octanoate, 283
Linalyl phenylacetate, 283
Linalyl propionate, 283
Linden leaves, 334
Lipase
  marketed in U.S., 71, 73, 74, 75, 76
  milk, 105
  pancreatic, 105, 106
  pregastric, 105—106

uses, 66, 68, 69, 105—107
Lipase protease, 76
Lipid oxidation, 21—24
Lipoxidase, 69, 107, 108
Lipoxygenase, see Lipoxidase
Lippia citriodora, 334
Liquidambar
orientalis, 335
styraciflua, 335
Liquipanol®, 73
Locust bean gum, 155, 160
Lo Han Kuo (dried fruit), 192
Lovage, 334
Lubricants, 327—328
Lungmoss, 334
Lungwort, 334
Lyases, 60
Lysozyme, 96

M

Mace, 232, 246
Maceration, definition, 323
Maidenhair fern, 334
Maillard reaction, definition, 323
Main note, see Middle note
Majorana hortensis, 232
Malic acid, 153, 283
Malt amylase, 71
Malt amylase PF, 77
Maltase, 88
Maltol, 283
Maltyl isobutyrate, 283
Mannitol, 150, 161
Maple, mountain, 334
Marigold, 335
Marjoram, 231, 232
Marla-Set®, 75
Marschall rennet, 75
Marsin®, 75
Mashase®, 76
Masking agent, definition, 323
Mastich, 38
Maxamyl®, 74
Maxa zyme NP, 74
Maxazyme P, 74
Maxilact®, 74
Maxinvert®, 74
Meat, 103—104
Meaty, 315
Meer papain, 74
Melaleuca
leucadendron, 333
species, 333
Melanol, 267
Melibiase, 88, 90
Menstruum, definition, 323
Mentha
arvensis, 248
piperata, 248
pulegium, 334
p-Mentha-1,8,dien-7-ol, 283, 284, 295
d,p-Mentha-1,8-diene, 283
Menthadienol, 283
p-Mentha-6,8-dien-2-ol, 263
6,8,(9)-*p*-menthadien-2-one, 263
Menthadienyl acetate, 283
p-Menthan-1,8 (10)dien-9-ol, 283
p-Menthan-1,8 (10) dien-9yl acetate, 283
p-Menthan-2-one, 283
p-Mentha-8 thiol-3-one, 283
p-Menth-1,4 (8)-diene, 302
p-Menth-1-en-3-ol, 284
p-Menth-1-ene-9-ol, 284
p-Menth-3-en-1-ol, 284
p-Menth-8-en-3-ol, 282
n-Menth-4(8)-en-3-one, 300
p-Menth-8-en-3-one, 282
8,9-*p*-Menthen-1-ol, 302
p-Menth-1-en-8-ol, 302
1-p-Menthen-4-ol, 263
1-*p*-Menthen-9-yl acetate, 284
Mentho furan, 303
Menthol, 238, 284, 316
D-neo-Menthol, 284
Menthone, 284
Menthyl acetate, 284
Menthyl isovaterate, 284
Menyanthes trifoliata, 333
Mercaptan, definition, 323
1/4 + Mercapto-3-butanol, 284
3-Mercapto-2-butanone, 284
4-Mercapto-2-butanone, 282
2-Mercaptomethyl pyrazine, 284
2-Mercaptonaphthalene, 292
3-Mercapto-2-pentanone, 284
2,3 or 10-Mercaptopinane, 284
Mercaptopropionic acid, 284
Mesityl oxide, 289
Methanthiol, 288
Methional, 36—37, 291, 316
D,L-Methionine, 284
Methocel®, 197
p-Methoxyacetophenone, 256
Methoxyalpha methyl cinnamaldehyde, 284
o-Methoxy benzaldehyde, 284
p-Methoxybenzaldehyde, 284
Methoxy benzene, 258
p-Methoxybenzyl alcohol, 258
p-Methoxybenzyl butyrate, 259
p-Methoxybenzyl formate, 259
p-Methoxy benzyl propionate, 259
o-Methoxycinnamaldehyde, 284
p-Methoxy enzyl acetate, 258
2-Methoxy-3 (5 and 6)-isopropylpyrazine, 284
2-Methoxy-4-methyl phenol, 284
2-Methoxy-3-(1-methylpropyl)-pyrazine, 285
2,5 or 6-Methoxymethyl pyrazine, 285
o-Methoxyphenol, 275
o-Methoxyphenyl acetate, 275
4-(*p*-Methoxyphenyl)-2-butanone, 285
1-(4-Methoxyphenyl)-4-methyl-1-pentene-3-one, 285

1 (*p*-Methoxyphenyl)-1-penten-3-one, 285
1-(*p*-Methoxyphenyl)-2-pronanone, 285
1-Methoxy-4-propenyl benzene, 258
Methoxy pyrazine, 285
3,5, or 6-Methoxypyrazine, 271
o-Methoxy toluene, 285
p-Methoxy toluene, 285
2-Methoxy-4-vinylphenol, 285
Methyl acetate, 285
4-Methylacetophenone, 285
1-Methyl-2-acetyl pyrrole, 285
Methyl alcohol, 327
2-Methylallyl butyrate, 285
Methyl allyl disulfide, 257
Methyl allyl trisulfide, 257
Methyl amyl ketone, 275, 316
Methyl anisate, 285
Methyl-*o*-anisate, 288
o-Methyl anisole, 285
p-Methyl anisole, 285
Methyl anthranilate, 285
Methyl anthronilote, 314
Methylated silica, 285
Methyl benzoate, 285
alpha-Methylbenzyl acetate, 285
alpha-Methylbenzyl alcohol, 285
alpha-Methylbenzyl butyrate, 285
Methyl benzyl disulfide, 285
alpha-Methylbenzyl formate, 286
alpha-Methylbenzyliso butyrate, 286
alpha-Methylbenzyl propionate, 286
4-Methylbiphenyl, 286
2-Methyl-1-butanethiol, 286
3-Methyl-2-butanethiol, 286
2-Methyl-2-butenal, 286
3-Methyl butylacetate, 313
2-Methylbutyl isovalerate, 286
3-Methylbutyl-2-methyl-butanoate, 286
2-Methylbutyl 3-methyl-butanoate, 286
2-Methylbutyl 2-methyl-butyrate, 286
3-Methylbutyl-2 methyl-propanoate, 286
Methyl beta-methyl thiopropionate, 315
2-Methyl butyraldehyde, 286
3-Methyl butyraldehyde, 286
Methyl butyrate, 286
2-Methyl-butyric acid, 286
Methyl cellulose, 286
Methyl chavicol, 268
alpha-Methylcinnaldehyde, 286
p-Methyl cinnamic aldehyde, 286
Methyl cinnamate, 286
6-Methyl coumarin, 286
3-Methylcrotonic acid, 286
2-Methyl-1, 3-cyclohexadine, 286
1-Methyl-2,3-cyclohexadione, 286
3-Methyl-2-cyclohexen-1-one, 286
3-Methyl-1-cyclopenta-decanone, 287
1-Methyl-1 cyclopentene-3-one, 287
Methylcyclopentenolone, 287
5H-5-Methyl-6, 7 dihydrocyclopenta(*P*)-pyrazine, 287
Methyl dihydrojasmonate, 287
Methyl 3,7-dimethyl-6-octeneoate, 287
Methyl disulfide, 287
Methyl dodecanoate, 288
Methylene chloride, 327
4-(3,4-Methylenedioxy-phenyl) 2-butanone, 287
Methyl ester of rosin, 287
Methylethyl acetaldehyde, 286
Methyl ethyl ketone, 261
2-Methyl-3-foranthiol, 287
5-Methyl furfural, 287
Methyl furfuryl disulfide, 287
2-Methyl-3,5 or 6-furfuryl thiopyrazine, 287
Methyl 2-furoate, 287
2 Methyl-3 furyl acrolein, 287
3-(5-Methyl-2-furyl)-butanal, 287
2Methyl-3-furyl disulfide, 260
2-Methyl-3-furyl tetrasulfide, 260
4-Methyl guaiacol, 284
6-Methyl-3, 5 heptadien-2-one, 287
Methyl heptanoate, 287
2 Methyl heptanoic acid, 287
6-Methyl-5-hepten-2-one, 287
Methyl heptine carbonate, 289
Methyl heptyl ketone, 293
5-Methyl-2,3 hexandione, 287
Methyl hexanoate, 287
2-Methylhexanoic acid, 287
Methyl 2-hexenoate, 287
Methyl 3-hexenoate, 287
5-Methyl-3-hexen-2-one, 287
5-Methyl-5-hexen-2-one, 287
Methylhexylacetaldehyde, 289
Methyl hexyl ketone, 294
Methyl *p*-hydroxy benzoate, 287
Methyl 3-hydroxyhexanoate, 288
alpha-Methyl-beta-hydroxypropyl alpha-methyl-beta mercaptopropyl sulfide, 288
Methyl alpha-ionone, 288
Methyl beta-ionone, 288
3-Methyl indole, 301
Methyl delta-ionone, 288
alpha iso-Methyl ionone, 288
Methyl isobutylate, 288
Methyl isobutyl ketone, 289
1-Methyl-4-iso-propenyl-cyclohexanane-2-one, 266
6-Methyl-3-iso-propenylcyclohexanol, 266
1-Methyl-4-isopropyl benzene, 265
1-Methyl-4-isopropyl-1, 4-cyclohexadiene, 302
1-Methyl-4-isopropyl-1,3-cyclohexadiene, 302
2-Methyl-5-isopropyl-1,3-cyclohexadiene, 296
1-Methyl-4-isopropyl-cyclohexan-3-ol, 284
1-Methyl-4-isopropyl-cyclohexan-3-one, 284
1-Methyl-4-isopropyl-1-cyclohexen-3-one, 299
1-Methyl-4-isopropylidene 1-cyclohexen-2-one, 299
5-Methyl-2-isopropyl phenol, 303
2-Methyl-3-(*p*-isopropyl phenyl) propionaldehyde, 288
Methyl isovalerate, 288
Methyl jasmonate, 288
Methyl laurate, 288

Methyl linoleate, 288
Methyl mercaptan, 288
Methyl *o*-methoxybenzoate, 288
1-Methyl-3 methoxy-4-isopropylbenzene, 288
Methyl methoxy pyrazine, 315
2-Methyl-5 methoxy thiazole, 288
Methyl n-methyl-anthranilate, 288
Methyl 2-methyl butyrate, 288
Methyl 3-methylbutyrate, 288
7 Methyl-3 methylene-1,6 octadiene, 292
Methyl-4-methylpentanoate, 289
Methyl-4-(methylthiol)-butyrate, 289
5-Methyl-5-(methylthiol)-furan, 288
Methyl-3-methylthio-propionate, 288
2-Methyl 3,5 or 6 (methylthiol) pyrazine, 289
Methyl 4 methylvalerate, 289
Methyl myristate, 289
1-Methyl naphthalene, 289
Methyl beta-naphthyl ketone, 289
Methyl nonanoate, 289
Methyl-2-nonenoate, 289
Methyl nonyl ketone, 305
Methyl 2 nonynoate, 289
2-Methyloctanal, 289
Methyl octanoate, 289
Methyl *cis*-4-octenoate, 289
Methyl octyne carbonate, 289
Methyl 2-octynoate, 289
Methyl parabin, 287
2-Methyl pentanal, 289
3-Methyl pentanoic acid, 289
4-Methyl pentanoic acid, 289
4 Methyl-2-pentanone, 289
4-Methyl-2-3-pentandione, 289
2-Methyl-2-pentenal, 289
4-Methyl-2-pentenal, 289
2-Methyl-2-pentenoic acid, 289
2-Methyl-3-pentenoic acid, 289
2-Methyl-4-pentenoic acid, 289
4-Methyl-3-penten-2-one, 289
3-Methyl-2-(2 pentenyl)-2-cyclopenten-1-one, 289
beta-Methylphenethyl alcohol, 290
alpha-Methylphenethyl butyrate, 290
Methyl phenethyl carbinyl isobutyrate, 290
Methyl phenethyl ether, 290
alpha-Methyl phenylacetaldehyde, 298
Methyl phenylacetate, 290
3-Methyl-4-phenyl-3-butene-2-one, 290
2-Methyl-4-phenyl-2-butylacetate, 290
2-Methyl-4-phenyl-2-butyl isobutyrate, 290
2-Methyl-4-phenyl-butyraldehyde, 290
3-Methyl-2-phenyl-butyraldehyde, 290
Methyl 4 phenylbutyrate, 290
5-Methyl-2-phenyl 2 hexenal, 290, 315
Methyl phenyl ketone, 256
4-Methyl-2-phenyl-m-dioxolane, 259
4-Methyl-1-phenyl-2-pentanone, 290
4-Methyl-2-phenyl-2-pentenal, 290
Methyl 3-phenlpropionate, 290
Methyl propionate, 290
2-Methyl-3-propronaldehyde, 313
5-Methyl-5-propyl-2 cyclohexen-1-one, 290
Methyl propyl disulfide, 290
Methyl propyl ketone 295
2 Methylpropyl-3-methylbutyrate, 290
2-(2-Methylpropyl)pyridine, 290
3-(2-Methylpropyl)pyridine, 290
2-(1-Methylpropyl)thiazole, 290
Methyl propyl trisulfide, 290
2-Methyl pyrazine, 291
Methyl pyrazinyl ketone, 256
Methyl-2-pyridyl ketone, 256, 315
Methyl 3-pyridyl ketone, 256
Methyl-2-pyrrole ketone, 291
6-Methyl quinoline, 291
5-Methyl quinoxaline, 291
Methyl salicylate, 291
Methyl sulfide, 291
Methyl-*p*-tertiary butylphenyl acetate, 286
Methyl tetradecanoate, 289
2-Methyl tetrahydrofuran-3-one, 291
2-Methyltetrahydro-thiophene-3-one, 291
Methyl theobromine, 262
4-Methyl-5-thiazole-ethanol, 291
5-Methyl-5-thiazole ethanol acetate, 291
2-Methyl thioacetaldehyde, 291
3-(Methylthio) butanal, 291
4-(Methylthio) butanal, 291
1-(Methylthio)-2-butanone, 291
4-(Methylthio)-2-butanone, 291
Methyl thiobutyrate, 291
Methyl thiofuroate, 291
3-Methylthio-1-hexanol, 291
4-(Methylthio)-4-methyl-2-pentanone, 291
Methyl (thio) methylpyrazine, 291
Methyl thiomethyl pyrazine, 315
5-(Methylthio)-2-(methylthio)methyl penten-2-al-1, 291
5-Methyl-2-thiophene carboxaldehyde, 291
o-(Methylthio) phenol, 291
2-Methyl thiophenol, 304
3-(Methylthio) propanol, 291
3-Methylthiopropionaldehyde, 291
3-Methylthiopropyliso-thiocyanate, 291
2-Methyl-3-tolyl-propionaldehyde, 291
Methyl trisulfide, 268
2-Methylundecanal, 291
Methyl-9-undecenoate, 292
Methyl undecyl ketone, 304
Methyl 2-undecynoate, 292
Methyl valerate, 292
2-Methylvaleric acid, 292
2-Methyl vinylpyrazine, 292
4 Methyl-5-vinylthiazole, 292
Metroclot®, 76
Mexican sarsaparilla, 335
Microbial rennet, 74
Micrococcus lysodeikticus, 111
Middle note, definition, 323
Miles fungal lactase, 75
Milezyme® AFP, 75
Milezyme® 8X, 75
Milk-clotting enzymes, 99—101
Mimosa, 238

Mimosa flowers, 334
Mint oils, 244
Miraculin, 189—190
Modifier, definition, 323
Momordica grosvenori, 192
Monellin, 188
Monosodium glutamate, 292
Morcurd®, 76
Mortierella vinacea var. *raffinoseutilizer*, 90
Mucor
  pusillus, 73, 100
  miehei, 75, 76, 77, 100
Mullein flowers, 334
Muscone, 287
Mushroom, 315
Musk ambrette, 292
Mustard seed, 232
Mycobacterium tuberculosis, 66
Mycolase®, 77
Mycozyme®, 76
Mylase®, 76
Myrcene, 292
Myristica fragrans, 232
Myristic acid, 292
Myristic aldehyde, 292
Myrothecium verrucaria, 95
Myroxyton balsamum, 335
Myrrh, 334
Myrtenal, 273
Myrtenol, 292
Myrtle leaves, 334
Myrtus communis, 334

N

2-Naphthalenthiol, 292
Naphthol Yellow, 340
beta-Naphthyl anthranilate, 292
beta-Naphthyl ethyl ether, 292
Naringinase, 93
Naringin DHC, 208
Naringin dihydrochalcone, 203
Natural flavors, definition, 323
Natural ingredient definition, 1
Natural product sweetners, see Sweetners, natural products
Nature identical flavoring substances, definition, 323
NDGA
  analytical methods, 46—47
  antioxidant activity, 41
  antioxidant for vegetable oils, 26—27
  chemical structure, 35
  food approved, 32, 35
Neoeriocitrin DHC, 208
Neohesperidin DHC, 208
β-Neohesperidin dihydrochalone (NDHC)
  physical properties, 204
  safety, 204
  synthesis, 204—207
delta-Neomenthol, 292
Nerol, 292
Nerolidol, 292
Nerolin, 292
Neryl acetate, 292
Neryl butyrate, 292
Neryl formate, 292
Neryl isobutyrate, 292
Neryl isovalerate, 292
Neryl propionate, 292
Neutrase®, 76
N.F., definition, 323
Nitroanilines, 216—218
Nitrous oxide, 292
Nona-2-*trans*-6-*cis*-dienal, 292, 313
2,4-Nonadienal, 293
t-2-cis-6-Nonadienal, 316
2,6-Nonadienal diethyl acetal, 292
2,6-Nonadien-l-ol, 292
delta-Nonalactone, 279
γ-Nonalactone, 292, 313, 314
Nonanal, 292
1,3-Nonanediol acetate, 292
1,9-Nonanedithiol, 292
Nonanoic acid, 292
2-Nonanol, 292
2-Nonanone, 292
3-Nonanone, 293
3-Nonanon-l-yl acetate, 293
Nonanoyl-4-hydroxy-3-methoxybenzylamide, 293
2-Nonenal, 293
trans-2-Nonen-1-ol, 293
cis-6-Nonen-1-ol, 293
Nonnutritive sweetners
  baked goods, 161—162
  canned fruits, 148—149
  carbonated beverages, 145—147
  dry beverage bases, 147—148
  frozen desserts, 159—161
  gelatin desserts, 149—152
  jam, jellies & preserves, 157—159
  principals of application, 142—145
  properties
    cyclamate, 126—128, 129, 130
    saccharin, 128, 131—137
  puddings, 152—154
  relative sweetness, 138—140
  safety, 131—132, 135—138
  salad dressings, 154—157
  sensory panel testing, 141—142
Nonyl acetate, 293
Nonyl alcohol, 293
Nonyl isovalerate, 293
Nonyl octanoate, 293
Noot katone, 293, 315
Nopinene, 298
Nordihydroquaiaretic acid, see NDGA
Note, definition, 324
Novozyme C, 75
Nutmeg, 231, 232

O

Oak, English, wood, 334

Oak moss, 334
Oak, white, chips, 334
Ocimene, 293
Octadecanoic acid, 302
9-Octadecenoic acid, 295
2-*trans*-6-*trans* Octadienal, 293
gamma-Octalactone, 294
delta-Octalactone octanal, 294
Octanol, 294
Octanal diethyl acetal, 294
1,8 Octanedithiol, 294
Octanoic acid, 294
1-Octanol, 294
3-Octanol, 294
2-Octanol methyl hexyl carbinol, 294
2-Octanone, 294
3-Octanone, 294
3-Octanon-1-ol, 294
2-Octenal, 294
1-Octen-3-ol, 294, 315
cis-3-Octen-1-ol, 294
3-Octen-2-one, 294
trans-2-Octen-l-yl-acetate, 294
1-Octen-3-yl-acetate, 294
trans-2-Octen-1-yl-butanoate, 294
Octyl acetate, 294
3-Octyl acetate, 294
Octyl butyrate, 294
Octyl formate, 294
Octyl-2-furoate, 294
Octyl gallate, 26—27
Octyl heptanoate, 295
Octyl isobutyrate, 295
Octyl isovalerate, 295
Octyl octanoate, 295
Octyl phenylacetate, 295
Octyl propionate, 295
Oil anise, 238
Oil bergamot, 238
Oil bitter almond, 238
Oil bois de vose, 238
Oil caraway, 238
Oil cinnamon, 238
Oil clove, 238
Oil Davana, 313
Oil geranium, 238
Oil, in salad dressing, 155
Oil Jasmin, 316
Oil lemongrass, 238, 248
Oil Neroli, 233, 236
Oil of niobe, 285
Oil Olibanum, 316
Oil peppermint, 238, 248
Oil Tagette, 313
Oil thyme, 238
Oil Yellow GG, 351, 355
Oleic acid, 295
Oleoresins
  definition, 324
  dry spice equivalents, 232-233
  flavor advantages, 245
  production, 244—245
  solubility, 246—247
Olibanum, 334
Onicus benedictus, 335
Onion, 232
Opopanax chironium, 334
Orange G, 351, 355
Orange GGN, 351, 355
Orange I, 340
Orange RN, 351, 355
Oregano, 231, 232
Origanum dictamnus, 334
Orris root, 334
Osladin, 191—192
Ovazyme®, 74
Oxanone, 279
Oxathiazinone dioxides, 219—221
Oxidoreductases
  catalase, 111—112
  classification, 60
  galactose oxidase, 110—111
  glucose oxidase, 107—110
  lipoxidase, 107, 108
  peroxidase, 111
Oximes, 222
3-Oxobutanal dimethyl acetal, 295
Oxygenases, see Polyphenol oxidase
Oxygenated compounds, definition, 324

P

Palmitic acid, 295
Pancreatin, 71
Pancreatin NF, 76
Pancrelipase NF, 76
Panel test, definition, 324
Paniplus amylase tablets, 76
Paniplus MLO, 76
Panol®, 73
Pansy, 334
Panzyme, 74
Papain, 71, 74, 73, 76, 103
Paprika, 232, 246
Parsley, 232
Parsley seed, 246
Passiflora incarnata, 334
Passion flower, 334
Patchouly, 334
Patent Blue V, 352, 357, 361
Paullinia cupana, 334
Peach, 314
Peach leaves, 334
Peanut, 315
Pear, 316
Pectic acid lyases (PAL), 91
Pectic enzymes
  nomenclature, 91
  uses, 92—93
Pectinases, 69, 73—77
Pectinesterase, see Pectin methylesterase
Pectinex®, 76
Pectin lyases (PL), 91

Pectin methylesterase (PME), 91-93
Pectinol®, 76
Pectins, 91, 155
Pektolase L-60, 74
Pelargonyl vanillyamide, 293
Penicillium
  amagasakiense, 107
  notatum, 107
  species, 112
Pennyroyal, American, 334
Pennyroyal, European, 334
omega-Pentadecalactone, 295
2,4-Pentadienal, 295
2,3-Pentandione, 295
1-Pentanol, 258
2-Pentanol, 295
2-Pentanone, 295
2-Pentenal, 295
4-Pentenoic acid, 295
1-Penten-3-ol, 295
1-Penten-3-one, 295
3-Penten-2-one, 295
Pentosanase, 74
Pentyl butyrate, 258
Pentyl formate, 258
2-Pentyl furan, 295
Pentyl-2-furoate, 258
Pentyl-2-furyl ketone, 295
Pentyl heptanoate, 258
2-Pentyl pyridine, 295
Peppermint, 316
Pepsin, 60, 73—76, 100, 103
Peptidase, 60
Percolation, definition, 324
Perilla namkinensis, 222
Perilla alcohol, 283
Perillaldehyde, 295
Perillyl acetate, 295
Perillyl alcohol, 284, 295
Peroxidase, 66, 69, 111
Peroxide formation, 31
Pet foods
  FD&C dyes, 368—369
  FD&C lakes, 375
Petitgrain, 233
Petroselinum sativum, 232
Peucedanum
  graveolons, 334
  ostruthium, 334
Peumus boldus, 76
Pfizer rennet, 76
PG
  addition limits, 51
  analytical methods, 44—47
  antioxidant activity, 41
  antioxidants, 16—17
  chemical structure, 16
  food approved, 25
  in vegetable oils, 26—27, 39
  iron discoloration, 32
  synthetic, 300
alpha-Phellandrene, 296
Phenethyl acetate, 296
Phenethyl alcohol, 296
β-Phenylethyl alcohol, 296
Phenethylamine, 296
Phenethyl anthranilate, 296
2-Phenylethyl anthranilate, 296
Phenethyl benzoate, 296
2-Phenylethyl benzoate, 296
Phenethyl butyrate, 296
2-Phenylethyl butyrate, 296
Phenethyl cinnamate, 296
Phenylethylene, 302
Phenethyl formate, 296
Phenethyl-2-furoate, 296
Phenethyl hexanoate, 296
Phenethyl isobutyrate, 296
Phenethyl isovalerate, 296
Phenethyl-2-methylbutyrate, 296
Phenethyl 3-methylerotonate, 297
Phenethyl octanoate, 296
Phenethyl phenylacetate, 296
Phenethyl propionate, 296
Phenethyl salicylate, 297
Phenethyl senecioate, 297
Phenethyl tiglate, 297
Phenol, 297
Phenolase, see Polyphenol oxidase
Phenolic oils, 244
Phenothiazine, 46
Phenoxyacetic acid, 297
2-Phenoxyethyl isobutyrate, 297
Phenylacetaldehyde, 297
Phenylacetaldhyde 2,3-butylene glycol acetal, 297
Phenylacetaldehyde di-isobutyl acetal, 297
Phenylacetaldehyde dimethyl acetal, 297
Phenylacetaldehyde glyceryl acetal, 297
Phenylacetic acid, 297
Phenylbenzene, 260
4-Phenyl-2-butanol, 297
2-Phenyl-2-butenol, 297
4-Phenyl-3-buten-2-ol, 297
4-Phenyl-3-buten-2-one, 297
4-Phenyl-2-butylacetate, 297
2-Phenyl-3-carbethoxy furan, 297
2-Phenyl-*m*-dioxan-5-ol, 259
Phenyl disulfide, 297
2-Phenylethyl acetate, 296
2-Phenylethyl cinnamate, 296
2-Phenylethyl formate, 296
2-Phenylethyl-2-furoate, 296
2-Phenylethyl hexanoate, 296
2-Phenylethyl isobutyrate, 296
2-Phenylethyl isovalerate, 296
2-Phenylethyl-2-methylbutyrate, 296
Phenylethyl methyl carbinol, 297
2-Phenylethyl octanoate, 296
2-Phenylethyl phenylacetate, 296
2-Phenylethyl propionate, 296
2-Phenylethyl salicylate, 297
2-Phenylethyl tiglate, 297
Phenyl-3-methyl-3-pentanol, 297
1-Phenyl-2-pentanol, 300

3-Phenyl-4-pentanol, 297
2-Phenyl-4-pentanaol, 298
1-Phenyl-1, 2 propandione, 298
1-Phenyl-1-propanol, 298
3-Phenyl-1-propanol, 298
2-Phenylpropionaldehyde, 298
3-Phenylpropionaldehyde, 298
3-Phenylpropionic acid, 298
3-Phenylpropyl acetate, 298
2-Phenylpropyl butyrate, 298
3-Phenylpropyl cinnamate, 298
3-Phenylpropyl formate, 298
3-Phenylpropyl hexanoate, 298
2-Phenylpropyl isobutyrate, 298
3-Phenylpropyl isobutyrate, 298
3-Phenylpropyl isovalerate, 298
3-Phenylpropyl propionate, 298
2-(3-Phenylpropyl)-tetrahydrofuran, 298
Phloridzin, 209
Phosphatase, 66
Phosphate salts, 154
Phospholipids, 14—15
Phosphoric acid, 26—27, 298
Phosphorylases, 60
Photo-oxidation, 23—24
Phyllodulcin, 192
Phyllodulcin analogues, 216
Pi butyl sulfide, 262
Picea
  glauca, 335
  mariana, 335
Picrasma excelsa, 334
Pigments, 38
Pimenta officinalis, 232
Pimenta berries, 247
Pimpinella anisum, 232
Pineapple, 315
Pineapple mercaptan, 289
Pine, dwarf, 334
Pine, Scotch, 334
Pine, white, bark, 334
Pine, white oil, 334
alpha-Pinene, 298
beta-Pinene, 298
2-Pinen-4-ol, 306
Pine tar 0.1, 298
Pinocarveol, 299
Pinocembrin DHC, 208
Pinus
  mugo Turra var. *pumilio*, 334
  palustris, 334, 335
  species, 334, 335
  strobus, 334
  sylvestris, 334
Piper
  cubeba, 232, 333
  nigrum, 232
Piperidine, 299
Piperine, 299
Piperitenone, 299
Piperitenone oxide, 299
d-Piperitone, 299
Piperonal, 299
Piperonyl acetate, 299
Piperonyl isobutyrate, 299
Pistacia lentiscus, 38
Plated flavors, definition, 324
Pogostemon
  cablin, 334
  heyneanus, 334
Poly AO-79®, 42, 43
Polydyes, 377—382
Polygalacturonases (PG), 91
Polylimonene, 299
Polymethylgalacturonases (PMG), 91
Polyoxyethylene, 160
Polyphenolase, see Polyphenol oxidase
Polyphenol oxidases, 69, 71
Polypodium vulgare, 191
Poly R™-478, 378—380
Poly R™-481, 378—380
Polysorbate 20, 299
Polysorbate 60, 299
Polysorbate 8, 299
Poly Y™-607, 378—380
Ponceau 3R, 340
Ponceau 4R, 350, 353, 361
Ponceau 6R, 351, 354
Ponceau SX, 340, 351, 354
Poncirin DHC, 208
Popcorn, 315
Popular buds, 334
Populus
  balsamifera, 334
  candicans, 334
  nigra, 334
  tacamahacca, 334
Potassium acetate, 299
Potassium sorbate, 299
Potato, 316
Prolase® 300, 77
Prolase® EB-21, 77
Prolase RH, 77
L-Proline, 299
1,2-Propanedithiol, 299
Propanethiol, 299
2-Propene-1-thiol, 257
Propenyl guaethol, 299
Propenyl propylsulfide, 299
Propionaldehyde, 299
Propionic acid, 299
Propiophenone, 299
Propyl acetate, 299
Propyl alcohol, 299
p-Propyl anisole, 299
Propyl benzoate, 299
Propyl butyrate, 299
Propyl cinnamate, 300
Propyl disulfide, 300
Propylene glycol, 17, 300
Propylene glycol alginate, 155—157, 300
Propylene glycol dibenzoate, 300
Propylene glycol stearate, 300
Propyl formate, 300

Propyl-2-furanacrylate, 300
Propyl-2-furoate, 300
Propyl gallate, see PG
Propyl heptanoate, 300
Propyl hexanoate, 300
Propyl-*p*-hydroxybenzoate, 300
Propylidene phthalide, 300, 315
Propyl isobutyrate, 300
Propyl isovalerate, 300
n-Propylmercaptan, 299
Propyl methyl trisulfide, 290
alpha-Propylphenethyl alcohol, 300
o-Propylphenol, 300
Propyl phenylacetate, 300
Propyl propionate, 300
Propyl thioacetate, 300
Propyl trisulfide, 268
Proteases
  classification, 68, 69
  in baking, 101—102
  in brewing, 102—103
  in food industry, 98—99
  marketed in U.S., 73—77
  meat tenderizing, 103—104
  minor applications, 104—105
  rennin, 99—100
Proteinases
  industrial nonfood uses, 106
temperature-activity curves, 63
Prunus
  avium, 333
  cerasus, 333
  laurocerasus, 333
  persica, 334
Pseudomonas
  hydrophila, 97
  myxogenes, 100
  species, 84
Pterocarpus san alinus, 334
Ptyalin, 60
Puddings, low calorie, 152—154
Pulegone, 300
Pullulanase, 73, 84
Pulluzyme®, 73
Pyrazinyl ethanethiol, 300
Pyrazinyl methyl sulfide, 300
Pyridine, 301
2-Pyridinemethanethiol, 301
Pyroligneous acid, 301
Pyrrole, 301
Pyrrolidine, 301
Pyruvaldehyde, 301
Pyravic acid, 301

Q

Quassia amara, 334
Quebrachia lorentzii, 334
Quebracho bark, 334
Quercus
  alba, 334
  occidentalis, 333
  robur, 334
  suber, 333
Quikset®, 73
Quillaia, 334
Quillaja saponaria, 334
Quinine bisulfate, 301
Quinine hydrochloride, 301
Quinine sulfate, 301
Quinoline, 301
Quinoline Yellow, 350, 351, 356, 361

R

Raspberry, 314, 316
Rebaudioside A & B, 190, 191
Rectified oils, see Redistilled oils
Red FB, 351, 354
Red 2G, 350, 351, 353, 361
Red 6B, 351, 353
Redistilled oils, definition, 324
Red sandalwood, see Red saunders
Red saunders, 334
Regulase®, 73
Release agents, 327-328
Rennet, 74
Rennilase®, 76
Rennin, 73—77, 99—100
Rennin pepsin, 75, 76
Resinoid, definition, 324
Rhamnus purshiana, 333
Rhatany root, 334
Rheum
  officinale, 334
  palmatum, 334
  rhaponticum, 334
Rhizopus
  arrhizus, 74
  arrhizus var. *delemar*, 107
  oligosporus, 38
Rhodinol, 301
Rhodinyl acetate, 301
Rhodinyl butyrate, 301
Rhodinyl formate, 301
Rhodinyl isobutyrate, 301
Rhodinyl isovalerate, 301
Rhodinyl phenylacetate, 301
Rhodinyl propionate, 301
Rhozyme, 76
Rhozyme HP, 76
Rhozyme® H-39, 76
Rhozyme P-11, 76
Rhozyme P-53, 76
Rhozyme 41, 76
Rhozyme 54, 76
Rhubarb, garden root, 334
Rhubarb root, 334
Ribes nigrum, 333
Ricinus communis, 333
Roccella fuciformis, 333
Rose Bengale, 351, 355
Roselle, 334
Rosemary, 39—40, 231, 232

Rose oxide, 303
Rosin, 334
Rosmarinus officinalis, 232
Rubus, 333
Rum ether, 301

S

Saccharin
  baked goods, 161—162
  canned fruits, 148—149
  carbonated beverages, 145—147
  dry beverage bases, 147—148
  flavor characteristics, 141
  frozen desserts, 159—161
  gelatin desserts, 149—152
  jams, jellies, and preserves, 157—159
  principles of application, 141—145
  properties
    density, 137
    solubility, 133—134
    titration, 132
    viscosity, 135, 136
  puddings, 152—154
  relative sweetness, 138—140
  safety, 132, 136, 138
  salad dressings, 154—157
  synthetic, 301
Saccharomyces cerevisiae, 76, 77
Saccharomyces yeast, 74
Safranal, 304
Sage, 39—40, 247
Saint John's bread, 251
St. Johnswort, 335
Salicylaldehyde, 301
Sambucus nigra, 334
Sandalwood, 335
Sandarac, 335
alpha-Santalol, 301
beta-Santalol, 301
Santalum album, 335
Santalyl acetate, 301
Santalyl phenylacetate, 301
Sarsaparilla, 335
Sassafras albidum, 335
Sassafras leaves, 335
Saussurea lappa, 351, 354
Scarlet GN, 351, 354
Schiff base, definition, 324
Scoville units, definition, 324
Seasoning, definition, 324
Senna, Alexandria, 335
Sensory panel testing, 141—142
Serpentaria, 335
Sesquiterpene, definition, 324
Sesquiterpeneless oil, definition, 324
Shelf-life, definition, 324
Simaruba amara, 335
Simaruba bark, 335
alpha-Sinensal, 267
Skatole, 301
Smilax
  aristolochiaefolia, 335
  febrifuga, 335
  regelii, 335
  species, 335
Smoke, 314
Snakeroot, Canadian, 335
Soapbark, see Quillaia
Society of Flavor Chemists Inc., definition, 325
Sodium acetate, 301
Sodium benzoate, 301
Sodium carboxymethyl cellulose (CMC), 144, 155
Sodium chloride, 151
Sodium citrate, 302
Sodium cyclamate, 127, 129, 130
Sodium hexametaphosphate, 302
Solubilized flavor, definition, 324
Solubilizer, definition, 324
Solvents, 324, 327—328
Sorbic acid, 302
Sorbitan monostearate, 302
Sorbitol, 150, 161
d-Sorbitol, 302
Spanish "origanum", see *Thymus capitatus*
Spark-L®, 75
Spartium junceum, 334
Spearmint, 233
Spices
  antioxidant, 38—39
  definition, 325
  essential oils equivalents, 234
  flavors, 230—233
  in salad dressing, 155
  oleoresin equivalents, 232—233
Spiro-(2,4-dithia-1-methyl-8-oxabicyclo(3,3,0)
    octane-3,3′, 268
Sponge process, 234
Spray drying, definition, 325
Spruce needles and twigs, 335
Squalene, 16
Standard of identity, definition, 325
Starch, salad dressing, 155
Starch-splitting enzymes, 80—88
Stearic acid, 302
Sterols, 16
Stevia rebaudiana, 190
Steviol glycosides, 190
Sticta pulmonacea, 334
Stokes Law, definition, 325
Storax, 355
Strawberry, 313, 315
Streblus asper, 100
Strecker degradation, definition, 325
Streptococcus mitis, 84
Streptomyces, 97, 112
Styrax
  benzoin, 333
  paralleloneurus, 333
  tonkinensis, 333
Styrene, 302
Styrollyl acetate, 285

Styrollyl alcohol, 285
Styrollyl butyrate, 285
Styrollyl formate, 286
Styrollyl isobutyrate, 286
Styrollyl propionate, 286
Substrate, 60
Succinic acid, 302
Sucrose
  relative sweetness, 138—140
Sucrose octaacetate, 302
Sucrovert®, 76
Sugar, 161—162
Sugar alcohols
  antioxidants, 38
Sulfatodihydrochalcones, 213
Sulfoalkyldihydrochalcones, 213
Sulfodihydrochalcones, 213
Sulfur dioxide, 302
Sunset Yellow, 340, 350, 361
Sunset Yellow FCF, 350, 351, 356
Sure-Curd®, 76
Sweetners
  alkoxyaromatic, new
    chlorogenic acid, 216
    dihydrochalcone, 207—215
    β-neohesperidin dihydrochalone, 203—207
    nitroanilines, 216—218
    phyllodulcin analogues, 216, 217
    structural considerations, 200—203
  L-aspartic acid dipeptide, new, 201
  dipeptide, new
    aspartame, 193—200
  miscellaneous, new
    chlorosucrose, 222
    oxathiazinone dioxides, 219—221
  natural products, new
    Lo Han kuo (dried fruit), 192
    miraculin, 189—190
    monellin, 188
    osladin, 191—192
    phyllodulcin, 192
    steviol glycosides, 190
    thaumatins, 188—189
    xylitol, 193
Sweetners, nonnutritive, see Nonnutritive sweetners
Sweetzyme®, 76
Swertia chirata, 333
Synergists, 26—27
Synsepalum dulcificum, 189
Synthetases, see ligases
Synthetic, definition, 325
Syrup production, 81—84

T

Tagetes
  erecta, 335
  glandulifera, 335
  minuta, 335
  patula, 335
Taka-Diastase®, 81
Takamine bromelian, 75
Takamine catalase L, 75
Takamine cellulase, 75
Takamine fungal amylase, 74
Takamine fungal protease, 75
Takamine lipase powders, 75
Takamine pancreatic lipase, 75
Takamine pancreatin, 75
Takamine papain, 75
Takamyl®, 74
Tanacetum vulgare, 335
Tannase, 114
Tannic acid, 302
Tansy, 335
Tarragon, 231
Tartaric acid, 26—27, 302
Tartrazine, 340, 350, 351, 355, 361, 255
Taste, definition, 325
TBHQ
  analytical methods, 46—47
  antioxidant activity, 41
  antioxidant for vegetable oil, 26—27
  antioxidants, 16—17
  chemical structure, 16
  in breakfast cereals, 33
  in lemon essence oil, 43
  in oils, 33
  in vegetable oils, 39
  iron discoloration, 32
  package liners, 34, 35
  physical properties, 28, 29
  solubility, 28, 29
Tempeh, 38
Tenase®, 74
Tenderay®, 67
Tendrin®, 75
Termamyl®, 75
Terpene, definition, 325
Terpeneless oils
  chemical composition, 240, 242—243
  definition, 325
  extraction, 238
  flavor advantages, 241—242
  flavor disadvantages, 242
  production
    alcohol solvent, 241
    distillation under vacuum, 239—241
    solvent extraction, 241—242
alpha-Terpinene, 302
gamma-Terpinene, 302
4-Terpinenol, 263
alpha-Terpineol, 302
beta-Terpineol, 302
Terpinolene, 302
Terpinyl acetate, 302
Terpinyl anthranilate, 302
Terpinyl butyrate, 302
Terpinyl cinnamate, 302
Terpinyl formate, 302
Terpinyl isobutyrate, 302
Terpinyl isovalerate, 302

Terpinyl propionate, 302
Tetraclinis articulata, 335
Tetradecanal, 292
Tetradecanoic acid, 292
4,5,6,7-Tetrahydro-3,6-dimethyl-benzofuran, 303
Tetrahydrofurfuryl acetate, 303
Tetrahydro furfuryl alcohol, 303
Tetrahydro furfuryl butyrate, 303
Tetrahydro furfuryl cinnamate, 303
Tetrahydrofurfuryl propionate, 303
Tetrahydro geraniol, 267
Tetrahydrolinalool, 303
Tetrahydro-4-methyl-2-(2-methyl propen-l-yl)-
 pyran, 303
Tetrahydropseudo ionone, 303
Tetrahydropyrrole, 301
5,6,7,8-Tetrahydro-quinoxaline, 303
Tetramethyl ethylcyclo-hexenone, 303
1,5,5,9-Tetramethyl-13-oxatricyclo (8.3.0.0) (4,9)
 tridecane, 303
2,3,5,6-Tetramethyl pyrazine, 303, 315
Teucrium
 chamaedrys, 334
 polium, 334
Thamnidium, 67
Thaumatins, 188—189
Thaumatococcus danielli, 188
THBP
 antioxidant for vegetable oils, 26—27
 chemical structure, 30
 food approved, 30—31
Thiamine hydrochloride, 303
2-Thienyl disulfide, 303
2-Thienyl mercaptan, 303
2,2′-(Thiodimethylene)-difuran, 303
Thiodipropionic acid, 17, 26—27
Thiogeraniol, 303
Thiophenol, 258
Thistle, blessed, 335
Throw, definition, 326
Thuja occidentalis, 333
4-Thujanol, 303
Thyme, 316
Thymol, 238, 303, 316
Thymus capitatus, 335
Tillia, 334
Tincture, definition, 325
α-Tocopherol
 antioxidant activity, 41
γ-Tocopherol antioxidant activity, 41
Tocopherols
 antioxidant, 17, 26—27
 in fats, 14
 food approved, 35—37
Tolu, 335
Tolualdehyde glyceryl acetal, 303
Tolualdehydes, 303
o-Toluenesulfonamide (OTS), 136
o-Toluenethiol, 304
p-Tolylacetaldehyde, 304
o-Tolyl acetate, 304
p-Tolyl acetate, 304
4-(*p*-Tolyl)-2-butanone, 304
p-Tolyl isobutyrate, 304
p-Tolyl laurate, 304
p-Tolyl 3-methylbutyrate, 304
p-Tolyl phenylacetate, 304
2-(*p*-Tolyl)-propionaldehyde, 304
Tomato, 314
Top note, definition, 326
Transferases, 60
Triacetin, 275, 304
Tributyl acetylcitrate, 304
Tributyrin, 275
Tricalcium phosphate, 304
Trichoderma
 koningii, 95
 viride, 95
Trichloroethylene, 327
2-Tridecanone, 304
2-Tridecenal, 304
Triethyl citrate, 304
2,4,5-Trihydroxybutyrophenone, see THBP
Trimethyl amine, 304
p-alpha, alpha-Trimethyl benzylalcohol, 304
2,6,6 Trimethylbicyclo-(3.1.1)-2-heptene, 298
4-(2,6,6-Trimethyl cyclohexa-1, 3-dienyl) butene-
 2-en-4-one, 304
2,6,6-Trimethyl cyclohexa-1, 3-dienyl methanol,
 304
2,6,6-Trimethyl-1-cyclo hexen-1-acetaldehyde,
 305
2,6,6-Trimethylcyclo hexa-2-one-1, 4-dione, 305
4-(2,6,6-Trimethylcyclo-hexa-1-enyl) but 2 en-4-
 one, 304
2,2,6-Trimethylcyclo hexanone, 304
4-(2,6,6-Trimethylcyclo hex-1-enyl) but-2-en-4
 one, 305
3,7,11-Trimethyl-2,6,10-dodecatrien-1-ol, 273
3,7,11-Trimethyl-1,6 dodecatrien-3-ol, 292
3,5,5-Trimethyl hexanol, 305
3,5,5-Trimethyl-1-hexanol, 305
d-1,3,3-Trimethyl-2-norbornanone, 273
1,3,3-Trimethyl-2-norbornanyl acetate, 305
2,2,4-Trimethyl-1,3-oxacyclopentane, 305
2,4,5-Trimethyl, delta-3-oxazoline, 305
2,6,10-Trimethyl-2,6,10-pentadecatrien-14-one,
 305
2,3,5-Trimethyl pyrazine, 305
2,4,5-Trimethyl thiazole, 305, 315
Triphenylphosphine, 40
Tripropionin, 275
Trithioacetone, 305
Trolox C, 40—42
True fruit extract, definition, 326
True fruit flavor WONF, definition, 326
Trypsin, 60, 71
Tsuga
 canadensis, 334
 heterophylla, 334
Turmeric, 247
Turnera diffusa, 333
Turpentine, 335
Tyrosinase, 69, 71

U

Ultrazym® 100, 73
Uncaria gambir, 334
2,4 Undecadienal, 305
2,3-Undecadione, 305
delta-Undecalactone, 279
gamma-Undecalactone, 305, 313, 314
Undecanaic acid, 305
Undecanal, 305
2-Undecanol, 305
2-Undecanone, 305
2-Undecenal, 305
9-Undecenal, 305
10-Undecenal, 305
10-Undecenoic acid, 305
Undecen-1-ol, 305
10-Undecen-l-yl acetate, 305
Undecyl alcohol, 306
Urease, 59, 60
U.S.D.A., definition, 326
USP, definition, 326

V

Valencene, 306
Valeraldehyde, 306
Valeriana officinalis, 335
Valerian rhizome and roots, 335
Valeric acid, 306
gamma-Valerolactone, 306
d,l-Valine, 306
Vanatrope, 299
Vanilla, 251, 253, 315, 336—337
Vanillin, 248, 306, 310, 315
Vanillin acetate, 306
Vegetables, 69—71, 112—114
Veratraldehyde, 306
Verbascum
  phlomoides, 334
  thapsiforme, 334
Verbena officinalis, 335
Verbenol, 306
Vernonia anthelmintica, 40
Veronica officinalis, 335
Verrain, European, 335
Vetiver, 335
Vetiveria zizanioides, 335
Viburnum prunifolium, 334
o-Vinylanisole, 306
p-Vinylguaiacol, 285
Viola
  calcarata, 335
  tricolor, 334
Violet BNP, 352, 358
Violet 6B, 352, 358
Violet leaf alcohol, 292
Violet, Swiss, 335
Virginia snakeroot, 235
Vitamin B_1, 303
Vitamin C, 258
Vitamin U, 306
Volatile oils, definition, 326

W

Wallerstein papain, 77
Walnut husks, 335
West Indian sandalwood, see Amyris
Wild ginger, 335
Woodruff, sweet, 335
World Health Organization (WHO), 1—2
Wormwood, see Artemisia

X

Xanthan gum, 155
Xanthine oxidase, 66
2,6-Xylenol, 268, 306
Xylitol, 193

Y

Yarrow, 335
Yellow AB, 340
Yellow OP, 340
Yellow 2G, 351, 356, 361
Yellow #5 Lake, 374
Yerba santa, 335
Ylang Ylang, 236
Yucca
  brevifolia, 335
  mohavensis, 335
  schidigera, 335
Yucca, Joshua-tree, 335
Yucca, Mohave, 335

Z

Zingerone 4, 306
Zingiber officinale, 232